Rising Temperatures and Plant Ecosystems

Amit Awasthi · Avneesh Kumar · Vajinder Kumar ·
Tanupriya Choudhury
Editors

Rising Temperatures and Plant Ecosystems

Effects and Mitigation

Springer

Editors
Amit Awasthi
Applied Science Cluster
University of Petroleum and Energy Studies
Dehradun, Uttarakhand, India

Avneesh Kumar
Department of Botany
Shri Ramasamy Memorial University
Gangtok, Sikkim, India

Vajinder Kumar
School of Chemistry and Astbury
Center for Structural Molecular Biology
University of Leeds
Leeds, UK

Tanupriya Choudhury
School of Computer Science
University of Petroleum and Energy Studies
Dehradun, Uttarakhand, India

ISBN 978-3-032-29476-0 ISBN 978-3-032-29477-7 (eBook)
https://doi.org/10.1007/978-3-032-29477-7

This Springer imprint is published by the registered company Springer Nature Switzerland AG
The registered company address is: Gewerbestrasse 11, 6330 Cham, Switzerland

Contents

Temperature Pattern and the Escalation of Heat Waves

Rahul Malik, Tanya Mittal, Divyanka Kanyal, Sapna Rana, Tanupriya Choudhury, Bahati Shabani Nzeyimana, Kassia T. T. Amesho, and Amit Awasthi

Abstract The frequency, intensity and duration of extreme heat events are escalating all over the world, but little has been integrated in literature regarding its multi-scale effects on plant ecosystems. This chapter summarizes temperature changes, heat wave processes and physiological ecosystem-level impacts. The study of the significant heat waves in history (2003, Europe; 2010, Russia; 2015–2022, India; 2022, China) shows the increasing human and environmental expenses. Heat stress and climate adaptation: Bibliometric analysis of 110,000 publications (2015–2025) shows an increasing interest in the research topic. At cellular scales, extreme heat (>40 °C) triggers protein denaturation, reactive oxygen species accumulation, and xylem embolism, disrupting photosynthesis, water transport, and reproduction. On an ecological scale, forests are no longer carbon sinks but carbon sources due to a change in community structure; biodiversity is lower than it would be in a healthy ecosystem; agricultural yield drops by 10–20% when temperatures go excessively high or excessively low. Local heating is increased by critical feedback loops, for example, decreased evapotranspiration, soil desiccation, and vegetation loss. We find knowledge gaps in climate-prone areas (Africa, Latin America, South Asia) and suggest interdisciplinary studies combining climate modeling, plant physiology, and policy formulation to make ecosystem more resilient to the further rise in temperature.

R. Malik (✉)
Uttarakhand Pollution Control Board, Dehradun, Uttarakhand, India
e-mail: rahulmalik078@gmail.com

T. Mittal · D. Kanyal · S. Rana · T. Choudhury
School of Computer Science, UPES, Dehradun, Uttarakhand, India

B. Shabani Nzeyimana
PG and Research Department of Environmental Sciences Bishop Heber College (Autonomous) affiliated to Bharathidasan University, Tiruchirappalli, India

K. T. T. Amesho
Department of Higher Education and Lifelong Learning, School of Education, Faculty of Education and Human Sciences, University of Namibia, Windhoek, Namibia

A. Awasthi
Department of Applied Science, School of Advanced Engineering, UPES, Dehradun, Uttarakhand, India

A. Awasthi et al. (eds.), *Rising Temperatures and Plant Ecosystems*,
https://doi.org/10.1007/978-3-032-29477-7_1

Keywords Temperature · Heat wave · Plant ecosystem

1 Introduction

Temperature is a key variable in climate science and ecology because it influences how energy is distributed and how air moves. It also helps define the thermal limits for life on Earth Over geological timescales, temperature has influenced evolutionary trajectories, reorganized ecosystems, and threatened the stability of human civilizations to support livelihoods, but now those changes are occurring at a rate not previously seen in human history (Gowdy, 2020). There is a continuous increase in global mean temperatures. Among these, we may observe heat waves (HW), severely hot spell caused by trapped air. There are few habitats more impacted than what HW are to the biosphere and specifically to plant systems which are especially responsive to temperature (Miracle et al., 2025).

The response of plant physiology to temperature shows that they are more susceptible to climate change. Plants are important sources of structure and function in ecosystems, primary production, carbon flux, and habitat complexity, but their mobility makes them vulnerable to climate change because they are constrained by such small thermal ranges in their physiological functions (e.g., photosynthesis, water relations, and reproduction) (Oishi et al., 2025). Even a slight change or "change to either side of their optimum physiological level will lower their efficiency and survival in the long run. In a very short time, once extreme events of heat exceed biological thermal limits, the repercussions can manifest themselves within a very short period and on a multitude of scales, which influence the performance of plants, the ecology, and systemic risks of communities, which rely on such ecosystems. Considering the centrality of such events in the development of ecological outcomes, it is necessary to have a clear definition of what a heat wave is (Weiskopf et al., 2020).

A heat wave scientifically is a phenomenon that is associated with extreme weather characterized by a prolonged episode of abnormal heat relative to the normal climatological average of that region. Definitions vary from institution to institution and region to region (Awasthi et al., 2022; Yadav et al., 2023). This is due to the differences in the average climatology or baselines for what is "normal." The World Meteorological Organization (WMO) defined a heat wave as at least five consecutive days when the daily maximum temperature is at least 5 °C above the average climatological norm (Boni et al., 2023). The India Meteorological Department (IMD) considers heat wave conditions in India when maximum temperatures reach 40 °C or above in the plains or 30 °C in hilly regions, with 2 consecutive days being the minimum duration of a HW.

The high temperature is not the significant factor in the HW, it is in relation to the local climate change and continuity (Awasthi et al., 2022; Ren et al., 2023). What could pass as a normal summer day in a certain area of the world could be a life threatening heat wave in another area. This relativity points out the significance of context. Equally, time is essential and when one day is hot, it could be stressful but the longer the time spent in the hot atmosphere, day after day, night after night, the worse it is. This continuation further increases the burden on living creatures since they do not have an opportunity to rest. In the case of plants, it implies a constant physiological stress in the most susceptible period of their development and reproduction.

Global warming is directly related to the rising rate and severity of HW. Human activities have increased the average temperature of the planet since the mid-20th century. It is not only that it is moving the average around, but it is also increasing the frequency of extreme heat events that were previously infrequent (Zhao et al., 2022). Consider it as loading the dice of the climate. The climate change has increased dramatically the chance of rolling an extreme heat event. This has been proved by climate studies which give numerous recent extreme HW such as the one in Europe in 2003 and in Russia in 2010 (Barriopedro et al., 2023). High temperature drives species to physiological limits, accelerates water loss, disrupts reproduction and in the worst scenario, causes mass death. Forests, especially, are at risk of a massive dieback when HW and drought coincide (Rahmstorf, 2024). The agricultural systems experience declining crop yields endangering world food security. Not even the plants in the urban environment, which are the necessary cooling and purifying air, can participate in long heat spells. Biological stressors do not act in a vacuum. The condition of stressor processes in time and space is cumulative in its character. Both the ecosystems and the species that are contained in them are becoming less resilient as the HW get longer and more frequent. Plants losing their role of reducing extremes in climate are collapsing, and all the feedback loops are enhanced everywhere: less shading, less evapotranspiration, less soil moisture enhances the effects of the heat (Yuan et al., 2024). This biodegradation is not only a science problem, but also in actual sense, it has imitative effects in human health and wellbeing, particularly to communities that are already the most vulnerable to climatic variations. The effects of these stresses spill over to the human societies, where weaker ecosystems become less efficient in capturing carbon, becoming less biodiverse and providing fewer life-sustaining facilities to societies.

Researching on HW in association with plants and ecosystems is significant due to a variety of reasons. First, it can determine the weak areas in agricultural and ecological systems and this information may be used to make more specific adaptations such as breeding heat resistant crops or reforming the management system such as irrigation or forest management (Singh et al., 2023). As an example, in semi-arid areas, farmers have already started to experiment with drought-resistant varieties and change the time of sowing. These practical modifications demonstrate how research can be directly applied into the field. Secondly, it will shed light on the global processes of water and carbon, as the stressed ecosystems will no longer be a sink of carbon and water and turn into a source of these substances. Such stress is starting to appear in the many forest biomes in India and elsewhere when they experience long HW; it means that this transition is no longer hypothetical (Ravi et al., 2025). Thirdly, it emphasizes the long-term resilience of ecosystems, which may be relevant to the policies concerning the loss of biodiversity and food security. Lack of this understanding can cause policy makers to respond too slowly or too quickly to crises in disaster relief rather than resilience. More to the point, the fact that plant systems become the basis of life in terrestrial ecosystems, implies that their reaction to extreme heat may serve as a leading indicator of an ecological reaction that much larger (Nguyen et al., 2023).

The structure of this chapter is such that, it forms a clear background to the realization of the temperature trends, and their impact on accelerating the HW in the climate change setting. It starts by discussing the nature of temperature dynamics and scientific and meteorological meaning of HW. Then there is the focus on the characteristics of HW and the climatic drivers like greenhouse gases emission, atmospheric circulations, soil–moisture interactions, and urban heat island effects that determine the frequency and severity of HW. In the next chapter, the author focuses on the effect of global warming on increasing HW using examples of historical and recent cases, such as the 2003 European heat wave, the 2010 Russian crisis, the 2015 Indian episode, and the 2022 heat wave in China. Subsequent sections concern the implications to the ecosystems and plant life, including physiological strain, species range changes, and ecological relationship disturbances. To place these discussions in the context of the wider field, a bibliometric review is given, which explains the current trends in research, contributions to the research on a regional level, and knowledge gaps present. A conclusion of the chapter contains a synthesis of the findings, which forms the basis of other chapters which will discuss plant physiology, adaptation and ecosystem management in the times of increasing temperature extremes.

In such a way, the chapter makes the urgency of the issue as well as the conceptual premises necessary to understand the increasing dangers that HW have on plant ecosystems and, by implication, on human societies. Despite this fact, the need to approach the problem of HW in terms of the climatological and ecological phenomena emerges. This introductory chapter attempts to offer that groundwork by investigating the increasing temperature patterns, meteorological and climatic processes involves in heating waves, and increasing evidence of their heightening in various areas. It also puts the scientific discourse in the context of a bibliographic review, which illustrates how academic research has reacted to extreme heat events. By doing it, the chapter preconditions the further sections that become more directly plant-oriented analyses.

2 Bibliographic Analysis of Heat Wave Research

2.1 Keywords Used

The bibliometric analysis revealed a wide range of recurring keywords within the literature, including circulation pattern, heat stress, spatial pattern, heat wave, temperature anomaly, climatic hazard, soil moisture, drought, climate model, and heat index. Additional emerging terms such as RCP4.5, RCP scenarios, machine learning, random forest, and ensemble simulation also appeared frequently, alongside related concepts like urban heat island, climate adaptation, vulnerability, and extreme temperature events. Together, these keywords represent the dominant thematic language used across studies addressing temperature patterns, HW, and associated climatic impacts.

2.2 *Methodology for Bibliometric Visualization*

The bibliometric data generated by Dimensions through search query Temperature Pattern and HW provided me with the visual representations applied in this review: bar charts, year-wise publication trends, network maps etc. After data extraction, distribution of the frequency of keywords was done and key word frequencies were plotted through spreadsheet and statistical software to generate clean bar charts to depict themes prominence in the literature. VOSviewer, which works with bibliographic metadata to produce network maps based on their strength of links, cluster pattern and node density, was used to generate co-authorship and keyword co-occurrence visualizations. These graphs are graphical representations of team relationships, subjects and trends in the study area. The line chart representing the yearly publications was created by summing on the number of publications in each year and plotting it so as to demonstrate how this changes with time. All these methodological measures, in combination, guarantee that the graphical results are analytically sound and visually clear to provide a true picture of the research environment in the context of the temperature patterns and HW.

2.3 *The Country Co-authorship Map in Heat Wave Research*

The Country Co-authorship Map in Figs. 1 and 2 highlights strong global collaboration in heat wave research, with China and the United States emerging as the primary leaders; among them, China shows the highest level of co-authorship, indicating its dominant contribution and extensive international partnerships. Two major collaborative clusters are visible. The Euro-Atlantic group comprising the United States, the United Kingdom, Germany, Canada, and Western European countries, focuses largely on public health impacts and policy responses. In parallel, an East Asian–Vulnerable Economies cluster connects China and South Korea with highly affected regions such as India, Iran, and Saudi Arabia, emphasizing adaptation strategies for densely populated and climate-sensitive environments. Overall, the network reflects a robust system of cross-border cooperation, underscoring that extreme heat is recognized globally as a shared scientific and policy challenge.

The bar chart in Fig. 3. entitled Top 10 Keywords in Heat Wave Research (2015–2025) is a visual representation of frequent themes in research publications across the world on the topic of temperature pattern and HW as it is found in the Dimensions.ai database. The information covers a 15-year period (2015–2025) and covers

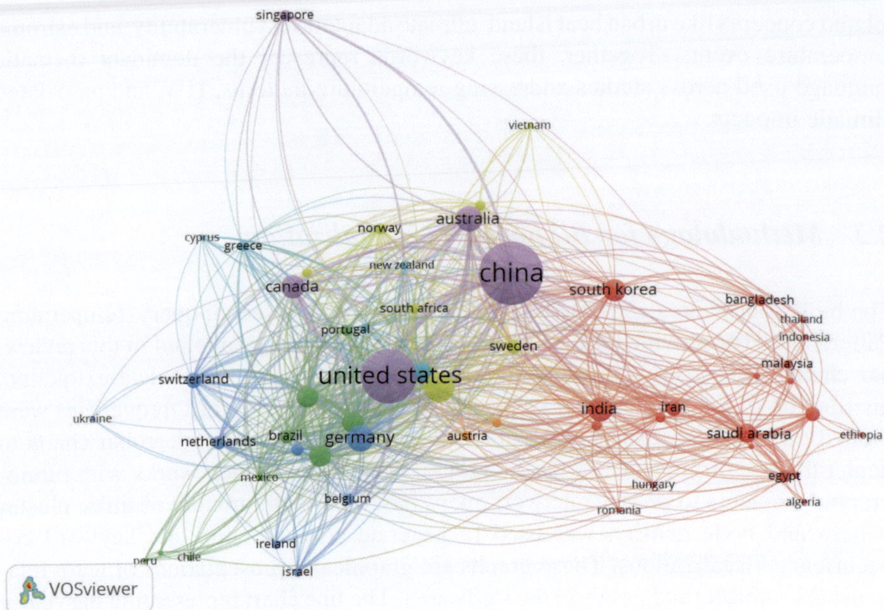

Fig. 1 Global country co-authorship network in heat wave research, highlighting major collaborative clusters between Euro-Atlantic and East Asian–vulnerable economy regions and the central role of China and the United States in international partnerships

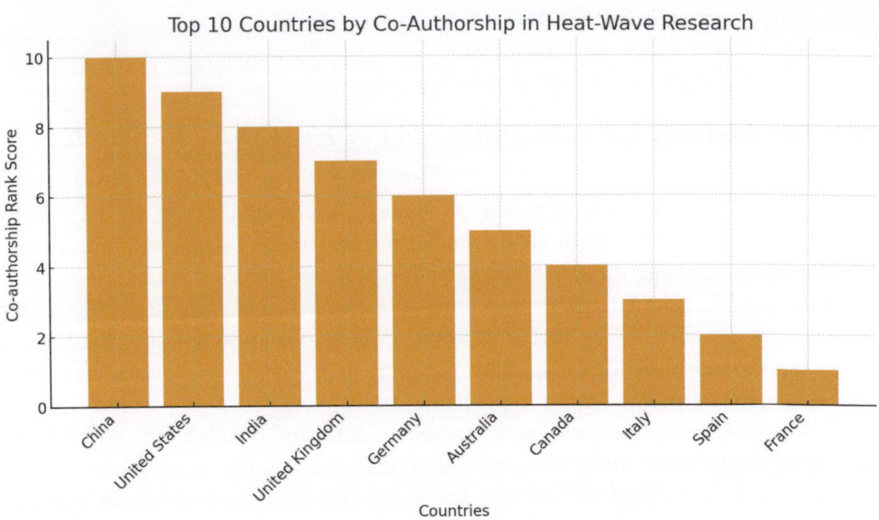

Fig. 2 Analysis of Top 10 countries by co-authorship in heat wave research

different scientific interests as much as atmospheric dynamics to biological responses and data modeling. The key word, which is most prevalent the circulation pattern, shows the main theme of climatic science to comprehend the impacts of large-scale atmospheric patterns in the initiation and continuation of HW. This is in line with the research done on the study of anomalies in temperature in the region and change in the behavior of the jet streams which dictate the occurrence and even the intensity of heat events. The prevalence of the term heat stress is an indication of the growing interest in the ecological, agricultural, and health effects of temperature change, as reflected by the interdisciplinary spread of climate sciences beyond meteorology as a field in its own right. Combining remote sensing, machine learning, and statistical climatology research has emerged as an important aspect of high-resolution heat risk assessment. Keywords like soil moisture and drought indicate a high degree of correlation between land surface processes and the intensity of heat, which indicates that the feedback of dry soils may increase local extremes of temperature. The methodological diversity is manifested in lower but nonetheless important terms such as temperature anomaly, RCP4.5, random forest, and heat index. They specify the joint application of emission scenario models, data-based algorithms, and physiological metrics in evaluating extreme heat conditions on the environmental and human levels.

The cloud of words in Fig. 4, presented in this study, provides a summary of most frequently reliant keywords, in the existing literature on the temperature tendencies, HW, and other similar climatic changes. With the help of determining the size of each word in accordance with its presence in the reviewed literature, the graphic concentrates the ideas with the highest level of scholarly interest. The terms such

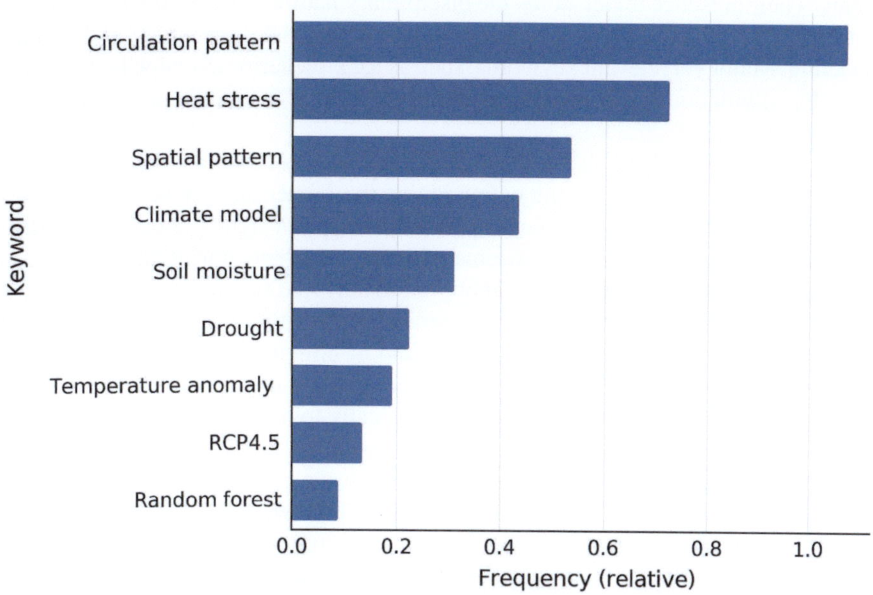

Fig. 3 Analysis of Top 10 keywords in heat wave research (2015–2025)

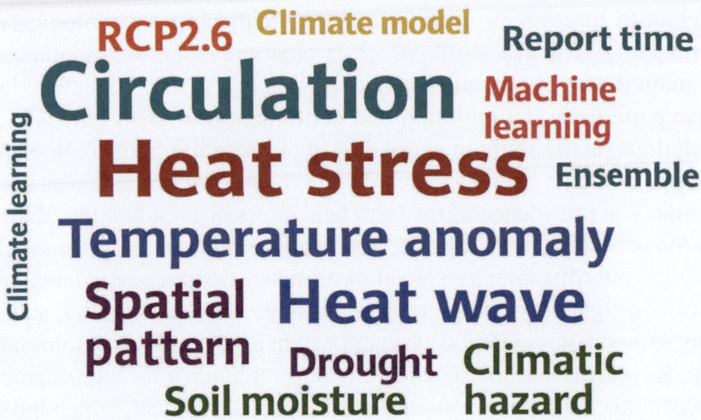

Fig. 4 Provides a summary of most frequently reliant keywords, in the existing literature on the temperature tendencies, HW, and other similar climatic changes

as heat stress circulation pattern, heat wave temperature anomaly and pattern are in larger fonts and that proves their significance in the themes of current research. The keywords used accentuate the necessity to understand the causes and consequences of heat episodes, climate variations and changes, in atmospheric dynamics, in the region.

The keywords of size like soil moisture drought climatic hazard and climate model are the auxiliary topics that are commonly used in discussions of heat-related climate events. Their presence underscores the disciplinary nature of the heat wave research that links disciplines such as meteorology, hydrology, risk assessment and climate modeling. Major yet relevant keywords, such as RCP scenarios machine learning and ensemble simulation imply the opportunities to develop the techniques and expand areas of interest.

The line chart in Fig. 5 shows the annual publication of temperature patterns and (HW) during 2015–2025, by the extraction of the Dimensions data on November 30, 2025. The general trend of the graph is that there is a significant and consistent increase in scientific interest in extreme temperature events during this decade. In 2015, the activity of research starts at approximately 46,000 publications. The next two years, 2016 and 2017 are only characterized by slight increases, which is a point of slow adoption as the knowledge in the field of climate-related studies was still growing but had not yet taken off. An evident change happens in the year 2018, and the number of publications increases drastically to approximately 60,000. This is possibly because of the heightened anxiety about the effects of global warming, and the rising number of reported extreme heat wave events in other parts of the world such as Europe, South Asia, and Australia. Following a modest downturn in 2019, there is again a rapid increase in research output in 2020, with an estimate of approximately 70,000 being reached. This is growth consistent with an increase in global discussion of climate resilience and health effects during and after the

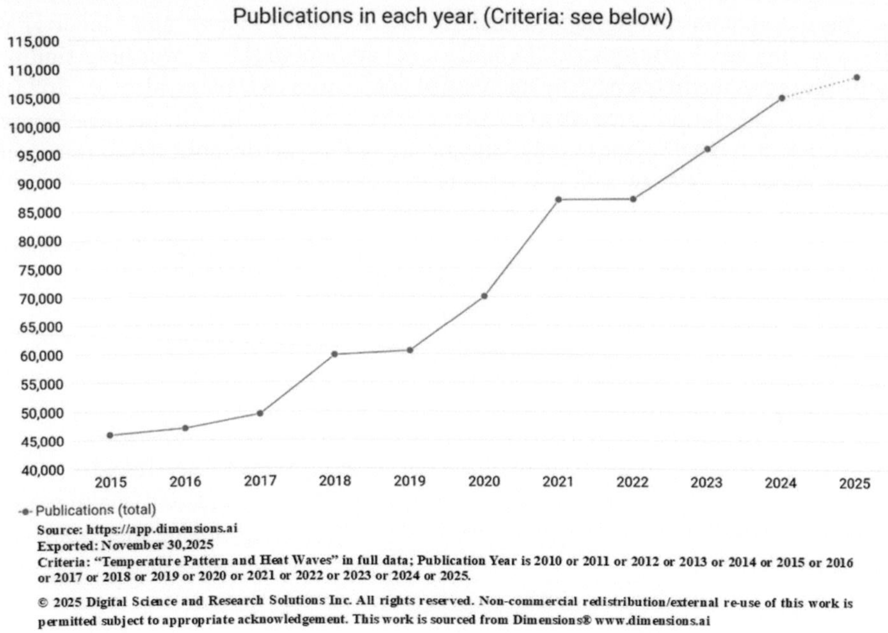

Fig. 5 Analysis of number of publications year-wise from 2015 to 2025

COVID-19 pandemic, when environmental vulnerabilities were put under a more critical focus. The number of publications increases to almost 87,000 in 2021 and 2022, indicating an increased scientific interest in the modeling of the heat extremes, the comprehension of the warming trends, and future temperature deviations. The trend is projected to rise until 2023 and 2024 reaching 105,000 publications. The estimated figure of 2025 of nearly 110,000 suggests that the trend is here to stay.

3 Global Temperature Trends

After 1970s there is a sharp increase in global average surface temperature which is due to greenhouse gases, that is, the gases in Earth's atmosphere which trap heat such as methane, carbon dioxide, methane, ozone and nitrous oxide whereas this is not the only because it is also linked from deforestation, industrialization and intensive warming. Nowadays industrialization plays a major role in impacting earth's overall climate but during the preindustrial era (1850–1900) temperature was relatively stable which became clearly detectable by mid-twentieth century period (Chakraborty et al., 2025; Kabir et al., 2023).

Warming is not uniform globally because it varies crucially from region to region due to factors like land versus ocean differences, latitude and geography. In Middle East and Arctic, the warming is almost twice as fast as global average influencing extreme weather in mid latitudes causing more frequent, intense and longer HW affecting the global atmospheric circulation. An additional 0.5 °C warming significantly increases the frequency and intensity of heat extremes (Pattnayak et al., 2023). South Asia and North Africa are experiencing sharp increase in extreme heat facing greater risk than other regions even India in 2022 saw one of the hottest pre-monsoon periods in over a century with temperature exceeding 45 °C which shows climate impacts are unevenly distributed (Sharma et al., 2022; Tandon et al., 2025).

Cities such as Delhi, Phoenix and Shanghai experience high night time temperatures due to Urban Heat Island effect which intensify heat because materials like concrete, asphalt, and glass absorb large amount of solar radiation during the day and remit it slowly as heat this reduces the ability of flora and fauna to recover from daytime heat stress resulting in urban areas facing higher heat wave impact than surrounding regions (Malik et al., 2025).

Climate projections (CMIP6) suggest that under high emission global temperature could rise by over 3 °C by 2100, in this alarming situation there is a need of adaptation strategies such as heat wave warning systems, cool infrastructure, and health plans as future HW will be more frequent, severe, and long lasting making them climate resilience challenge.

4 Characteristics of HW

To fully understand HW, it is important to know about the characteristics of it. A heat wave is a prolonged period of excessively hot weather, accompanied by high humidity and stagnant winds, usually lasting for around two or more days. Due to this phenomenon, there is seen a significant rise in temperature, above the historical average for a specific region (Barriopedro et al., 2023).

Though the definition being relative to both by different regions and institutions such as World Metrological Organization (WMO), which defines it broadly as a period of 2–3 consecutive days temperature increasing the local historical average by 5 °C (Tao et al., 2025). Whereas Indian Metrological Department (IMD) categorizes it on absolute temperatures, that is, the maximum temp for cities must reach above 40 °C, while for hilly areas the threshold is above 30 °C (Arya & Rao, 2023). As for United States National Weather Service (NWS), they rely upon heat index combining both air temperature and humidity and the duration of heat events, issuing heat warnings when the temp hits above 105 °F (Hawkins et al., 2017).

Despite these differences, all institutions agree over the fact that HW are a relative phenomenon as geographical conditions differ across the world. This makes global comparison difficult and in regions like Northern Europe a temp of 35 °C might be considered severe but for arid regions like the Middle East it is considered fairly normal (Klingelhöfer et al., 2023).

Apart from differences in definition, HW also differ in duration and intensity as they can range from 2–3 days to over a week (Faye et al., 2021). In recent years, an increase in nighttime temperature has been seen leading to increased risk of cardiovascular failure and chronic stress in humans. While in plants this has led to disruption in respiration and carbon balance, ultimately leading to reduced crop yield and retardation in growth (Chen et al., 2023).

As HW depend upon geographical factors, their seasonality is also shaped by it. In tropical and subtropical regions HW generally occur around the pre monsoon or dry season, while in temperate regions they are majorly a mid-summer phenomenon (Ravindra et al., 2024). In places such as southern Africa and Australia which are categorized as vast and dry continents having very long summers causes more than one-time seasonal occurrence of HW (Meque et al., 2022).

Apart from all these characteristics, HW have proven to have devastating potential leading to many historical records that have occurred in the last two decades. The 2003 European heat wave that killed over 70,000 people and caused widespread forest and crop losses. The 2010 Russian heat wave destroyed 25% of the nation's crops and claims over 50,000 lives. In Southeast Asia, the 2015 Indian heat wave which was recorded as the fifth deadliest in history took over 2,500 lives with temperature reaching 48 °C. While in 2022 Pakistan recorded prolonged temperature above 45 °C in March causing Glacial Lake Outburst floods that destroyed the infrastructure in the north. In the same year China experienced the most severe heat wave ever recorded in global history for duration of more than 70 days leading to drought, energy shutdowns and eventually the cut down of the global chain supply.

Table 1 shows a reconstructed summary of the key historical and current heat wave phenomena, focusing on their geographical coverage, strength, duration and respective effects of the event on human and ecological systems. The table has compared landmark episodes including the 2003 European heat wave, the 2010 Russian heat wave, the 2015–2022 Indian heat waves, and the 2022 heat wave in China in a systematic way, showing the increasing frequency and severity of heat waves over the past decades. In every event, it describes notable meteorological features, such as maximum temperature deviation relative to climatological normals and the duration of extreme conditions on successive days, and how the deviation surpasses critical thermal limits of the effects on human health and plant activity. Besides physical parameters, the table is also a compilation of human casualties, economic losses, and recorded impacts on agriculture and ecosystems, including the decline in crop yields, forest dieback, and carbon sink to carbon source behavior. This multi-dimensionalities highlights the fact that heat waves are not single meteorological events but multifaceted socio-ecological crises that spread throughout sectors. Table 1 also helps highlight regional susceptibility by comparing what is happening in the various climate zones and developmental settings, especially in climate-sensitive regions like South Asia, where the issue of high population density and a weak adaptive capacity increases the risk factor. Generally, the table serves as a brief empirical basis of the chapter, supporting the accelerated pace of the heat wave intensity and giving a comparative context to further deliberations of the physiological stress of plants and an unstable ecosystem in general and the necessity of combined adaptive and policy measures.

Ultimately, the cases in Table 1 reveal that HW are not just a public health crisis but also pose a greater threat to ecological and economic stability. To fully understand

Table 1 Major twenty-first-century HW and their climatic drivers, impacts on agriculture and ecosystems, and associated mortality across Europe, Russia, India, and China

Heat wave event	Year/period	Region	Time	Peak temperature (°C)	Primary causes	Key impacts	Deaths
European heat wave	July–August 2003	France, Central/ Western Europe	~4–6 weeks	35–40	Warm Sahara air masses, stationary high-pressure systems	70,000 deaths, reversed carbon sink role, reduced photosynthesis over millions of hectares, agricultural losses (fruits, maize), power shortages, drought impact on water supply	~70,000
Russian heat wave	July–August 2010	Western Russia (Moscow region)	4–6 weeks	38–40	High-pressure blocking systems trapping hot dry air	55,000 deaths, massive wildfires destroying boreal forests, wheat export ban (Aug 2010–July 2011), grain harvest dropped 25%, biodiversity decline, rural destruction	~55,000

(continued)

Table 1 (continued)

Heat wave event	Year/period	Region	Time	Peak temperature (°C)	Primary causes	Key impacts	Deaths
Indian heat wave (I)	Late May–June 2015	Andhra Pradesh, Telangana, Bihar, Odisha, Delhi	3–4 weeks	45–50	Dry continental winds, delayed monsoons, persistent high pressure	2,500 deaths (2,000 in AP/Telangana), melted roads, agricultural losses, water shortage, power cuts, infrastructure strain	~2,500

(continued)

Table 1 (continued)

Heat wave event	Year/period	Region	Time	Peak temperature (°C)	Primary causes	Key impacts	Deaths
Indian heat wave (II)	March–May 2022	North/Central India (Delhi, Punjab, Haryana, UP, Rajasthan, MP, Gujarat)	2–3 months (early onset)	44–49	Persistent high-pressure systems, weak Western Disturbances, climate change influence	Hottest March in 122 years, wheat yield reduction (10–20%), wheat export ban, record electricity demand, coal shortage, forest fires in Uttarakhand/HP, hundreds of suspected heat-related deaths	Hundreds (estimated)
Chinese heat wave	June–Late August 2022	Southern and Central China	70+ days (longest recorded)	40–45	Persistent subtropical high-pressure system, reduced rainfall	Dual stress (heat + water scarcity) on rice and maize crops, food security concerns, global food price impact, power restrictions on factories, public cooling shelters opened	Not clearly documented

this growing threat, we must now look upon the various climate drivers of HW int the next section, which are responsible for generating and amplifying HW.

5 Climate Drivers of HW

High-pressure systems such as anticyclones are the causes of HW, which are usually viewed as occurrences in isolation, since this causes the air to sink to the ground where it is compressed by atmospheric pressure. This pressure forms a top to the rising heat that traps the upshots close to the surface and does not allow the infiltration of the colder fresh air. This impact is also referred to as the Heat Dome Effect that causes HW (McGregor et al., 2024). A case in point of the Russian heat wave in the year 2010 which was instigated by a blocking high that hovered over western Russia causing the entrapment of hot air worsening the drought conditions over a period of weeks indicates that understanding the underlying conditions is essential not only in scientific terms, but also in designing proper predictive models and developing effective mitigation measures.

A simultaneous factor driving HW is ocean drivers such as El Nino and La Nina. While El Nino effect which tends to release an immense amount of heat from the Pacific Ocean into the atmosphere is major cause of disruption in global weather patterns and increased likelihood of HW in regions of South Asia, Australia, and North America; La Nina despite having its general cooling effect, shifts jet streams to cause severe drought driven HW through prolonged rainfall in Southern USA and South America. Even though both phases disrupt the global weather circulation, El Nino remains the primary driver of increased global temperature spikes (Murari et al., 2016).

Another global factor, which has remained majorly responsible for rising temperatures and HW, is the 'greenhouse effect'. Due to excessive release of heat trapping gases like carbon dioxide, methane and nitrous oxide, there has been a significant rise in the planet's average temperature. Moreover, these gases create an insulating blanket that traps heat preventing it from going out into the space, thus weakening jet streams, causing them to stall over a particular region instead of moving away as they should, which as a result creates heat domes over regions thus preventing the air from cooling down even at night, making extreme HW event even more frequent (Chakraborty et al., 2025).

A major amplifier of HW is the Urban Heat Island Effect (UHI). In today's modern world, excessive use of asphalt, concrete, etc., is causing a rise in temperatures by absorbing and retaining more heat than natural landscapes. As a result of both anthropogenic sources such as air conditioners, vehicles, industrial exhausts and unplanned infrastructure, UHI is acting as a force multiplier of HW, making cities 3–5 °C hotter than their surrounding neighborhoods lying in the outskirts of the city, thus pushing temperatures into a dangerous territory (He et al., 2021).

The soil moisture levels act as a surface amplifier for HW, as under normal conditions the moisture in soil helps in actively cooling the air above it through the process of evaporation which turns water into water vapor. But in conditions where the soil moisture level is depleted, the sun's energy is forced to increase the ground temperature and the air directly above it instead of evaporating water, which leads to pumping heat upward in the atmosphere. This results in a feedback loop as rising temperatures dry out the soil further, which in turn intensifies the heat, thus causing more pronounced HW. This effect is especially pronounced in landlocked regions, as these regions primarily depend upon evapotranspiration for regulating the surface temperatures (Gurung & Chen, 2024).

Ultimately, climate change has shifted the global baseline. Transforming extreme heat events which were once considered an anomaly in the mid-twentieth century are now being recognized as a recurring reality. Therefore, understanding this baseline shift in the growing threats posed by HW has now become much more crucial.

6 Escalation of HW Under Climate Change

Mounting evidences from advanced climate models, long-term studies and observations points to the reality that these once in a decade events which were once considered an anomaly are now becoming more frequent and intense as global heating continues, thus occurring with alarming frequency worldwide due to which the regions of South Asia, the Middle East, Southern Europe and parts of North America are especially being majorly affected by HW (Hansen et al., 2025).

Now, (HW) are not just frequent; their intensity and duration have also increased, creating a great difference between now and as it was 50 years back. According to IPCC AR6, extreme heat events that used to occur once every ten years are now more likely to happen four times per decade. Under a high emission scenario (SSP5-8.5), HW could become two to three times more frequent by the century's end. Even under more moderate pathways (SSP2-4.5), the planet is more likely to warm up by approximately 2.7 °C by 2100, which will in turn affect regions having dense population and fragile ecosystem to bear the consequences of these changes (de Bont et al., 2024).

As previously discussed, the European heat wave of 2003 was initially viewed as an unexpected event. But now it is well known for the effect of human-induced warming. Similarly, the 2010 Russian heat wave which disrupted the nations grain crops and overwhelmed public health systems causing more than 50,000 deaths was traced back as the result driven by human activities. In south Asia, India has seen a significant rise in pre-monsoon HW over the past two decades. While China experienced the most severe heat wave in 2022 which stretched across multiple provinces for over two months. This resulted in drying up the Yangtze River, causing droughts and reduced hydropower output. These prolonged events show that the danger lies not only in rising temperatures, but in their persistence that turns HW into a disaster forcing ecological, agricultural, and infrastructural systems past their breaking points.

Apart from this, escalating heat has affected our ecosystem in a great way. Our terrestrial ecosystem, which consists of forests, has stopped acting as carbon sinks and has started releasing carbon, which in turn is accelerating planetary heating. While the marine ecosystem consists of oceans which absorb 90% of excess planetary heat, is causing marine HW, which have increased by 50% over the past decade. This has led to the devastation of the coral reefs, and the reduction in oxygen concentration which in turn is causing mass die offs of fishes thus disrupting the food webs of the ocean (Hansen et al., 2025).

For human societies, the rise in HW has affected human health, causing increase in death rates from cardiovascular and kidney failures, while also affecting the infrastructure and economy by degrading physical assets such as melting asphalt roads, deforming railway tracks, and destroying transformers causing billions of economic losses. Overall HW have emerged as one of the most urgent and the deadliest faces of our changing climate (Rao et al., 2023).

7 Implications of HW for Ecosystems and Plant Systems

The very simple concept of HW is that it increases the temperature, yet it does not limit itself to that. The effects of HW on our ecosystem particularly the plant system are enormous. Their biological roles, development cycles, distributions pattern and ecological relationship all rely heavily on the temperature of the surrounding in that when the temperature exceeds the capacity of the surrounding, an under spread disturbance of the ecological system is caused.

On the cellular scale, extreme heat often greater than 40 °C causes essential proteins and enzymes, especially those crucial for photosynthesis and respiration to lose their shape and efficiency. There has also been seen reduced photosynthesis during HW as the enzyme responsible for it starts losing its shape and specificity which causes the enzyme to bind oxygen instead of carbon dioxide. This process is termed photorespiration, which is stated as a wasteful metabolic process. It reduces photosynthesis efficiency by burning energy and releasing already fixed carbon, while producing zero sugar in return, ultimately reducing crop yield.

HW also cause the buildup of toxic molecules such as reactive oxygen species (ROS) inside the plant cells often called 'free radicals'. This sudden growth of ROS inside the plants overwhelms their natural antioxidant system which leads to the destruction of vital parts of cell like the cell membrane, proteins and DNA overall leading to scorched leaves, yellowing, premature aging and ultimately tissue death.

Due to rapid increase in temperature by HW, water evaporation from lakes, reservoirs and soil takes place rapidly, which causes water scarcity. This situation makes it increasingly difficult for plants to retain moisture, therefore, to conserve water, they tend to close their stomata. But this defense mechanism limits their breathing process and the intake of carbon dioxide during photosynthesis. Furthermore, under extreme conditions, the need to transpire becomes overwhelming, which causes too much pressure upon xylem, forming an embolism that blocks the tube. This is known as xylem embolism, which can turn fatal for plants if not resolved.

Apart from this, HW can also disrupt the entire plants' reproductive system. During the flowering stage extreme heat can sterilize pollen, thus reducing pollen viability, preventing it from growing further. As a survival defense, the plant has to lose its flowers and fruits to conserve the resources for the parent plant. Overall, this leads to a reduction in yields and empty grains especially in staple crops like wheat, rice and maize; all of it resulting as pollination failure.

As a result, there's been seen migration in different species of plants due to die-off of plants in their original homelands, they are now shifting toward higher altitude to follow their ideal temperature range as it moves, thus leading to a generational shift in them. However, these migrations often fail because of geographical barriers like cities and farms, leading to higher risks of local extinction.

Ecosystems may enter into a completely new condition by prolonged durations of repeated heating. Forests where dieback occurs frequently can become shrublands or grasslands and this will reduce the carbon storage capacity and change the surface reflectivity. Grasslands, in their turn, can present the transition toward the more heat- and drought-tolerant species, which can influence the forage quality of herbivores. Agricultural systems would restrict farmers using traditional crops by forcing them to adopt heat-tolerant crops which in most cases are less nutritious.

Finally, the ecological effects of HW are felt on a molecular scale, even in the cell, up to the macro level, such as landscape changes. These occurrences are not isolated cases but they alter the very core of ecosystems and tend to have long lasting effects which may be hard to reverse.

Table 2 is a systematic, scale-oriented summary of the translation of extreme heat into cascading effects of cellular processes into entire ecosystems and agriculture as a whole. It classifies the information into six interrelated dimensions of scale of impact, physiological or biophysical mechanism, specific process affected, observed outcomes, ecological consequences, and implications of agriculture and ecosystem services, pegged with recent references. At the cellular level, it emphasizes the denaturation of proteins and enzymes, the buildup of reactive oxygen species, dysfunction of the xylem, and increased photorespiration and demonstrates that the high temperatures above a critical point leads to the impairment of photosynthesis, respiration and water movement and lowers primary productivity and biomass. At the organism-level, the table highlights stomatal closure, and reproductive failure in heat stress as a response, as well as associating water-conservation trade-offs and pollen sterility to growth restriction, seed or fruit abortion and crop failure in major staples, such as wheat, rice, and maize. Population and community level rows define the range changes to cooler temperature and the shift in species structure, such as the substitution of sensitive species with C4 grasses and xerophytes that are able to survive the heat, reducing genetic diversity, altering biodiversity and reducing the quality of forage. At the ecosystem scale, the table reports forest die back, carbon sink-source transitions, and declining evaporation showing how repeated occurrences of heat and drought events have caused canopy loss, net carbon release, and enhanced local heat extremes. Lastly, it describes the destabilization of crop phenology, pollination systems, and soil health as immediate dangers to food security, ecosystem services, and land sustainability in the long run, suggesting the systemic quality of heat wave effects at scales.

8 Expanded Case Studies of Heat Wave Impacts

Case Study 1: The European Heat Wave of 2003

Occurred in late July to mid-August 2003 which was triggered by warm Sahara air masses and stationery high-pressure systems. It caused nearly 70,000 deaths across Europe with France impacted the worst, temporarily reversed Europe's role as carbon sink, drought lowered major rivers impacting the water supply and transport leading to several agricultural losses especially in fruit orchards, maize, and livestock fodder. It also led to reduced photosynthesis over millions of hectares, power shortages due to increased electricity demand in the country. This incident exposed Europe's lack of preparedness for extreme heat.

Case Study 2: The Russian Heat Wave of 2010

This heat wave had a major global economic impact It is one of the most intense HW in Russia's history occurred in July–August 2010 and majorly affected the western part of Russia especially Moscow. They were triggered by high-pressure systems, creating a blocking event that trapped hot and dry air. It caused nearly 55,000 deaths with Moscow experiencing major deaths due to smoke inhalation and heat stroke. Large areas of boreal forest destroyed by massive wildfires. Wheat export ban (Aug 2010–July 2011) leading to higher global food prices whereas grain harvest dropped. This heat wave caused a decline in biodiversity, fires altering forest structure and destruction of many rural homes due to which government had to take major steps such as creating firebreaks around villages and emergency medical support.

Case Study 3: The Indian Heat Wave of 2015 and 2022

The 2015 Heat Wave

India is already at risk of HW because of its dense population and reliance on agriculture. It occurred in late May–June 2015. The main affected areas were Andhra Pradesh, Telangana, Bihar, Odisha, and Delhi. Temperatures reached an all-time high of 45–48 °C, with some areas even exceeding 50 °C. The cause was said to be dry continental winds, delayed monsoons, and obstinate high-pressure conditions. Estimated deaths were around 2,500, with around 2,000 in Andra Pradesh and Telangana alone. The high mortality was closely linked to poverty and lack of cooling access. Roads melted in the cities; the agriculture suffered, water shortage intensified, and power cuts were frequent as the demand for cooling surged. Government and health systems were on high alert. Water distribution, activation of disaster management

Table 2 Multi-scale physiological and ecological impacts of HW on plant systems

Scale of impact	Physiological/ cellular mechanism	Specific process affected	Observed outcomes	Ecological consequences	Agricultural/ ecosystem service impact	References
Cellular level	Protein and enzyme denaturation	Photosynthesis & Respiration (>40°C)	Loss of enzyme shape and specificity; reduced efficiency	Cell dysfunction; reduced primary productivity	Decreased crop yield; reduced carbon sequestration	Zahra et al. (2023)
Cellular level	Reactive Oxygen Species (ROS) accumulation	Antioxidant system overwhelmed	Scorched leaves; yellowing; premature aging; tissue death	Cell death; loss of photosynthetic tissue	Reduced biomass; decreased productivity	Fortunato et al. (2023), Samat et al. (2025)
Cellular level	Xylem dysfunction	Water transport under transpiration stress	Xylem embolism; blockage of water-conducting vessels	Loss of hydraulic conductivity; wilting	Crop failure; tree death; ecosystem collapse	Bai et al. (2023)
Cellular level	Photorespiration activation	Enzyme binds O_2 instead of CO_2	Wasteful metabolic process; energy loss without carbon fixation	Reduced net photosynthesis; carbon loss	Reduced crop yield; weakened plant vigor	Samat et al. (2025)
Organismal level	Stomatal closure	Water conservation response	Reduced CO_2 intake; limited photosynthesis	Growth limitation; developmental stress	Reduced biomass accumulation; stunted growth	Kurmanbayeva et al. (2022)
Organismal level	Reproductive failure	Pollen sterilization during flowering	Reduced pollen viability; pollination failure	Seed/fruit abortion; generational decline	Crop failure (wheat, rice, maize); seed security loss	Palermo et al. (2025)

(continued)

Table 2 (continued)

Scale of impact	Physiological/ cellular mechanism	Specific process affected	Observed outcomes	Ecological consequences	Agricultural/ ecosystem service impact	References
Population level	Range shift/ Migration	Upslope/ latitudinal movement toward cooler areas	Geographic redistribution of species	Fragmented populations; isolation by barriers	Reduced genetic diversity; local extinction risk	Ullah et al. (2022)
Community level	Species composition change	Replacement of sensitive species by heat-tolerant species	Shift toward C_4 grasses & xerophytic vegetation	Altered biodiversity; reduced nutritional quality	Lower forage quality; reduced ecosystem services	Khan et al. (2022)
Ecosystem level	Forest dieback	Repeated heat + drought; pathogen activation; insect outbreaks	Death of dominant tree species (e.g., eucalyptus); canopy loss	Conversion of forest to shrubland/ grassland	Reduced carbon storage; altered albedo; loss of habitat	Arimoto et al. (2023)
Ecosystem Level	Carbon sink → source transition	Reduced photosynthesis + decomposition from dead biomass	Net carbon release from previously stable systems	Positive feedback to atmospheric CO_2; warming acceleration	Climate feedback amplification; reduced climate buffering	Cheung et al. (2023)
Ecosystem level	Evapotranspiration reduction	Reduced vegetative cover; soil moisture feedback	Intensified local heating; reduced cooling capacity	Enhanced heat dome effects; ecosystem stress amplification	Reduced resilience; increased drought risk	Mahecha et al. (2024)

(continued)

Table 2 (continued)

Scale of impact	Physiological/ cellular mechanism	Specific process affected	Observed outcomes	Ecological consequences	Agricultural/ ecosystem service impact	References
Agricultural system	Crop phenology disruption	Early flowering + adverse conditions during grain fill	Reduced grain quality; empty seeds; premature maturation	Reduced harvest yields (10–20% in major producer regions)	Food security threats; economic losses	Jia et al. (2024)
Ecosystem service	Pollination network disruption	Insect pollinator stress; flower/ fruit abortion	Reduced seed set; reproduction failure	Breakdown of plant–pollinator mutualisms	Loss of ecosystem services; reduced plant diversity	Duneau and Ferdy (2022)
Ecosystem service	Soil health degradation	Reduced litter quality; altered microbial communities; nutrient cycling disruption	Nitrogen immobilization; reduced soil fertility	Long-term productivity decline; ecosystem resilience loss	Reduced agricultural productivity; land degradation	Fikri et al. (2021)

teams, and restriction of outdoor work during peak heat hours were some of the actions taken by the government.

The 2022 Heat Wave

The heat wave of 2022 occurred in March–May. Due to its early onset, this heat wave was unique. Mainly affecting North India, Central India, and parts of Pakistan, the worst hit regions included Delhi, Punjab, Haryana, Uttar Pradesh, Rajasthan, Madhya Pradesh and Gujarat. March 2022 being India's hottest March in over 122 years, the temperatures reached 44–49 °C across the plain areas. The causes were persistent high-pressure systems, weak Western Disturbances, and climate change influence. The early onset had an impact on the reproductive stage of wheat and led to the reduction of yields by 10–20% in major producer states. This prompted a ban on wheat exports and sparked worries about food security. Electricity demands hit record-breaking highs, consequently leading to a shortage of coal and power cuts. There was also an increase in forest fires, especially in Uttarakhand and Himachal Pradesh. Death estimates vary; however, hundreds are suspected as heat related. Government actions included national and state advisories, implementation of Heat Action Plans, increased crop monitoring, and emergency water supply and cooling centers in urban areas.

Case Study 4: China's Heat Wave of 2022

Occurred during the summer 2022, from June to late August, it was one of the longest HW in China's record history (70 plus days long). The southern and central China were especially affected. The cause was said to be a persistent, powerful subtropical high-pressure system and reduced rainfall. The yields of crops such as rice and maize suffered because of experiencing dual stress from heat and water scarcity. The agricultural industry suffered and hence raised concerns about food security and global food prices. The government opened underground bomb shelters as public cooling shelters. Power-usage restrictions were imposed on factories to maintain power supply to residential areas. The 2022 event is described as the most sever heat wave China has ever recorded.

Case Study 5: Forest Dieback in Australia

Forest dieback is the decline and death of trees, especially eucalyptus across Australia. First dieback events were recorded in the 1970s–1980s and drastically intensified over the years. Worst affected zones include NSW Northern Rivers, Victorian Central Highlands, WA Jarrah Forest and dry inland eucalyptus woodlands. They are driven by multiple causes. Extreme heat being the major cause, other causes include prolonged drought, insect outbreaks, nutrient loss from soils, salinity of soils, pathogens, etc.

9 Broader Discussion: Linking Temperature Patterns, Plant Ecosystems, and Human Systems

The intricate linkage between climatic activities, vegetative ecosystems and human livelihood is why there are rising temperatures, and frequent HW. Among these, the most important is the plant ecosystems because they do not only react and signal the climate change but they also contribute toward it. The alteration in climatic pattern has an impact on the biology and composition of plants which produces a chain reaction within the ecosystem eventually impacting on the human livelihoods. The links between temperature patterns, flora, and human machineries need to be studied to gain a complete insight into the key impacts and design solutions (Khan et al., 2025).

As stated in the previous sections of this paper, in 1.7, the influence of heat on plants is pivotal to this correlation. The increased temp does not only interfere with the major metabolic activities such as photosynthesis, respiration and transpiration, but also interferes with the plants by ruining tissues, cells and enzymes thereby compromising the overall plant health and growth. Such transformations modify the biodiversity and ecosystem service over stability and resilience (Li et al., 2021).

Though thinking plants are just passive recipients of climate change would be unjustified as they help in regulating the climate by acting as carbon sinks, pulling massive amounts of climate warming carbon dioxide during photosynthesis. Even though natural vegetation such as forests and grasslands etc., acts as natural climate buffers they die back during prolonged heat events, resulting in reduced carbon absorption which leads to intensifying greenhouse gas, increasing overall global warming (Zhou et al., 2024).

Furthermore, this can be seen in urban environments as well. Even though vegetation in urban areas lowers the rising temperature and helps in purifying air, overall improving the UHI effect, these ecosystems are vulnerable to extreme heat, water scarcity and pollution. This results in a decrease in the ability to act as carbon sinks and cool the air. Ultimately, this leads to a reduction in air quality, posing greater health risks upon humans.

As agriculture and forestry are highly sensitive to extreme heat, it leads to reduced crop yield, drying out the vegetation which results in megafires. Therefore, in areas depending upon rainwater farming, prolonged heat stress increases food and economic insecurity. Apart from food, plants also support water cycles, soil health and pollination, which are all necessary for human survival.

However, the human response to climate change and prolonged HW shapes the adaptation strategies. When we respond to grid failures and deaths in vulnerable societies due to prolonged exposure to HW, we build mitigation strategies to lessen the impact of it. But in this process, our actions sometimes affect the plant ecosystem as well by overusing and introducing non-native species. Therefore, the decisions taken today must be made keeping all aspects in mind as they will eventually determine the future of both ecological and human systems in the face of heat stress.

Thus, it is needed that there is a system-based holistic approach. Climate, ecosystems and human societies should be viewed as self-reliant components of a wider social and ecological system. Plant ecosystems are also essential to this system between our atmosphere and our society as they do not only control the climate, they also sustain the biodiversity and livelihoods of a considerable number of the society. Their vulnerability to heat, which is aggravated by the demands of global warming, requires us to address it by formulating useful policies, management structures and research need to be conducted on it.

To conclude, the connection between the patterns of temperatures, human systems, and plant ecosystems is a vicious circle: climate change influences the dynamics of the plants which in their turn affect the wellbeing of people; and human activity influences the resilience of plant ecosystems. The knowledge of this cycle is essential to developing climate strategies that will emphasize the protection and adaptation of plant ecosystems.

10 Conclusion

The HW symbolize the intersection of climatic forces, nature vulnerability, and human health. This chapter highlights the critical role played by extreme heat in the formation of plant ecologies as well as the environment by analyzing their trends, reasons, and effects. Observational data, climate forecasts, and historic records can demonstrate that there is a definite tendency toward the fact that the number of HW grows more frequent, more intense, and longer-lasting as a result of climate change caused by humanity. Their influence is quite extensive: they interfere with the physiology of plants, reduce the agricultural output, trigger the death of forests, cause the migration of species, and destabilize ecological interactions.

The bibliographic review indicates that the world has made significant progress in researching the issue of HW, but there are still gaps in knowledge especially in the regions of the world that are susceptible to climate change like Africa, Latin America and some parts of Asia. Enhancing the interdisciplinary partnership that cuts across climate change, plant biology, socio-economic analysis, and policy development will be essential in enhancing the understanding and developing viable adaptation strategies.

Finally, it demonstrates that HW are not single-phenomena but structural stressors that result in cascading effects within the ecosystems and societies. The ability of plant ecosystems to withstand extreme heat will decisively determine future ecological stability, food security, and climate feedback processes in the decades to come as the following will access in more detail.

References

Arimoto, K. I., Miyauchi, S., Troutman, T. D., Zhang, Y., Liu, M., Stoner, S. A., & Zhang, D. E. (2023). Expansion of interferon inducible gene pool via USP18 inhibition promotes cancer cell pyroptosis. *Nature Communications, 14*(1), 251. https://doi.org/10.1038/s41467-022-35348-5

Arya, E. V., & Rao, M. S. (2023). Assessing Climatic Anomalies and Trends: A Comprehensive Study of rainfall indices, minimum and maximum temperature. https://doi.org/10.21203/rs.3.rs-3324202/v1

Awasthi, A., Vishwakarma, K., & Pattnayak, K. C. (2022). Retrospection of heatwave and heat index. *Theoretical and Applied Climatology, 147*(1), 589–604. https://doi.org/10.1007/s00704-021-03854-z

Bai, F., Ma, H., Cai, Y., Shahid, M. Q., Zheng, Y., Lang, C., & Wang, L. (2023). Natural allelic variation in GRAIN SIZE AND WEIGHT 3 of wild rice regulates the grain size and weight. *Plant Physiology, 193*(1), 502–518. https://doi.org/10.1093/plphys/kiac405

Barriopedro, D., García-Herrera, R., Ordóñez, C., Miralles, D. G., & Salcedo-Sanz, S. (2023). HW: Physical understanding and scientific challenges. *Reviews of Geophysics, 61*(2), e2022RG000780. https://doi.org/10.1029/2022RG000780

Boni, Z., Bieńkowska, Z., Chwałczyk, F., Jancewicz, B., Marginean, I., & Serrano, P. Y. (2023). What is a heat (wave)? An interdisciplinary perspective. *Climatic Change, 176*(9), 129. https://doi.org/10.1007/s10584-023-03562-0

Chakraborty, M., Singh, R., Awasthi, A., Kumar, V., Kumar, A., & Kaur, S. (2025). Assessment of greenhouse gas emissions in Delhi landfills: Analyzing factors affecting emissions and environmental implications. *Journal of Material Cycles and Waste Management, 27*(1), 330–342. https://doi.org/10.1007/s10163-024-02114-2

Chen, M., Zhang, T. L., Hu, C. G., & Zhang, J. Z. (2023). The role of drought and temperature stress in the regulation of flowering time in annuals and perennials. *Agronomy, 13*(12), 3034. https://doi.org/10.3390/agronomy13123034

Cheung, W. W., Maire, E., Oyinlola, M. A., Robinson, J. P., Graham, N. A., Lam, V. W., & Hicks, C. C. (2023). Climate change exacerbates nutrient disparities from seafood. *Nature Climate Change, 13*(11), 1242–1249. https://doi.org/10.1038/s41558-023-01800-0

de Bont, J., Nori-Sarma, A., Stafoggia, M., Banerjee, T., Ingole, V., Jaganathan, S., & Ljungman, P. (2024). Impact of heatwaves on all-cause mortality in India: A comprehensive multi-city study. *Environment International, 184*, Article 108461. https://doi.org/10.1016/j.envint.2024.108461

Duneau, D., & Ferdy, J. B. (2022). Pathogen within-host dynamics and disease outcome: What can we learn from insect studies? *Current Opinion in Insect Science, 52*, Article 100925. https://doi.org/10.1016/j.cois.2021.100925

Faye, M., Dème, A., Diongue, A. K., & Diouf, I. (2021). Impact of different heat wave definitions on daily mortality in Bandafassi, Senegal. *Plos One, 16*(4), Article e0249199. https://doi.org/10.1371/journal.pone.0249199

Fikri, M., Joulian, C., Motelica-Heino, M., Norini, M. P., & Hellal, J. (2021). Resistance and resilience of soil nitrogen cycling to drought and heat stress in rehabilitated urban soils. *Frontiers in Microbiology, 12*, Article 727468. https://doi.org/10.3389/fmicb.2021.727468

Fortunato, S., Lasorella, C., Dipierro, N., Vita, F., & de Pinto, M. C. (2023). Redox signaling in plant heat stress response. *Antioxidants, 12*(3), 605. https://doi.org/10.3390/antiox12030605

Gowdy, J. (2020). Our hunter-gatherer future: Climate change, agriculture and uncivilization. *Futures, 115*, Article 102488. https://doi.org/10.1016/j.futures.2019.102488

Gurung, T. R., & Chen, L. (2024). Understanding the influence of soil moisture on heatwave characteristics in the contiguous United States. *Environmental Research Letters, 19*(6), Article 064070. https://doi.org/10.1088/1748-9326/ad4ec3

Hansen, J. E., Kharecha, P., Sato, M., Tselioudis, G., Kelly, J., Bauer, S. E., Ruedy, R., Jeong, E., Jin, Q., Rignot, E., Velicogna, I., Schoeberl, M. R., von Schuckmann K., Amponsem J., Cao J., Keskinen A., Li J., & Pokela, A. (2025). Global warming has accelerated: Are the United Nations

and the public well-informed? *Environment: Science and Policy for Sustainable Development, 67*(1), 6–44.
Hawkins, M. D., Brown, V., & Ferrell, J. (2017). Assessment of NOAA national weather service methods to warn for extreme heat events. *Weather, Climate, and Society, 9*(1), 5–13. https://doi.org/10.1175/WCAS-D-16-0042.1
He, B. J., Wang, J., Liu, H., & Ulpiani, G. (2021). Localized synergies between HW and urban heat islands: Implications on human thermal comfort and urban heat management. *Environmental Research, 193*, Article 110584. https://doi.org/10.1016/j.envres.2020.110584
Jia, B., Cui, X., Zhang, Z., Li, X., Hou, Y., Luo, J., & Guo, W. (2024). Arbuscular mycorrhizal fungi regulate amino acid metabolism, phytohormones and glycolysis pathway to promote the growth of Suaeda salsa under combined Cd and NaCl stresses. *Plant Physiology and Biochemistry, 214*, Article 108921. https://doi.org/10.1016/j.plaphy.2021.02.003
Kabir, M., Habiba, U. E., Khan, W., Shah, A., Rahim, S., De los Rios-Escalante, P. R., & Shafiq, M. (2023). Climate change due to increasing concentration of carbon dioxide and its impacts on environment in 21st century; a mini review. *Journal of King Saud University-Science, 35*(5), 102693. https://doi.org/10.1016/j.jksus.2023.102693
Khan, I., Wu, J., & Sajjad, M. (2022). Pollen viability-based heat susceptibility index (HSIpv): A useful selection criterion for heat-tolerant genotypes in wheat. *Frontiers in Plant Science, 13*, 1064569. https://doi.org/10.3389/fpls.2022.1064569
Khan, R., Wheeler, P., & Gowing, D. (2025). Characterising heatwave responses and climate driver impacts using multicollinearity-controlled generalised linear mixed models in urban and forest trees (2018–2023). *Earth Systems and Environment, 1–19*. https://doi.org/10.1007/s41748-025-00559-3
Klingelhöfer, D., Braun, M., Brüggmann, D., & Groneberg, D. A. (2023). Heatwaves: Does global research reflect the growing threat in the light of climate change? *Globalization and Health, 19*(1), 56. https://doi.org/10.1186/s12992-023-00912-w
Kurmanbayeva, A., Bekturova, A., Soltabayeva, A., Oshanova, D., Nurbekova, Z., Srivastava, S., & Sagi, M. (2022). Active O-acetylserine-(thiol) lyase A and B confer improved selenium resistance and degrade l-Cys and l-SeCys in Arabidopsis. *Journal of Experimental Botany, 73*(8), 2525–2539. https://doi.org/10.1093/jxb/erab564
Li, G., Chen, T., Feng, B., Peng, S., Tao, L., & Fu, G. (2021). Respiration, rather than photosynthesis, determines rice yield loss under moderate high-temperature conditions. *Frontiers in Plant Science, 12*, Article 678653. https://doi.org/10.3389/fpls.2021.678653
Mahecha, M. D., Bastos, A., Bohn, F. J., Eisenhauer, N., Feilhauer, H., Hickler, T., & Quaas, J. (2024). Biodiversity and climate extremes: Known interactions and research gaps. *Earth's Future, 12*(6), e2023EF003963. https://doi.org/10.1029/2023EF003963
McGregor, G. (2024). Synoptic scale atmospheric processes and heatwaves. In *Heatwaves: Causes, Consequences and Responses* (pp. 207–259). Cham: Springer International Publishing. https://doi.org/10.1007/978-3-031-69906-1_6
Malik, R., Rehman, F., Dhupper, R., Nadeem, B., Jain, H., Bhardwaj, A., & Choudhury, T. (2025). Intelligent predictive systems for modeling PM10, ozone, and temperature dynamics: A forecasting approach for urban air quality in Delhi. *Aerosol Science and Engineering, 1–11*,. https://doi.org/10.1007/s41810-025-00301-8
Meque, A., Pinto, I., Maúre, G., & Beleza, A. (2022). Understanding the variability of heatwave characteristics in southern Africa. *Weather and Climate Extremes, 38*, Article 100498. https://doi.org/10.1016/j.wace.2022.100498
Miracle, F. A., Nathan, M. E., Aransiola, A. B., Adewoye, P. O., Dahunsi, A. A., Aransiola, S. A., & Maddela, N. R. (2025). HW in wetlands ecosystem biodiversity and bioresources. In *Wetland Ecosystems: Conservation Strategies, Policy Management and Applications* (pp. 149–163). Cham: Springer Nature Switzerland. https://doi.org/10.1007/978-3-031-91982-4_8
Murari, K. K., Sahana, A. S., Daly, E., & Ghosh, S. (2016). The influence of the El N iño Southern Oscillation on HW in India. *Meteorological Applications, 23*(4), 705–713. https://doi.org/10.1002/met.1593

Nguyen, T. T., Grote, U., Neubacher, F., Do, M. H., & Paudel, G. P. (2023). Security risks from climate change and environmental degradation: Implications for sustainable land use transformation in the Global South. *Current Opinion in Environmental Sustainability, 63*, Article 101322. https://doi.org/10.1016/j.cosust.2023.101322

Oishy, M. N., Shemonty, N. A., Fatema, S. I., Mahbub, S., Mim, E. L., Raisa, M. B. H., & Anik, A. H. (2025). Unravelling the effects of climate change on the soil-plant-atmosphere interactions: A critical review. *Soil & Environmental Health, 100130*. https://doi.org/10.1016/j.soilenh.2025.100130

Palermo, N., Resentini, F., Moscatelli, S., Mendes, M. A., Costa, A., & Janni, M. (2025). Heat stress phenotyping: A key for plant selection in a warming climate. *Plant Science, 112675*. https://doi.org/10.1016/j.plantsci.2025.112675

Pattnayak, K. C., Awasthi, A., Sharma, K., & Pattnayak, B. B. (2023). Fate of rainfall over the North Indian states in the 1.5 and 2 C warming scenarios. *Earth and Space Science, 10*(2), e2022EA002671. https://doi.org/10.1029/2022EA002671

Rahmstorf, S. (2024). Climate and weather at 3 degrees more: Earth as we don't (want to) know it. In *3 Degrees More: The Impending Hot Season and How Nature Can Help Us Prevent It* (pp. 3–17). Cham: Springer Nature Switzerland. https://doi.org/10.1007/978-3-031-58144-1_1

Rao, I. R., Bangera, A., Nagaraju, S. P., Shenoy, S. V., Prabhu, R. A., Rangaswamy, D., & Bhojaraja, M. V. (2023). Chronic kidney disease of unknown aetiology: A comprehensive review of a global public health problem. *Tropical Medicine & International Health, 28*(8), 588–600. https://doi.org/10.1111/tmi.13913

Ravi, A., Pillai, D., & Deshpande, M. V. (2025). Declining net carbon sequestration of west-central Indian ecosystem in response to frequently occurring drought: Inference from satellite measurements and modeling. *Ecological Informatics, 103386*. https://doi.org/10.1016/j.ecoinf.2025.103386

Ravindra, K., Bhardwaj, S., Ram, C., Goyal, A., Singh, V., Venkataraman, C., & Mor, S. (2024). Temperature projections and heatwave attribution scenarios over India: A systematic review. *Heliyon, 10*(4). https://doi.org/10.1016/j.heliyon.2024.e26431

Ren, Y., Liu, J., Zhang, T., Shalamzari, M. J., Arshad, A., Liu, T., & Wang, T. (2023). Identification and analysis of heatwave events considering temporal continuity and spatial dynamics. *Remote Sensing, 15*(5), 1369. https://doi.org/10.3390/rs15051369

Samat, A. T., Soltabayeva, A., Bekturova, A., Zhanassova, K., Auganova, D., Masalimov, Z., & Kurmanbayeva, A. (2025). Plant responses to heat stress and advances in mitigation strategies. *Frontiers in Plant Science, 16*, 1638213. https://doi.org/10.3389/fpls.2025.1638213

Sharma, A., Andhikaputra, G., & Wang, Y. C. (2022). Heatwaves in South Asia: Characterization, consequences on human health, and adaptation strategies. *Atmosphere, 13*(5), 734. https://doi.org/10.3390/atmos13050734

Singh, D., Yadav, A., Singh, S., Yadav, R. K., Sen, M., Yadav, A. K., & Singh, A. K. (2023). HW and its impact on crop production and mitigation techniques: A review. *International Journal of Environment and Climate Change, 13*(9), 377–382. https://doi.org/10.9734/ijecc/2023/v13i930575

Tandon, A., Awasthi, A., & Pattnayak, K. C. (2025). Efficacy of machine learning in simulating precipitation and its extremes over the capital cities in North Indian states. *Scientific Reports, 15*(1), 10345. https://doi.org/10.1038/s41598-024-84360-w

Tao, J., Xu, Z., Ho, H. C., Zheng, H., Xing, X., Hu, J., & Cheng, J. (2025). Quantifying localized heatwave impact on mortality: A multi-country modeling study in the Asia-Pacific region. *The Lancet Regional Health-Western Pacific, 61*,. https://doi.org/10.1016/j.lanwpc.2024.101062

Ullah, A., Nadeem, F., Nawaz, A., Siddique, K. H., & Farooq, M. (2022). Heat stress effects on the reproductive physiology and yield of wheat. *Journal of Agronomy and Crop Science, 208*(1), 1–17. https://doi.org/10.1111/jac.12542

Weiskopf, S. R., Rubenstein, M. A., Crozier, L. G., Gaichas, S., Griffis, R., Halofsky, J. E., & Whyte, K. P. (2020). Climate change effects on biodiversity, ecosystems, ecosystem services, and natural resource management in the United States. *Science of the Total Environment, 733*, Article 137782. https://doi.org/10.1016/j.scitotenv.2020.137782

Yadav, N., Rajendra, K., Awasthi, A., Singh, C., & Bhushan, B. (2023). Systematic exploration of heat wave impact on mortality and urban heat island: A review from 2000 to 2022. *Urban Climate, 51*, Article 101622. https://doi.org/10.1016/j.uclim.2023.101622

Yuan, X., Li, S., Chen, J., Yu, H., Yang, T., Wang, C., & Ao, X. (2024). Impacts of global climate change on agricultural production: A comprehensive review. *Agronomy, 14*(7), 1360. https://doi.org/10.3390/agronomy14071360

Zahra, N., Hafeez, M. B., Ghaffar, A., Kausar, A., Al Zeidi, M., Siddique, K. H., & Farooq, M. (2023). Plant photosynthesis under heat stress: Effects and management. *Environmental and Experimental Botany, 206*, Article 105178. https://doi.org/10.1016/j.envexpbot.2022.105178

Zhao, Q., Yu, P., Mahendran, R., Huang, W., Gao, Y., Yang, Z., & Guo, Y. (2022). Global climate change and human health: Pathways and possible solutions. *Eco-Environment & Health, 1*(2), 53–62. https://doi.org/10.1016/j.eehl.2022.05.002

Zhou, M., Wan, N., Zheng, T., Xu, H., Yang, L., & Wang, T. (2024). New insights into global warming: End-to-end visual analysis and prediction of temperature variations. *arXiv preprint* arXiv:2409.16311. https://doi.org/10.48550/arXiv.2409.16311

Impact of Variation in Temperature on Plant Reproduction and Growth

Phogat Ishu, Soni Anchal, Sunita, Bagri Sandeep, Anjali, Debashis Paul, Dutta Suman, and Rakesh Kumar

Abstract The atmospheric temperature is an important environmental factor that dictates the developmental process of a plant at the beginning to the end, and it has great influence on growth as well as reproduction of plants. In this chapter, the multi-directional impacts of temperature change on major processes in the plant life cycle are concentrated. Low and high temperatures can interfere with the basic cellular activities. Lowering temperatures can also damage the membranes and deactivate enzymes and decrease the photosynthetic activity and consequently slow down growth and even kill the plant. Conversely, heat stress may cause protein denaturation, elevate respiration rates at the cost of photosynthesis as well as be associated with oxidative damage. There is also optimum temperature that is crucial in signaling flowering time which is vernalization. The non-ideal variations have the ability to influence fruitful pollination activities by postponing or avoiding flowering. The ambient temperature of the air greatly influences pollen and its viability, the stigma receptivity, and germination; heat stress may render pollen sterile, and cold stress may reduce the length of life of the pollen. Similarly, temperature deviations tend to frustrate seed development and fruit set. The high temperature levels at these stages may lead to poor quality fruit, reduced germination percentages and smaller seeds. Understanding of these complex interactions is important in predicting the way plants will be affected by climate change, designing crop types that will be resilient to climate and ensuring food security in the world. The chapter shall also dwell on the

P. Ishu · S. Anchal · Anjali
Chaudhury Devi Lal University, Sirsa, Haryana, India

Sunita · B. Sandeep
ICAR-Central Institute for Cotton Research, Regional Station, Sirsa, Haryana, India

D. Paul (✉)
ICAR-Central Institute for Cotton Research, Nagpur, Maharashtra, India
e-mail: deba.13.paul@gmail.com

D. Suman
Ramakrishna Mission Vivekananda Educational and Research Institute, Howrah, West Bengal, India

R. Kumar
Graphic Era School of Management, Graphic Era Deemed to Be University, Dehradun, India

A. Awasthi et al. (eds.), *Rising Temperatures and Plant Ecosystems*,
https://doi.org/10.1007/978-3-032-29477-7_2

molecular and physiological processes of temperature-mediated responses, which will give an overall picture of temperature stress tolerance in the plant scheme.

Keywords Pollen viability · Stigma receptivity · Seed development · Protein synthesis · Temperature stress

1 Introduction

Temperature is a very crucial ecological parameter that affects all the possible metabolic processes of the organisms practically (Aslam et al., 2021). Global atmosphere change, which is characterized by a sharp rise in average global temperatures since 1880 and the early 2000s, which can directly be associated with the observed principal effects of heatwaves, drought, flooding, and glacier melting (Fig. 1). The effects of temperature on living things have been found to be many; including cells, morphology, physiology, behavior, growth, and the spread of plants and animals, among others (Patel et al., 2023). It influences nearly all aspects of plant development, progress and survival thus a factor of high priority in plant biology. The temperature is a major predictor in determining plant biology, such that the temperature affects physiological processes (e.g., photosynthesis, respiration), phenological processes (e.g., germination, reproduction), and macro-processes such as crop distribution and yield (Zinn et al., 2010). The main functions of plants are directly regulated by temperature, such as activity of enzymes, photosynthesis, water uptake, flowering and resilience to stress. Even small changes would affect the productivity and health of the plants (Chaturvedi et al., 2021).

Despite immense gains, there remain enormous gaps in the knowledge of molecular and metabolic basis of thermotolerance. There is also lack of enough research on the interaction of temperature stress with other abiotic factors, including salinity and drought. The long-term adaptations of perennial plants have not been known well in comparison to annual crops. More so, there is inadequacy in translating laboratory findings into field-level procedures in developing crops over the long term. The chapter is critical as it provides a summary of the knowledge on the effect of temperature extremes in vegetative and reproductive phases of plants, critical regimes where reproductive biology of the plants is susceptible to climatic stress and provide the future research direction on crop adaptation to heat stress. The present chapter provides a thorough study of the impact of changes in temperature on the growth and reproduction of plants thanks to the association of physiological, morphological, and molecular reactions to heat stress. It further examines the effects of heat and cold stress on pollen production, flowering, growth, fruit set and seed production with a long-term perspective on their effects on ecological balance and yield stability. It promotes breeding programs and agricultural methods that enhance resilience to climate changes, on top of technologies that are needed to offer food security amid global warming. The present scientific knowledge on the resilience breeding and

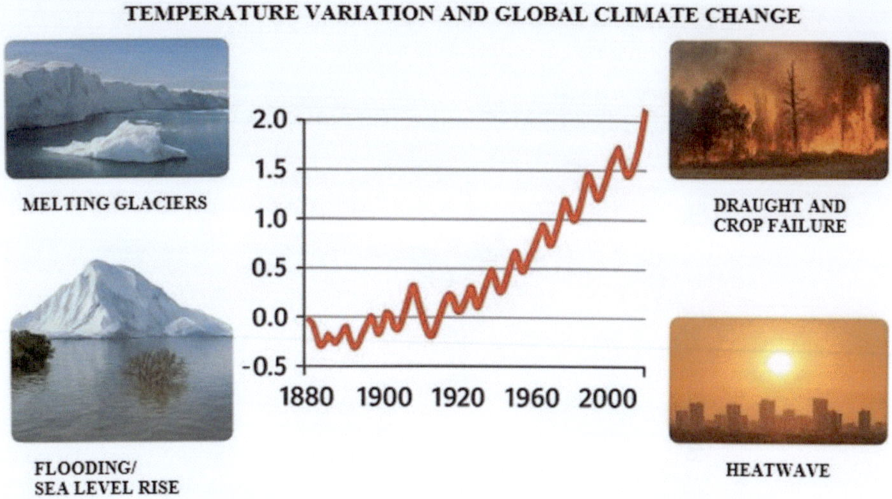

Fig. 1 Effect of successive temperature variations on global climate

supporting management practices is integrated in this chapter, to realize its inferences in the future.

2 Effect of Temperature Stress on Plant Physio-Chemical Response

Light intensity and temperature are two major environmental factors which influence the growth and development of the plants (Asada, 2006). Any fluctuations in their quantity, term or scope may initiate acclimation, bring about cell damage, which may lead to the death of plants (Sanghera et al., 2011). Plants have evolved several effective defense strategies to survive in the harsh extreme conditions of light and temperature. These systems are mainly made up of the xanthophyll cycle, the photorespiratory route and the electron transport system of photosynthesis (Asada, 2006). Stressed plants generate a higher level of reactive oxygen species (ROS) (Ahmad et al., 2010; Garg et al., 2024). In addition to damaging, the ROS are also vital as signaling molecules (Baxter et al., 2014). Plants respond to this by the production of various low-mass antioxidants and antioxidative enzymes. The intensity and the length of the stress defines the quantity of these antioxidants they contain (Asada, 2006). The concentrations of antioxidants, such as ascorbate and glutathione, are often decreased in short yet severe light or temperature stress, which indicates that optimum ROS scavenging is necessary (Baxter et al., 2014). Nevertheless, with a long-term exposure to adverse environments, the level of antioxidants grows slowly due to acclimation (Fig. 2). Studies of plant antioxidant responses have shown that much of this defense

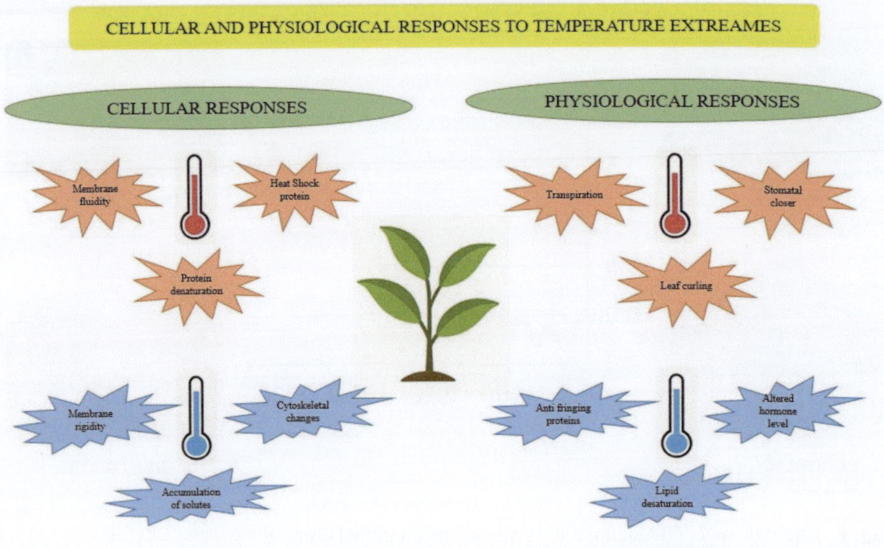

Fig. 2 Cellular and physiological response of the plant to extreme temperatures

process operates in the chloroplasts in which the components operate in close concert (Baxter et al., 2014). The intensity of the antioxidant response also depends on the species as well as is associated with the natural capability of the plant to endure stress. Thus, the examination of a large variety of plant species can be an important source of information concerning the natural adaptations of antioxidant systems to the high-light and temperature stress (Baxter et al., 2014).

2.1 Cold Stress: Chilling and Freezing Injuries

Tropical and subtropical plants, including many vegetable species, are susceptible to chilling damage and even death as a result of the biochemical disturbances that low temperatures (1–10 °C) generate in their cells (Alam & Jacob, 2002). Low temperatures affect plants' essential biological processes, including respiration, photosynthesis, mineral uptake, water balance, and general metabolism (Bloom et al., 2004). The reduction in metabolic activity during chilling is a complex outcome influenced by both the temperature level and the duration of exposure. In plants, cold affects the gene expression of a number of low-mass proteins (Alam & Jacob, 2002). Several strategies to enhance cold tolerance into several groups: thermal treatments (such as the effects of heat stress, intermediate warming, thermal conditioning, and low-temperature hardening), chemical treatments (including trace elements, synthetic growth regulators, and antioxidants), and genetic and cell engineering approaches were already reported (Chen et al., 2014).

2.2 *Heat Stress: Oxidative Damage and Denaturation of Proteins*

All living things encounter heat stress (HS) when temperatures rise over optimal ge. Due to HS's disruption of cellular homeostasis, plants frequently experience severe delays in growth and development and, in the worst situations, die (Allakhverdiev et al., 2008). Because they are sedentary, plants are always subject to abiotic stimuli like temperature variations (Asthir, 2015). Globally, large agricultural losses are caused by heat, frequently combined with drought or other stress factors (Asthir, 2015). The accumulation of heat shock proteins (HSPs), which are regulated by heat stress transcription factors (HSFs), is considered to be a significant component of the heat stress response (HSR) and acquired thermotolerance in plants and other organisms. In several recent reviews, our current knowledge of HSP and HSF functions has been comprehensively updated (Kumar et al., 2024). Attempts to boost thermotolerance by overexpressing a single HSF or HSP gene have not been very successful because of the genetic complexity of the HSR. Our knowledge of the plant HSR has significantly increased as a result of genome-wide expression investigations conducted on *Arabidopsis* under heat stress or in cultivars lacking thermotolerance. Similar patterns of gene expression changes, impacting roughly 2% of the genome, are shown when comparing expression profiles from different plant species under varied heat stress situations, such as distinct tissues, developmental stages, growth settings, or stress treatment regimens (Allakhverdiev et al., 2008). Recent research has demonstrated the molecular roles of *HSPs* and the complex regulatory network of HSFs with new findings related to other molecular processes involved in heat perception and survival, coupled with the interplay between heat stress and other abiotic stresses (Asthir, 2015).

Changes in temperature impact negatively on several morphological and physiological characteristics of plants (Zinn et al., 2010). There is a need to understand the reaction of plants to warming or cooling to ensure that there are sustainable agricultural conditions. There is a serious effect of the current increase in the global average temperature on the productivity of crops (Chaturvedi et al., 2021). Both high and low temperatures which are influenced by the terrain, climate change, and land to sea temperature differentials may stop plant growth and development. Biological processes can accelerate and decelerate because of temperature changes, and it often leads to the denaturation of enzymes (Baxter et al., 2014). The temperatures exceeding the optimum levels due to stress length and intensity may create reversible and irreversible damages in plants with sometimes resulting in death of plants. By studying the morphological, biochemical, and physiological responses of plants to various conditions, it is possible to assess the degree of resilience of the plant (Patel et al., 2023). Plants possess defense systems to survive unpredictable temperature stress thereby reducing the harmful consequences of physiological and metabolic processes (Asthir, 2015).

3 Impact of Temperature Variation on Plant Reproductive Physiology

One of the most important factors in finding the effectiveness of plant reproduction is temperature. In comparison to the vegetative growth, any damage caused during the reproductive stage (e.g., flower initiation, pollen development, fertilization, and fruit/seed set) is irreparable, which directly causes reduced productivity (Chaturvedi et al., 2021; Zinn et al., 2010). The importance of reproductive physiology response to temperature stress is significant to achieve food security since extreme temperatures are increasing in frequency due to atmospheric transformation (Awasthi et al., 2022; Sage & Kubien, 2015). The change in temperature in an agricultural habitat with regard to the plant is believed to be imperative to understand. Temperature changes cause enzyme denaturation, which exponentially rises forward and reverse bio reactions (Aslam et al., 2021). Plants may be negatively impacted by heat and cold temperature stressors at any stage of their growth. The response to stress on different plant species to reproductive (gametophytic) will be crucial to future management of agricultural production because sexual reproduction in flowering plants would generate most of our food supply. Hot and cold temperature stressors result in reactive oxygen species or ROS. The concentrations of these ROS may become so large that they cause oxidative damages and even cell death. Among the most obvious effects of temperature stress on photosynthetic tissues (Fig. 3) is the suppression of photosynthesis. High temperatures also can cause stress that changes the production and transmission of hormones (Kotak et al., 2007). High temperature stress, for example, stimulates ethylene signaling, salicylic acid and abscisic acid. ABA signaling can also be used to determine low temperature tolerance (Thomashow, 1999). Moreover, cold stress signaling is thought to be associated with phytohormones auxin, GA (gibberellins) and ethylene. Heat stress induces heat stress transcription factors expression which induces the expression of many other stress transcripts (Beniwal et al., 2004). The noteworthy group of up-regulated genes encode heat shock proteins some of which act as chaperones to inhibit protein denaturation. Among the cold-specific transcription factors are CBF/DREB1 that is up-regulated during cold stress responses (Zinn et al., 2010).

3.1 Temperature-Mediated Flowering

Temperature is an important factor in flowering which is an important developmental transition that is influenced by both intrinsic and extrinsic factors. Long-day plants such as wheat and Arabidopsis flower faster in high temperatures and is regulated by FLOWERING LOCUS T (FLT) (Kumar et al., 2012), a mobile signal that interacts with thermal and photoperiod signals. Some crops such as wheat, barley and canola, however, require that they be exposed to vernalization, or exposure to cold, in order

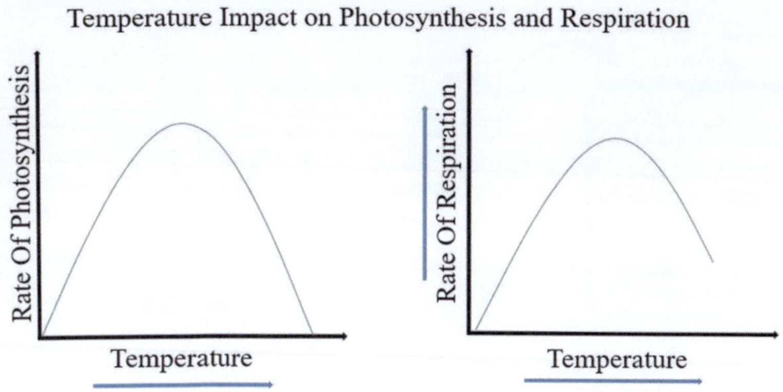

Fig. 3 Temperature impact on photosynthesis and respiration

to suppress the FLOWERING LOCUS C, a flower suppressor, and ensure that flowering occurs only after winter (Amasino, 2010). Two of the epigenetic modifications applied by vernalization and that irreversibly repress FLC are histone methylation and chromatin remodeling (Berry & Dean, 2015). This shows how plants grow to accommodate seasonal variations in temperature to attain optimal conditions in relation to reproduction. Reproductive outcomes are also influenced by the diurnal temperature changes in daytime and evening or thermoperiodism. An example is that tomatoes produce improved fruit sets when they are subjected to normal day/night cycles (25/15 °C) instead of subjected to high temperatures (Peet et al., 1998). Daytime stress causes plants to recover under such cyclic variation and increases their reproductive performance (Awasthi et al., 2024).

3.2 *Vernalization: Role of Cold in Flowering*

Extended cold stress promotes plant blossoming through a process called vernalization (Fig. 4). The vernalization genes (VRN1, VRN2, and VIN3) are activated by cold exposure. When floral integrator genes SOC1 and FT (blooming LOCUS T) are connected, their expression is inhibited. FLOWERING LOCUS C, which normally acts as a strong blossoming repressor, is suppressed as a result. When FLC is suppressed, FT and SOC1 are expressed, which triggers flowering. Thus, after being exposed to winter cold, vernalization guarantees that plants flower at the appropriate period. The method by which heat stress causes plants to blossom is depicted in Fig. 4. The transcription factor PHYTOCHROME INTERACTING FACTOR 4 is activated by heat stress. Inside the nucleus, activated PIF4 binds directly to the FLOWERING LOCUS T gene's promoter region, increasing the expression of the gene. Following their interaction, the FT protein and FD in the shoot apical meristem activate the genes that identify floral meristems. As a result, under heat

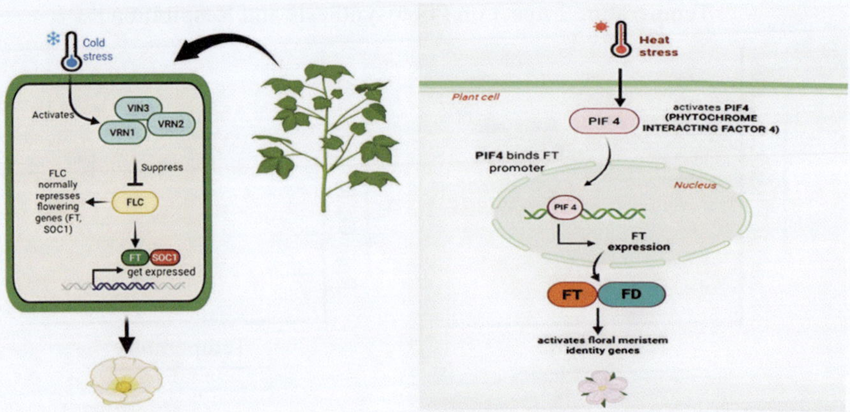

Fig. 4 Cold and heat stress mediated signaling for flowering

stress, flowering begins. PIF4 thus serves as a crucial regulator that connects signals of high temperature to the induction of flowering.

3.3 Thermoperiodism: Daily Temperature Cycles

The process of thermoperiodism, in which day/night temperature cycles control plant growth and flowering. Sensors such heat shock proteins (HSPs), ELF3, and phytochrome B (PhyB) are the first to detect temperature changes. The plant's circadian clock, which consists of morning genes (CCA1, LHY) and evening genes (TOC1, ELF3, ELF4, LUX), is integrated with these signals. Cell elongation is promoted by the circadian system working in tandem with temperature signals to maintain PIF4, which in turn activates the YUCCA genes that are in charge of auxin production. Plant hormones are also influenced by temperature cycles, which maintain ABA balance and raise auxin and gibberellins (GA20ox, GA3ox). Developmental transitions are supported by these hormonal shifts. In the end, temperature-regulated pathways cause the production of FT (blooming LOCUS T), which travels to the shoot apical meristem (SAM) and initiates blooming through SOC1 and AP1. In order to maximize flowering time and growth, thermoperiodism combines temperature cycles, circadian rhythm, hormones, and transcriptional regulators (Fig. 5).

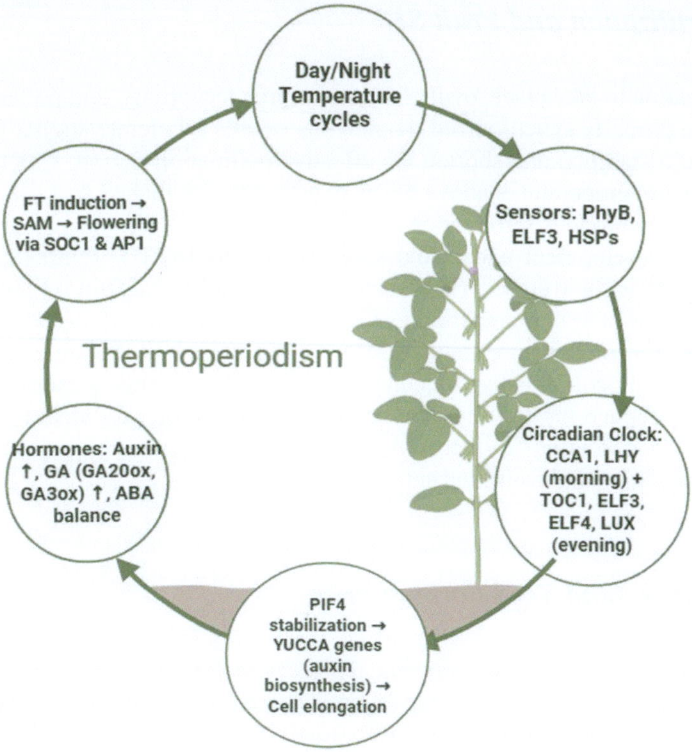

Fig. 5 Thermoperiodism: regulation of daily temperature cycles

3.4 Pollen Development and Viability Under Thermal Stress

Among the structures, most susceptible to temperature changes are pollen grains. Grain shriveling or sterility results from heat stress (>35 °C), which also affects meiosis, tapetum function, and pollen starch buildup (Prasad et al., 2017). Cellular structures are further harmed by the buildup of reactive oxygen species, and sterility is made worse by inadequate antioxidant defenses (Chaturvedi et al., 2021). Cold stress (<15 °C), however, hinders the formation of pollen tubes and delays pollen germination, decreasing the likelihood of successful fertilization (Sharma et al., 2016). At high temperatures, heat shock proteins and transcription factors are quickly activated to stabilize membranes and preserve proteins (Grover et al., 2013). Pollen grains are deformed as a result of cold stress, which interferes with callose deposition and tapetal growth. Remarkably, under cold stress, exogenous phytohormones like auxin and brassinosteroids increase pollen fertility, pointing to possible mitigation techniques.

3.5 Fertilization and Fruit Set

For fertilization to take place, ovule lifetime, stigma receptivity, and pollen viability must all be properly synchronized. High temperatures accelerate stigma withering, reduce ovule viability, and shorten the effective pollination period. Chickpeas that are exposed to temperatures above 35 °C experience diminished pod set and ovule abortion. Asynchronous fertilization and pod abortion result from soybeans' slower embryo sac development under cold stress (less than 18 °C). Temperature has a significant impact on fruit set. The tomato exhibits nearly total fruit set failure when subjected to constant hot temperatures (32–36 °C) (Peet et al., 1998). High temperatures (>38 °C) and low temperatures (<15 °C) can cause boll shedding and embryo abortion, making cotton balls particularly sensitive. Temperatures exceeding 35 °C significantly reduce the fertility of rice spikelets. Therefore, an ideal temperature window for coordinating gametophyte and sporophyte processes is necessary for proper fertilization and fruit/seed set.

3.6 Embryo and Seed Formation

Embryogenesis and seed development represent the culmination of reproductive success. Heat stress accelerates ovule degeneration and causes early embryo abortion (Farooq et al., 2011). Even brief exposure to 40 °C in soybeans lowers the number of seeds per pod by 50%. In cereals like wheat and maize, cold stress slows down fertilization rates and delays the maturity of embryo sacs. According to recent studies, heat and cold stress both interfere with hormonal signaling, especially that of auxin, gibberellins, and abscisic acid, which results in abnormal growth regulation and poor embryonic development.

3.7 Hormonal and Nutritional Modulation for Thermotolerance

Auxins, abscisic acid, brassinosteroids, and salicylic acid are examples of phytohormones that can be applied externally to improve a plant's resistance to heat during reproduction. These substances protect cell membranes, control the expression of genes related to reproduction, and improve the ability of reactive oxygen species to be scavenged. Similarly, by maintaining ovule activity and improving pollen viability, minerals including calcium, potassium, magnesium, and nitrogen help plants become more resilient to stress. Furthermore, new studies demonstrate that biostimulants and hormone priming methods can improve crop reproductive success under both cold and heat stress (Li et al., 2025).

4 Genetic Mechanisms and Molecular Regulation

Key components of the complex, hierarchical genetic basis of plant thermotolerance have been found in both main crops and model plants. This system includes quick transcriptional reprogramming, immediate protein protection, and developmental pathway modification (Table 1). The immediate defense at the cellular level depends on molecular chaperones, particularly the heat shock proteins, which include the tiny HSPs in Arabidopsis, Hsp100/ClpB, Hsp90/HtpG, Hsp70/DnaK, and Hsp60/GroEL. Additionally, sHSPs are crucial for defending photosystem II against heat stress, which protects photosynthesis machinery. These heat shock proteins stabilize protein structure and quality by avoiding denaturation and aiding in the refolding of heat-damaged proteins (Kotak et al., 2007; Kumar et al., 2024). Because bZIP TFs like bZIP28 and bZIP60 in *Arabidopsis* act as ER (Endoplasmic Reticulum) stress regulators to remove misfolded proteins collected in the Endoplasmic Reticulum under heat stress, they promote this essential process of protein homeostasis (Deng et al., 2011; Howell, 2013).

The transcriptional machinery is the one in charge of the entire defense and is mostly involved in the synthesis of the HS (heat shock)-sensitive genes and is controlled by the heat stress transcription factors of tomatoes, including; HsfA1a, HsfA2, and HsfB1 (Kotak et al., 2007). In addition to the acute stress response, thermotolerance is highly associated with development in Arabidopsis thaliana that are controlled by PIF4, HY5, DELLAs, and HFR1. They mediate environmental temperature signals to growth and development by regulating protein breakdown and making sure that cellular processes are affected by average temperature (Casal & Balasubramanian, 2019). Lastly, in staple crops (rice, wheat and tomatoes), Quantitative Trait Loci (QTLs) have been discovered to help in improving agriculture. These genetic regions are also important in regulating grain filling, spikelet fertility, and yield stability to heat stress and complex heat tolerance traits (Jagadish et al., 2010; Ye et al., 2015). All these genes and factors, in totality, constitute a multi-layered genetic process to guarantee the survival of plants in addition to sustaining agronomic yield faced with increased global temperatures.

A signaling pathway that is generally activated in plant cells under conditions of heat and cold stress is shown in Fig. 6, showing how plants use Ca^{2+} signaling and variations in plasma membrane fluidity/rigidity to detect and combat to heat and cold stress responses. Stress causes Ca^{2+} influx, which in turn activates proteins that bind calmodulin and set off the MAPK cascade. The signals are further amplified by secondary messengers such as ROS, H_2O_2, NO, ABA, and phosphatidic acid. These pathways trigger the expression of HSP genes, which encode molecular chaperones that preserve protein folding and stability, by activating Heat Shock Transcription Factors (HSFs) in response to heat stress. Osmo protection, membrane integrity, and stress adaptation are provided via the upregulation of COR and LEA genes by CBF/CRT transcription factors, which are activated by the MAPK cascade under cold stress. Therefore, plants use different transcription factors (HSF for heat, CBF/CRT for cold) and stress-responsive genes (HSP vs. COR/LEA) to attain overall

Table 1 Genetic basis of thermotolerance

S.N	Crop	Gene and factors involved	Gene functions/ regulators	Mechanism in thermotolerance	References
1	Arabidopsis	*bZIP* TFs (e.g., *bZIP28, bZIP60)*	ER stress regulators	Heat stress triggers the unfolded protein response	Deng et al. (2011)
2	Arabidopsis	Eukaryotic/ Escherichia coli nomenclature for small HSP (sHSP) proteins, Hsp90/ HtpG, Hsp70/ DnaK, Hsp60/ GroEL, and Hsp100/ClpB	Molecular chaperones in the regulation of protein quality	Photosystem II is protected by *sHSP* in stressful situations	Kotak et al. (2007)
3	Tomato	*HsfA1a*, *HsfA2* and *HsfB1*	Nuclear retention factor	Accountable for Heat shock responsive gene expression	Kotak et al., (2007)
4	*A. thaliana*	*PIF4, HY5, DELLAs, HFR1*	Protein degradation	Cellular processes are affected by average temperature	Casal and Balasubramanian (2019)
5	Arabidopsis	*bZIP* TFs (e.g., bZIP28, bZIP60)	ER stress regulators	Unfolded protein response activation during heat stress	Deng et al. (2011), Howell (2013)
6	Rice, wheat, tomato	QTLs for heat tolerance	Quantitative trait loci	Under heat stress, control yield stability, spikelet fertility, and grain filling	Jagadish et al. (2010), Ye et al. (2015)

stress tolerance, even if they use identical upstream signaling events (Ca^{2+} influx and MAPK cascades).

5 Phytohormones' Function in Temperature Response

Phytohormones are endogenous messenger molecules that are essential for plant development and stress response (Table 2). Several prior research provided solid evidence that phytohormones considerably influence how plants respond to heat stress (Ahammed et al., 2016).

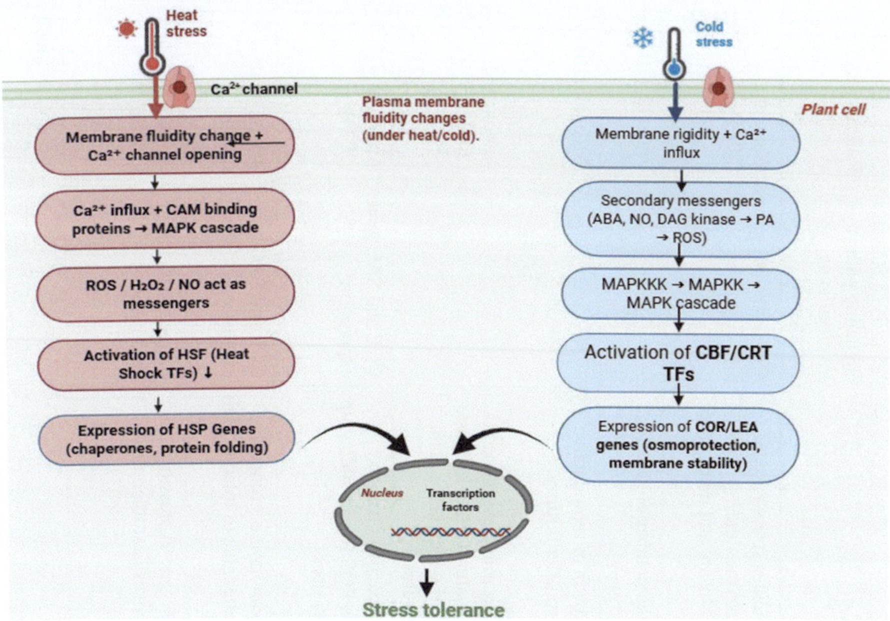

Fig. 6 Gene expression changes under temperature stress

6 Strategies for Enhancing Crop Thermotolerance

For healthy growth and development, each genotype of a plant, or more accurately, each species, needs a certain range of temperatures. When temperatures fall outside this range, the resulting stress conditions can negatively affect the genotype's performance in various ways. The two main environmental stressors that restrict crop productivity and geographic spread are heat and cold. Through both physiological barriers and evolutionary processes, plants have evolved a variety of adaptations over time to endure these climatic variations (Viswanathan & Zhu, 2002). Current adaptations to temperature stress through technical and management practices are insufficient to maintain consistent crop performance. As a result, there is a growing demand for breeding temperature-tolerant crop varieties (Driedonks et al., 2016). However, developing thermotolerant crops is challenging due to the high variability in temperature sensitivity across different developmental stages. When plants experience heat stress, a number of reactions are set off, including the synthesis of stress-sensitive genes. The underlying genetic mechanisms and the breadth of genetic variation related to heat and cold tolerance in plants are still mostly unknown, despite tremendous advancements in the breeding of tolerant lines.

Table 2 Phytohormones' function in temperature response

S.N	Crop	Phytohormone	Role in heat/cold stress	Signaling pathways involved/gene expressions	References
1	Rice (*Oryza sativa*)	Abscisic Acid (ABA)	Accumulates under cold stress, upregulates stress-responsive (COR) genes, enhances antioxidant enzymes and osmoprotectants, modulates stomatal closure and membrane stability	CBF and COR genes	Wang et al. (2024)
2	Rice (*Oryza sativa*)	Cytokinin and Abscisic Acid	Modify the floret differentiation and the number of spikelets per panicle	–	Wu et al. (2019)
3	Rice (*Oryza sativa*)	Indole-3-Acetic Acid (IAA) And Gibberellin (GA)	actively involved in controlling the growth of reproductive organs, pollination, fertilization, and spikelet fertility	–	Wu et al. (2019)
4	Arabidopsis and Ryegrass	Methyl Jasmonic Acid	alters the antioxidant defense system, lessens heat-induced chlorophyll loss, preserves the plant's natural water balance, and stops crop electrolyte leakage, all of which helps plants become more tolerant of high temperatures	–	Angon et al. (2024)
5	*Arabidopsis*, Grapes, Kiwi, Barley, Rice, Ryegrass and Chinese Cabbage	Melatonin	Melatonin, a potent antioxidant and (Reactive Oxygen Species) ROS scavenger, keeps excess ROS from building up and stops plants from senescing due to stress	CCE and SAG gene expression	Angon et al. (2024)
6	Tomato	Melatonin	Tomato seedlings treated with melatonin have a stronger antioxidant defense system and are better able to tolerate high temperatures	CCE and SAG gene expression	Angon et al. (2024)
7	Potato, Mustard, Tobacco, Tomato, Bean, and Arabidopsis Thaliana	Salicylic acid	Through the induction of specific HSP gene expression, exogenous salicylic acid therapy improves heat tolerance	SA-accumulating cpr5 mutants	Ahammed et al. (2016)

(continued)

Table 2 (continued)

S.N	Crop	Phytohormone	Role in heat/cold stress	Signaling pathways involved/gene expressions	References
8	Wheat	Salicylic acid	Salicylic acid, prior to treatment, prevents the D1 protein from degrading under heat and intense light exposure, in addition to increasing protein kinase activity	Degradation of D1 protein	Ahammed et al. (2016)
9	Grape vine	Salicylic acid	By maintaining a greater ribulose-1,5-bisphosphate carboxylase/oxygenase activation state and boosting photosystem II efficiency, salicylic acid pretreatment lessens heat stress-induced photosynthetic inhibition	Accumulation of HSP21	Ahammed et al. (2016)

6.1 Breeding for Heat Resistance

Conventional breeding is a tried-and-true way to increase temperature tolerance and lessen the detrimental effects of heat on crop output. Typically, breeding activities are conducted in climacteric regions that are comparable to the desired producing environments. Therefore, breeding lines destined for production in warmer climates are chosen in high-temperature environments (Mickelbart et al., 2015). In order to purposefully produce unique crop genotypes that can tolerate heat, conventional breeding has also been used. By maturing sooner, an enhanced broccoli variety, for instance, avoids exposure to high temperatures later in the season, particularly during the heat-sensitive stage of blossom initiation, resulting in higher-quality heads (Farnham & Bjorkman, 2011). Furthermore, novel cowpea varieties grown in hotter climates with longer photoperiods during the reproductive phase have been shown to increase average grain yields (Ehlers & Hall, 1998). The wild progenitor "*Triticum tauschii*" has also been successfully used as a gene donor in recurrent selection to improve wheat. As a result, BC_1F_6 plants now have bigger grain sizes and higher grain filling rates (Gororo et al., 2002). With yield increases of up to 37.8%, three cycles of recurrent selection for heat tolerance in potato breeding resulted in significant genetic benefits (Benites & Pinto, 2011).

Plant heat tolerance seems to be a polygenic characteristic, which probably adds to our incomplete knowledge of its genetic foundation (Driedonks et al., 2016). One of the main goals of efforts to improve our genetic understanding of thermotolerance has been the identification of quantitative trait loci in segregating mapping populations. A thorough list of QTLs linked to heat tolerance was compiled by (Jha et al., 2014) for a variety of plant species, including *Arabidopsis*, beans, Brassica, barley, maize, cowpea, potato, rice, tomato, sorghum, and wheat. Heat tolerance qualities have been linked to a number of QTLs. These include QTLs associated with reduced canopy temperature and increased chlorophyll in wheat during the vegetative and reproductive stages (Vijayalakshmi et al., 2010). Higher chlorophyll fluorescence implies heat-resilient photosynthetic efficiency, whereas lower canopy temperature indicates increased water uptake capacity (Wahid et al., 2007). Htg6.1, a key QTL associated with high-temperature seed germination in lettuce, was found to co-locate with $LsNCED_4$, a gene encoding a temperature-sensitive enzyme involved in the synthesis of abscisic acid (Argyris et al., 2011). Five and six QTLs for pollen viability and pollen tube growth, respectively, have been identified in maize with high heritability estimates of 0.64 and 0.68, respectively. It's important to remember that these evaluations were carried out in vitro, they might not accurately reflect the physiological reactions under field conditions (Frova & Sari-Gorla, 1994).

6.2 *Breeding for Cold Resistance*

Long-term evolutionary adaptation has revealed that plants often develop freezing resistance and other temperature stress tolerances as a result of exposure to low but non-freezing temperatures, a process known as cold acclimation (Chinnusamy et al., 2007). However, little is known about the fundamental processes underpinning this adaptive response. Therefore, elucidating the genomic basis of cold tolerance in plants holds significant importance for advancing both fundamental knowledge and applied crop improvement (Chen et al., 2014). Conventional breeding approaches, particularly those involving interspecific or intergeneric hybridization, have achieved limited success in enhancing cold tolerance in major crop species (Ma et al., 2024). But recent developments in mutagenesis, transgenic plant research, whole-genome profiling, and sequencing have greatly expanded our knowledge of the intricate transcriptional networks triggered by cold stress. Significant changes in gene expression and the accumulation of hundreds of metabolites are brought on by low temperatures; some of these metabolites provide protection from the negative effects of the cold. Numerous cold-inducible genes, many of which are involved in cold tolerance, have been discovered to be present in plants. A portion of these genes are controlled by dehydration and C-repeat binding factors. CBF/DREB1, two essential transcription factors that support cold stress responses, are responsive element-binding proteins. The process of cold acclimatization involves a variety of physiological and molecular alterations, indicating that cold resistance is a complex characteristic governed by multiple routes rather than a single regulatory mechanism (Sanghera et al., 2011). Crop tolerance to cold stress is improved by using molecular techniques to incorporate cold-regulated genes into crop hybrids. Many proteins, such as molecular chaperones, antifreeze proteins, and other proteins thought to offer resistance to freezing stress-induced dehydration, as well as enzymes involved in respiration and the metabolism of lipids, carbohydrates, phenylpropanoids, and antioxidants, are encoded by cold-responsive genes (Sanghera et al., 2011; Thomashow, 1999). During cold acclimation, a developmental process that increases a plant's resistance to low temperatures, several gene expression alterations take place (Steponkus, 1993).

6.3 *Management Practices to Mitigate the Effect of Temperature on Crops*

Crops can be spared extreme heat by changing the sowing dates. This can react well with major staples such as wheat, rice, and maize (Mahajan et al., 2025). Conservation agriculture enhances the heat resistance of crops, delays crop senescence, maintains soil moisture, and regulates the soil temperature using methods such as surface mulching and minimal tillage (Mahajan et al., 2025). Mulching and row covering substance, including used plastic film and biomass, enable crops such as melons to manage the needed ambient soil temperature, particularly in periods of extreme temperature (Ferrante & Mariani, 2018). Row covers and mulches also aide in the growth and establishment of plants by supporting, warming the soil and microclimate around plants during cold weather (Ferrante & Mariani, 2018). Moisture stress that is caused by high temperature is reduced through effective watering. This includes water-saving techniques, effective management of water, and drip irrigation (Ahmad et al., 2019). Foliar sprays (e.g., KNO3, urea, salicylic acid, GA3, and micronutrients) enhance photosynthesis, maintain membrane integrity, and reduce the impact of heat stress, which have been used successfully on wheat. The physical resilience of seeds is increased by the use of compounds like thiourea to prime seeds; in the case of camelina, thiourea seed priming increased seed yield by around 15 percent during heat stress (Raza et al., 2023).

Heat-tolerant cultivars can be developed by breeding and biotechnology. These processes are the reproductive stages which can be overexpressed with the use of the marker-assisted genetic selection and the cold tolerance breeding (Soualiou et al., 2022). Approaches such as genomic selection, editing through CRISPR, or marker-assisted selection can be focused on such traits as heat shock protein genes (HSFA2), deep root, or wax deposition (Mahajan et al., 2025). Nanotechnology enhances water retention and cell integrity at the presence of heat and fortifies antioxidant defense, including silicon or zinc oxide nanoparticles (Mahajan et al., 2025). Microbial bio stimulants, such as endophytes and mycorrhizae, and advantageous rhizobacteria have the ability to improve the root development, nutrient and water uptake, and stress-response mechanisms and, thus, increase thermotolerance (Bhagchandka, 2025). Early detection of thermal stress can be achieved through timely monitoring through thermal infrared cameras equipped on drones or handheld cameras, which means that prompt and swift actions can be undertaken (Ferrante & Mariani, 2018). Heat stress in the fields can be real-time monitored and proactively controlled through sensors, Artificial Intelligence (AI) technologies, satellite imaging and remote sensing (Bhagchandka, 2025). Crop-specific cold acclimation reduces damage caused during the occurrence of freezing or chilling by inducing biochemical and morphological changes, including accumulation of hormones, osmolytes, and protective proteins (Soualiou et al., 2022).

7 Conclusion and Prospects for the Future

The complex interactions between temperature change and plant reproduction are of greater importance than ever in the times of global climate change (Maity et al., 2023). Temperature variations have impacts on most physiological, biochemical, and molecular processes of plants, including respiration and photosynthesis as well as reproduction and ultimate production (Aslam et al., 2021). The creation of climate-resistant varieties of crops and methods of adaptive management has emerged as an objective on sustainable agricultural practices as temperatures increase globally, and the number of extreme cases of temperatures increase (Biswas et al., 2025). Therefore, the research needs to combine the fields of physiology, molecular, and genetic data and create a more advanced concept of plant thermotolerance in the future (Kumar et al., 2024). Important regulatory genes and pathways of temperature resilience can be identified with the help of more advanced molecular techniques, such as transcriptome profiling, metabolomic analyses, and CRISPR/Cas-based genome editing (Dutta et al., 2024). Interesting ways to enhance tolerance to heat and cold include the study of the regulatory processes of heat shock proteins and antioxidant defense systems and membrane-stabilizing molecules (Dutta et al., 2025; Kumar et al., 2024). Similarly, analyzing the genetic basis of thermotolerance by studies of genome-wide association mapping and QTL sequencing will assist breeders to create marker-assisted selection models that suit certain temperature stresses (Gain et al., 2025). One of the research frontiers is the temperature-mediated control of reproductive growth and development. Among the outcomes of the temperature stress are pollen sterility, failure of fertilization processes, and reduce seed set, which ultimately have their effects on the final crop productivity (Bhattacharya, 2022). The future research direction must be on the mechanisms that control the flower induction, pollen viability and embryo development in the case of different temperature regimes. The knowledge of the interaction between thermoperiodism and vernalization in flowering regulation can potentially provide control over the phenological characteristics to ensure enhanced alignment of reproductive stages with optimal temperature-related factors (Naik et al., 2025). Stressing physiological and biochemical level, it is necessary to study hormonal signaling pathways, such as ethylene, ABA, and salicylic acid interaction with nutrient dynamics to enhance thermotolerance (Shah et al., 2025). Improving plant metabolic processes during temperature stress through the development of bio-stimulant-based interventions will help stabilize the process and supplement genetic improvement strategies. In order to accelerate the development of cultivars with heat and cold tolerance, molecular breeding, high-throughput phenotyping, and speed breeding should be used in breeding programs (Kallugudi et al., 2025). The management strategies such as the use poly-mulches, micro-irrigation, micro-environment controlled agriculture and the accurate weather based advisory to combat temperature extremes at the field level should also be accorded the first priority. Finally, resilience of crops to temperature stress requires an interdisciplinary approach that cuts across the various fields of physiology, seed science, molecular biology, genetics, plant breeding, and agronomy. The AI-driven climate prediction models, remote stress monitoring based on remote sensing, and decision-support system could transform the breeding and management of thermotolerant. By integrating genetic innovation with sustainable field practices, it will be simple to safeguard global sustainable food systems to the emergent challenges of the unpredictability of temperatures and sustainability, and attain productivity, consistency, and food security in the climate change scenario.

Acknowledgements Not Applicable

Authors Contribution DP and SD: Conceptualization and supervision; IP, AS, SB, Sunita: Data curation, preparation of tables and figures; IP, AS, DP, Sunita, SB: Preparation of first draft manuscript; All authors: editing and proofreading.

Consent for Publication
Agreed by all the authors.

Conflict of Interests
Authors declare no conflict of Interests.

Funding Not applicable.

References

Ahammed, G. J., Li, X., Zhou, J., Zhou, Y. H., & Yu, J. Q. (2016). Role of hormones in plant adaptation to heat stress. In *Plant hormones under challenging environmental factors* (pp. 1–21). Springer, Netherlands.

Ahmad, I., Ullah, A., Rahman, M. H. U., Ahmad, B., Wajid, S. A., Ahmad, A., & Ahmed, S. (2019). Climate change impacts and adaptation strategies for agronomic crops. In *Climate change and agriculture.*

Ahmad, P., Jaleel, C. A., Salem, M. A., Nabi, G., & Sharma, S. (2010). Roles of enzymatic and nonenzymatic antioxidants in plants during abiotic stress. *Critical Reviews in Biotechnology, 30*(3), 161–175.

Alam, B., & Jacob, J. (2002). Overproduction of photosynthetic electrons is associated with chilling injury in green leaves. *Photosynthetica, 40*(1), 91–95.

Allakhverdiev, S. I., Kreslavski, V. D., Klimov, V. V., Los, D. A., Carpentier, R., & Mohanty, P. (2008). Heat stress: An overview of molecular responses in photosynthesis. *Photosynthesis Research, 98*(1), 541–550.

Amasino, R. (2010). Seasonal and developmental timing of flowering. *The Plant Journal, 61*(6), 1001–1013.

Angon, P. B., Das, A., Roy, A. R., Khan, J. J., Ahmad, I., Biswas, A., & Yeasmin, S. T. (2024). Plant development and heat stress: Role of exogenous nutrients and phytohormones in thermotolerance. *Discover Plants, 1*(1), 17.

Argyris, J., Truco, M. J., Ochoa, O., McHale, L., Dahal, P., Van Deynze, A., & Bradford, K. J. (2011). A gene encoding an abscisic acid biosynthetic enzyme (LsNCED4) collocates with the high temperature germination locus Htg6. 1 in lettuce (Lactuca sp.). *Theoretical and Applied Genetics, 122*(1), 95–108. https://doi.org/10.1007/s00122-010-1425-3

Asada, K. (2006). Production and scavenging of reactive oxygen species in chloroplasts and their functions. *Plant Physiology, 141*(2), 391–396.

Aslam, M. A., Ahmed, M., Hassan, F. U., Afzal, O., Mehmood, M. Z., Qadir, G., Asif, M., Komal, S., & Hussain, T. (2021). Impact of temperature fluctuations on plant morphological and physiological traits. *Building climate resilience in agriculture: Theory, practice and future perspective* (pp. 25–52). Springer International Publishing.

Asthir, B. (2015). Protective mechanisms of heat tolerance in crop plants. *Journal of Plant Interactions, 10*(1), 202–210.

Awasthi, A., Vishwakarma, K., & Pattnayak, K. C. (2022). Retrospection of heatwave and heat index. *Theoretical and Applied Climatology, 147*(1), 589–604.

Awasthi, R., Devi, P., Jha, U. C., Sharma, K. D., Roorkiwal, M., Kumar, S., & Nayyar, H. (2024). Exploring the synergistic effects of drought and heat stress on chickpea seed development: Insights into nutritional quality and seed yield. *Plant Stress, 14*, Article 100635.

Baxter, A., Mittler, R., & Suzuki, N. (2014). ROS as key players in plant stress signalling. *Journal of Experimental Botany, 65*(5), 1229–1240.

Benites, F. R. G., & Pinto, C. A. B. P. (2011). Genetic gains for heat tolerance in potato in three cycles of recurrent selection. *Crop Breeding and Applied Biotechnology, 11*, 133–140. https://doi.org/10.1590/S1984-70332011000200005

Beniwal, S. K., Bharti, K., Chan, K. Y., Fauth, M., Ganguli, A., Kotak, S., Mishra, S. K., Nover, L., Port, M., Scharf, K. D., Tripp, J., Weber, C., Zielinski, D., & Von, Koskull-Doring, P. (2004). Heat stress response in plants: a complex game with chaperones and more than twenty heat stress transcription factors. *Journal of Biosciences, 29*, 471–487.

Berry, S., & Dean, C. (2015). Environmental perception and epigenetic memory: Mechanistic insight through FLC. *Plant Journal, 83*(1), 133–148.

Bhagchandka, H. V. (2025). Managing crop stress: Techniques to protect crops from biotic & abiotic stress. *Global Agriculture.*

Bhattacharya, A. (2022). Effect of low temperature stress on photosynthesis and allied traits: A review. In *Physiological processes in plants under low temperature stress*, pp.199–297.

Biswas, A., Sarkar, S., Das, S., Dutta, S., Choudhury, M. R., Giri, A., Bera, B., Bag, K., Mukherjee, B., Banerjee, K., & Gupta, D. (2025). Water scarcity: A global hindrance to sustainable development and agricultural production—A critical review of the impacts and adaptation strategies. *Cambridge Prisms: Water, 3*, Article e4.

Bloom, A. J., Zwieniecki, M. A., Passioura, J. B., Randall, L. B., Holbrook, N. M., & St. Clair, D. A. (2004). Water relations under root chilling in a sensitive and tolerant tomato species. *Plant, Cell & Environment*, *27*(8), 971–979.

Casal, J. J., & Balasubramanian, S. (2019). Thermomorphogenesis. *Annual Review of Plant Biology, 70*(1), 321–346.

Chaturvedi, S. (2021). Heat stress response mechanisms in pollen development. *New Phytologist.*

Chen, L. J., Xiang, H. Z., Miao, Y., Zhang, L., Guo, Z. F., Zhao, X. H., & Li, T. L. (2014). An overview of cold resistance in plants. *Journal of Agronomy and Crop Science, 200*(4), 237–245.

Chinnusamy, V., Zhu, J., & Zhu, J.-K. (2007). Cold stress regulation of gene expression in plants. *Trends in Plant Science, 12*(10), 444–451.

Deng, Y., Humbert, S., Liu, J. X., Srivastava, R., Rothstein, S. J., & Howell, S. H. (2011). Heat induces the splicing by IRE1 of a mRNA encoding a transcription factor involved in the unfolded protein response in Arabidopsis. *Proceedings of the National Academy of Sciences, 108*(17), 7247–7252.

Driedonks, N., Rieu, I., & Vriezen, W. H. (2016). Breeding for plant heat tolerance at vegetative and reproductive stages. *Plant Reproduction, 29*(1), 67–79.

Dutta, S., Islam, Z., Das, S., Barman, A., Chowdhury, M., Mondal, B. P., Ajnabi, J., & Manna, D. (2025). Harmonizing plant resilience: Unveiling the symphony of membrane lipid dynamics in response to abiotic stresses: A review. *Discover Plants, 2*(1), 61.

Dutta, S., Paul, D., Das, A., Kumar, S., Roy, A., Barua, A., & Chinnusamy, V. (2024). Directed mutagenesis in plants through genome editing using guide RNA library. *The Nucleus, 67*(1), 61–73.

Ehlers, J. D., & Hall, A. E. (1998). Heat tolerance of contrasting cowpea lines in short and long days. *Field Crops Research, 55*(1–2), 11–21. https://doi.org/10.1016/S0378-4290(97)00055-5

Farnham, M. W., & Bjorkman, T. (2011). Breeding vegetables adapted to high temperatures: A case study with broccoli. *HortScience, 46*(8), 1093–1097.

Farooq, M., Bramley, H., Palta, J. A., & Siddique, K. H. (2011). Heat stress in wheat during reproductive and grain-filling phases. *Critical Reviews in Plant Sciences, 30*(6), 491–507.

Ferrante, A., & Mariani, L. (2018). Agronomic management for enhancing plant tolerance to abiotic stresses: High and low values of temperature, light intensity, and relative humidity. *Horticulture, 4*(3), Article 21.

Frova, C., & Sari-Gorla, M. (1994). Quantitative trait loci (QTLs) for pollen thermotolerance detected in maize. *Molecular and General Genetics MGG*, *245*(4), 424–430. http://www.ncbi.nlm.nih.gov/pubmed/7808391

Gain, N., Muthusamy, V., Chhabra, R., Kyada, A. D., Dutta, S., Chand, G., Kasana, R. K., Sarika, K., Devi, E. L., Madhavan, J., & Subramani, R. (2025). Accelerated development of MTL and DMP gene-based subtropically-adapted maternal haploid inducer lines in maize using molecular breeding. *Planta, 262*(4), 88.

Garg, T., Kumar, A., Joshi, A., Awasthi, A., Rana, A., Kumar, V., & Kaur, S. (2024). The role of phytohormones in reducing the arsenic-induced stress in plants. *South African Journal of Botany, 168*, 296–315. https://doi.org/10.1016/j.sajb.2024.03.020

Gororo, N. N., Eagles, H. A., Eastwood, R. F., Nicolas, M. E., & Flood, R. G. (2002). Use of Triticum tauschii to improve yield of wheat in low-yielding environments. *Euphytica, 123*(2), 241–254. https://doi.org/10.1023/A:1014910000128

Grover, A., Mittal, D., Negi, M., & Lavania, D. (2013). Generating high temperature tolerant transgenic plants: Achievements and challenges. *Plant Science, 205–206*, 38–47.

Howell, S. H. (2013). Endoplasmic reticulum stress responses in plants. *Annual Review of Plant Biology, 64*(1), 477–499.

Jagadish, S. V. K., Cairns, J., Lafitte, R., Wheeler, T. R., Price, A. H., & Craufurd, P. Q. (2010). Genetic analysis of heat tolerance at anthesis in rice. *Crop Science, 50*(5), 1633–1641.

Jha, U. C., Bohra, A., & Singh, N. P. (2014). Heat stress in crop plants: Its nature, impacts and integrated breeding strategies to improve heat tolerance. *Plant Breeding, 133*(6), 679–701. https://doi.org/10.1111/pbr.12217

Kallugudi, J., Lata, C., Thribhuvan, R., Ravikiran, K. T., Dutta, S., Krishnan, P. A., Patel, C., Thakur, A., Manjul, A. S., Prasad, P., & Gangwar, O. P. (2025). Advanced strategies for crop improvement against abiotic stresses: An integrated view from breeding to genomics. *Cutting Edge Technologies for Developing Future Crop Plants*, 83–123.

Kotak, S., Larkindale, J., Lee, U., von Koskull-Döring, P., Vierling, E., & Scharf, K. D. (2007). Complexity of the heat stress response in plants. *Current Opinion in Plant Biology, 10*(3), 310–316.

Kumar, P., Paul, D., Jhajhriya, S., Kumar, R., Dutta, S., Siwach, P., & Das, S. (2024). Understanding heat-shock proteins' abundance and pivotal function under multiple abiotic stresses. *Journal of Plant Biochemistry and Biotechnology, 33*(4), 492–513.

Kumar, S. V., et al. (2012). Transcription factor PIF4 controls the thermosensory activation of flowering. *Nature, 484*(7393), 242–245.

Li, J., et al. (2025). Optimization of crop response to climatic stress via bio stimulants and hormones. arXiv Preprint.

Ma, S., Huang, X., Zhao, X., Liu, L., Zhang, L., & Gan, B. (2024). Current status for utilization of cold resistance genes and strategies in wheat breeding program. *Frontiers in Genetics, 15*, 1473717.

Mahajan, S., Thakur, P., Das, S., Sharma, R. P., Manuja, S., Jha, P. K., Saini, A., Sahoo, C., & Fayez Zadeh, M. R. (2025). Impression of contemporary heat stress complexities in agricultural crops: A review. *Plant Growth Regulation.*

Maity, A., Paul, D., Lamichaney, A., Sarkar, A., Babbar, N., Mandal, N., Dutta, S., Maity, P. P., & Chakrabarty Chakrabarty, S. K. (2023). Climate change impacts on seed production and quality: Current knowledge, implications, and mitigation strategies. *Seed Science and Technology, 51*(1), 65–96.

Mickelbart, M. V., Hasegawa, P. M., & Bailey-Serres, J. (2015). Genetic mechanisms of abiotic stress tolerance that translate to crop yield stability. *Nature Reviews Genetics, 16*(4), 237–251. https://doi.org/10.1038/nrg3901

Naik, Y. D., Bahuguna, R. N., Garcia-Caparros, P., Zwart, R. S., Reddy, M. S., Mir, R. R., Jha, U. C., Fakrudin, B., Pandey, M. K., Challabathula, D., & Sharma, V. K. (2025). Exploring the multifaceted dynamics of flowering time regulation in field crops: Insight and intervention approaches. *The Plant Genome, 18*(2), Article e70017.

Patel, A., Kumar, K., & Yadav, A. S. (2023). Effects of temperature on plants and animals. *Krishi Science, 4*(10), 21–24.

Peet, M. M., Sato, S., & Gardner, R. G. (1998). Comparing heat stress effects on male-fertile and male-sterile tomatoes. *Plant, Cell & Environment, 21*(2), 225–231.

Prasad, P. V. V., Bheemanahalli, R., & Jagadish, S. V. K. (2017). Field crops and the fear of heat stress. *Field Crops Research, 200*, 114–121.

Raza, A., Wang, D., Zou, X., & Prakash, C. S. (2023). Developing temperature-resilient plants: A matter of present and future concern for sustainable agriculture [Editorial]. *Agronomy, 13*(4), Article 1006.

Sanghera, G. S., Wani, S. H., Hussain, W., & B. Singh, N. (2011). Engineering cold stress tolerance in crop plants. *Current Genomics, 12*(1), 30–43.

Sage, T. L., & Kubien, D. S. (2015). The temperature response of C3 and C4 photosynthesis. *Plant, Cell & Environment, 28*(1), 68–96.

Shah, S. A., Arshad, M., & Aslam, S. (2025). Comprehensive review on the role of exogenous phytohormones in enhancing temperature stress tolerance in plants. *Journal of Crop Health, 77*(4), 136.

Sharma, K. D., Nayyar, H., & Bains, T. S. (2016). Cold stress and reproductive success in maize. *Plant Stress Physiology, 5*(2), 40–49.

Soualiou, S., Duan, F., Li, X., & Zhou, W. (2022). Crop production under cold stress: An understanding of plant responses, acclimation processes, and management strategies. *Plant Physiology and Biochemistry, 190*, 47–61.

Steponkus, P. L. (1993). A contrast of the cryostability of the plasma membrane of winter rye and spring oat-two species that widely differ in their freezing tolerance and plasma membrane lipid composition. *Advances in Low Temperature Biology, 2*, 211–312.

Thomashow, M. F. (1999). Plant cold acclimation: Freezing tolerance genes and regulatory mechanisms. *Annual Review of Plant Biology, 50*(1), 571–599.

Vijayalakshmi, K., Fritz, A. K., Paulsen, G. M., Bai, G., Pandravada, S., & Gill, B. S. (2010). Modeling and mapping QTL for senescence-related traits in winter wheat under high temperature. *Molecular Breeding, 26*(2), 163–175. https://doi.org/10.1007/s11032-009-9366-8

Viswanathan, C., & Zhu, J. K. (2002). Molecular genetic analysis of cold–regulated gene transcription. *Philosophical Transactions of the Royal Society of London. Series B: Biological Sciences, 357*(1423), 877–886.

Wahid, A., Gelani, S., Ashraf, M., & Foolad, M. R. (2007). Heat tolerance in plants: An overview. *Environmental and Experimental Botany, 61*(3), 199–223.

Wang, H., Charagh, S., Dong, N., Lu, F., Wang, Y., Cao, R., & Wei, X. (2024). Genome-wide analysis of heat shock protein family and identification of their functions in rice quality and yield. *International Journal of Molecular Sciences, 25*(22), 11931.

Wu, C., Tang, S., Li, G., Wang, S., Fahad, S., & Ding, Y. (2019). Roles of phytohormone changes in the grain yield of rice plants exposed to heat: A review. *PeerJ, 7*, Article e7792.

Ye, C., Tenorio, F. A., Redoña, E. D., Morales–Cortezano, P. S., Cabrega, G. A., Jagadish, K. S., & Gregorio, G. B. (2015). Fine-mapping and validating qHTSF4. 1 to increase spikelet fertility under heat stress at flowering in rice. *Theoretical and Applied Genetics, 128*(8), 1507–1517.

Zinn, K. E., Tunc-Ozdemir, M., & Harper, J. F. (2010). Temperature stress and plant sexual reproduction. *Plant Physiology, 154*(2), 634–640.

Effects of Rising Temperature on Plant Water Relations and Hydraulics

D. P. Singh, Varun Vijay Savadi, Siya Tuteja, and Kailash Pandey

Abstract Rising average global temperature due to climate change has a strong influence on plant water relations and hydraulic function, thereby impacting plant physiology, productivity, and ecosystem resilience. This chapter explains how high temperatures affect important processes including transpiration, stomatal control, incorporating physiological processes, cell responses, ecological effects, xylem conductivity, and susceptibility to cavitation and involvement of the physiological processes involved in plant responses to rising vapor pressure deficits, shifts in water potential gradients, and the threat of hydraulic failure. Further, this study also explains structural adaptations, that is, xylem anatomical changes and root/shoot ratios, explained within the frames of short-term acclimation and long-term adaptation. In addition, an interspecific variation of hydraulic approaches, the consequences for agricultural systems and forest dynamics and possibilities of mitigation measures such as breeding and agronomic practices have also been explained. This chapter highlights the need to take into consideration plant hydraulics in a warming world to provide sustainable ecosystem function and crop resilience through the integration of existing research and knowledge gaps.

Keywords Climate change · Heat stress · Plant water relations · Plant hydraulics · Vapor pressure deficit · Xylem conductivity · Cavitation vulnerability · Crop and ecosystem resilience

D. P. Singh (✉) · S. Tuteja · K. Pandey
Department of Physics, SoAE, UPES, Dehradun, Uttarakhand, India
e-mail: devendrapal@ddn.upes.ac.in

V. V. Savadi
Agnira Sanlayan Private Limited, Industrial area, Gandhinagar, Gujarat, India

A. Awasthi et al. (eds.), *Rising Temperatures and Plant Ecosystems*,
https://doi.org/10.1007/978-3-032-29477-7_3

1 Introduction

Increasing global average temperature resulting from modern-day human needs has resulted in great impact on plant water relations and hydraulic functioning. Overall impact leads to decline in productivity and reduced yields affecting the entire ecosystems which pose a serious threat to food security for people (Robinson et al., 2019; Smith et al., 2015; Sperry & Love, 2015a). Temperature is an important determinant of physiological functions in plants and affects transpiration, stomatal conductance, vapor pressure deficit (VPD), plant water potential, and xylem vessel hydraulic efficiency (Grossiord et al., 2020; López et al., 2021). Increased temperatures not only elevate evaporative demand and water losses for plants, but may also compromise hydraulic balance, thus compromising photosynthesis, growth, and even survival in plants (Dos Santos et al., 2022; Sadok et al., 2021a). In addition to the aforementioned environmental stressors, there is a greater risk of cavitation and embolism occurring in xylem vessels, which lowers hydraulic conductance and increases the potential for hydraulic failure (Venturas et al., 2017a). Simultaneously, plants can show remarkable elasticity to regulate water balance through immediate and short-term physiological adjustments, for example, regulating water-use efficiency via stomatal regulation, osmotic adjustment, hormonal response, etc., but also through longer-term structural adjustments, for example, root-to-shoot ratios, leaf structure, and xylem anatomy acclimation (Lipiec et al., 2013; Venturas et al., 2017a). However, the responses to higher temperatures vary widely across species. Species will vary in hydraulic strategy, determining whether a species has more conservative traits saving water or whether the species has risk-prone traits that are more water-efficient. Such differences in hydraulic traits would affect community composition and ecosystem functioning under climate-related stressors. In forest ecosystems, this would change community structure, carbon sequestration capacity, and resilience to mortality events and in agricultural production systems, the impacts would dictate crop productivity, water-use efficiency, and irrigation use projections (Mbava et al., 2020). Consequently, the downstream implications require appropriate adaptation and mitigation strategies. Breeding and biotechnology are rapidly advancing in developing varieties that would have greater heat and drought safety margins, while agronomic and silvicultural practices (e.g., precision irrigation, canopy management, and agroforestry) are proven to be suitable strategies to maintain and improve productivity and ecosystem stability (Adeyemi et al., 2017). A study domain exhibits a knowledge deficiency on the thresholds of hydraulic susceptibility and the interplay of rising temperatures with other abiotic stressors, including drought, increased CO_2 concentrations, and nutrient imbalances (N. G. McDowell, 2011; Sperry & Love, 2015b). Future studies should take a more holistic view by integrating molecular, physiological and ecosystem scales, supported with new relevant technologies in imaging, modeling, and remote sensing to improve the predictive capacity of climate impact on plants, as well as methods to develop climate-resilient management strategies (Costa et al., 2019). This chapter focuses on understanding various aspects of different interaction phenomena because of increasing temperature in plants. Key highlights are utilization of water loss through evapotranspiration, functioning of stomata for balancing water flow and how xylem conducts the water flow (Roy et al., 2024). It also presents an overview of these complex interactions for different plant species and how it affects the entire ecosystem. Furthermore, it emphasizes new research ideas and techniques which help in understanding the problem caused by temperature effects on entire plant ecosystems and highlights potential ways in which this critical challenge can be tackled by protecting plants and ecosystems from climate change.

2 Temperature and Plant Water Relations

Some of the key interactions in plant ecosystems include conductance efficacy of water by stomata, rate of evapotranspiration and behavior of water potential gradients within plant tissues with respect to increase in temperature. Rise in temperature often leads to increased loss of water due to increase in evaporation and transpiration process (Dai et al., 2018; Will et al., 2013a). This loss in water happens through the pores on leaves (stomata) which let out water vapor. If water retention capacity in the soil is reduced, it leads to drying of plant cells and induces stress finally causing damage (Asbjornsen et al., 2011; Sperry & Love, 2015c). Evapotranspiration and VPD are significant pathways in which temperature has direct effects on plant water relationships (Lambers and Oliveira, 2019; Massmann et al., 2019). Each increase in temperature raises the VPD by increasing the gradient between water vapor pressure in the leaf and the surrounding atmosphere. Subsequently, this creates a stronger driving force for transpiration. As a consequence, higher temperatures combined with increased VPD can cause rates of water loss from the plant to exceed root uptake during drought or limited soil moisture conditions. Although some increases in the evapotranspiration can be beneficial to cooling and positively influencing nutrient mobility when temperature increases are moderate, increased VPD likely leads to stomatal closure, reduced carbon fixation, and increased hydraulic stress. Therefore, the relationship between evapotranspiration and VPD can serve as an important framework through which to understand plant responses to temperature increases either directly or indirectly (as a complex influence will obviously have multiple effects).

- As temperature increases, VPD also increases, through an increase in the difference in the water vapor pressure inside the leaf and that of the surrounding atmosphere.
- Increased VPD increases transpiration, and overall evaporative demand on the plant increases.
- Increased temperature, and thus increased VPD, leads to increased depletion of soil moisture and increased risk of hydraulic failure.

In terms of plant adaptation to higher temperatures, without a doubt, leaf water potential (LWP) and osmotic adjustment are significant traits for plant adaptation to high temperatures and increased risk of water stress (Blum, 2017; Chen and Jiang, 2010). Generally speaking, the increase in temperature leads to lowered leaf water potential as transpiration rates and evaporative demand increase, creating a challenge for keeping turgor pressure and metabolic rate of the plant. The plants challenge of keeping turgor pressure under conditions of water limitation leads to osmatic, which is often through the process (accumulation) of growth promoting compatible solutes (e.g., proline and sugars), which can also promote water retention during stressful circumstances, which means that biological sanctions are a key factor regulating the water-hydraulic balance of the plant at a warmer climate.

- Increased temperature typically leads to a lower LWP as transpiration rates increase even if leaf water saturation is achieved.
- Osmotic adjustment of plants is generally through the accumulation of compatible solutes (like sugars) to maintain turgor pressure during reduced water availability.
- Long exposure to increased temperatures will cause permanent wilting and decreased cellular growth.

Figure 1 illustrates a schematic flow process how rise in temperature leads to different interactions in plant ecosystems. It also highlights limitations resulting in temperature effects and some of key ideas and future research directions which can be adopted to mitigate these effects.

3 Hydraulic Conductance and Temperature

Temperature has its direct impact on hydraulic conductance efficiency on any plant ecosystems. This can be understood with the fact that as the temperature increases, water content in the plant system proportional starts getting decreased, as described in Fig. 2. This leads to reduced viscosity of water through the plant system to compensate for the rising temperatures (Hatfield & Dold, 2019a; Sadok et al., 2021b; Will et al., 2013b). Elevated temperatures for longer durations result in formation of air bubbles inside the xylem conduits, which leads to blockages in water flow to entire plant system. This phenomenon shuts the plants' ability of transportation of water to its leaves which causes closure of stomata as result carbon intake reduces leading to possible damage. Identifying temperature dependent impacts in plant systems is important in order to understand the different stresses imposed in plant systems and address them in a systematic manner in the future to mitigate such losses. Further, the effect of tree water status on key eco-physiological parameters is given in Fig. 3.

Xylem Hydraulic Conductivity, as a variable of plant water transport potential, may respond significantly to changes in temperature (Gleason et al., 2012; Kim et al., 2014; Venturas et al., 2017b). Raising temperature can also slightly reduce water viscosity and increase xylem conductivity, thus enhancing whole-plant water

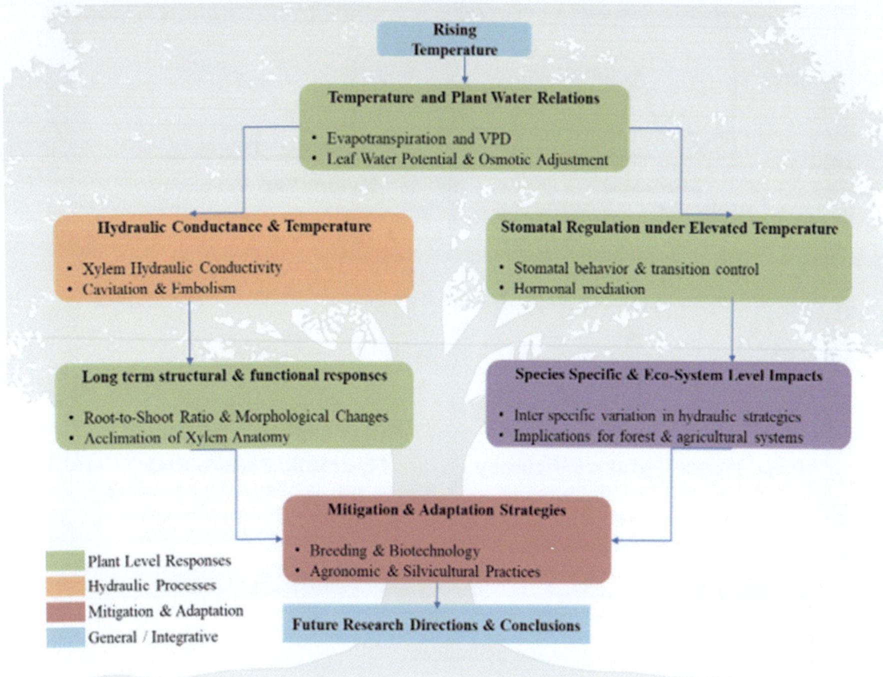

Fig. 1 Conceptual framework: effects of rising temperature on plant relations and hydraulics

transport efficiency through the plant vascular system. However, prolonged higher temperatures increase transpiration demand, thereby increasing xylem tension and amplifying the risk of cavitation and embolism in xylem vessels, which can rapidly and substantially decrease hydraulic conductivity, thereby halting the continuous movement of water from roots to leaves. Thus, maintaining xylem hydraulic stability and reducing embolism risk is fundamental to maintaining plant water status and photosynthetic performance under thermal stress or mortality.

- Hydraulic conductivity increases only slightly with higher temperature due to lower water viscosity.
- However, xylem architecture including conduit size and pit membrane porosity is likely the major factor regulating the efficiency of water transport flow.
- Increased flow due to warming is offset by a concomitant increase in embolism risk and cavitation, especially at increased VPD.

Cavitation and Embolism Formation are one of the key limitations affecting the overall functioning of plant water transport system triggered by rise in temperature effects (Cochard et al., 2013; Savi et al., 2015; Venturas et al., 2017c). Rupture of water column in the plant system is driven by cavitation phenomenon which results through formation of air bubbles caused by high water losses and dry air through stress induced from strong pulling forces in plants water pipe (xylem). Continued

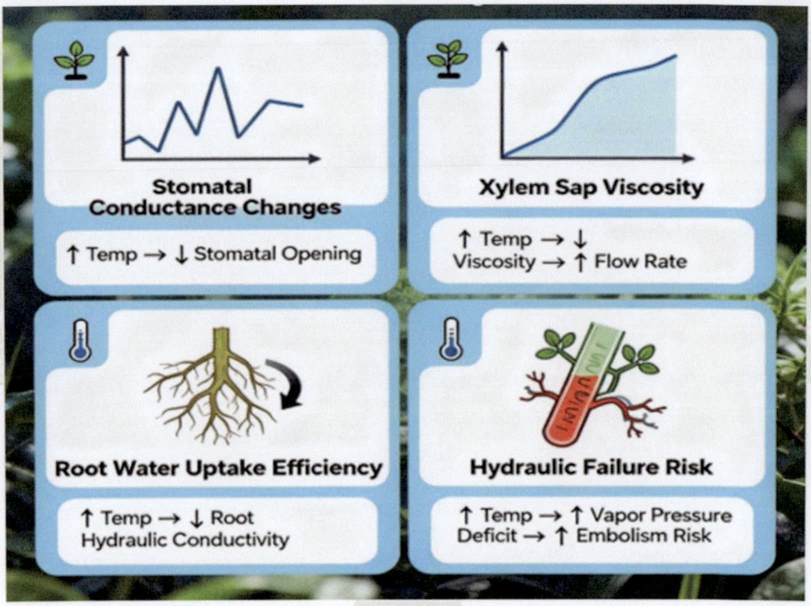

Fig. 2 Temperature effects on plant water hydraulics

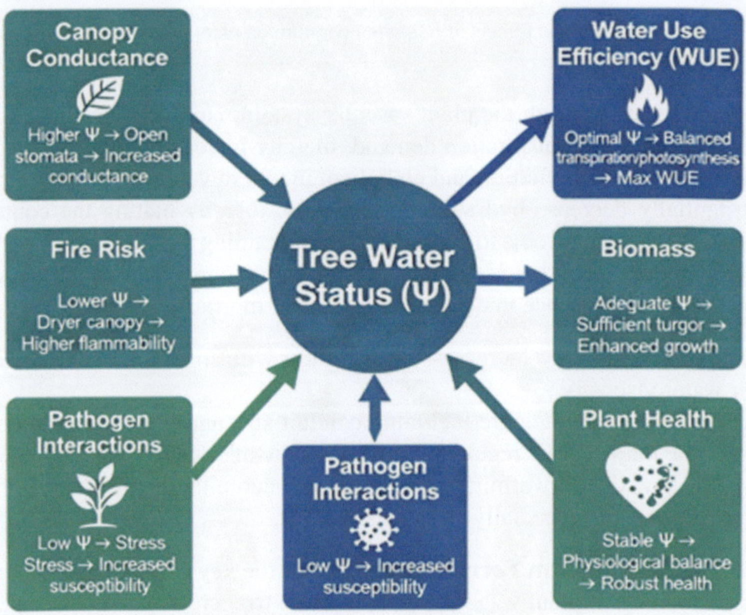

Fig. 3 Effect of tree water status on key eco-physiological parameters

stress due to rise in temperature leads to embolism formation through air bubbles hindering the flow of water in the entire plant system. At elevated temperatures plants become water deficit and soil loses its moisture content, where chances of these blockages are more prominent. Embolism formation ceases water movement leading to closure of stomata which drops the photosynthesis effect and overall plant growth (Brodersen & McElrone, 2013; Lens et al., 2013). Various plant species have their unique threshold capacity of fixing these losses occurred due to formation of air bubbles depending upon available moisture content levels. Repetitive heat stresses caused by prolonged exposure to higher temperatures results in permanent damage.

The interaction of high temperature and water deficit increases the probability of cavitation or air bubble formation, which obstructs water transport.

- Embolisms reduce xylem water permeability, thereby reducing performance of water transport to leaves, which can result in hydraulic failure all (Çeperlo et al., 2013).
- Species with larger diameter xylem vessels (e.g., ring-porous trees) are generally more vulnerable to cavitation than species with smaller diameter xylem vessels (e.g., conifers).

4 Stomatal Regulation at High Temperatures

Stomatal regulation in response to elevated temperatures, is an important adaptive mechanism to respond to the competing demands of water conservation and carbon acquisition (Chaves et al., 2016; Hatfield & Dold, 2019). As the temperature rises, transpiration occurs at an increasing rate due to an increased vapor pressure deficit, and consequently, the stomata will undergo either partial closure or complete closure to limit excess water loss. This response improves the leaf water potential and prevents hydraulic failure, but the trade-off is that CO_2 cannot enter the leaf impacting the efficiency of photosynthesis and biomass accretion. Elevated temperatures can also change stomatal sensitivity to environmental signals like humidity, light, and internal CO_2 which may result in a change in stomatal closure with extreme heat stress which is less conservative. For some species increased thermal hormonal signaling pathways especially increased abscisic acid (ABA) increased stomatal closure while other species will allow for partial or nearly full opening to remove evaporative cooling at the cost of water. The efficacy of stomatal regulation for species is dependent on their inherent characteristics, the amount of water available for utilization, and the duration of exposure to increased heat. Overall, stomatal regulation in higher temperatures represents an important mechanism for plant water use efficiency, hydraulic stress capacity, and adaptability/change in a warming environment (Chaves et al., 2016; Hatfield & Dold, 2019).

Stomatal behavior and transpiration control are critical for plant survival under increased temperature, since they regulate the exchange of water vapor and carbon dioxide between the leaf and atmosphere (Zandalinas et al., 2016). First, as temperature increases, stomata will open wider first to mediate evaporative cooling through transpiration, so to protect cellular and enzymes structures from thermal injury (Keller, 2005; Tomasella et al., 2019). However, vapor pressure deficit increases, and continued stomatal conductance can result in extreme water loss and significant declines in leaf water potential resulting in hydraulic failure. In this respect, plants will dynamically tend to adjust stomatal aperture such as closing or restricting stomatal pores in reduction to CO_2 acquisition for photosynthesis and decreasing transpiration. This is a trade-off between precisely calibrated regulation that is subject to numerous cues, including the water status of the leaf, abscisic acid accumulation, and CO_2 dependent feedback from photosynthesis. In arid conditions, stomata can be more conservative to favor water, despite low transpiration rates, and in more saturated conditions, can remain partially open for cooling purposes. Under elevated temperature conditions, stomatal behavior is thus a highly regulatory dynamic that integrates environmental and intrinsic physiological signals to optimize the regulation of transpiration dependent on physiological needs and risk tolerance in part directly impacting hydraulic safety for the plant and productivity overall.

- Stomata act as central mediators and regulate trade-offs between water loss and carbon gain.
- At high temperatures and VPD, stomata can be closed to conserve water, and photosynthesis can decline.
- There are plants that exhibit isohydric types (tight stomatal control), and plants with anisohydric types (less stomatal control, greater risk).

Hormonal Mediation is important to stomatal regulation at high temperatures as it acts as an internal signal network that integrates stress signals originating from the environment with physiological responses in plants (Garg et al., 2024; Li et al., 2021; Zhao et al., 2021). Abscisic acid (ABA), in particular, is the primary hormone regulating stomatal closing as it accumulates rapidly in leaves when plants experience heat and drought stress, which underlies stomatal closure to reduce transpiration water loss (Bharath et al., 2021; Muhammad Aslam et al., 2022). The expectation is that higher air and leaf temperatures will increase root-to-leaf transport of ABA and biosynthesis paced by high temperatures to facilitate hydraulic safety in plants while moderating carbon assimilation. In addition to ABA there are many other hormones that work together, some with synergistic effects, to regulate stomatal response under heat stress including jasmonic acid (JA), ethylene, and salicylic acid (SA). Ethylene is shown to enhance the sensitivity of stomata to ABA while JA has been found to promote stomatal closure under high-heat drought stress. Cytokinins could have an opposite effect in that they may promote stomatal opening, worked counter to the effects of ABA on WUE. Importantly, the signal trade-offs that come with hormonal cross-talk supply a greater indication that stomatal response to heat is not simply reactive but is part of a complex approach that coordinates water demand for conservation, cooling, and photosynthesis. Thus, hormonal mediation could be considered an integral mechanism employed by plants to modulate stomatal control in response to increasing temperatures and the affiliate hydraulic stress.

- ABA plays a central role in mediating stomatal closure during drought and heat stressed plants.
- Increased temperature will affect ABA induces biosynthesis rate or affect ABA signaling or affect the plant's ability to manage water loss effectively.

5 Long-Term Structural and Functional Responses

Long-term physiological and structural adjustments to increasing temperatures are reflective of the capacity of plants to acclimate and remodel their hydraulic plant structure and water relations with chronic stress (Chapin III et al., 1996; Soudzilovskaia et al., 2013). Physically, plants can acclimate by reducing leaf area, changing leaf morphology by forming thicker cuticles or smaller stomatal size, and changing root/shoot ratios in ways that increase water uptake and minimize transpirational loss. Xylem anatomy may change, for instance, smaller vessel diameters, or strengthened cell walls, which inherently limit the risk of cavitation but also lower maximum hydraulic conductivity. Functionally, plants will also increase their osmotic adjustment capability, increase compatible solute accumulation, and increase stomatal sensitivity to correspond to the increased vapor pressure deficit that is occurring in the long term. Such acclimations are often paired with permanent hormonal reprogramming, alterations to metabolic pathways, and changes to photosynthetic pathways that improve water-use efficiency. In perennials, the same types of responses can include phenological shifts such as early flowering or inside growing habits, which would allow them to escape heat stress. Collectively, structural and functional acclimations afford plants the opportunity to maintain hydraulic integrity and survive chronic heat, but at the trade-off of growth and development. Understanding these long-term responses will ultimately prepare us to predict plant resilience and corresponding ecosystem processes under climate warming.

The root-to-shoot ratio and morphological modifications are potentially the most important long-term adaptive responses of plants to high temperature and associated water stress (Calleja-Cabrera et al., 2020; Chaves et al., 2002; Kou et al., 2022). With continued heat stress, most plants would show an increase in the root-to-shoot ratio by allocating additional biomass to roots compared to above ground structures. This modification facilitates a greater ability to extract water from deeper soil layers and retain transpiration and leaf hydration under high evaporative demand. Simultaneously, shoot growth is often mitigated, resulting in a reduction of leaf area, thicker cuticles, and adjustments to leaf angle that help reduce losses to transpiration. Stomata size and density are also prone to alteration; whereby smaller or reduced stomata develop as a structural adaptation to improve water-use efficiency over time. Some species also can develop various modifications in xylem anatomy, such as less vessel size, to reduce the risk of cavitation under high tension, but these potential modifications reduce maximum hydraulic conductivity. Collectively, these morphological changes represent long-term adaptations developed through time to moderate water acquisition and conservation so that hydraulic stability and survival under chronic thermal stress are sustained along the journey, albeit at the expense of growth and yield potential.

- High temperatures typically cause changes in biomass allocation. For example, the allocation of resources toward the roots will allow for an increased root-to-shoot ratio that enhances water acquisition.
- Leaves may become smaller, thicker, or more reflective (e.g., greater pubescence or wax deposition) that limits transpirational demand.

Changes in xylem anatomy can be an important longer-term strategy that allows plants to maintain hydraulic integrity in high temperatures and concomitant water stress (Hacke & Sperry, 2001a; Maurel & Nacry, 2020). Long-term heat exposure often increases transpirational demands and ultimately increases the chance for cavitation and embolism to occur in xylem conduits. Then, most plants would exhibit anatomical changes, like decreases in vessel diameter, increases in vessel wall thickness, and increases in the frequency of smaller conduits that contribute to a greater resistance to embolism and hydraulic safety in the vessel structure (Hacke & Sperry, 2001; Sperry, 2003; Venturas et al., 2017d). These changes may increase resistance to water transport and reduce maximum hydraulic conductivity but represent a trade-off for survival over efficient water transport. Similarly, plants can also modify the structure or density of the pit membrane to mitigate the risk of air-seeding and embolism transfer. Likewise, acclimation can alter xylem density and arrangement, with relatively more fibers and supportive tissues protecting the vascular system from thermal and hydraulic stress in the xylem capacity. Likewise, changes in allocation of carbon can accompany acclimation in xylem anatomy, with resources devoted to constructing less efficient but safer hydraulic architecture. In summary, xylem adjustment is an adaptive and plastic response that allows plants to cope with chronic heat stress, as well as sustain hydraulic transport and resilience from climate induced hydraulic stress.

- Plants can acclimate to extended heating with change in the xylem anatomy, for example, smaller vessels and thicker walls that support cavitation.
- These changes are typically associated with lower hydraulic efficiency.

6 Impacts Specific to Species and Ecosystems

Impacts of increased temperatures on plant hydraulics and water relations can be species-specific, and delivery of impacts can be at the ecosystem scale, reflecting the complex interactions between physiological traits at the individual level and ecological processes at the scale of the community and ecosystem. Certain plants and species will be more or less sensitive and resistant to thermal and hydraulic stress, typically representing combinations of anatomical, physiological, and evolutionary adaptations (De Micco & Aronne, 2012; Volaire, 2018a). Drought resistant species may apply conservative water use strategies in addition to having shorter xylem vessel diameters, greater osmotic adjustment ability, but as with thermal sensitive plant species, those potentially sensitive to drought may experience acute decreases in hydraulic conductance and survival under sustained heat. If these differences are at the species level, assessment of heat and drought allow for approximation of such greater community scale changes that may be beneficial to heat and/or drought tolerant plant species but would limit sensitive species and yield shifts in diversity. Diversity shifts can in turn alter ecosystem process, such as rates of evapotranspiration, carbon cycling, and nutrient cycling, ultimately also effecting productivity and resiliency of the ecosystem. At the level of the forest, xylem embolism or death of sensitive species can lead to a thinning canopy, larger-scale declines in carbon storage or sequestration, and even provide feedback at the regional and local climate scale. In agricultural systems, species and cultivar-level variation in hydraulics determines productivity, water use efficiency, and food security with a warming climate. Therefore, increased temperatures may have cascading effects not only for individual species but can range across entire ecosystems that will rearrange plant-water relationship and functioning in response to increased temperatures.

Inter specific variation in hydraulic strategies refers to the different plant species' response to warmer temperatures and increased water stress and ultimately how this affects their survivability, distribution, and ecological roles (N. McDowell et al., 2008; Volaire, 2018b). There is a great range in the balance of hydraulic safety and hydraulic efficiency among species. Some species will adopt a conservative strategy, often indicated by narrower xylem vessels and wide radial pathways to move water, high resistance to cavitation, and stomatal behavior that restricts transpiration. On the other hand, most species will use risky but effective water transport strategy which can lead to growth potential when the plant is well watered, however, it increases the chance of embolism caused by heat or drought. For instance, desert shrubs and drought tolerant trees have high hydraulic safety margins through osmotic adjustment and conservative water transport, while fast growing temperate trees can be more efficient but have a higher likelihood of thermal and hydraulic failure through cavitation. Differences between species can also be observed through root architecture, leaf morphology, and hormonal control traits that lead to varying levels of hydraulic risks and performance in response to high vapor pressure deficit and historical warming. Furthermore, hydraulic variability does influence competitive interactions and community dynamics; i.e., plastic species can dominate less plastic and more sensitive species under prolonged periods of extreme heat stress, which can alter the vegetation structure and ecosystem processes associated with that vegetation. It is useful to recognize that hydraulic strategy variations provide essential biological context for anticipating species' adaptive abilities and for developing management practices that aim to sustain some level of ecosystem resilience under a warmer climate.

- Drought Resistant species usually have associated traits of deeper roots, smaller leaf area, and more robust xylem.
- Tropical species, which have evolved within a restricted suite of temperatures, may have a particularly low tolerance for hydraulic failure associated with warming.
- Xerophytes could exhibit some degree of tolerance related to acclimatization strategies that are structurally and biochemically based.

The consequences for Forest and Agricultural Systems from rising temperatures should be substantial because of the cascading changes expected to occur to plant water relations and hydraulic function that would have direct impacts on productivity, resilience, and sustainability in the longer-term. In forested systems, rising temperatures not only enhance evaporative demand, but also increase risks of cavitation and embolism resulting in high degrees of hydraulic failure, especially in sensitive tree species (Tyree & Sperry, 1989). These events can lead to mortality events across the landscape, changes in species composition and succession, and reduced capacity for long-term carbon sequestration, weakening forest ecosystem function as climate regulators. Distance-drawn changes in hydraulic function in forests could also mean reduced forest canopy density, with consequences accruing to biodiversity response, nutrient cycling, and the availability of water resources to neighboring ecosystems. Similarly, in agricultural systems, crop species too stand to experience hydraulic stress as rising temperatures translate into increased water demand and subsequently transpiration consumption resulting in reduced soil moisture storage, destabilizing stomatal control, and ultimately reducing yields and water use efficiency. Sensitivity to heat drew vegetable crops such as wheat, rice, and maize experience greater damage to water transport due to hydraulic stress, whereas others have cultivars bred for better osmotic adjustments and more conservative water-use conductance to help comparable productivity as temperatures increase. Also, the need for irrigation will increase significantly, increasing pressures on freshwater resources, and exacerbating competition among agriculture, industry, and domestic consumption. Collectively, these effects represent a clear and impending need for breeding and management strategies to develop cultivated and natural systems with hydraulic resilience to sustain forests and crops productivity amid accelerating climate warming.

- Increasing temperatures could change tree species composition in the forest as more heat-and drought-adapted species become more dominant.
- Crop production could be markedly reduced from damage to hydraulic transport, especially in areas that experience heavy thermal loads.
- Hydraulic failure is listed as a leading cause of tree death as global temperatures continue to climb.

7 Mitigation and Adaptation Strategies

The continuing rise in temperature will have adverse implications for plants' water relations and hydraulic functioning. Therefore, it is very important to initiate actions for mitigation and adaptation with respect to ecosystems resiliency and sustainable agriculture (Srivastav et al., 2021). Genetic breeding and biotechnology initiatives are adaptation actions intended to increase heat and drought tolerance and include attempts to develop cultivars that show greater osmotic adjustment, improved stomatal control, and xylem that resist cavitation. Agronomic management strategies can reduce the effects by improving irrigation scheduling and with the incorporation of mulches and organic matter when cropping systems to reduce evaporative loss and to improve moisture availability in hot temperatures. On the forestry side—assisted migration of heat tolerant species, reforestation of degraded land, and increased genetic diversifications among species, will be important adaptation actions to improve hydraulic stability and long-term adaptation. On a larger scale, combining strategies for climate-resilient agricultural systems with ways to improve water management strategies like rainwater harvesting, drip-irrigation, and precision agricultural technologies will assist balanced water use under increased water use demand. Finally, global action on reducing greenhouse gas emissions from all sources is vital, of course, because minimizing the risk of potential catastrophic warming scenarios is imperative to not further challenge plant limits on adaptive responses. As a result, these strategies taken together reflect a recognized need to consider the intersecting physiological, ecological, and technological implications of managing plant hydraulic function to be able to sustain productivity under increasingly variable climate.

Both Breeding and Biotechnology represent strong paths toward developing plant types with greater tolerance to the hydraulic constraints of increased heat (Munawar et al., 2020). Conventional genetic selection programs have sought traits such as deeper root systems, reduced stomatal density, and increased osmotic adjustment for the purposes of optimizing water use efficiency and hydraulic safety during heat and drought-stress. With modern advances in molecular breeding and genomics, scientists can now more precisely identify quantitative trait loci and genes related to xylem anatomy, stomatal control and plant hormone signaling, which should make it possible to breed climate and drought tolerant cultivars. Moreover, biotechnology expands options through genetic manipulation and genome editing processes such as CRISPR-Cas, enabling efficient and precise changes to processes such as ABA biosynthesis, aquaporin function, and embolism resistance. Biotechnology could also be used to enhance hydro-protectant synthesis and signaling pathways that would protect hydraulic function during elevated heat stress. To add to studies focused on single relevant species to climate change, there appears to be an increasing consideration of marker-assisted selection and transgenic approaches in forestry species to increase the likelihood of long-term hydraulic stability in natural forests. By integrating conventional breeding methods with modern biotechnology, the time for developing plant types that will tolerate higher vapor pressure deficits, lessen cavitation risk and safeguard productivity will be shortened under climate scenarios to sustain food security and ecological health.

- Selection and breeding of thermo-tolerant varieties to accelerate water use efficiency and xylem resistance.
- Genetic modification to increase ABA sensitivity, antioxidant synthesis, and accumulation of osmo-protectants, to calculate hydraulic function.

Agronomic and Silvicultural Practices represent an essential means to reduce the impact of global warming on plant-water relations and hydraulic systems by improving soil structure, water supply, and reducing hydraulic stress in agricultural and forest ecosystems. Farming practices such as conservation tillage, mulching, cover cropping, and organic matter addition in soils will maintain soil moisture, keep evaporation low, and increase water holding capacity; this in effect, provides a buffer to the crops, particularly during periods of high evaporative demand (Abdallah et al., 2021; El-Beltagi et al., 2022). Precision irrigation systems—including drip and sprinkler systems—that utilize smart scheduling based on soil moisture and climate information will maximize water-use efficiency and reduce hydraulic stress. In relation to forestry, silvicultural practices—such as thinning, mixed-species stands, and canopy manipulation—will also minimize water competition, improve microclimate conditions, and enhance the hydraulic stability of tree populations with respect to cavitation and embolism. Soil conservation practices and assisted regeneration using drought-and heat-tolerant species further increases the hydraulic stability of forest ecosystems. Agroforestry, which involves tree/crop combinations, provides another effective adaptation strategy by improving control of microclimate, decreasing evapotranspiration, and increasing root zone water uptake. The combination of these agronomic and silvicultural practices represent ecologically sustainable approaches of adaptation to temperature influenced hydraulic issues, while maintaining the long-term productiveness of agricultural landscapes and climate stability of forested ecosystems.

- Adjusting planting and irrigation timing, and fertilization of soil to reduce temperatures impact.
- Agroforestry and mixed-species planting to increase resilience, and reduction of water stress.

Quantitative summary of the effect of rising temperature on significant plant-water relations and hydraulic metrics, relative to representative physiological responses of a typical C_3 species in controlled settings are given in Table 1. Values are approximate C_3 plant responses (e.g., soybean, wheat) to increased temperature. Arrows provide direction and magnitude of the change in response to soil temperature (25 °C).

Table 1 Quantitative Effects of Rising Temperature on Plant Water Relations and Hydraulic Parameters for C_3 plant (Anfodillo et al., 2013; Chaves et al., 2003; Cochard et al., 2021)

Parameter	Units	Optimal temperature (25 °C)	Moderate heat (35 °C)	Severe heat (45 °C)	% Change (45 °C versus 25 °C) (%)	Remarks
Leaf water potential (Ψleaf)	MPa	−0.6	−1.1	−1.8	↓ 70	Steep decline due to excessive transpiration
Relative water content (RWC)	%	88	76	62	↓ 30	Progressive tissue dehydration
Stomatal conductance (gs)	mol H_2O m^{-2} s^{-1}	0.35	0.21	0.10	↓ 71	Stomatal closure limits CO_2 uptake
Transpiration rate	mmol H_2O m^{-2} s^{-1}	4.2	6.3	5.1	↑ 21	Initial rise, then decline under stress
Photosynthetic rate (an)	μmol CO_2 m^{-2} s^{-1}	22.5	14.2	6.8	↓ 70	Reduced due to stomatal and metabolic inhibition
Xylem hydraulic conductivity	% of control	100	78	46	↓ 54	Cavitation and embolism formation
Root hydraulic conductivity	% of control	100	85	60	↓ 40	Membrane damage and aquaporin inhibition
Osmotic potential (Ψs)	MPa	−1.2	−1.5	−1.8	↓ 50	Accumulation of osmolytes helps osmotic adjustment
Vapor pressure deficit (VPD)	kPa	1.5	2.8	4.6	↑ 207	Drives enhanced transpiration rate
Cavitation incidence	% of xylem area affected	5	18	46	↑ 820	Strongly temperature-dependent xylem failure
Hydraulic safety margin	MPa	0.8	0.4	0.2	↓ 75	Narrower safety range increases vulnerability

8 Future Research Directions

Future research should focus on a mechanistic understanding of how increased temperature effects plant hydraulic and water relations by incorporating physiological, molecular, and ecosystem-based methods to predict plant resilience in a changing climate. A critical area of focus should be to explain the genetic and molecular basis of hydraulic traits (e.g., xylem vulnerability to cavitation, stomatal sensitivity, osmotic adjustment), as these can be investigated for breeding and biotechnology applications for resilient climate species. Use of techniques such as micro-CT scanning, isotopic tracing, and process-based hydraulic models will be valuable not only to explore the mechanistic understanding of plant hydraulic responses to temperature stress, but to also be able to capture the "real-time" dynamics of water transport under different thermal and moisture conditions. Long-term field studies in combination with remote sensing approaches at an ecosystem scale, will be needed to understand whether inter-specific differences in hydraulic strategies will yield differences in the community composition, carbon, water fluxes, and ecosystem service function. Future research also needs to encompass cumulative effects of increased temperature with other stresses (drought, high CO_2, nutrient deficiency, etc.) as those additive stress are more likely to align with climate change. Particularly in agricultural and forestry research, candidate research information coupled with socio-economic factors will inform targeted adaptation measures. Finally, inter-disciplinary research at the intersection of plant physiology, plant genetics, plant ecology, and climate sciences will be necessary to define hydraulic thresholds to avert catastrophic mortality events in natural and agricultural food production systems.

- High-resolution models that couple hydraulics, carbon balance and climatic drivers will be needed.
- We will need to know some of the molecular control of thermal hydraulic responses.
- We will need opportunities to increase knowledge on synergisms (temperature + CO_2 + drought).

9 Conclusion

Increasing temperatures have serious and potentially widespread consequences for plant water relations and hydraulic functioning, modifying how plants take up, transport, and store water in increasingly stressful climatic conditions. Warmer climates promote atmospheric vapor pressure deficit and increased evapotranspiration, both of which result in reduced leaf water potential, impact stomatal control, and increase the likelihood of cavitation and embolism which decreases xylem hydraulic conductivity. Under these conditions, plants respond in the short-term with physiological responses such as stomatal closure, osmotic adjustment and hormonal regulation, while over longer time frames they may adopt strategies which alter their root-to-shoot ratios, leaf form, or xylem anatomy which can compromise hydraulic safety and efficiency. There are species specific responses to these conditions, resulting in a range of hydraulic strategies within individual species, potentially leading to variation in intra-specific hydraulic strategies with ecological implications ranging from forest die back and shifting community composition, through to reduced agricultural yields and increased irrigation demands. The downstream impacts emphasizes the need for adaptation and mitigation strategies including breeding and biotechnology tools to produce hydraulic debtor species, adoption of agronomic and silviculture best practices to enable maximal water use, and ecosystem level management strategies to provide resilience and stability in these challenging climates. Future research should connect approaches such as molecular, physiological and ecological and be able to predict threshold hydraulic dysfunction more accurately for guidelines on unmanaged ecosystems and managed agricultural systems in warmer climates. Collectively, the key findings suggest that while plants have a great deal of plasticity and adaptive capacity, thermal exposure remains a primary threat to plant hydraulic integrity. Future work addressing these threats will be essential for conserving biodiversity, ecosystem service function, and global food security.

Increases in temperature can strongly affect how plants maintain their water uptake and the function of their hydraulic system. These water relations can then directly affect individual plant performance, which in turn can indirectly affect performance across whole ecosystems. Responses of plants to increases in temperature are highly species dependent, and complex based on anatomical, physiological, and environmental interactions. Anticipating and mitigating these responses is important to maintain ecosystem resilience and productivity in a warming world.

References

Abdallah, A. M., Jat, H. S., Choudhary, M., Abdelaty, E. F., Sharma, P. C., & Jat, M. L. (2021). Conservation agriculture effects on soil water holding capacity and water-saving varied with management practices and agroecological conditions: A review. *Agronomy, 11*(9), 1681.

Adeyemi, O., Grove, I., Peets, S., & Norton, T. (2017). Advanced monitoring and management systems for improving sustainability in precision irrigation. *Sustainability, 9*(3), 353.

Anfodillo, T., Petit, G., & Crivellaro, A. (2013). Axial conduit widening in woody species: A still neglected anatomical pattern. *IAWA Journal, 34*(4), 352–364.

Asbjornsen, H., Goldsmith, G. R., Alvarado-Barrientos, M. S., Rebel, K., Van Osch, F. P., Rietkerk, M., Chen, J., Gotsch, S., Tobon, C., & Geissert, D. R. (2011). Ecohydrological advances and applications in plant–water relations research: A review. *Journal of Plant Ecology, 4*(1–2), 3–22.

Bharath, P., Gahir, S., & Raghavendra, A. S. (2021). Abscisic acid-induced stomatal closure: An important component of plant defense against abiotic and biotic stress. *Frontiers in Plant Science, 12*, 615114.

Blum, A. (2017). Osmotic adjustment is a prime drought stress adaptive engine in support of plant production. *Plant, Cell & Environment, 40*(1), 4–10.

Brodersen, C. R., & McElrone, A. J. (2013). Maintenance of xylem network transport capacity: A review of embolism repair in vascular plants. *Frontiers in Plant Science, 4*, 108.

Calleja-Cabrera, J., Boter, M., Oñate-Sánchez, L., & Pernas, M. (2020). Root growth adaptation to climate change in crops. *Frontiers in Plant Science, 11*, 544.

Chapin, F. S., III, Bret-Harte, M. S., Hobbie, S. E., & Zhong, H. (1996). Plant functional types as predictors of transient responses of arctic vegetation to global change. *Journal of Vegetation Science, 7*(3), 347–358.

Chaves, M. M., Costa, J. M., Zarrouk, O., Pinheiro, C., Lopes, C. M., & Pereira, J. S. (2016). Controlling stomatal aperture in semi-arid regions—The dilemma of saving water or being cool? *Plant Science, 251*, 54–64.

Chaves, M. M., Maroco, J. P., & Pereira, J. S. (2003). Understanding plant responses to drought—From genes to the whole plant. *Functional Plant Biology, 30*(3), 239–264.

Chaves, M. M., Pereira, J. S., Maroco, J., Rodrigues, M. L., Ricardo, C. P. P., Osório, M. L., Carvalho, I., Faria, T., & Pinheiro, C. (2002). How plants cope with water stress in the field? Photosynthesis and growth. *Annals of Botany, 89*(7), 907.

Chen, H., & Jiang, J.-G. (2010). Osmotic adjustment and plant adaptation to environmental changes related to drought and salinity. *Environmental Reviews, 18*, 309–319.

Cochard, H., Badel, E., Herbette, S., Delzon, S., Choat, B., & Jansen, S. (2013). Methods for measuring plant vulnerability to cavitation: A critical review. *Journal of Experimental Botany, 64*(15), 4779–4791.

Cochard, H., Pimont, F., Ruffault, J., & Martin-StPaul, N. (2021). SurEau: A mechanistic model of plant water relations under extreme drought. *Annals of Forest Science, 78*(2), 55.

Costa, J. M., Marques da Silva, J., Pinheiro, C., Barón, M., Mylona, P., Centritto, M., Haworth, M., Loreto, F., Uzilday, B., & Turkan, I. (2019). Opportunities and limitations of crop phenotyping in southern European countries. *Frontiers in Plant Science, 10*, 1125.

Dai, A., Zhao, T., & Chen, J. (2018). Climate change and drought: A precipitation and evaporation perspective. *Current Climate Change Reports, 4*(3), 301–312.

De Micco, V., & Aronne, G. (2012). Morpho-anatomical traits for plant adaptation to drought. In *Plant responses to drought stress: From morphological to molecular features* (pp. 37–61). Springer.

Dos Santos, T. B., Ribas, A. F., de Souza, S. G. H., Budzinski, I. G. F., & Domingues, D. S. (2022). Physiological responses to drought, salinity, and heat stress in plants: A review. *Stress, 2*(1), 113–135.

El-Beltagi, H. S., Basit, A., Mohamed, H. I., Ali, I., Ullah, S., Kamel, E. A. R., Shalaby, T. A., Ramadan, K. M. A., Alkhateeb, A. A., & Ghazzawy, H. S. (2022). Mulching as a sustainable water and soil saving practice in agriculture: A review. *Agronomy, 12*(8), 1881.

Garg, T., Kumar, A., Joshi, A., Awasthi, A., Rana, A., Kumar, V., & Kaur, S. (2024). The role of phytohormones in reducing the arsenic-induced stress in plants. *South African Journal of Botany, 168*, 296–315. https://doi.org/10.1016/j.sajb.2024.03.020

Gleason, S. M., Butler, D. W., Ziemińska, K., Waryszak, P., & Westoby, M. (2012). Stem xylem conductivity is key to plant water balance across Australian angiosperm species. *Functional Ecology, 26*(2), 343–352.

Grossiord, C., Buckley, T. N., Cernusak, L. A., Novick, K. A., Poulter, B., Siegwolf, R. T. W., Sperry, J. S., & McDowell, N. G. (2020). Plant responses to rising vapor pressure deficit. *New Phytologist, 226*(6), 1550–1566.

Hacke, U. G., & Sperry, J. S. (2001). Functional and ecological xylem anatomy. *Perspectives in Plant Ecology, Evolution and Systematics, 4*(2), 97–115.

Hatfield, J. L., & Dold, C. (2019). Water-use efficiency: Advances and challenges in a changing climate. *Frontiers in Plant Science, 10*, 103.

Keller, M. (2005). Deficit irrigation and vine mineral nutrition. *American Journal of Enology and Viticulture, 56*(3), 267–283.

Kim, H. K., Park, J., & Hwang, I. (2014). Investigating water transport through the xylem network in vascular plants. *Journal of Experimental Botany, 65*(7), 1895–1904.

Kou, X., Han, W., & Kang, J. (2022). Responses of root system architecture to water stress at multiple levels: A meta-analysis of trials under controlled conditions. *Frontiers in Plant Science, 13*, 1085409.

Lambers, H., & Oliveira, R. S. (2019). Plant water relations. In *Plant physiological ecology* (pp. 187–263). Springer.

Lens, F., Tixier, A., Cochard, H., Sperry, J. S., Jansen, S., & Herbette, S. (2013). Embolism resistance as a key mechanism to understand adaptive plant strategies. *Current Opinion in Plant Biology, 16*(3), 287–292.

Li, N., Euring, D., Cha, J. Y., Lin, Z., Lu, M., Huang, L.-J., & Kim, W. Y. (2021). Plant hormone-mediated regulation of heat tolerance in response to global climate change. *Frontiers in Plant Science, 11*, 627969.

Lipiec, J., Doussan, C., Nosalewicz, A., & Kondracka, K. (2013). Effect of drought and heat stresses on plant growth and yield: A review. *International Agrophysics, 27*(4).

López, J., Way, D. A., & Sadok, W. (2021). Systemic effects of rising atmospheric vapor pressure deficit on plant physiology and productivity. *Global Change Biology, 27*(9), 1704–1720.

Massmann, A., Gentine, P., & Lin, C. (2019). When does vapor pressure deficit drive or reduce evapotranspiration? *Journal of Advances in Modeling Earth Systems, 11*(10), 3305–3320.

Maurel, C., & Nacry, P. (2020). Root architecture and hydraulics converge for acclimation to changing water availability. *Nature Plants, 6*(7), 744–749.

Mbava, N., Mutema, M., Zengeni, R., Shimelis, H., & Chaplot, V. (2020). Factors affecting crop water use efficiency: A worldwide meta-analysis. *Agricultural Water Management, 228*, 105878.

McDowell, N. G. (2011). Mechanisms linking drought, hydraulics, carbon metabolism, and vegetation mortality. *Plant Physiology, 155*(3), 1051–1059.

McDowell, N., Pockman, W. T., Allen, C. D., Breshears, D. D., Cobb, N., Kolb, T., Plaut, J., Sperry, J., West, A., & Williams, D. G. (2008). Mechanisms of plant survival and mortality during drought: Why do some plants survive while others succumb to drought? *New Phytologist, 178*(4), 719–739.

Muhammad Aslam, M., Waseem, M., Jakada, B. H., Okal, E. J., Lei, Z., Saqib, H. S. A., Yuan, W., Xu, W., & Zhang, Q. (2022). Mechanisms of abscisic acid-mediated drought stress responses in plants. *International Journal of Molecular Sciences, 23*(3), 1084.

Munawar, S., M. T., u. Q., Mustafa, G., Khan, M. S., & Joyia, F. A. (2020). Role of biotechnology in climate resilient agriculture. In *Environment, climate, plant and vegetation growth* (pp. 339–365). Springer.

Robinson, D. A., Hopmans, J. W., Filipovic, V., van der Ploeg, M., Lebron, I., Jones, S. B., Reinsch, S., Jarvis, N., & Tuller, M. (2019). Global environmental changes impact soil hydraulic functions through biophysical feedbacks. *Global Change Biology, 25*(6), 1895–1904.

Roy, A., Mandal, M., Das, S., Popek, R., Rakwal, R., Agrawal, G. K., & Sarkar, A. (2024). The cellular consequences of particulate matter pollutants in plants: Safeguarding the harmonious integration of structure and function. *Science of the Total Environment, 914*, 169763. https://doi.org/10.1016/j.scitotenv.2023.169763

Sadok, W., Lopez, J. R., & Smith, K. P. (2021). Transpiration increases under high-temperature stress: Potential mechanisms, trade-offs and prospects for crop resilience in a warming world. *Plant, Cell & Environment, 44*(7), 2102–2116.

Savi, T., Bertuzzi, S., Branca, S., Tretiach, M., & Nardini, A. (2015). Drought-induced xylem cavitation and hydraulic deterioration: Risk factors for urban trees under climate change? *New Phytologist, 205*(3), 1106–1116.

Smith, S. J., Edmonds, J., Hartin, C. A., Mundra, A., & Calvin, K. (2015). Near-term acceleration in the rate of temperature change. *Nature Climate Change, 5*(4), 333–336.

Soudzilovskaia, N. A., Elumeeva, T. G., Onipchenko, V. G., Shidakov, I. I., Salpagarova, F. S., Khubiev, A. B., Tekeev, D. K., & Cornelissen, J. H. C. (2013). Functional traits predict relationship between plant abundance dynamic and long-term climate warming. *Proceedings of the National Academy of Sciences, 110*(45), 18180–18184.

Sperry, J. S. (2003). Evolution of water transport and xylem structure. *International Journal of Plant Sciences, 164*(S3), S115–S127.

Srivastav, A. L., Dhyani, R., Ranjan, M., Madhav, S., & Sillanpää, M. (2021). Climate-resilient strategies for sustainable management of water resources and agriculture. *Environmental Science and Pollution Research, 28*(31), 41576–41595.

Tomasella, M., Petrussa, E., Petruzzellis, F., Nardini, A., & Casolo, V. (2019). The possible role of non-structural carbohydrates in the regulation of tree hydraulics. *International Journal of Molecular Sciences, 21*(1), 144.

Tyree, M. T., & Sperry, J. S. (1989). Vulnerability of xylem to cavitation and embolism. *Annual Review of Plant Physiology and Plant Molecular Biology, 40*(1), 19–36.

Venturas, M. D., Sperry, J. S., & Hacke, U. G. (2017). Plant xylem hydraulics: What we understand, current research, and future challenges. *Journal of Integrative Plant Biology, 59*(6), 356–389.

Volaire, F. (2018). A unified framework of plant adaptive strategies to drought: Crossing scales and disciplines. *Global Change Biology, 24*(7), 2929–2938.

Will, R. E., Wilson, S. M., Zou, C. B., & Hennessey, T. C. (2013). Increased vapor pressure deficit due to higher temperature leads to greater transpiration and faster mortality during drought for tree seedlings common to the forest–grassland ecotone. *New Phytologist, 200*(2), 366–374.

Zandalinas, S. I., Rivero, R. M., Martínez, V., Gómez-Cadenas, A., & Arbona, V. (2016). Tolerance of citrus plants to the combination of high temperatures and drought is associated to the increase in transpiration modulated by a reduction in abscisic acid levels. *BMC Plant Biology, 16*(1), 105.

Zhao, B., Liu, Q., Wang, B., & Yuan, F. (2021). Roles of phytohormones and their signaling pathways in leaf development and stress responses. *Journal of Agricultural and Food Chemistry, 69*(12), 3566–3584.

Crop Responses to Elevated Temperatures: Cellular Mechanisms and Opportunities for Genetic Improvement

Indu Rialch, Sunayana, Gurpreet Kaur, and Akshita Awasthi

Abstract Heat stress impairs vital physiological and reproductive functions leading to diminished photosynthetic efficiency, reduced pollen viability, limited seed development, and accelerated ageing particularly during germination, flowering, and seed formation. Plant responses include morpho-physiological adjustments, antioxidant defence activation, osmotic regulation, and transcriptional reprogramming. However, this widespread substitution of traditional landraces with genetically homogeneous, high-yielding varieties threatens crop genetic diversity. Conventional breeding programmes through QTL mapping, marker-assisted selection, and genomic selection have been employed for enhancing various traits of interest, including seed yield, stay green capacity, osmolyte accumulation, and antioxidant activity. Recent advances in omics technologies, transgenic approaches, and genome editing provide accurate tools for manipulating heat-responsive genes, while complementary approaches such as high-throughput phenomics, systems biology, and exogenous protectants enhance adaptive capacity. Integrating genomics, transcriptomics, proteomics, and precision breeding offers a pathway to develop thermotolerant cultivars.

Keywords Heat stress · Crop diversity · Plant defence · Omics · Stress management

I. Rialch (✉) · G. Kaur · A. Awasthi
Department of Plant Breeding and Genetics, Punjab Agricultural University, Ludhiana, Punjab, India
e-mail: indudehal@pau.edu

Sunayana
Regional Research Station, Punjab Agricultural University, Abohar, Punjab, India

A. Awasthi et al. (eds.), *Rising Temperatures and Plant Ecosystems*,
https://doi.org/10.1007/978-3-032-29477-7_4

1 Introduction

Global warming has accelerated faster than ever in recent years. From June 2023 to May 2024, the average global surface temperature reached a record 1.63 °C above the 1850–1900 pre-industrial baseline. According to climate predictions, globally average temperatures above 1.5° over pre-industrial levels are expected to be exceeded in 2024–2028, illustrating just how quickly the world is approaching critical thresholds of change. Of all the sectors, agriculture is severely threatened by climate change because crop growth, development, and productivity are under strict control by temperature, precipitation, and other weather variables. Empirical evidence, as well as model-based predictions, indicates that mean annual temperatures in the major food-producing regions are rising, with large consequences for crop yields and stability of food systems around the world (Zhao et al., 2017). This means it is a severe threat to food security globally, as food demand is expected to rise by more than 50% by 2050 (Falcon et al., 2022). In addition to that, water scarcity is a basic and significant compounding stressor. Globally, some 70% of total freshwater withdrawals have been used for the production of irrigated crops, accounting for up to 40% of the world's total food calorie supply. With global freshwater demand from domestic and industrial sectors expected to grow significantly by 2050, competition for limited water supplies will increase, which reduces agriculture's potential ability to meet the increased demand under such conditions (Luck et al., 2015).

Heat stress is a critical abiotic constraint in agriculture; exposure to above-optimal temperatures affects crop performance. The extent of crop loss is determined by the intensity and interval of high temperature exposure, crop species, and developmental stages. High impacts are revealed via quantitative evaluations: Additionally, a global meta-analysis found that heat stress was the most detrimental abiotic factor lowering agricultural productivity, with an average yield drop of 33% (Cohen et al., 2021). This abiotic stress inhibits crop growth globally, resulting in lower productivity. This necessitates management strategies that are climate-resilient. Farmers' adaptive reactions include using early-maturity or heat-tolerant cultivars, improving watering schedules and modifying planting windows. However, underlying thermotolerance processes that rely on molecular and genetic factors are not addressed by these reactions. These lead to deterioration of crop genetic variety, which is a necessary precondition for long-term climate variability resilience. Thus, germplasm conservation and exploitation continue to be essential to the development of climate-resilient and sustainable agricultural systems.

Other climate variables including heat waves, cyclones, droughts, and floods have become more frequent and intense during the past few decades. In addition to reducing agricultural output, these shocks also cause productive assets to deteriorate, rural livelihoods to become unstable, poverty to worsen, and overuse of natural reservoirs to accelerate (World Bank, 2013). Due to their heavy reliance on agriculture, smallholder farming systems, and lack of access to institutional, financial, and technological shock-smoothing mechanisms, developing nations including India are especially vulnerable. One of the main causes of low agricultural output, slow economic growth, and ongoing poverty is thought to be frequent exposure to unfavourable climate shocks (Carter et al., 2014). An integrated and multifaceted strategy that combines indigenous knowledge systems, adaptive on-farm methods, and contemporary developments in agricultural sciences, meteorology, and ICTs is required to address such difficulties. Since roughly 45% of India's cropland is still rainfed, production methods are extremely susceptible to variations in rainfall and temperature extremes (Birthal et al., 2014; Easterling et al., 2007; Tandon et al., 2025). Therefore, coordinated farm, community and institutional actions for risk reduction, resource efficiency, and bolstering adaptive ability in the face of rapid climate change are necessary to increase resilience.

Rising Temperatures and Heat Stress

The frequency of heat stress in ecosystems can be correlated with temperature increases in a linear fashion. One of the fundamental climatic factors that governs biological processes is temperature. The current increase in global temperatures causes heat stress in humans, animals, crops, and natural ecosystems further impairing ecosystem stability and efficiency (Ponsonby & Corleto, 2024; Yang et al., 2023). Concurrently, the frequency of extreme weather occurrences has already significantly increased, resulting in extensive devastation and overall economic instability (Awasthi et al., 2025; Newman & Noy, 2023). Heat stress exposure is expected to rise as a result of ongoing climate change-related warming (Pattnayak et al., 2023). Crop growth, development, and plant metabolism are all significantly impacted by atmospheric temperature. At the agroecosystem level, heat conditions have a significant impact on the activities of plants and soil biota. Empirical evidence shows that seasonal temperature variability drastically alters plant growth, vegetation dynamics, and seawater thermal regimes. In addition, rising temperatures increase food insecurity, restrict the availability of water supplies, and worsen air quality all of which increase hazards to human health, livelihoods, and death. Over the past century, near-surface temperature has significantly increased globally, and, in recent decades, this trend has accelerated. According to the Intergovernmental Panel on Climate Change (IPCC), human activity has clearly warmed the climate, with the average global surface temperature from 2011 to 2020 being roughly 1.1 °C higher than the pre-industrial baseline from 1850 to 1900. Global surface temperatures are predicted to rise further and beyond 1.5 °C once more in the upcoming decades under numerous emissions scenarios. With a global mean near-surface temperature of about 1.45 ± 0.12 °C above the 1850–1900 baseline, year 2023 was the warmest year on record, according to the World Meteorological Organization (WMO) (2024). NASA estimates the 2024 annual anomaly at about + 1.47 °C relative to the mid-nineteenth-century average.

1.1 *Heat Stress and Food Security*

The intensification of heat waves driven by climate change poses significant short- and long-term risks to global food security (Behera et al., 2023). Moreover, multiple environmental stressors, including pollution, drought, and elevated temperatures, threaten food security at a large scale (Bhardwaj et al., 2022; Ramya et al., 2023). Among different stresses, heat stress is the most devastating stress which reduces the agricultural productivity at large. Very high temperatures affect the crops, wild plant species, aquaculture systems, and livestock in a negative manner, thereby compromising food production across sectors. The heat-induced reductions in productivity and quality are a crucial indicator that climatic variability affects global food security as demonstrated by the relationship between climate change and agricultural systems.

2 Mechanism Through Which Heat Stress Impacts Plants

The plant morphology, physiology, metabolism, molecular pathways, and rhizosphere interactions are affected by heat stress. One of the earliest indications that a plant is under heat stress could be abnormalities in its morphology. Numerous tolerance pathways, including the control of ion channels, heat shock proteins, osmolytes, and signalling pathways involving kinases and transcription factors (TF), are activated at the cellular and molecular levels. The ability of plants to detect temperature cues, transduce these signals, and modify their gene and metabolic expressions in order to maintain homeostasis makes them resilient to thermal stress.

Protein synthesis, metabolite buildup, and heat shock gene induction are all essential for fortifying cell structures and offering thermo-protection. Recent advancements made in molecular biology, genomics, and biotechnologies have significantly helped unravel the pathways of regulation of temperature tolerance and enabled the development of crop varieties resistant to thermal stress. Plant responses based on morpho-physical, biochemical, and molecular changes due to heat tend towards indicating an interesting aspect of the complexity and adaptability of plant survival responses under high temperature conditions. High temperature effects at phenological, physiological, biochemical, and molecular levels are represented graphically in Fig. 1. Crop plants have embraced stress signalling pathways and transcriptional controllers for countering heat stress damage. Heat stress is mediated by changes in cell membrane fluidity and ion homeostasis, and these lead to signalling pathways that turn on heat shock transcription factors, stress proteins, and thermotolerance genes. Moreover, small RNAs, especially microRNAs, have recently been recognized as crucial regulators of transcriptional and post-transcriptional responses to heat stress.

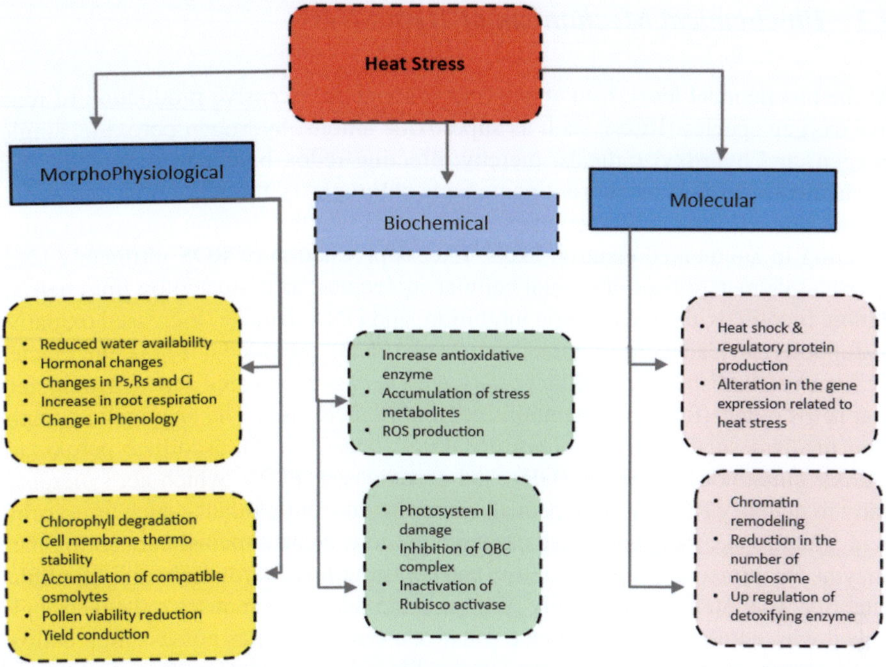

Fig. 1 Different heat stress-induced effects at morpho-physiological, biochemical, and molecular levels

2.1 *Physiological Mechanism of Heat Stress*

Heat stress restricts germination, vegetative growth, and reproductive development by disrupting cell division, turgor maintenance, and assimilate allocation. In its early phase, heat stress strongly inhibits germination, seedling establishment, and growth of radicles and plumules. During the vegetative phase, reduced cell division, inhibited expansion, and turgor loss limit biomass accumulation. Additionally, morphological changes of leaf scorch, sunburn, early senescence, and loss of green leaf area further decrease assimilatory surface and growth space. Among physiological transformations of plant growth, photosynthesis is most sensitive when subjected to increased temperature concentrations. Heat stress can create chloroplast ultrastructural alterations, reduce chlorophyll content, and impair thylakoid structures, inhibiting photochemical activity of photo-system two (PSII). Simultaneously, stomata close, reducing carbon dioxide uptake, thereby strongly restricting growth and development due to increased CO_2 content restrictions and increased inhibition of whole-plant photosynthesis activity as well. Reproductive stages are highly sensitive because heat reduces pollen viability, impairs anther dehiscence, and shortens grain filling.

2.2 Biochemical Mechanism of Heat Stress

At the biochemical level, heat shock brings about an excessive production of reactive oxygen species (ROS), such as superoxide anions, hydrogen peroxide, singlet oxygen, and hydroxyl radicals, thereby affecting redox balance inside cells. The production of ROS is most prominent inside chloroplasts, mitochondria, and peroxisomes due to elevated temperatures that can easily alter electron transport chains, resulting in augmented electron leaks. Excess production of ROS ultimately leads to oxidative metabolism of crucial cellular molecules, as indicated by lipid peroxidation, protein oxidation, enzyme inhibition, and DNA damage. Increased oxidative disruptions can lead to programmed cell death (PCD), provided it is not counteracted by some mechanism or other. Plant cells have a very sophisticated antioxidant system that helps reduce ROS and maintains cellular homeostasis. The enzymatic component involves superoxide dismutase (SOD), catalase (CAT), ascorbate peroxidase (APX), glutathione reductase (GR), and peroxidases (POX), which act synergistically to detoxify ROS. Simultaneously, non-enzymic antioxidants, such as ascorbic acid, glutathione, tocopherol, and carotenoids, complement mainstream antioxidant enzymes by protecting cells from ROS and contributing towards maintaining cellular integrity. The intensity and ability of these antioxidants' responses remain strongly linked with natural thermotolerance. Heat-resistant genotypes possess high antioxidant enzyme activities and higher compatible solute content, such as proline and glycine betaine, playing a crucial role within them, thereby determining thermotolerance and protecting them against HS. All these biochemical responses remain major contributors towards thermotolerance of plants against HS, as they help them combat oxidative stress, osmotic stress, and cell damage due to HS.

2.3 Molecular Mechanism of Heat Stress

Heat stress response of plants is mediated through molecular recognition, based on changes that occur on the plasma membrane of plants due to fluidity and ionic equilibrium. These alterations activate complex intracellular signalling cascades, prominently mediated by calcium-dependent protein kinases (CDPKs), mitogen-activated protein kinases (MAPKs), nitric oxide, and phytohormonal regulators including abscisic acid, salicylic acid, and jasmonic acid. A pivotal component of thermotolerance is production of heat shock proteins (HSPs). Under elevated temperatures, after release from cytosolic repressor complexes, HSFs are activated and translocate to the nucleus, binding to heat shock elements in promoters of heat shock genes. It promotes high-level expression of molecular chaperones such as HSP70, HSP90, HSP100, HSP60, and small HSPs. Collectively, these HSPs preserve proteostasis by stabilizing denatured proteins, preventing irreversible aggregation, refolding misfolded polypeptides, and maintaining membrane integrity. Moreover, *HSPs*, other stress-responsive genes, are triggered that encode detoxifying enzymes, osmolyte biosynthesis pathways, ion transporters, and regulatory proteins. Exogenous application of phytohormones, osmoprotectants (e.g., proline, glycine betaine), polyamines, nitric oxide, and selenium has been shown to modify gene expression and augment antioxidant defences. These interventions further enable crosstalk among stress pathways and promote the establishment of stress memory, conferring adaptive advantages during recurrent thermal episodes.

3 Role of Heat Stress in Affecting Crop Growth and Development

High temperature is a devastating abiotic stress that negatively affects agricultural productivity, particularly in semi-arid and dry areas. The balance of physiological, metabolic, and molecular processes necessary for regular plant functioning is disturbed by high temperatures. Thus, crop plants are being negatively impacted in a number of ways by the rising temperature. High temperatures affect cell division and elongation activities during the vegetative phase, which prevents seed germination, lowers seedling vigour, and hinders root and shoot elongation. Heat stress-induced loss of membrane integrity and enzyme stability further limits photosynthetic efficiency and nutrient uptake. Excessive heat not only speeds up leaf senescence and chlorophyll degradation, but it also decreases carbon assimilation and light interception.

Heat stress affects critical activities including pollen generation, fertilization, and grain filling; hence, it is particularly detrimental during the reproductive phase. By reducing pollen viability, hindering stigma receptivity, and impairing anther dehiscence, high temperatures result in poor seed set and sterility. Furthermore, heat-induced alterations in hormonal balance, especially in auxins and gibberellins, impair the development of reproductive organs. Elevated temperatures shorten the duration of the grain filling phase, leading to insufficient starch production and reduced grain weight. Figure 2 shows how heat stress affects crop growth. Understanding the complicated impacts of thermal stress on various crop species is crucial for maintaining agricultural systems in light of rising average and maximum temperatures.

Various biochemical and physiological processes control the different stages of plant development. These activities are very sensitive to temperature. Level of damage or resistance is strongly dependent on intensity, duration, and frequency of exposure to high temperatures, species, and genotypic tolerance levels. Maintenance of crop productivity under high-temperature stress conditions demands a comprehensive understanding of adaptability, avoidance, and acclimatization techniques of crop plants. High-temperature stress adversely affects crop plants during their development phases, of which the reproduction phase is predominantly observed during

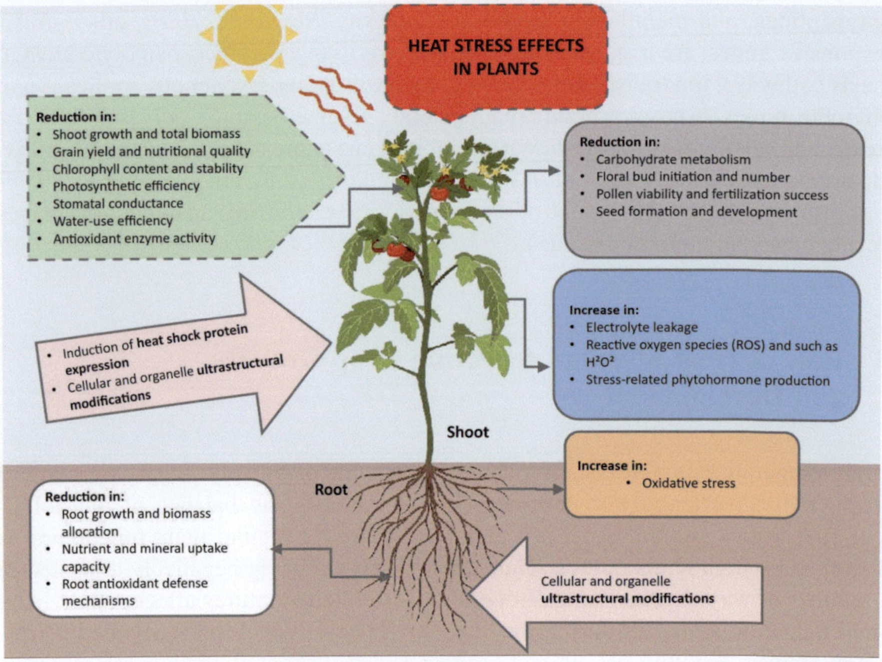

Fig. 2 Heat stress effects on crop growth and development

anthesis, and the seed development phase is considered most susceptible. Effect of heat stress on different crop plants is briefed in Table 1. An essential phenomenon of thermotolerance is associated with transporters present on cell membranes, indicating the role of transporters of carbohydrates, calcium, polyamines, osmolytes, and other cellular metabolites. Inactivation of these transporters due to HS indicates disturbed metabolic functions and a lack of sensitivity towards stress conditions. In addition, the adverse effects of HS on different crop plants are explained in Table 1, and molecular details required for the counteraction of HS are explained in Table 2.

4 Effect of Heat Stress on Crop Diversity

Selection pressure imposed by elevated temperatures alters allele frequencies and accelerates genetic erosion (Fowler & Mooney, 1990). Climate change has worsened these threats, contributing towards genetic erosion of local varieties and influencing the survival of wild species. Many existing cultivars will be replaced by improved cultivars or alternative landraces that are better adapted to the new climatic scenarios due to fluctuating environmental conditions. It is clear that a number of centres of crop diversity are situated in areas that are extremely susceptible to climatic variability, even though the exact effects of heat stress on the diversity of primitive cultivars

Table 1 Effect of heat stress on different crop species

Crops	Phase	Developmental phase	Optimum temperature (°C)	Adverse effects	References
–Wheat	Vegetative	Emergence to tillering	20–25	Reduced germination, biological yield, and tiller formation; reduction in spikelet/pollen fertility and viability, decreased grain number per spike; shortened grain-filling duration; reduced grain weight as per disturbed source sink balance; damages photosystem II, decreases Rubisco activity, increases respiration rate and oxidative stress	Chavan et al. (2019), Wang et al. (2011)
	Reproductive	Double ridge to booting	17–21		Draeger et al. (2020)
	Seed development	Seed set	14–18		Abdelrahman et al. (2020), Kino et al. (2020)
	Biochemical process	Photosynthesis/ respiration	25/14–15		Djanaguiraman et al. (2018), Impa et al. (2019), Posch et al. (2019)

(continued)

Table 1 (continued)

Crops	Phase	Developmental phase	Optimum temperature (°C)	Adverse effects	References
Barley	Vegetative	Emergence to tillering	15–22	Decreases seedling emergence and leaf area, poor tiller initiation and early senescence; reduced pollen viability, spikelet fertility, grain number per spike; pollen sterility; shorten flowering period and reduces grain size and weight. reduced endosperm filling and starch biosynthesis; premature ripening leading to shrivelled grains; damages chlorophyll, decreases photosynthetic rate and increases membrane lipid peroxidation, reduced endosperm filling and starch biosynthesis, shrivelled grains; damages chlorophyll, decreases photosynthetic rate and increases membrane lipid peroxidation	Ford et al. (2016)
	Reproductive	Double ridge to booting	15–20		Ejaz and von Korff (2017)
	Heading to flowering	–	20		Ejaz and von Korff (2017)
	Seed development	Seed set	13–18		Cochrane et al. (2000), Howard et al. (2012)
	Biochemical process	Photosynthesis/ respiration	25/15		Jedmowski and Brüggemann (2015), Oukarroum et al. (2016)

(continued)

Table 1 (continued)

Crops	Phase	Developmental phase	Optimum temperature (°C)	Adverse effects	References
Maize	Vegetative	Emergence to leaves	30–32	Impairs leaf expansion, chlorophyll biosynthesis, and seedling vigour; delays early growth; reduces pollen viability; asynchrony pollen shed and silk emergence leads to poor fertilization; reduces kernel number and grain; accelerates endosperm development, reduces starch accumulation; reduced photosynthetic efficiency, disruption of PSII, and increased respiration rates lead to energy imbalance	Singh et al. (2014)
	Reproductive	Initiation of Tassel and ear; silking of ear	26–28		Begcy et al. (2019)
	Flowering		28–33		Sánchez et al. (2014)
	Development of Kernel	Seed set	24–29		Commuri and Jones (2001), Sánchez et al. (2014)
	Biochemical level	Photosynthesis/ Respiration	38–40/23–26		Wang et al. (2020)
Rice	Vegetative ara>	Emergence to tillering	25–31	Early growth increased tillers; at later stage decline in tillers, spikes, and florets; reduced seed germination, radicle and plumule growth; floret sterility; poor pollen shedding and germination, anther dehiscence, pollen tube elongation; lipid peroxidation of thylakoid membranes; declined chlorophyll; disruption of oxygen-evolving complex; possible cell death	Krishnan et al. (2011), Oh-e et al. (2007)
	Reproductive	Initiation of Panicle and booting	26–30		Essemine et al. (2010), Kumar et al. (2019), Martínez-Eixarch and Ellis (2015), Sánchez et al. (2014)

(continued)

Table 1 (continued)

Crops	Phase	Developmental phase	Optimum temperature (°C)	Adverse effects	References
	Heading to flowering	–	30–33		Bheemanahalli et al. (2017), Shi et al. (2018), Zhang et al. (2018)
	Seed development	Seed set	27–28		Nakata et al. (2017), Xu et al. (2020a)
	Biochemical level	Photosynthesis/ respiration	30–35/23		Bahuguna et al. (2017), Mohammed and Tarpley (2010)
Soybean	Vegetative	Development of leaf and stem	28–30	Reduces leaf expansion, photosynthetic activity, and stem elongation; pollen viability, flower number, and pod initiation; pod set, seed number, and seed weight; photosynthetic rate and enzyme activity; oxidative damage to membranes, increased flower abscission ROS production and photorespiration	Alsajri et al. (2019), Jumrani and Bhatia (2018), Lyu et al. (2020)
	Reproductive	Flower initiation to anthesis	26–30		Djanaguiraman et al. (2011), Wiebbecke et al. (2012)

(continued)

Table 1 (continued)

Crops	Phase	Developmental phase	Optimum temperature (°C)	Adverse effects	References
	Post-anthesis	Formation of pod to seed set	22–31		Jumrani and Bhatia (2018), Siebers et al. (2015), Wei et al. (2020)
	Biochemical process	Photosynthesis/ Respiration	28–35/18–22		Alsajri et al. (2019), Djanaguiraman et al. (2011)
Tomato	Vegetative	Leaves and stem development	25–28	Causes leaf scorching, reduced vegetative growth, fruit-bearing potential, pollen viability, inhibits anther dehiscence and poor fruit set; reduced fruit quality; reduced PSII efficiency and chlorophyll content; imbalance between photosynthesis and respiration under prolonged heat	Havko et al. (2020), Lu et al. (2017), Tsai et al. (2019),
	Reproductive	Inflorescence formation to flower development	21–25		Ayenan et al. (2019), Paupière et al. (2017), Xu et al. (2017)
	Flowering/ post-anthesis	Fruit set	–		Lu et al. (2017), Wiebbecke et al. (2012)
	Biochemical process	Photosynthesis/ respiration	26–30/15		

Table 2 Molecular mechanism to combat the heat stress

Stress/crop	Adverse effect	References
Molecular		
Tomato	Heat shock proteins form cytoplasmic granules to protect protein biosynthesis machinery	Jagadish et al. (2010), Miroshnichenko et al. (2005)
Low molecular weight HSPs	Maintain membrane integrity; localize in chloroplasts to protect PSII; role in electron transport	Barua et al. (2003)
HSP100 with sHSPs	Molecular chaperone complex enhances acclimation and immunity	Bosl et al. (2006)
TFs and HSFs (soybean, rice, wheat, maize, tomato)	Heat sensing and signalling; regulate HSP gene expression; conserved mechanisms	Lavania et al. (2018), Ohama et al. (2017)

are still unknown because there are few empirical studies on the subject. In order to ensure adaptive capacity, maintain production system resilience, and meet the food demands of an expanding population, modern agriculture is globally interdependent and depends on crop genetic diversity from various regions.

At the international level, germplasm exchange is required due to changing climates, necessitating facilitated access to exotic varieties and landraces from analogous agroecological zones. This will likely increase global demand for genetic resources, particularly those that provide resilience against climate change (Lane & Jarvis, 2007). National and international breeding programmes have already initiated efforts to produce cultivars with improved acceptance to such pressures. However, the natural distribution for crop wild relatives is often geographically limited to centres of origin, many of which are confined to specific sub-regions. Many crops showed a remarkable decline in productivity due to high temperature stress. In South Asia and parts of Africa, each 1 °C increase in temperature above the threshold level in wheat during grain filling leads to 6–10% reduction in yield, causing a shift in traditional cultivars (Asseng et al., 2015). In rice, high temperatures above 35 °C also led to the displacement of landraces from eastern India and Southeast Asia (Jagadish et al., 2007; Kumar et al., 2019), while in maize, temperature stress reduces kernel set, causing a reduction up to 15–20% in yield (Lobell et al., 2011). Potato leads to reduced tuber formation due to heat stress above 25 °C, which shifts its production towards cooler zones (Hijmans, 2003). Advancement in molecular breeding and screening technologies allow quick evaluation and germplasm use. In a similar way, contemporary approaches like transgenic breeding, marker-assisted selection, and genome editing may minimize the range of variety needed as they permit the precise introgression of single genes instead of complete trait complexes. Furthermore, climate change may modify the significance of underutilized or minor crops. Species that are presently regarded as marginal, such as those suitable for biofuel production or resilient crops adapted to extreme conditions, may gain increasing global importance as climate-resilient alternatives. This diversity offers crucial opportunities for enhancing the adaptive capacity of global agriculture under changing climatic conditions. Subsequent studies should consequently focus on comprehensive analyses to evaluate heat stress effects and other climate-related challenges on landrace diversity, to formulate effective strategies for conservation and utilization.

5 Strategies to Improve Heat Stress Tolerance

The identification of phenological, physiological, and molecular responses towards heat stress led to extensive investigation into how crops withstand high temperatures by innate genetic dissimilarity or developing non-natural deviations by using genome editing and mutational breeding. To strengthen the progress of the work on heat stress tolerance in plants, there is a requirement to amalgamate heat stress recovery.

5.1 Conventional Approaches

Tolerance to heat stress is usually coupled with restricted interruption of photosynthetic machinery, along with improved formation of protective compounds. Photosynthesis as well as reproductive development are vulnerable to elevated temperatures. Thus, under elevated temperature a variety that is tolerant to high temperature should have superior membrane thermostability, efficiency of photosynthetic machinery with good fruit setting (Craita & Tom, 2013). For the development of stress-tolerant varieties, conventional breeding plays an important role. Since cultivars originating from cooler regions are less heat-tolerant than those from warmer regions, the development of yield-efficient promising genotypes was more fruitful by conventional breeding (Kugblenu et al., 2013; Mickelbart et al., 2015). Heat-tolerant lines are successfully created in key crops, like *Oryza sativa*, *Triticum aestivum*, *Zea mays*, *Solanum tuberosum*, and *Solanum lycopersicum*. Breeding for altered phenology, such as earlier flowering, helps avoid peak thermal periods. Introgression breeding in *Oryza sativa* contributed to the transition of heat stress-tolerant 'N22' into 'Xieqingzao B' line to create 'BC1F8'generation (Jiang-lin et al., 2011). Gayabyeo/N22 cross led to the development of heat tolerance line along with high yielding (Jiang-lin et al., 2011). Through backcrossing, *Aegilops tauschii* was used as a donor in wheat to incorporate heat-tolerant significant traits, including cell membrane stability along with chlorophyll retention in the PBW550 cultivar (Sehgal et al., 2011). In the rice crop, the main aim is to classify QTLs and genes that are linked with tolerance to heat stress. Among many selection methods, both marker-assisted recurrent selection and genomic selection were used across several crop plants to develop new, improved varieties (Varshney & Dubey, 2009). Although conventional breeding has developed valuable heat-tolerant lines, the underlying genetic basis remains complex (Uday et al., 2014).

5.2 Biotechnology Approaches

Biotechnology offers advanced ways to improve the yield, growth, and development of crop plants in today's climate, which is changing rapidly. It provides insight by identifying key components like genes, promoters, transcription factors, chromosome locations, hormone signalling pathways, and molecular interactions for addressing heat stress. These approaches deepen the understanding of plants' physiology for the progress of heat-stress-tolerant crop plants by stabilizing yields in stressful circumstances.

5.2.1 Marker-Assisted Selection

As a high-throughput technique, marker-assisted selection (MAS) has proven to be very effective in accelerating breeding programmes through increased yield (Zafar et al., 2018). Crop improvement has made use of marker-assisted backcrossing (MABC), marker-assisted recurrent selection (MARS), genome-wide association studies (GWAS), and genomic selection (GS) (Singha et al., 2022).

5.2.2 Genomic Selection

Genomic selection (GS) is a novel approach in plant breeding which utilizes data from DNA markers to forecast breeding value for quantitative traits (Tayade et al., 2018). Minor and major alleles are effectively identified by the GS method as compared to MAS. Cairns and Prasanna (2018) reported that high-density single-nucleotide polymorphism and climate-resilient wheat varieties may be produced using the GS approach to understand the complicated genome of *Triticum aestivum.* In wheat crop, GS increased the resilience to biotic and abiotic stresses along with producing higher yielding cultivars (Bhat et al., 2016; Cairns & Prasanna, 2018; Juliana et al., 2019; Mondal et al., 2021; Saini et al., 2022). GS can boost genetic gain for complex characteristics by screening genetic resources for advantageous variation of quantitative variables. The genomic selection has been found very effective to hasten breeding efficiency for tolerance to heat stress (Mondal et al. 2021; Wang et al., 2018a).

5.2.3 Genomics

Genomics includes many approaches like association mapping (AM), genome-wide association studies (GWAS), next generation sequencing, and single nucleotide polymorphism (SNP) for the analysis of the genetic basis of many traits, as well as genotypic and phenotypic relationships through high-resolution mapping. Among them, GWAS and other QTL mapping coupled with heat-tolerant phenomenon helped in the production of cultivars suited for high-temperature conditions (Paliwal et al., 2012). Across multiple environments, many QTLs have been identified as constitutive QTLs that can be exploited for developing heat-tolerant landraces. Contrarily, QTLs detected under designed environments known as adaptive QTLs can be used for areas which are heat-stress-specific (Collins et al., 2008). High-resolution QTL mapping and GWAS identify genomic loci associated with membrane stability, chlorophyll retention, and reproductive thermotolerance (Langridge et al., 2015). Since heat tolerance is a complex trait that is governed by numerous minor-effect QTLs, dissecting the loci vital for genetic improvement (Bohnert et al., 2006). Heat stress tolerance-related QTLs in different crop species are given in Table 3. For heat tolerance; availability of reference genome sequences in numerous crops and map-based cloning has advanced considerably for major QTLs. Many adaptive mechanisms are governed by multiple genes, which also regulate the heat tolerance mechanism, ultimately providing higher yield. Additionally, genomics approaches have acknowledged candidate genes responsible for heat stress tolerance.

5.2.4 Transcriptome

This technique has emerged as a trustworthy tool for the identification of heat-tolerant genes considering the discrepancy of expression and redeployment of pathways associated with heat stress tolerance. On different developmental phases, transcriptomics provided important insight into gene structure, which genes are linked to heat stress tolerance mechanisms, along with bioinformatics. HSFs being prominent take part in a central function in signal transduction and heat shock protein's activation (Jie et al., 2014; Wentao et al., 2019). Additionally, autophagy has been identified as another protective mechanism with studies of Arabidopsis and tomato species (Gonul et al., 2019). Plants show differential responses to abiotic stress at diverse stages of development due to changes in transcriptional reprogramming. For abiotic stresses, temperature shift changes in metabolite concentration, including sugar, hormones, and calcium ions, influence transcriptional activity modulating circadian oscillator genes (Khurram et al., 2019). Genes associated with polyamine biosynthesis have shown tolerance by supporting physiological processes like photosynthesis, defence molecules against oxidative stress, osmoregulation, and membrane structure stabilization (Marco et al., 2011). Aquaporin brainwashing genes were acknowledged, as 35 in Arabidopsis, 31 in maize, 33 in rice, 47 in tomato, 66 in soybean, 71 in cotton, 41 in *Solanum tuberosum,* and 3 in *Cucumis sativus*, for utilization in crop plants in future breeding programmes (Ranganathan et al., 2018; Yong et al., 2019).

Table 3 QTLs related to heat tolerance across crop species

Crop	Trait	QTL name/ chromosome	Linked marker/ region	References
Wheat	Seed yield	qSYHS-A03	SSR marker BrSSR123	Talukder et al. (2014)
	Days to flowering	qDFHS-C02	SNP marker BnSNP567	Wang et al. (2018b)
	Membrane stability index	qMSIHS-A05	SSR marker BRMSI45	Zhang et al. (2019a)
	Pollen viability	qPVHS-C09	SNP marker BnPV89	Li et al. (2021a)
	Chlorophyll content	qCHLHS-A02	SSR marker BRCHL112	Kumar et al. (2022a)
Rice	Spikelet fertility under heat	qSFHS-5	SNP marker RM287	PS et al. (2017)
	Chlorophyll content	qCHLHS-9	SNP marker RM1000	PS et al. (2017)
	Flowering	*OsGSK1*	–	Koh et al. (2007)
	Seedling, flowering, filling	*TT1*	–	Li et al. (2015)
	Seedling	*OsHTAS*	–	Liu et al. (2016)
	Seedling, flowering, fertilization	*OsFBN1*	–	Li et al. (2019)
	Seedling, flowering	*SLG1*	–	Xu et al. (2020b)
	Seedling	*HTS1*	–	Chen et al. (2021)
	Seedling	*HTH5*	–	Cao et al. (2022)
	Seedling	*TT3.1*	–	Zhang et al. (2022)
	Seedling	*TT3.2*	–	Zhang et al. (2022)
	Seedling	*OsSGS3*	–	Gu et al. (2023)
	Seedling, booting, fertilization	*OsGRP3*/ *OsGRP162*	–	Yang et al. (2024)
Maize	Grain yield	qGYHS-4	SSR marker ZmSSR123	El Sappah et al. (2022)
	Anthesis-silking interval	qASIHS-2	SNP marker ZmSNP456	El-Sappah et al. (2022)
Chickpea	Seed yield per plant	Meta-QTL1 (CaLG01)	SNP marker CaSNP789	Kumar et al. (2023a)
	Days to flowering	Meta-QTL2 (CaLG03)	SNP marker CaSNP234	Kumar et al. (2023b)

(continued)

Table 3 (continued)

Crop	Trait	QTL name/ chromosome	Linked marker/ region	References
Lentil	Seed yield	qSYHS-7	SSR marker LSSY123	Singh et al. (2020a)
	Days to maturity	qDTHS-3	SNP marker LSNM456	Singh et al. (2020b)
Rapeseed-Mustard	Seed yield	qSYHS-A03	SSR marker BrSSR123	Rahman et al. (2020)
	Days to flowering	qDFHS-C02	SNP marker BnSNP567	Wang et al. (2018c)
	Membrane stability index	qMSIHS-A05	SSR marker BRMSI45	Zhang et al., 2019b
	Pollen viability	qPVHS-C09	SNP marker BnPV89	Li et al. (2021b)
	Chlorophyll content	qCHLHS-A02	SSR marker BRCHL112	Kumar et al. (2022b)

Annexins are a conserved multigene family of calcium and phospholipid-binding proteins resulting in plant development and resistance to heat. Annexins' transcriptomics have been done on many crops like Arabidopsis, Indian Mustard, cabbage, and rapeseed (Xin et al., 2020). Transcriptomic analysis continues to uncover novel stress-responsive genes harnessed through molecular breeding strategies or genetic engineering for the generation of heat stress-resilient varieties.

5.2.5 Proteomics

Proteomics unravels the proteins that mediate heat stress response because of changes in gene expression. Since post-transcriptional and post-translational modifications alter protein activity, proteomic analyses offer a more direct view of cellular mechanisms under stress. Proteomics has been used for investigating the response to heat stress in *Oryza sativa, T. aestivum, H. vulgare, Zea mays,* and *alfalfa* (Qi et al., 2018). Proteomics unravels the method implicated in heat stress adaptation in developing pollen grains. Using gel LC–MS fractionation, proteins implicated in oxidative stress protection were upregulated during heat exposure (Sridharan et al., 2018).

5.2.6 Genetic Engineering

Genetic engineering offers an influential platform for the development of heat-tolerant crop plants by manipulating the complex traits. This approach enables modifications by manipulating the complex genetic traits, like increasing photosynthetic activity, restrictive RuBP oxygenation pathway, and increasing the enhanced accumulation of compatible solutes like proline, glycine, betaine, and pigment that augment resilience to heat stress. Genetic engineering for resilience to heat stress has been executed in many crops (Sotirios et al., 2015). In potato, annexins are implicated in heat tolerance and mediate protection against photo-oxidative damage (Michal et al., 2015). Efforts have also been made to transfer C4 photosynthetic characteristics into C3 plants for improving yield and sustainability under heat stress. For instance, introducing maize (C4) enzymes such as NADP-malic enzyme, PEP carboxylase, and pyruvate orthophosphate dikinase enhanced photosynthetic capacity and limited photo-oxidative damage in rice leaves (Maurice et al., 2000). The integration of soybean and rice promoters with Streptococcus pyogenes Cas9 (SpCas9) and alternative systems like CRISPR/Cas12a has driven significant advancements in crop engineering. CRISPR/Cas9's success has facilitated the adoption of alternative gene editing systems, including CRISPR/Cas12a, to broaden genetic engineering strategies (Soumen et al., 2019).

6 Management of Increased Temperatures and Heat Stress

6.1 Timely Sowing and Water Management

The implementation of appropriate agronomic practices such as timely sowing, optimum seed rates, proper spacing, optimum sowing depth, and treatment of seeds or seedlings improves crop establishment, enhances the resource-use efficiency, and increases overall survival of crop plants. Early sowing is considered to be effective in wheat crop and is an economical approach to mitigate terminal heat stress (Kumar & Rai, 2014a, 2014b). In addition to this, supplemental irrigation during the critical reproductive stages can significantly enhance yield under heat stress conditions. The transplanting of rice in time during summers leads to flowering before the onset of the peak heat period avoiding the heat stress (Prasad et al., 2017). Maize crop showed higher yields when sown in the cooler months or early in the summer (Shah et al., 2011). Irrigation during critical reproductive stage, during seed-filling and flowering stages in wheat, rice, and maize crop, leads to higher yields along with better grain quality under high temperature stress (Gupta et al., 2010; Singh et al., 2019).

6.2 Mulching

Mulching acts as an insulating layer which moderates soil temperature and reduces the heat stress throughout the crop growing season. It has shown to decrease soil temperature by 1–2 °C which lowers extreme summer temperatures and increases the winter soil warmth and prevents excessive soil heating (Awasthi et al., 2023).

There is an increase in crop yields, improvement in water-use efficiency (WUE) and enhancement in nitrogen-use efficiency (NUE) by 60% as compared to conditions without mulch in maize, tomato, and capsicum (Qin et al. 2015; Zhao et al., 2019). Therefore, under heat stress conditions, these enhancements lead to higher plant growth, production, and improved quality.

6.3 *Use of Osmoprotectants*

Osmoprotectants are an easy and affordable way to reduce the heat stress in crops (Bakhtavar et al., 2015). For example, the foliar application of salicylic acid enhances leaf number and growth in wheat and maize plants by inducing thermotolerance (Afzal et al., 2020; Hussein et al., 2007). The plant-derived natural growth enhancers such as sorghum water extract (SWE) and moringa leaf extract (MLE) in rice and sunflower are documented as affordable and efficient bio-stimulants (Latif et al., 2018; Rehman et al., 2017).

6.4 *Suitable Crop and Cultivar Selection*

In late-sown conditions HS has a negative impact on crop plants as it reduces yield. The floral organs are exposed to higher microclimatic temperatures when HS occurs concurrently with flower initiation (Prasad et al., 2017). In order to avoid this, crops and cultivars that flower in the early morning or late evening when the temperature is lower must be selected. These avoidance mechanisms are exhibited by heat-tolerant genotypes of crops like wheat (HD 2967), sorghum (CSV27 & SPV1616), and pearl millet (Chiluwal et al. 2020; Sehgal et al., 2020). Similar adaptive behaviour has been shown by other crop species including *Oryza australiensis* and other wild rice relatives (Jagadish et al., 2015).

7 Conclusions and Future Prospects

The high temperature stress is one major factor limiting the overall growth and development of crops. Stress leads to oxidative damage, interferes with metabolic processes, and alters cellular structures. These factors work together to shorten crop growth duration, speed up senescence, and decrease biomass accumulation all of which eventually lower yield potential. Conventional breeding methods rely on taking advantage of genetic diversity which needs improvements in phenotyping and genotyping. By modification of heat-responsive genes and QTLs, genetic engineering enables improvement in crop performance under terminal heat stress. The fast and accurate measurement of complex polygenic traits that supplement traditional breeding methods has been made possible by recent advancements in high-throughput phenomics and systems biology techniques. Furthermore, omics technologies speed up breeding for sustainable agriculture and provide mechanistic understanding of plant responses. Gene mapping, introgression, and trait identification have all been successfully accomplished with breeding techniques such as QTL mapping, MABC, MARS, and GS. But when it comes to delineating the physiological and molecular mechanisms of stress tolerance, these methods offer little resolution. On the other hand, although their high cost prevents widespread use particularly for resource-poor farmers the high-throughput sequencing and proteomic technologies allow for a more better understanding of underlying mechanisms of heat stress. Although integrating with break-through technologies like CRISPR/Cas genome editing, transgenic techniques, and molecular engineering into large-scale breeding is still difficult, they present promising paths for improving crop heat tolerance. Therefore, combining traditional and contemporary methods is a sensible way to increase breeding efficiency and accessibility for heat-tolerant cultivars. Furthermore, by improving stress tolerance at the physiological level, exogenous applications of protective proteins and osmolytes offer an extra layer of resilience. Despite the considerable advancements achieved in comprehending and alleviating heat stress, additional research is essential to completely clarifying its fundamental mechanisms. Future generations will have access to food if new technologies and traditional farming practices are combined to produce climate-resilient crops.

References

Abdelrahman, M., Burritt, D. J., Gupta, A., Tsujimoto, H., & Tran, L. S. P. (2020). Heat stress effects and tolerance in wheat and barley: Integrating transcriptomics and metabolomics. *Frontiers in Plant Science, 11*, 732.

Afzal, I., Basra, S. M. A., Farooq, M., Nawaz, A., & Alghamdi, S. S. (2020). Salicylic acid seed priming and foliar spray improves wheat growth and yield under heat stress conditions. *Journal of Agronomy and Crop Science, 206*(4), 452–463.

Alsajri, F. A., Singh, B., Bhatia, V. S., Kaushik, P., & Sita, K. (2019). Heat stress effects on soybean physiology and yield. *Legume Research, 42*(5), 715–721.

Asseng, S., Ewert, F., Martre, P., Rötter, R. P., Lobell, D. B., Cammarano, D., Kimball, B. A., et al. (2015). Rising temperatures reduce global wheat production. *Nature Climate Change, 5*(2), 143–147.

Awasthi, A., Pattnayak, K. C., Tandon, A., Sarkar, A., & Chakraborty, M. (2023). Implications of climate change on surface temperature in North Indian states: Evidence from CMIP6 model ensembles. *Frontiers in Environmental Science, 11*, 1264757. https://doi.org/10.3389/fenvs.2023.1264757

Awasthi, A., Pattnayak, K. C., & Tiwari, P. R. (2025). Application of artificial intelligence-supported process-based climate models to understand the atmosphere/weather patterns and their prediction. *Frontiers in Environmental Science, 12*, 1543951. https://doi.org/10.3389/fenvs.2024.1543951

Ayenan, M. A. T., Danquah, A., & Danquah, E. Y. (2019). Genetic improvement for heat tolerance in tomato (Solanum lycopersicum): Progress and prospects. *Plant Breeding, 138*(3), 389–400.

Bahuguna, R. N., Solis, C. A., Shi, W., & Jagadish, S. V. K. (2017). Post-flowering night respiration and heat tolerance in rice. *Functional Plant Biology, 44*(9), 1036–1044.

Bakhtavar, M. A., Afzal, I., Basra, S. M. A., Ahmed, A., & Noor, M. (2015). Physiological strategies to improve the performance of spring maize (*Zea mays* L.) planted under early and optimum sowing conditions. *PLoS ONE, 10*(4), e0124441.

Barua, D., Heckathorn, S. A., & Coleman, J. S. (2003). Variation in heat-shock proteins (HSPs) among populations of Solidago altissima: Evidence for ecotypic variation in HSP expression. *American Journal of Botany, 90*(5), 729–735. https://doi.org/10.3732/ajb.90.5.729

Begcy, K., Sandhu, J., Walia, H., & Jagadish, S. V. K. (2019). Transient heat stress during early reproductive stages alters maize kernel set. *Plant Physiology, 181*(2), 503–516.

Behera, B., Haldar, A. & Sethi, N. (2023). Agriculture, food security, and climate change in south Asia: A new perspective on sustainable development. *Environment Development Sustainability. 26*(9), 22319–22344. https://doi.org/10.1007/s10668-023-03552-y

Bhardwaj, A., Devi, P., Chaudhary, S., Rani, A., Jha, U. C., Kumar, S., Bindumadhava, H., Prasad, P. V. V., Sharma, K. D., & Siddique, K. H. M. (2022). 'Omics' approaches in developing combined drought and heat tolerance in food crops. *Plant Cell Reports, 41*, 699–739.

Bhat, J. A., Ali, S., Salgotra, R. K., Mir, Z. A., Dutta, S., & Jadon, V. (2016). Genomic selection in the era of next generation sequencing for complex traits in plant breeding. *Frontiers in Genetics, 7*, 221. https://doi.org/10.3389/fgene.2016.00221

Bheemanahalli, R., Sathishraj, R., Manoharan, M., et al. (2017). High temperature effects on rice: Reproductive stage is most sensitive. *Plant Physiology Reports, 22*, 103–112.

Birthal, P. S., Joshi, P. K., & Roy, D. (2014). Weather shocks and farm household vulnerability to poverty in India. *Agricultural Economics Research Review, 27*(1), 25–36.

Bohnert, H. J., Gong, Q., Li, P., & Ma, S. (2006). Unraveling abiotic stress tolerance mechanisms—Getting genomics going. *Current Opinion in Plant Biology, 9*(2), 180–188. https://doi.org/10.1016/j.pbi.2006.01.003

Bosl, B., Grimminger, V., & Walter, S. (2006). The molecular chaperone Hsp104—A molecular machine for protein disaggregation. *Journal of Structural Biology, 156*(1), 139–148. https://doi.org/10.1016/j.jsb.2006.03.003

Cairns, J. E., & Prasanna, B. M. (2018). Developing and deploying climate-resilient maize varieties in the developing world. *Current Opinion in Plant Biology, 45*, 226–230. https://doi.org/10.1016/j.pbi.2018.05.004

Cao, Z. B., Tang, H. W., Cai, Y. H., Zeng, B. H., Zhao, J. L., Tang, X. Y., Lu, M., Wang, H. M., Zhu, X. J., Wu, X. F., Yuan, L. F., & Wan, J. L. (2022). Natural variation of *HTH5* from wild rice, *Oryza rufipogon* Griff., is involved in conferring high-temperature tolerance at the heading stage. *Plant Biotechnology Journal, 20* (8), 1591–1605.

Carter, M., et al. (2014). Poverty traps and climate risk: Evidence from India. *World Bank Economic Review, 28*(2), 243–275. https://doi.org/10.1093/wber/lht028

Chavan, S. G., Kadam, R. A., & More, A. D. (2019). Heat stress and its management in wheat. *International Journal of Chemical Studies, 7*(2), 1276–1280.

Chen, F., Dong, G. J., Wang, F., Shi, Y. Q., Zhu, J. Y., Zhang, Y. L., Ruan, B. P., Wu, Y. P., Feng, X., Zhao, C. C., Yong, M. T., Holford, P., Zeng, D. L., Qian, Q., Wu, L. M., Chen, Z. H., & Yu, Y. C. (2021). A β-ketoacyl carrier protein reductase confers heat tolerance via the regulation of fatty acid biosynthesis and stress signaling in rice. *New Phytologist, 232*(2), 655–672.

Chiluwal, A., Bheemanahalli, R., Kanaganahalli, V., Boyle, D., Perumal, R., Pokharel, M., Oumarou, H., & Jagadish, S. K. (2020). Deterioration of ovary plays a key role in heat stress induced spikelet sterility in sorghum. *Plant, Cell & Environment, 43*(2), 448–462.

Cochrane, M. P., Duffus, C. M., & Hedley, C. L. (2000). The effect of heat stress on barley grain development. *Journal of Cereal Science, 32*(2), 115–126.

Cohen, I., Zandalinas, S. I., Huck, C., Fritschi, F. B., & Mittler, R. (2021). Meta-analysis of drought and heat stress combination impact on crop yield and yield components. *Physiologia Plantarum, 171*, 66–76.

Collins, N. C., Tardieu, F., & Tuberosa, R. (2008). Quantitative trait loci and crop performance under abiotic stress: Where do we stand? *Plant Physiology, 147*(2), 469–486. https://doi.org/10.1104/pp.108.118117

Commuri, P. D., & Jones, R. J. (2001). High temperatures during endosperm cell division in maize: A genotypic comparison under in vitro and field conditions. *Plant Physiology, 125*(1), 292–298.

Craita, E. B., & Tom, G. (2013). Plant tolerance to high temperature in a changing environment: Scientific fundamentals and production of heat stress-tolerant crops. *Frontiers in Plant Science, 4*, 273.

Djanaguiraman, M., Boyle, D. L., Welti, R., Jagadish, S. V. K., & Prasad, P. V. V. (2018). Decreased photosynthetic rate under high temperature in wheat is due to lipid desaturation. *Plant, Cell & Environment, 41*(5), 1079–1091.

Djanaguiraman, M., Prasad, P. V. V., Boyle, D. L., & Schapaugh, W. T. (2011). Soybean pollen anatomy, viability and pod set under high temperature stress. *Journal of Agronomy and Crop Science, 197*(1), 92–99.

Draeger, T., Goswami, S., & Moore, G. (2020). Heat stress during meiosis in wheat: Mechanisms and breeding implications. *Journal of Experimental Botany, 71*(13), 3712–3725.

Easterling, W. E., Aggarwal, P. K., Batima, P., Brander, K., Erda, L., Howden, S. M., & Tubiello, F. N. (2007). Food, fibre and forest products. In M. L. Parry, O. F. Canziani, J. P. Palutikof, P. J. van der Linden, & C. E. Hanson (Eds.), *Climate change 2007: Impacts, adaptation and vulnerability. Contribution of Working Group II to the Fourth Assessment Report of the Intergovernmental Panel on Climate Change* (pp. 273–313). Cambridge University Press.

Ejaz, M., & von Korff, M. (2017). Heat stress responses and tolerance mechanisms in barley. *Plant, Cell & Environment, 40*(10), 2396–2411.

El-Sappah, A. H., Rather, S. A., Alotaibi, S. S., Alharbi, B. M., & Ibrahim, M. E. H. (2022). Identification of QTLs for heat stress tolerance in maize (*Zea mays* L.) through association mapping and genomic prediction. *Frontiers in Genetics, 13*, 869923. https://doi.org/10.3389/fgene.2022.869923

Essemine, J., Ammar, S., & Bouzid, S. (2010). Impact of heat stress on germination and growth in higher plants: Physiological, biochemical and molecular repercussions and mechanisms of defence. *Journal of Biological Sciences, 10*(6), 565–572. https://doi.org/10.3923/jbs.2010.565.572

Falcon, W. P., Ratna, S., & Kumar, S. (2022). Food security and global crop demand to 2050: Will climate change undermine it? *Agricultural Systems, 191*, Article 103181. https://doi.org/10.1016/j.agsy.2021.103181

Ford, B., Deng, W., Clausen, J., Oliver, S., Boden, S., Hemming, M., & Trevaskis, B. (2016). Barley (*Hordeum vulgare*) circadian clock genes can respond rapidly to temperature in an EARLY FLOWERING 3-dependent manner. *Journal of Experimental Botany, 67*(18), 5517–5528.

Fowler, C., & Mooney, P. (1990). *The threatened gene: Food politics and the loss of genetic diversity*. UK, Lutworth Press.

Gonul, D., Zhenhua, S., Nahoko, H., Mami, K., Masanori, I., & Atsushi, H. (2019). Autophagy mitigates high-temperature injury in pollen development of Arabidopsis thaliana. *Developmental Biology, 456*, 190–200.

Gu, X. T., Si, F. Y., Feng, Z. X., Li, S. J., Liang, D., Yang, P., Yang, C., Yan, B., Tang, J., Yang, Y., Li, T., Li, L., Zhou, J. L., Li, J., Feng, L. L., Liu, J. Y., Yang, Y. Z., Deng, Y. W., Wu, X. N., & Liu, J. Z. (2023). The OsSGS3-tasiRNA- OsARF3 module orchestrates abiotic-biotic stress response trade-off in rice. *Nature Communication, 14*(1), 4441.

Gupta, N. K., Agarwal, S., Agarwal, V. P., Nathawat, N. S., Gupta, S., & Singh, G. (2010). Effect of water stress on physiological attributes and their relationship with growth and yield in wheat cultivars at different growth stages. *Field Crops Research, 118*(3), 219–225.

Havko, N. E., Das, M. R., McClain, A. M., Kapali, G., Sharkey, T. D., & Howe, G. A. (2020). Heat stress uncouples growth-defense tradeoffs in tomato. *Proceedings of the National Academy of Sciences, 117*(13), 7510–7519.

Hijmans, R. J. (2003). The effect of climate change on global potato production. *American Journal of Potato Research, 80*(4), 271–279.

Howard, T. P., et al. (2012). Grain development under high temperature in barley. *Plant Physiology, 159*, 171–183.

Hussein, M. M., Balbaa, L. K., & Gaballah, M. S. (2007). Salicylic acid and salinity effects on growth of maize plants. *Research Journal of Agriculture and Biological Sciences, 3*(4), 321–328.

Impa, S. M., Sunoj, V. S. J., Krassovskaya, I., et al. (2019). Carbon balance and yield in wheat under high temperature stress. *Plant, Cell & Environment, 42*(4), 1233–1246.

Jagadish, S. V. K., Craufurd, P. Q., & Wheeler, T. R. (2007). High temperature stress and spikelet fertility in rice (*Oryza sativa* L.). *Journal of Experimental Botany, 58*(7), 1627–1635.

Jagadish, S. V. K., Craufurd, P. Q., & Wheeler, T. R. (2010). High temperature stress and spikelet fertility in rice (Oryza sativa L.). *Journal of Experimental Botany, 61*(1), 191–203. https://doi.org/10.1093/jxb/erp316

Jagadish, S. V. K., Murty, M. V. R. & Quick, W. P. (2015). Rice responses to rising temperatures–challenges, perspectives and future directions. *Plant, Cell & Environment 38*(9), 1,686–1,698.

Jedmowski, C., & Brüggemann, W. (2015). Photosynthesis and respiration responses of barley to heat stress. *Plant Biology, 17*(1), 5–12.

Jiang-lin, L., Hong-yu, Z., Xue-lian, S., Ping-an, Z., & Ying-jin, H. (2011). Identification for heat tolerance in backcross recombinant lines and screening of backcross introgression lines with heat tolerance at milky stage in rice. *Rice Science, 18*, 279–286.

Jie, Z., Jian, W., Jing, Q. Y., & Zhixiang, C. (2014). Role and regulation of autophagy in heat stress responses of tomato plants. *Frontiers in Plant Science, 5*, 174. https://doi.org/10.3389/fpls.2014.00174

Juliana, P., Montesinos-López, O. A., Crossa, J., Mondal, S., González Pérez, L., & Poland, J. (2019). Integrating genomic-enabled prediction and high-throughput phenotyping in breeding for climate-resilient bread wheat. *Theoretical and Applied Genetics, 132*(1), 177–194. https://doi.org/10.1007/s00122-018-3206-3

Jumrani, K., & Bhatia, V. S. (2018). Impact of elevated temperatures on soybean productivity. *Agricultural and Forest Meteorology, 253–254*, 94–104.

Khurram, B., Akihiro, M., Sultana, R., & Motoaki, S. (2019). Recent advances in the characterization of plant transcriptomes in response to drought, salinity, heat, and cold stress. *F1000 Research.* https://doi.org/10.12688/f1000research.18424.1.

Kino, H., et al. (2020). Heat tolerance mechanisms in wheat grain filling. *Frontiers in Plant Science, 11*, 1–12.

Koh, S., Lee, S. C., Kim, M. K., Koh, J. H., Lee, S., An, G., Choe, S., & Kim, S. R. (2007). T-DNA tagged knockout mutation of rice *OsGSK1*, an orthologue of *Arabidopsis BIN2*, with enhanced tolerance to various abiotic stresses. *Plant Molecular Biology, 65*(4), 453–466.

Krishnan, P., Ramakrishnan, B., Reddy, K. R., & Reddy, V. R. (2011). High temperature effects on rice growth, yield, and grain quality. *Advances in Agronomy, 111*, 87–206.

Kugblenu, Y. O., Oppong Danso, E., Ofori, K., Andersen, M. N., Abenney Mickson, S., Sabi, E., et al. (2013). Screening tomato genotypes in Ghana for adaptation to high temperature. *Acta Agriculturae Scandinavica Section B: Soil and Plant Science, 63*, 516–522.

Kumar, A., Dixit, S., Ram, T., Yadaw, R. B., Mishra, K. K., & Mandal, N. P. (2019). Breeding high-yielding drought-tolerant rice: Genetic variations and conventional and molecular approaches. *Journal of Experimental Botany, 70*(4), 1247–1266. https://doi.org/10.1093/jxb/erz010

Kumar, A., Sharma, R., & Singh, R. K. (2022a). Genetic dissection of chlorophyll stability and photosynthetic efficiency under heat stress in wheat (*Triticumaestivum* L.). *Frontiers in Plant Science, 13*, 828445. https://doi.org/10.3389/fpls.2022.828445

Kumar, S., et al. (2022b). Identification of QTLs associated with chlorophyll content under heat stress in Brassica juncea. *Journal of Plant Physiology, 267*, Article 153556. https://doi.org/10.1016/j.jplph.2021.153556

Kumar, R. et al. (2023a). High confidence QTLs and key genes identified using Meta-QTL analysis for heat tolerance in chickpea. *Frontiers in Plant Science, 14*, 1274759. Link

Kumar, S., Sandhu, J. S., Singh, I., & Bharadwaj, C. (2023b). Meta-QTL analysis and candidate genes for heat stress tolerance in chickpea (*Cicerarietinum* L.). *BMC Plant Biology, 23*(1), 1–15. https://doi.org/10.1186/s12870-023-04012-7

Kumar, R. R., & Rai, G. K. (2014a). Breeding strategies for improving heat tolerance in wheat: A review. *Journal of Wheat Research, 6*(2), 1–12.

Kumar, R.R., & Rai, R.D. (2014b). Can wheat beat the heat: under standing the mechanism of thermos tolerance in wheat (Triticumaestivum L.) a review? *Cereal Research Communication, 42*(1), 1–18.

Lane, A. & Jarvis, A. (2007). Changes in climate will modify the geography of crop suitability: Agricultural biodiversity can help with adaptation. Paper presented at ICRISAT/CGIAR 35th Anniversary Symposium. In *Climate-Proofing Innovation for Poverty Reduction and Food Security*, 22–24 November 2007, ICRISAT, Patancheru, India. Available at: http://www.icrisat.org/Journal/SpecialProject/sp2.pdf. Accessed 8 March 2010. 12 pp.

Langridge, P., & Reynolds, M. P. (2015). Genomic tools to assist breeding for drought tolerance. *Current Opinion in Biotechnology, 32*, 130–135. https://doi.org/10.1016/j.copbio.2014.11.027

Latif, S., Ullah, F., & Ayub, G. (2018). Moringa leaf extract improves growth, physiology, and yield of sunflower under heat stress. *South African Journal of Botany, 115*, 130–137.

Lavania, D., Dhingra, A., Siddiqui, M. H., Al-Whaibi, M. H., & Grover, A. (2018). Current status of the production of high temperature tolerant transgenic crops for cultivation in warmer climates. *Plant Physiology and Biochemistry, 131*, 191–204. https://doi.org/10.1016/j.plaphy.2018.04.033

Li, J. J., Yang, J., Zhu, B. H., & Xie, G. S. (2019). Overexpressing *OsFBN1* enhances plastoglobule formation, reduces grain-filling percent and jasmonate levels under heat stress in rice. *Plant Science, 285*, 230–238.

Li, X. M., Chao, D. Y., Wu, Y., Huang, X. H., Chen, K., Cui, L. G., Su, L., Ye, W. W., Chen, H., Chen, H. C., Dong, N. Q., Guo, T., Shi, M., Feng, Q., Zhang, P., Han, B., Shan, J. X., Gao, J. P., & Lin, H. X. (2015). Natural alleles of a proteasome α2 subunit gene contribute to thermotolerance and adaptation of African rice. *Nature Genetics, 47*(7), 827–833.

Li, J., et al. (2021a). Genetic dissection of pollen viability under heat stress in Brassica species using QTL mapping. *Frontiers in Plant Science, 12*, Article 640490. https://doi.org/10.3389/fpls.2021.640490

Li, Y., Wang, J., Xia, X., & He, Z. (2021b). Identification of QTLs for heat tolerance during the reproductive stage in wheat. *Theoretical and Applied Genetics, 134*(2), 373–385. https://doi.org/10.1007/s00122-020-03699-4

Liu, J. P., Zhang, C. C., Wei, C. C., Liu, X., Wang, M. G., Yu, F. F., Xie, Q., & Tu, J. M. (2016). The RING finger ubiquitin E3 ligase OsHTAS enhances heat tolerance by promoting H_2O_2-induced stomatal closure in rice. *Plant Physiology, 170*(1), 429–443.

Lobell, D. B., Schlenker, W., & Costa-Roberts, J. (2011). Climate trends and global crop production since 1980. *Science, 333*(6042), 616–620.

Lu, Y., Li, R., Mageney, V., et al. (2017). Heat stress responses in tomato at vegetative stage. *BMC Plant Biology, 17*(1), 1–16.

Luck, M., et al. (2015). Increasing pressure on freshwater resources due to human population growth and changing consumption patterns. *Nature Climate Change, 6*(8), 617–621.

Lyu, J., et al. (2020). Genetic basis of soybean response to heat stress. *Plant Biotechnology Journal, 18*(8), 1859–1875.

Marco, F., Altabella, T., Alca´zar, R., Cuevas, J., Bortolotti, C., Gonza´lez, M. E., et al. (2011). Transcriptome analysis of polyamine overproducers reveals activation of plant stress responses and related signaling pathways tolerance in plants. *Omics and Plant Abiotic Stress Tolerance, 2011*, 82–90.

Martínez-Eixarch, M., & Ellis, R. H. (2015). Rice reproductive development under high temperature. *Field Crops Research, 182*, 138–147.

Maurice, K., Cho, D., Ranade, U., & Hsu, T.-P. (2000). Photosynthetic performance of transgenic rice plants overexpressing maize C4 photosynthesis enzymes. *Studies in Plant Science, 7*, 193–204.

Michal, S., Barbara, S., Wojciech, R., et al. (2015). Potato annexin STANN1 promotes drought tolerance and mitigates light stress in transgenic Solanum tuberosum L. Plants. *PLOS ONE*. https://doi.org/10.1371/journal.pone.0132683

Mickelbart, M. V., Hasegawa, P. M., & Bailey-Serres, J. (2015). Genetic mechanisms of abiotic stress tolerance that translates to crop yield stability. *Nature Reviews Genetics, 16*, 237–251. https://doi.org/10.1038/nrg390

Miroshnichenko, D., Becker, T., & Schierle, C. (2005). Heat shock proteins and stress tolerance in plants. *FEBS Letters, 579*(30), 5832–5838. https://doi.org/10.1016/j.febslet.2005.08.054

Mohammed, A. R., & Tarpley, L. (2010). Effects of high night temperature on crop physiology and productivity: A focus on rice. *Field Crops Research, 119*(2–3), 197–207. https://doi.org/10.1016/j.fcr.2010.07.005

Mondal, S., Sallam, A., Sehgal, D., Sukumaran, S., Farhad, M., Navaneetha Krishnan, J., Kumar, U., & Biswal, A. (2021). Advances in breeding for abiotic stress tolerance in wheat. In C. Kole (Ed.), *Genomic designing for abiotic stress resistant cereal crops* (pp. 71–103). Springer International Publishing. https://doi.org/10.1007/978-3-030-75875-2-4

Nakata, M., et al. (2017). Grain setting in rice under heat stress. *Plant Production Science, 20*(2), 145–158.

Newman, R., & Noy, I. (2023). The global costs of extreme weather that are attributable to climate change. *Nature Communication, 14*, 6103.

Ohama, N., Sato, H., Shinozaki, K., & Yamaguchi-Shinozaki, K. (2017). Transcriptional regulatory network of plant heat stress response. *Trends in Plant Science, 22*(1), 53–65. https://doi.org/10.1016/j.tplants.2016.08.015

Oh-e, M., Saitoh, K., & Kuroda, T. (2007). Effects of high temperature on growth, yield and dry-matter production of rice grown in the paddy field. *Plant Production Science, 10*(4), 412–422. https://doi.org/10.1626/pps.10.412

Oukarroum, A., et al. (2016). Heat stress effects on photosynthesis in barley. *Plant Biology, 18*(3), 431–443.

Pattnayak, K. C., Awasthi, A., Sharma, K., & Pattnayak, B. B. (2023). Fate of rainfall over the North Indian states in the 1.5 and 2 C warming scenarios. *Earth and Space Science, 10*(2), e2022EA002671. https://doi.org/10.1029/2022EA002671

Paliwal, R., Röder, M. S., Kumar, U., Srivastava, J. P., & Joshi, A. K. (2012). QTL mapping of terminal heat tolerance in hexaploid wheat (Triticum aestivum L.). *Theoretical and Applied Genetics, 125*(3), 561–575. https://doi.org/10.1007/s00122-012-1853-3

Paupière, M. J., et al. (2017). Responses of tomato reproductive development to high temperature. *Frontiers in Plant Science, 8*, 1–18.

Ponsonby, W., & Di Corleto, R. (2024). Climate change and heat stress. *Occupational Medicine, 74*, 138–139.

Posch, B. C., Zangerl, P., & Zeeman, S. C. (2019). Wheat photosynthesis and respiration under heat stress. *Plant Physiology, 181*(3), 1095–1107.

Prasad, P. V., Bheemanahalli, R., & Jagadish, S. K. (2017). Field crops and the fear of heat stress opportunities, challenges and future directions. *Field Crops Research, 200*, 114–121.

PS, B., Swamy, B. P. M., Vikram, P., Ahmed, H. U., Dixit, S., Cruz, M. T. S., & Kumar, A. (2017). QTLs and candidate genes for heat stress tolerance identified in rice (*Oryza sativa* L.) using linkage mapping, GWAS, and transcriptomic approaches. *Plant Molecular Biology, 93*(4–5), 327–340. https://doi.org/10.1007/s11103-016-0566-2

Qi, Z., Wenxin, C., Jiayi, B., Hao, X., Ying, L., et al. (2018). Proteomics and phosphoproteomics of heat stress-responsive mechanisms in Spinach. *Frontiers in Plant Science*. https://doi.org/10.3389/fpls.2018.00800

Qin, W., Hu, C. & Oenema, O. (2015). Soil mulching significantly enhances yields and water and nitrogen use efficiencies of maize and wheat: A meta-analysis. *Scientific Reports, 5*(1), 16,210.

Rahman, H., Mamidi, S., Rahman, M., & Mason, A. S. (2020). Genetic analysis and QTL mapping of heat stress tolerance in *Brassica juncea* (Indian mustard). *Theoretical and Applied Genetics, 133*(9), 2605–2617. https://doi.org/10.1007/s00122-020-03604-1

Ramya, A., Dhevagi, P., Poornima, R., Avudainayagam, S., Watanabe, M., & Agathokleous, E. (2023). Effect of ozone stress on crop productivity: A threat to food security. *Environmental Research, 236*, Article 116816.

Ranganathan, K., Maryam, V., & Janusz, J. Z. (2018). Regulation of aquaporins in plants under stress. *Biological Research, 51*, 4. https://doi.org/10.1186/s40659-018-0152-0

Rehman, H. U., Basra, S. M. A., Rady, M. M., Ghoneim, A. M., & Wang, Q. (2017). Moringa leaf extract improves wheat growth and productivity by delaying senescence and source-sink relationship. *International Journal of Agriculture & Biology, 19*(3), 479–484.

Saini, D. K., Chopra, Y., Singh, J., Sandhu, K. S., Kumar, A., & Bazzer, S. (2022). Comprehensive evaluation of mapping complex traits in wheat using genome-wide association studies. *Molecular Breeding, 42*, 1. https://doi.org/10.1007/s11032-022-01253-2

Sánchez, B., Rasmussen, A., & Porter, J. R. (2014). Temperatures and growth responses of maize and rice. *Plant, Cell & Environment, 37*(1), 20–30.

Sehgal, A., Sita, K., Kumar, J., Kumar, S., Singh, S., Siddique, K. H. M., & Nayyar, H. (2020). Effects of drought, heat and their interaction on the growth, yield and photosynthetic function of lentil (Lens culinaris Medikus) genotypes varying in heat tolerance. *Theoretical and Applied Genetics, 133*(4), 1263–1281.

Sehgal, S. K., Kaur, S., Gupta, S., Sharma, A., Kaur, R., & Bains, N. S. (2011). A direct hybridization approach to gene transfer from Aegilops tauschii Coss. to Triticum aestivum L. *Plant Breeding, 130*(1), 98–100.

Shah, F., Huang, J., Cui, K., Nie, L., Shah, T., Chen, C., & Wang, K. (2011). Impact of high-temperature stress on rice plant and its traits related to tolerance. *Field Crops Research, 124*(3), 279–290.

Shi, W., et al. (2018). High temperature effects on rice flowering and grain filling. *Journal of Agronomy and Crop Science, 204*(2), 141–150.

Siebers, M. H., et al. (2015). Soybean reproductive responses to high temperature. *Global Change Biology, 21*(1), 214–226.

Singh, A. K., et al. (2014). Heat stress effects on maize growth and productivity. *Maydica, 59*, 92–99.

Singh, D., Kumar, J., Basandrai, A. K., Singh, R., & Malhotra, N. (2020a). QTL mapping for heat tolerance in lentil (*Lens culinaris*Medik.). *Euphytica, 216*(7), 107. https://doi.org/10.1007/s10681-020-02630-2

Singh, R., et al. (2020b). Mapping QTLs for heat stress tolerance in lentil. *Journal of Plant Breeding and Crop Science, 12*(4), 89–98.

Singha, D. L., Das, D., Paswan, R. R., Chikkaputtaiah, C., & Kumar, S. (2022). Novel pproaches: Advanced molecular techniques for crop improvement. In *Plant-microbe interactions* (pp. 1–27). CRC Press.

Sotirios, F., Sascha, R., Enrico, S., Klaus-Dieter, S., et al. (2015). Prospects of engineering thermotolerance in crops through modulation of heat stress transcription factor and heat shock protein networks. *Plant, Cell and Environment, 38*, 1881–1895. https://doi.org/10.1111/pce.12396

Soumen, N., Bhuvan, P., Shan, Z., & Vibha, S. (2019). Heat-shockinducible CRISPR/Cas9 system generates heritable mutations in rice. *Plant Direct*. https://doi.org/10.1002/pld3.145

Sridharan, J., Palak, C., Arindam, G., et al. (2018). Proteomics of heat-stress and ethylene-mediated thermo tolerance mechanisms in tomato pollen grains. *Frontiers in Plant Science, 9*, 1558. https://doi.org/10.3389/fpls.2018.01558

Talukder, S. K., Babar, M. A., Vijayalakshmi, K., Poland, J., Prasad, P. V. V., Bowden, R., & Fritz, A. (2014). Mapping QTL for the traits associated with heat tolerance in wheat (*Triticumaestivum* L.). *BMC Genetics, 15*(97), 1–11. https://doi.org/10.1186/s12863-014-0097-4

Tandon, A., Awasthi, A., & Pattnayak, K. C. (2025). Extreme precipitation projections in Asian subregions regions: Study on CMIP6 development, geographic patterns, and research gaps. *Physics and Chemistry of the Earth, Parts a/b/c, 104148*. https://doi.org/10.1016/j.pce.2025.104148

Tayade, R., Nguyen, T. D., Oh, S. A., Hwang, Y. S., Yoon, I. S., & Deshmuk, R. (2018). Effective strategies for enhancing tolerance to high-temperature stress in rice during the reproductive and ripening stages. *Plant Breeding and Biotechnology, 6*(1), 1–18. https://doi.org/10.9787/PBB.2018.6.1.1

Tsai, C. J., et al. (2019). Molecular regulation of tomato under high temperature stress. *Plant Physiology and Biochemistry, 139*, 379–390.

Uday, C. J., Abhishek, B., & Narendra, P. S. (2014). Heat stress in crop plants: Its nature, impacts and integrated breeding strategies to improve heat tolerance. *Plant Breeding, 133*, 679–701.

Varshney, R. K., & Dubey, A. (2009). Novel genomic tools and modern genetic and breeding approaches for crop improvement. *Journal of Plant Biochemistry and Biotechnology, 18*, 127–138. https://doi.org/10.1007/BF03263311

Wang, X., Cai, J., Jiang, D., Liu, F., Dai, T., & Cao, W. (2011). Pre-anthesis high-temperature acclimation alleviates damage to the flag leaf caused by post-anthesis heat stress in wheat. *Journal of Plant Physiology, 168*(6), 585–593. https://doi.org/10.1016/j.jplph.2010.09.016

Wang, X., Cai, X., Xu, C., Wang, Q., & Dai, S. (2018a). Drought- and heat-stress–associated gene expression in wheat during grain filling. *Journal of Agronomy and Crop Science, 204*(2), 141–152. https://doi.org/10.1111/jac.12249

Wang, X., et al. (2018b). Genome-wide association mapping of heat tolerance-related traits in Brassica napus. *BMC Plant Biology, 18*, 145. https://doi.org/10.1186/s12870-018-1398-y

Wang, X., Xu, Y., Hu, Z., & Xu, C. (2018c). Genomic selection methods for crop improvement: Current status and prospects. *The Crop Journal, 6*(4), 330–340. https://doi.org/10.1016/j.cj.2018.03.001

Wang, Y., et al. (2020). Kernel set under heat stress in maize. *Plant Physiology, 183*(2), 950–964.

Wei, S., et al. (2020). Soybean seed setting under high temperature. *Frontiers in Plant Science, 11*, 1–11.

Wentao, G., Xi, L., Shaoying, G., Yunfeng, L., Faqiang, L., & Qingjun, X. (2019). Autophagy in plant: A new orchestrator in the regulation of the phytohormones homeostasis. *International Journal of Molecular Sciences, 20*, 2900. https://doi.org/10.3390/ijms20122900

Wiebbecke, A., et al. (2012). Soybean and tomato photosynthesis under heat stress. *Journal of Experimental Botany, 63*(3), 1219–1232.

World Bank. (2013). *Turn down the heat: Climate extremes, regional impacts, and the case for resilience.* World Bank.

World Meteorological Organization (WMO). (2024). *State of the global climate 2023.* WMO-No. 1315. Geneva: World Meteorological Organization. https://library.wmo.int/doc_num.php?explnum_id=11511

Xin, H., Li, L., Sai, X., Min, Y., Pan, X., Wei, L., et al. (2020). Comprehensive analyses of the annexin (ANN) gene family in Brassica rapa, Brassica oleracea and Brassica napus reveals their roles in stress response. *Scientific Reports, 10*, 4295. https://doi.org/10.1038/s41598-020-59953-w

Xu, Y. F., Zhang, L., Ou, S. J., Wang, R. C., Wang, Y. M., Chu, C. C., & Yao, S. G. (2020). Natural variations of *SLG1* confer high-temperature tolerance in *indica* rice. *Nature Communication 11*(1), 5441.

Xu, J., et al. (2017). Inflorescence and flower development in tomato under heat stress. *Plant Reproduction, 30*(2), 95–105.

Xu, Y., et al. (2020). Rice seed development under high temperature stress. *Plant Biotechnology Journal, 18*(4), 1244–1255.

Yang, C., Luo, A. N., Lu, H. P., Davis, S. J., & Liu, J. X. (2024). Diurnal regulation of alternative splicing associated with thermotolerance in rice by two glycine-rich RNA-binding proteins. *Science Bulletin, 69*(1), 59–71.

Yang, M., & Wang, G. (2023). Heat stress to jeopardize crop production in the US corn belt based on downscaled CMIP5 projections. *Agricultural Systems, 211*, Article 103746.

Yong, X. Z., Lei, Y., Ning, L., et al. (2019). Genome-wide identification, structure characterization, and expression pattern profiling of aquaporin gene family in cucumber. *BMC Plant Biology, 19*, 345.

Zafar, S. A., Hameed, A., Nawaz, M. A., Ma, W., Noor, M. A., & Hussain, M. (2018). Mechanisms and molecular approaches for heat tolerance in rice (Oryza sativa L.) under climate change scenario. *Journal of Integrative Agriculture, 17*(3), 726–738. https://doi.org/10.1016/S2095-3119(17)61774-1

Zhang, H., Zhou, J. F., Kan, Y., Shan, J. X., Ye, W. W., Dong, N. Q., Guo, T., Xiang, Y. H., Yang, Y. B., Li, Y. C., Zhao, X. Y., Yu, H. X., Lu, Z. Q., Guo, S. Q., Lei, J. J., Liao, B., Mu, X. R., Cao, Y. J., Yu, J. J., Lin, H. X. (2022). A genetic module at one locus in rice protects chloroplasts to enhance thermotolerance. *Science, 376*, 1293–1300.

Zhang, C., et al. (2018). Rice spikelet fertility under heat stress. *Plant Physiology and Biochemistry, 132*, 213–221.

Zhang, H., et al. (2019a). Mapping of QTLs for membrane stability index under heat stress in rapeseed (*Brassica napus* L.). *Plant Breeding, 138*(5), 570–579. https://doi.org/10.1111/pbr.12692

Zhang, H., Mittal, N., Leamy, L. J., Barazani, O., & Song, B. H. (2019b). Back to the wild: Important role of wild species in wheat breeding. *Molecular Plant Breeding, 12*(5), 619–631. https://doi.org/10.1016/j.molp.2019.03.016

Zhao, C., Liu, B., Piao, S., Wang, X., Lobell, D. B., Huang, Y., & Zhu, Y. (2017). Temperature increase reduces global yields of major crops in four independent estimates. *Proceedings of the National Academy of Sciences of the United States of America, 114*(35), 9326–9331. https://doi.org/10.1073/pnas.1701762114

Zhao, J., Zhang, Y., Li, Y., & Zhao, W. (2019). Effects of straw mulching on soil temperature, water content and yield of tomato under plastic greenhouse conditions. *Scientia Horticulturae, 257*, Article 108732.

Impact of Extreme Temperatures on Plant–Microbe Interactions: Mechanisms, Consequences, and Resilience Strategies

Sapna Baghel and **Kailash Pandey**

Abstract Climatic variability and global warming are also leading to extreme temperatures that are impacting the health and efficiency of plants by altering the plant–microbe interactions. These symbiotic relationships between plants and microbes are important ecological interactions, which have immense ecological and biotechnological consequences. In biotic and abiotic stress, soil microbes enhance resilience in plants, and promote growth and offer biocontrol benefits. Hot spells (over 40) and cold spells (0–10) may suppress the positive relationship between beneficial associations (like nitrogen fixation by rhizobia and uptake of nutrients by arbuscular mycorrhizal fungi). They are also capable of lowering the yield of crops and promoting the growth of some harmful microorganisms. Extreme temperatures influence rhizosphere, phyllosphere, and endosphere, as it changes the composition, diversity, and metabolic activity of the microbial populations. Heat stress and cold stress may have adverse effects on the activity of microorganisms and symbiosis and restrict the availability of resources to plants. Thermal stress physiological alterations in plants including changes in root exudation can inhibit the recruitment and symbiosis of microbes. The current review highlights the underlying molecular mechanisms of plant and microbial response to extreme temperature stress, and, as such, their interactions are a complex and dynamic phenomenon. It also reviews how extreme temperatures impact the environment of plant–microbe symbionts and also examines how these symbionts help increase the tolerance and resilience of plants. An explanation of these symbiotic processes can facilitate the creation of novel solutions to facilitate climate-resistant and eco-friendly agricultural systems.

Keywords Plant stress · Microbiome · Heat stress · Cold stress · Holobiont

---

S. Baghel (✉)
Department of Biotechnology, S.S.E.S. A's Science College, Congress Nagar, Nagpur, Maharashtra, India
e-mail: sapna.baghel153@gmail.com

K. Pandey
Department of Applied Science, School of Advanced Engineering, UPES, Dehradun, Uttarakhand, India

A. Awasthi et al. (eds.), *Rising Temperatures and Plant Ecosystems*,
https://doi.org/10.1007/978-3-032-29477-7_5

1 Introduction

Changing climate is widely considered a major environmental challenge of the twenty-first century. It influences not only climate patterns but also air quality, water availability, habitability, and food security of entire areas. Recently, extreme weather events have been occurring more frequently and intensely, with numerous temperature records being broken worldwide (Awasthi et al., 2025; Ibáñez et al., 2023; Tandon et al., 2025). Thermal stress refers to the adverse effects caused by exposure to extremely high or low temperatures, where sudden fluctuations or prolonged stress impair physiological, biochemical, and molecular processes in plants and associated microbes. However, each plant species and microbial community possesses specific temperature tolerance limits and may activate rapid defence responses or long-term adaptation mechanisms to survive cold or heat stress (Ahmad et al., 2022). Exposure to high temperature (40 °C) significantly affects the plant and plant-associated microbes' growth, while lower temperatures 0–10 °C slow down the growth and development of the plant and even cause death (Khan et al., 2025; Mitra et al., 2021).

Microorganisms have been universal for 3.5 billion years and play multidimensional roles in the ecosystem (Parasar et al., 2024). One gram of soil can contain various bacteria, fungi, and many other microorganisms like protozoa, epiphytes, archaea, viruses, ectomycorrhizae, and nitrogen-fixers, which all play important roles in plant health, particularly under adverse conditions. These microbes have various plant benefits, like defence mechanisms, synthesis of metabolites, regulation of nutrient dynamics, and alleviation of stressors (Dastogeer et al., 2022). Plant–microbe interactions or plant microbiome are also known as a symbiotic relationship or as a "holobiont," which significantly contributes to plant ecology and physiology, and is critical for plant well-being and productivity (Hassani et al., 2018). Among all environmental factors, temperature plays a major role in microbial community, composition, and function. It influences the allocation, metabolism, development, community structure, and genetic evolution among microbes, including genetic variation, species evolution, and biotic interactions.

Climate changes can have a profound impact on plant physiology such as the synthesis and constituent of root exudates. Such temperature variations, be it negative or positive, may alter plant immune response in either a positive or a negative manner, depending on the stress level and stress duration. High temperatures, which are even marginally above the optimal temperatures, have the capacity to repress Effector-Triggered Immunity (ETI) and salicylic acid (SA)-dependent signalling in plants, which is of concern since most crop species use these signalling pathways to defend against major pathogens (Cheng et al., 2019). ETI can be affected by high temperature since it destabilizes resistance (R) proteins, which provide a second line of defence against pathogens. This destabilization impairs the capacity of R proteins to be effective in recognition of pathogen effectors, hence diminishing immune activation and downstream defence signalling (Lee et al., 2012). Simultaneously, high temperature can either suppress SA production or impair its signalling pathway. SA is a key defence hormone and part of Pattern-Triggered Immunity (PTI), ETI, and

Systemic Acquired Resistance (SAR). Nevertheless, the immune response, which depends on SA, is very vulnerable to high temperatures, whereas the jasmonic acid (JA) and ethylene (ET) signalling are usually more active when exposed to high temperatures, meaning that hormonal priorities change when temperature is used as a stressor (Son & Park 2022). The combination of the effector recognition interference by the weakened ETI and of the SA-mediated defence response leads to a dual susceptibility at which effector recognition is impaired, as well as the activation of an effective immunity. In this case, the plants would lose disease resistance despite having functional resistance genes since the immune machinery fails to work effectively in a stressful environment of heat. In addition to plant physiology, heat disturbances like wildfire may also have additional impacts on ecosystem resilience by decreasing microbial activity and changing the microbial community structure, which can have effects on plant–microbe interactions (Koster et al, 2021). Extreme summer events are becoming more and more important restrictions on the productivity of plants, especially with the global temperatures estimated to rise by 1.5–2.0 by the year 2050 (Awasthi et al., 2023; Irfan et al., 2023). These temperature changes do not only impact biodiversity and ecosystems, but they also disrupt soil microbes. Such disruptions of temperature propagate within microbial networks, changing microbial functioning, ecological interaction, and community structure with a final detrimental effect on the ability of soil microbes to sustain plant growth and ecosystem functioning (Hacquard et al., 2022). Microbes in soils are influential controllers of ecosystem operation, which play a role in the cycling of nutrients, carbon fixation, degradation of organic matter, plant growth stimulation, and natural biocontrol (Chen et al., 2024). Therefore, minor yet consistent thermal variations may disrupt the whole plant–soil complex. Plants increasingly become unavailable to important microbial partners in the form of nitrogen fixation, phosphate solubilization, hormone regulation, and pathogen suppression as microbial functioning becomes less efficient or changes in nature. This erosion by microorganisms can over the years change into decreased agricultural productiveness, lower ecological stability, loss of biodiversity, and ultimate instability of the terrestrial ecosystems. Consequently, the knowledge of temperature variations in altering plant–microbiome interactions is important in anticipating the reaction of ecosystems to emerging climatic circumstances. Such understanding will aid us to determine vulnerable systems and shape the process of creating climate-resilient practices that safeguard plant productivity and ecosystem stability at both ends of the heat and cold. It has been reported by research papers that some bacterial families are tolerant to high temperatures or low temperatures, and they are able to increase in number after extreme stress on temperatures. Other bacteria that are either heat-sensitive or cold-sensitive and slow-growing bacteria would not survive under extreme temperatures. The current review is aimed at learning how extreme temperatures affect the plant–microbe associations and the modified symbiotic associations. Thermotolerant microbial inoculants and adaptive agronomic measures to develop plant resistance to extreme thermal conditions should also be discussed.

2 An Overview of Plant–Microbial Association

Depending on the type of symbiotic association between plants and microbes, plant–microbe interactions can be described as mutualism, commensalism, and parasitism. Mutualistic interactions are beneficial to the plant and the microbe as they lead to relationships that lead to the well-being of the plant and the functioning of the ecosystem. Examples are rhizobia, mycorrhizal fungi, and phosphate-solubilizing bacteria, which serve to improve the growth of plants by improving nutrient acquisition, especially phosphorus and nitrogen, transforming them into forms usable by plants, and increasing the ability of roots to penetrate the soil, hold water, and overall nutrient uptake (Iqbal et al., 2023; Oubohssaine et al., 2025). Commensalistic relationships are those where the relationship is mutually advantageous to one partner but to which the other partner is not much affected. An example is the phyllosphere bacteria which can be found on the leaves, stems, and other surface organs of the plant. To protect the plant against invasive microbes, these microbes synthesize antimicrobial substances (Iqbal et al., 2023). Contrarily, parasite relations are the pathogenic bacteria, fungi, or viruses that can colonize in plants and damage the host benefitting themselves, which in many cases leads to necrosis (Newton et al., 2010). These forms of interactions reveal the active and heterogeneous relationship between plants and microbes which can be beneficial or harmful depending on the ecological factors and the evolutionary compatibility.

3 Plant-Associated Microbes: Rhizosphere, Phyllosphere, and Endosphere

Plants form networks with a number of useful microorganisms such as bacteria, fungi, archaea, and actinomycota that promote plant growth, development, and tolerance to stress. These microbes aid in the movement of nutrients, hormone regulation, and improved biotic and abiotic stress defence (Ortiz & Sansinenea 2022). They inhabit different plant compartments according to their habitat like rhizosphere, phyllosphere, and endosphere, and they have different ecological functions.

3.1 Rhizosphere

The rhizosphere refers to a thin layer of soil around the roots of the plants, where interactions among the roots, soil, and communities of microorganisms are dynamically experienced (Kumar & Dubey, 2020). The microorganisms of this zone are important in the cycling of nutrients, enhancement of the soil structure, and generation of phytohormones and bioactive compounds that lead to the growth and defence of plants. Plant Growth-Promoting Rhizobacteria (PGPR) are of special importance because they can increase nutrient uptake, promote root growth, and control soil-borne pathogens due to several different mechanisms, including competitive exclusion, production of antibiotics, and induced systemic resistance. The region also releases carbohydrates, amino acids, vitamins, and secondary metabolites, which improve the plant defences to biotic or abiotic stresses by rhizodeposition (Upadhyay et al., 2022). PGPR improves the growth of plants by fixing nitrogen, solubilizing phosphorus, and producing cytokinins, auxin, and gibberellins hormones (Dong et al., 2019). Taxa like *Rhizobium, Azospirillum, Acinetobacter, Paenibacillus, Klebsiella, Pseudomonas,* and *Bacillus* have been well known to enhance nutrient uptake and stress tolerance (Kumar & Dubey 2020). *Rhizobium* develops root nodules in legumes and transforms nitrogen in the atmosphere to ammonia, which helps in the production of vital biomolecules, nucleotides (Arora et al., 2022; Brito-Lopez et al., 2025). Nitrification is also enhanced by other microbes like *Nitrosomonas*, which transform ammonium into nitrite (NO_2^-), which enhances further availability of nitrogen (Zayed et al., 2023; Fig. 1). *Arbuscular mycorrhizal* also promote the absorption of phosphorus, nitrogen, and micronutrients by extending the hyphae further into the soil layers especially when under stress (Valenzuela-Aragon et al., 2025). Moreover, certain microbes generate growth-regulating substances, including the cytokinin of *Azotobacter chroococcum* (Bonartsev et al., 2016) or indole-3-acetic acid (IAA) of *Bacillus* spp., and they can improve the root architecture (Erturk et al., 2010).

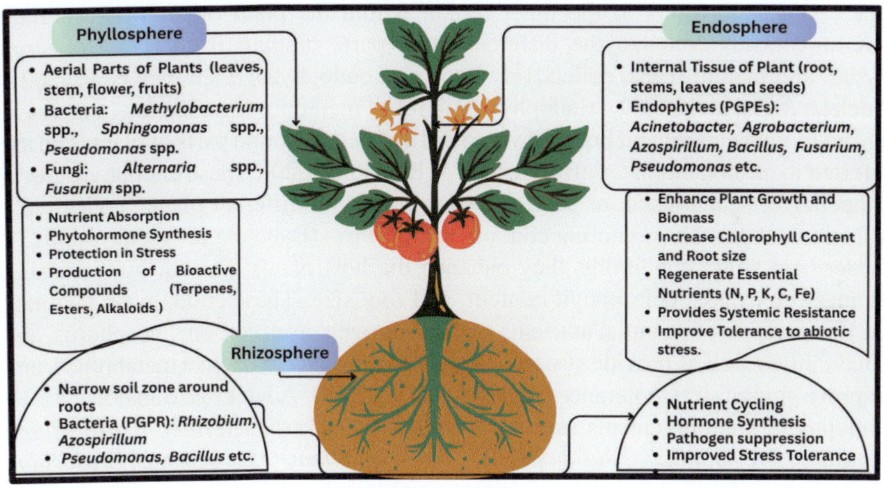

Fig. 1 Overview of plant-associated microbial habitats—Rhizosphere, Phyllosphere, and Endosphere—and their key roles in nutrient cycling, growth promotion, and stress tolerance

3.2 Phyllosphere

The phyllosphere comprises the aerial parts of a plant, including leaves, stems, flowers (such as pollen and nectar), and fruits (Bashir et al., 2022). These microbes include epiphytic bacteria, archaea, fungi, algae, protozoa, oomycetes, and nematodes that live both in the upper region (phylloplane) and the internal tissues of those parts (endosphere) (Krishna et al., 2024). It is one of the major terrestrial habitats next to soil (rhizosphere), approximately 60% but less nutrient-rich than rhizosphere, and faces much environmental pressure (Sohrabi et al., 2023). Phyllosphere microbes promote plant development by nutrient absorption, phytohormone synthesis, and supporting plants in acclimatizing to abiotic stress. These microbes release terpenes, esters, alkaloids, and various other bioactive compounds that modulate the composition and diversity of plant metabolites (Shi et al., 2024b). Some beneficial phyllosphere microorganisms are *Methylobacterium* spp., *Sphingomonas* spp., *Pseudomonas* spp., etc., whereas fungi are *Alternaria* spp., *Fusarium* spp., etc. (Huang et al., 2023, Fig. 1).

3.3 Endosphere

The endosphere refers to the inner region within the plant organ where certain microorganisms colonize the different plant parts ranging from roots to aerial tissues and mature seeds, collectively known as endophytes (Compant et al., 2021; Adeleke & Babalola 2022). Endophytic microbes including bacteria, fungi, archaea, and actinomycetes inhabit both aboveground and belowground parts of plants and are referred to as bionanomaterials (Adeleke & Babalola 2020). These endophytes form cooperative, antagonistic, or neutral associations with different plants, collectively termed plant growth-promoting endophytes (PGPEs) (Babalola & Adedayo, 2023). Under heat stress conditions, they enhance the host plants' growth by increasing biomass, dry mass, chlorophyll content, and root size. These endophytes facilitate the regeneration of essential nutrients such as nitrogen, iron, carbon, phosphorus, and potassium as well as provide systemic resistance, produce bioactive metabolites, and improve abiotic stress tolerance (Adegbeye et al., 2020; Adeleke & Babalola, 2021). Endophytic microorganisms such as *Acinetobacter, Agrobacterium, Azospirillum, Bacillus, Brevibacillus, Burkholderia, Herbaspirillum, Micrococcus, Fusarium, Xylaria,* and *Pseudomonas* are commonly employed as PGPEs. *Pseudomonas* and *Bacillus* species promote plant growth by dissolving insoluble phosphate through the production of gluconic acid, phosphatases, and phytases (Babalola & Adedayo, 2023; Fig. 1).

4 Extreme Temperature Stress and Its Effects on Plants

Extreme temperature is highly sensitive to plant development and efficiency. Both high and low heat stress affect growth, development, and metabolism by damaging biochemical, physiological, and molecular processes in the plant (Sidhu et al., 2024).

4.1 *Heat Stress: Effects on Plants*

The effects of heat stress are sunburn and premature ageing of leaves, twigs, and blotching of fruits and leaves (Mondal et al., 2023). It also increases transpiration and stomatal conductance as well as reduces photosynthesis and compromising electron transport. According to one of the researchers, high temperature changes the effector-triggered immunity (ETI) pathway and predisposes the pathogen *Pseudomonas syringae* pv. *tomato* to the pathogen (Huot et al., 2017; Table 1). Heat stress causes the activation of oxidative stress, which in turn causes the breakdown of membrane lipids, essential leakage of vital electrolytes, and denaturation of essential biomolecules (Pramanik et al., 2023). Stress due to high temperature influences the metabolic activity and absorption of nutrients that cause decreased nutrient concentration in the shoot system (Mishra et al., 2023). Proteins are destabilized by heat stress which affects the activity of vital enzymes, eventually disrupting the metabolic process. High temperature may cause changes in the conformation of the enzyme, which may decrease its catalytic activity, particularly in enzymes of the defence response, cellular receptor activation, and secondary metabolite production (Hasanuzzaman et al., 2013). Besides, concentration of amino acids is upsurged and that of starch, sugar, and lipids is reduced during heat stress. The heat stress results in the change in the protein structure that impairs protein production and the production of the gene encoding heat shock proteins (HSPs) at the molecular level (Dos Santos et al., 2022). This condition worsens the generation of reactive oxygen species (ROS) leading to oxidative injury, membrane depolarization, and activation of a programmed cell death pathway. Moreover, it also decreases the life cycle, decreases chlorophyll biosynthesis and phytohormones and antioxidants production, changes the chloroplast structure and organization, and disrupts hormonal homeostasis (Sidhu et al., 2024; Table 1). Stomatal closure and CO_2 availability inhibition coupled with inhibition of enzymes of carbon fixation are heat stress effects that suppress plant growth. Heat stress also suppresses the ATP production, slow leaf growth, and accelerates ageing (Dos Santos et al., 2022). When plants are exposed to heat stress, they change the expression of the genes to enable them to survive at the molecular level. Such operative genes control the synthesis of osmolytes, transporters, detoxifying enzymes, and different regulatory proteins (Hasanuzzaman et al., 2013). Salicylic acid (SA) is important in guarding against disease in plants. It reduces biosynthesis and signalling of plants under heat stress, thereby exposing them to disease (Choi & Lee, 2024). Conversely, ethylene signalling and histone modifications contribute to plants to develop heat tolerance through the establishment of so-called stress memory, thereby being more responsive to subsequent stress (Tiwari et al., 2024; Table 1).

4.2 Cold Stress: Effects on Plants

Cold stress limits plant performance and production through physiological, anatomical, morphological, and metabolic changes. These include osmotic damage, desiccation, stomatal dysfunction, pigment alterations, reduced fluorescence, and chloroplast impairment. Extremely low temperatures also alter absorption of nutrients, structure of roots, photosynthesis, water transport, cell multiplication, and signalling via plan hormone in plants. However, short exposure to cold stress helps plants develop early resistance against infections. For example, in *Arabidopsis,* it activates defence-related genes that enable the plant's immune system to respond more rapidly to bacterial attacks (Wu et al., 2019; Table 1).

It also causes yellowing and curling of leaves, dwarfed seedlings, localized tissue lesions, rupturing stems, low germination, weak growth, leakage of metabolites, and tissue injury (Adhikari et al., 2022). Cold stress causes intracellular ice formation and metabolic disruptions, reduces chlorophyll biosynthesis, and impairs photosynthetic efficiency (Adam & Murthy, 2013). It causes nutritional disorders, ROS accumulation, and hormonal imbalance. Moreover, it altered membrane fluidity by damaging the cell membrane, decreasing enzyme activity, lowering water potential, disrupting lipid composition, reducing ATP supply, and resulting in the buildup of toxic compounds (Li et al., 2024). Beneficial microbes promote the biosynthesis of cytokinin, auxin, and gibberellins as well as stress hormones to enhance plant tolerance (Qian et al., 2024; Table 1, Fig. 2)

5 Effect of Extreme Temperatures on Soil Microbial Communities

5.1 Heat Stress: Effects on Soil Microbial Communities

Vastly high temperatures influence the nutrient cycles of the microbial community of soil by dissolving soluble metal ions of decomposed organic matter, causing lysis of microbial cells and soil aggregation. It often restrains the nutrition supply, which leads to lowering the microbial biomass and abundance of microorganisms, which ultimately destabilize the structures of communities (Fadiji et al., 2023). The initial effect of high temperature is the increase in enzyme action and acceleration of the decomposition of organic matter, the enhancement of the metabolic activity of microorganisms, and the production of enzymes to a certain extent, after which the enzyme activity, including urease and phosphatase, decreases (Mutambara &

Table 1 Impacts of extreme temperature stress on the structure and function of plant–microbe associations

Sr. no.	Parameter	Under heat stress	Under cold stress	References
	Plant immunity & defence response	Heat stress affects effector-triggered immunity (ETI) pathway, making plants more prone to infection by *Pseudomonas syringae* pv. *tomato*	Short cold stress help plants to build early resistance against infections. In *Arabidopsis*, it activates defence genes which make plant's immune system to respond faster to bacterial attack	Huot et al. (2017), Wu et al. (2019)
	Rhizosphere structure	High temperature disrupts microbial population by altering their abundance, diversity, and functionality, leading to decline in temperature-sensitive microbes and an increased dominance of thermotolerant microbes such as *Bacillus, Pseudomonas, Nitrospira,* and fungal species	Cold stress decreases cell division and growth of microbes, enriches cold tolerant microbes like *Piriformospora, Agrobacterium, Bacillus, Pseudomonas, Arthrobacter, Trichoderma,* and fungal species	Fadiji et al. (2023), Shi et al. (2024a), Zhao et al. (2024), Schnecker et al. (2023)
	Symbiotic efficiency	Inhibited symbiotic nitrogen fixation by reducing nitrogenase activity, supressing root hair formation, delaying nodulation, decreasing the number of nodule numbers, and accelerating nodule death	Decreased cell motility, delayed nodulation, impaired nitrogen fixation, and symbiotic association with plants	Zhang et al. (2024), Janczarek et al. (2024), Abd-Alla et al. (2023)
	Microbial diversity and community composition	Heat stress reduced microbial variety, activity, and AMF diversity, shifted the community structure towards heat-tolerant taxa	Cold stress may reshape the microbial community around plant roots by altering the populations of specific bacterial groups	Yu et al. (2024), Xu et al. (2022), Qiu et al. (2021); Li et al. (2024), Schnecker et al. (2023)

(continued)

Table 1 (continued)

Sr. no.	Parameter	Under heat stress	Under cold stress	References
	Plant–microbe signalling	Alters root, exudates influence microbial attraction, and communication. Enhances heat tolerance through ethylene signalling and histone changes, creating stress memory	Under cold stress, plants and microbes use a two-compartment system to sense temperature changes and activate stress-responsive genes and cold shock proteins to tolerate cold stress	Tiziani et al. (2022), Tiwari et al. (2024), Fadiji et al. (2023)
	Nutrient uptake & enzyme activity	Elevated temperature stress, reduces metabolic activity, nutrient uptake, leading to lower nutrient concentration in the shoot system. While optimum temperature promotes enzyme activity, excessive heat eventually inhibits it	Cold stress reduces nutrient absorption and enzyme activity while cold-tolerant microbes enhance uptake by solubilizing P, K and fixing N	Mishra et al. (2023), Schnecker et al. (2023), Wang et al. (2023), Li et al. (2024), Mutambara and Nkongolo (2025), Expósito et al. (2022)
	Phytohormone regulation	Heat stress increases phytohormone such as ABA, SA levels, while plant growth-promoting bacteria enhance IAA, GA, and cytokinin production and reduce ABA and SA	Cold stress induces ABA, JA, SA, and Brs to enhance plant tolerance, whereas beneficial microbes promote the biosynthesis of cytokinin, auxin, and gibberellin to mitigate stress effects	Khan et al. (2020) a & b, Meza et al. (2025), Yadav et al. (2015), Qian et al. (2024)
	Oxidative status and antioxidant response	Heat stress causes the accumulation of ROS, leading to oxidative stress, membrane depolarization, and activation of antioxidant enzymes such as SOD, APX, and CAT	Cold stress leads to the accumulation of ROS, causing damage to membrane, protein, and nucleic acids while promoting antioxidant enzyme activity	Parasar et al. (2024), Hasanuzzaman et al. (2013), Ali et al. (2011), Ullah et al. (2025), Abdel Latef and Chaoxing (2011)

(continued)

Table 1 (continued)

Sr. no.	Parameter	Under heat stress	Under cold stress	References
	Photosynthetic efficiency	High temperature disrupts electron transport chain in photosystem II, reduces chlorophyll, PS II stability, and CO_2 fixation, beneficial microbes enhance photosynthetic pigments and protect PS II	Low temperature disturbs pigment composition, reduces chlorophyll fluorescence and chloroplast development, lowering photosynthetic efficiency	Mathur and Jajoo (2020), Issa et al. (2018), Abdel Latef and Chaoxing (2011), Li et al. (2024) Caradonia et al. (2019)
	Osmoregulation & metabolite accumulation	Heat stress triggers the accumulation of osmolytes (proline, sugars, glycine, betaine, and sugar alcohols including inositol, mannitol, and trehalose) that help retain water, maintain cell turgidity, enhance antioxidant activity, and maintain cellular balance	Cold stress builds up osmolytes that help cells to preserve water balance and protect cellular structures without interfering with normal metabolism	Jha and Mohamed (2023), Dastogeer et al. (2022), Hasanuzzaman et al. (2013)

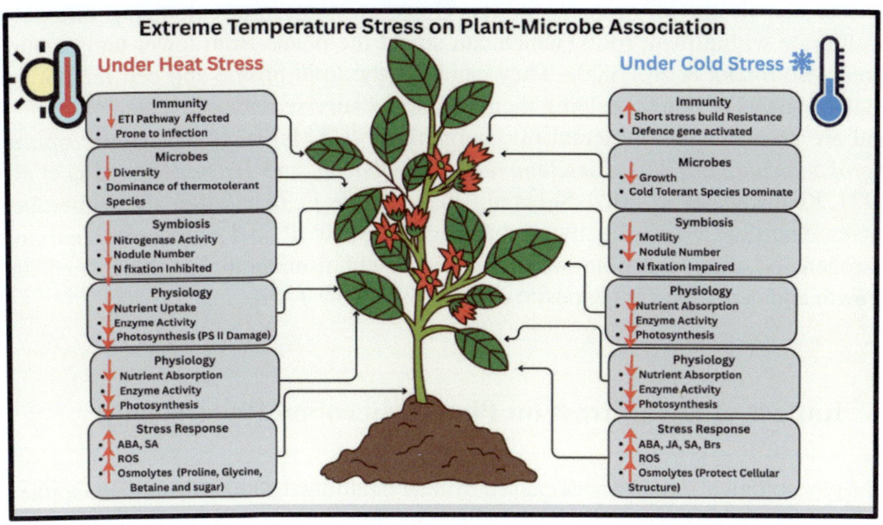

Fig. 2 Impact of extreme temperature stress on plant–microbe association: Heat and Cold stress differentially affect plant immunity, microbial diversity, symbiosis, and physiological functions

Nkongolo, 2025). The heat stress has great effects on the structure of the soil microbial community as it causes changes in the soil chemistry and availability of resources. High temperature speeds up the drying of soil, enhances the oxygen diffusion, and initiates reactions leading to the co-precipitation of phosphorus with the organic matter and the transformation of soluble ferrous iron to insoluble ferric complexes. When the nutrients are immobilized, the soil becomes a sink of nutrients, which limit the availability of vital resources needed in metabolism by the microorganisms (Parasar et al., 2024). These limitations cause changes in the microbial composition, as the population of useful microbes usually decreases and the opportunities or heat-adaptive taxa acquire a competitive edge. There are certain groups of microorganisms, which cannot grow or stay active until the temperature reaches a certain level, and such are referred to as thermotolerant microorganisms (Zhao et al., 2024; Table 1, Fig.2).

5.2 *Cold Stress: Effects on Soil Microbial Communities*

Extreme cold stress significantly affects soil microbes, reducing enzyme activity, growth, and cell division of microbes (Schnecker et al., 2023). Cold stress forms ice crystals in the cells, which cause dehydration and leakage of electrolytes that further lead to cell death (Mitra et al., 2021). However, some root-colonizing microbes proliferate within plant roots, which can shield the plants from lower temperature stress (Beirinckx et al., 2020). They can alter the lipid profile and cell membrane characteristics of plants, helping them adapt and survive under adverse conditions, and are known as cold-tolerant microorganisms (CTMs) such as *Piriformospora, Agrobacterium, Bacillus, Pseudomonas, Arthrobacter,* and *Trichoderma* (Jha et al., 2021; Kushwaha et al., 2020; Shi et al., 2024a; Table 1). In addition, these microbes increase nutrient availability by solubilizing phosphate (P) and potassium (K), fixing nitrogen (N), and producing hormones and essential minerals beneficial for plant growth and development (Expósito et al., 2022; Table 1, Fig. 2).

6 Impact of Heat Stress on Plant–Microbial Relationship

The physiological disturbances caused by heat mentioned above are not only applied to the plant but also have a direct effect on plant–microbe interactions stability. When internal plant environment is stressed, the ratio of useful microbes against harmful one changes, which results in a lack of symbiosis and inability to resist harmful factors. Low photosynthetic rates and ATP deficiency inhibit the movement of photosynthates and sugars in the vegetation. This causes a decrease in the number of root exudates that can be released to the rhizosphere and the release of beneficial microorganisms, including mycorrhizae, rhizobia, and plant growth-promoting bacteria whose nutrients are required (Mishra et al., 2023). Disruption of cellular membranes, disruption of receptor signalling, and hormonal imbalance also undermine the process of cellular membrane signalling, which complicates the process of root colonization and sustaining symbiotic relationships further. Meanwhile, repressed defence mechanisms make the host more vulnerable, thus permitting the growth of pathogenic microorganisms. Thus, the physical and metabolic stress of plants under heat stress was transferred to evident changes in the abundance, composition, and functional activity of the surrounding soil and tissue of plants (Tokas et al., 2023; Table 1, Fig. 2).

7 Impact of Cold Stress on Plant–Microbial Relationship

Cold stress interferes with the physiological and metabolic functions of plants and has a cascading impact on the microbial communities associated with the plants. A decrease in photosynthesis, slower metabolism, a slowing down of nutrient transport, and water control make compounds of carbon, sugars, and root exudates that support beneficial rhizosphere microbes less accessible (Ullah et al., 2025). Rigidity of membranes, hormonal disequilibrium, and the buildup of reactive oxygen species also undermine the plant–microbe communication, so the symbiotic microbes cannot easily form or sustain effective relationships. Consequently, although most microbes are impacted negatively, cold-tolerant microorganisms (CTMs) can be used to colonize roots and benefit plants by increasing nutrient uptake, regulating hormonal equilibrium, and stabilizing membrane and cellular activities, thus alleviating the negative effect of low temperatures (Shi et al., 2024a; Table 1, Fig. 2).

8 Mechanisms of Plant and Microbial Response to Thermal Stress

8.1 *Plant Responses*

Under heat stress, plant responds via different coping mechanisms to mitigate their effects. A plant went through physical adaptations in its body by generating signals that alter its metabolism. To cope with high-temperature stress, plants adjust their metabolism in several ways. They produce compatible solutes that stabilize proteins and cellular structures, regulate osmotic balance to keep cells turgid, and boost their antioxidant activity to prevent damage and maintain cellular balance (Hasanuzzaman et al., 2013). Moreover, plants can cool their leaves through transpiration by opening their stomata, leaf orientation changes in membrane lipid composition, and the xylem system for survival (Ding & Yang 2022). To protect from heat stress, the plant altered its metabolism of carbon assimilation. They utilize their chloroplastic starch reserves, breaking them down into energy, sugars, and metabolites, and preventing further damage (Dos Santos et al., 2022; Table 2, Fig. 3).

Plants cope with heat stress through alterations in gene expression at the molecular level. These include genes responsible for the expression of osmotic stabilizers (proline, sugars, glycine, betaine, and sugar alcohols, including inositol and mannitol, and trehalose). To prevent oxidative damage, plants produce molecular chaperons (heat shock proteins) and activate antioxidative enzymes (Buttar et al., 2020). These enzymes and proteins protect cells from heat damage by keeping damaged proteins stable and assisting in their proper refolding (Zhao et al., 2020). Moreover, plants produce enzymatic antioxidants such as superoxide dismutase (SOD), ascorbate peroxidase (APX), monodehydroascorbate reductase (MDHAR), dehydroascorbate reductase (DHAR), glutathione reductase (GR), catalase (CAT), glutathione peroxidase (GPX), and glutathione S-transferase (GST) that neutralize ROS and aid in reducing the negative effects of extreme temperature. Tocopherol, flavanones, carotenoids, and anthocyanins also help to mitigate oxidative stress (Eggersdorfer & Wyss, 2018; Khanzada et al., 2024) (Fig. 3). Numerous phytohormones like abscisic acids (ABA), gibberellic acids (GA), jasmonic acids (JA), and salicylic acid (SA), and brassinosteroids (BRs) help the plant to withstand extreme temperatures (Wang et al., 2025). Plants produce some signalling molecules, including nitric oxide, polyamines such as putrescine, spermidine, and spermine, and trace elements, e.g., selenium, silicon, and nutrients like nitrogen, phosphorus, potassium, and calcium, which play an important role in alleviating high-temperature stress-induced damage in plants (Hasanuzzaman et al., 2013; Table 2, Fig. 3).

8.2 Microbial Responses

The high-temperature stressed microorganisms use various structural, biochemical, and molecular methods to withstand thermal stresses. They avoid stressors of extreme temperatures using molecular chaperones, specialized lipids, transporters, and stress-adapted protein sequences (Seeger et al., 2022). They express certain polypeptides known as heat shock proteins (HSPs), which assist in stabilizing existing proteins, refold denatured proteins, prevent aggregation, and create biofilm in physical shedding to act as protective clothing (Yin et al., 2019). There are bacteria cells which give out exopolysaccharides (EPSs) which are used to retain water, protect those cells against oxidative injury, and cementing properties (Fig. 3). Some thermophiles (heat-loving microorganisms) alter their cell membranes by replacing the ester-linked lipids with ether-linked ones. Ether linkages create stronger bonds that are more chemically stable as compared to the ester bonds, which are prone to hydrolysis and oxidation during high temperatures. This enhances the strength of the bond, and eliminates membrane degradation and breakage in uncongenial thermal conditions. They tend to be monolayered instead of bilayered, increasing the rigidity and thermostability of the membrane and shielding them against oxidative damage in particular archaea (Rekadwad et al., 2023). Also, certain thermosensors, such as

Table 2 Overview of 41 investigations highlighting the tolerance mechanism of plant-associated microbes in plants under extreme temperature stress conditions

S. no.	Microbes	Temperature stress	Plant species	Tolerance mechanism	References
	Actinobacteria (Streptomycetaceae, Micrococcaceae, and*Mycobacteriaceae) Acidobacteria (Solibacteriaceae taxa,)*	Heat stress	Beet root *(Beta vulgaris)*	Various bioactive compounds including antibacterial and antifungal compounds are produced to prevent illness	Van der Voort et al. (2016)
	Arbuscular mycorrhiza	Heat stress	Maize (*Zea mays)*	Improve soil quality, prevent photosynthetic apparatus damage, enhance photosynthesis, and crop growth	Mathur et al. (2021)
	Arbuscular mycorrhiza	Heat stress	Maize (*Zea mays*)	Increased antioxidant enzyme activity and osmolyte levels, reduced membrane lipid peroxidation and membrane permeability	Zhu et al. (2010)
	Arbuscular mycorrhiza	Heat stress	Wheat (*Triticum aestivum* L.)	Enhance tolerance via carbon sink strength and grain-filling, alter nutrient allocation, and lower K/Ca ratio	Cabral et al. (2016)
	Arbuscular mycorrhiza	Heat stress	Maize (*Zea mays*)	Enhance root system for better water uptake, boost photosynthetic efficiency, and maintain the stability of photosynthetic unit under stress	Mathur and Jajoo (2020)

(continued)

Table 2 (continued)

S. no.	Microbes	Temperature stress	Plant species	Tolerance mechanism	References
	Arbuscular mycorrhiza (*Glomus fasciculatum*)	Heat stress	Cyclamen (*Cyclamen persicum* Mill.)	Antioxidative enzyme activity, ascorbic acid and polyphenol contents increased, promoting radical scavenging activity	Maya and Matsubara (2013)
	Arbuscular mycorrhiza (*Glomus fasciculatum*)	Cold stress	Tomato (*Lycopersicon esculentum* cv Zhongzha105)	Reduce membrane lipid peroxidation while increase chlorophyll content and build up osmolyte, and antioxidant enzyme activity	Abdel Latef and Chaoxing (2011)
	Arthrobacter nicotianae, Brevunimonas terrae, Paenibacillus tylopili, and Pseudomonas cedrina	Cold stress	Wheat (*Triticum aestivum* L.)	Act as a matrix for the biofilm formation, produce ACC deaminase to tolerate stress, synthesize IAA and siderophore, and solubilize phosphorus	Yadav et al. (2015)
	Aspergillus aculeatus	Heat stress	Perennial ryegrass (*Lolium perenne*)	Regulate hormone homeostasis, prevent damage of photosynthetic system, reduces oxidative damage, and modulate gene expression levels	Li et al. (2021)

(continued)

Table 2 (continued)

S. no.	Microbes	Temperature stress	Plant species	Tolerance mechanism	References
	Aspergillus japonicus	Heat stress	Sunflower & Soybean (*Helianthus annuus and Glycine max*)	Produce IAA responsible for enhancing chlorophyll amount, length of root and shoot along with higher dry weight. Regulates stress-related hormones (ABA) and antioxidant enzymes	Ismail et al. (2018)
	Azospirillum brasilense	Cold stress	Brachypodium (*Brachypodium distachyon*)	Improve the root and shoot development, promote stress endurance	Schillaci et al. (2021)
	Bacillus amyloliquefaciens	Heat stress	Wheat (*Triticum aestivum* L.)	Reduce generation of ROS thereby less cell damage, activate heat shock protein to tolerate stress	Abd El- Daim et al. (2014)
	Bacillus amyloliquefaciens and *Brevibacillus laterosporus*	Cold stress	Rice (*Oryza sativa* L.)	Promote plant health by enhancing proline, chlorophyll amount, and antioxidant enzymes while reducing membrane electrolyte efflux and monodehydroascorbate (MDHA) content	Kakar et al. (2016)
	Bacillus cereus	Heat stress	Soybean (*Glycine max*)	Improves, photosynthetic pigments, antioxidant levels, enhances the activation of molecular chaperon and stress-responsive genes	Chouhan et al. (2023)

(continued)

Table 2 (continued)

S. no.	Microbes	Temperature stress	Plant species	Tolerance mechanism	References
	Bacillus cereus	Heat stress	Soybean (*Glycine max*)	Enhance chlorophyll content and biomass due to the production of phytohormone such as SA, GAs, IAA, and organic acids and reduce plant stress hormone ABA	Khan et al. (2020b)
	Bacillus cereus	Heat stress	Tomato *Solanum lycopersicum*	Breakdown of ACC enzyme to tolerate stress, synthesize IAA, improve the morpho-physiological and biochemical parameters	Mukhtar et al. (2020)
	Bacillus cereus	Heat stress	Mugwort (*Artemisia vulgaris*)	Produce phytohormones which enhance the growth, differentiation, cell expansion, and stress resistance	Shaffique et al. (2023)
	Bacillus cereus + humic acid	Heat stress	Tomato (*Lycopersicon esculentum)*	Improve IAA, SA, and organic acid production, increase antioxidant enzymes, and enhance iron, phosphorus, and potassium intake	Khan et al. (2020a)
	Bacillus cereus, Myroides odoratimimus, and *Providencia rettgeri Bacillus cereus, Myroides odoratimimus,* and *Providencia rettgeri*	Heat stress	Sorghum*(Sorghum bicolor)*	Increase antioxidant activity and lower proline and MDA levels and promote the phytoremediation of chromium-contaminated soil	Bruno et al. (2020)

(continued)

Table 2 (continued)

S. no.	Microbes	Temperature stress	Plant species	Tolerance mechanism	References
	Bacillus proteolyticus and *Bacillus safensis*	Heat stress	Common bean (*Phaseolus vulgaris* L.)	Promotes root development and alleviates oxidative stress, synthesizes auxin (IAA), increases radicle and plant length, enhances antioxidant responses, and stabilizes MDA levels under heat thus reducing lipid peroxidation	Meza et al. (2025)
	Bacillus safensis	Heat stress	Wheat (*Triticum aestivum* L.)	Enhance antioxidant, and oxidoreductase activity, accumulate osmolytes. Preserve cell health and chloroplast morphology minimizing chloroplast damage	Sarkar et al. (2018)
	Bacillus velezensis	Heat stress	Wheat (*Triticum aestivum* L.)	Produce stress defence and energy-related proteins, modulate amino acids pathway. Accumulate GABA in leaves to balance carbon–nitrogen content. Modulate amino acids and carbohydrates metabolism	Abd El-Daim et al. (2019)
	Bacillus, Bacterovorax, Caulobacter, Chitinophaga, Dyadobacter, Halomonas, Mesorhizobium, Micromonospora, Niastella, Pseudonocardia, Rhizobacter	Heat stress	Maize (*Zea mays*)	Stimulate the production of phenolic compounds, benzoic acid, and malate thus enhancing antioxidant capacity	Tiziani et al. (2022)

(continued)

Table 2 (continued)

S. no.	Microbes	Temperature stress	Plant species	Tolerance mechanism	References
	Burkholderia phytofirmans	Cold stress	Grapevines (*Vitis vinifera* L.)	Cold tolerance is promoted through carbon fixation and buildup of osmoprotective compounds including carbohydrates starch, proline, and phenolic metabolites	Theocharis et al. (2012)
	Burkholderia phytofirmans PsJN	Cold stress	Thale Cress (*Arabidopsis thaliana*)	Strengthening cell wall to prevent freezing damages and collapse. Enhance pigment synthesis and activate cold stress signalling pathway	Su et al. (2015)
	Burkholderia sp. and *Saitozyma podzolica*	Heat stress	Chinese ginseng (*Panax notoginseng*)	Stimulates nucleotide metabolism, increases the release of purine and pyrimidine metabolites in roots thereby enhancing the plant heat tolerance	Liu et al. (2025)
	Enterobacter sp.	Heat stress	Wheat (*Triticum aestivum* L.)	Activates heat stress memory genes and heat shock factor via ethylene signalling and transcriptional control	Shekhawat et al. (2021)
	Funneliformis mosseae and Paraburkholderia graminis	Cold stress	Tomato (*Solanum lycopersicum* L.)	Reduce cell membrane injuries by lowering electrolyte leakage and boost photosystem II activity	Caradonia et al. (2019)

(continued)

Table 2 (continued)

S. no.	Microbes	Temperature stress	Plant species	Tolerance mechanism	References
	Lysinibacillus and *Lysinibacillus sphaericus*	Cold stress	Maize (*Zea mays* L.)	Increase phytohormones, antioxidant enzymes, total phenolic content and osmolytes to enhance plant resistance	Jha et al. (2023)
	Methylotrophic bacteria (Methylobacterium brachiatum, M. thiocyanatum, and Methylorubrum populi)	Heat stress	Black Gram (*Vigna mungo L.*)	Enhance chlorophyll content, carotenoid and antioxidant enzymes. Reduce oxidative stress and superoxide radicals	Gamit et al. (2024)
	Nitrospira, Novosphingobium, Penicillium, Phenylobaterium, and *Acidothermus*	Heat stress	Rhododendron (*Rhododendron hainanense*)	Improve soil fertility through higher level of soil nitrogen, carbon from organic matter and microbial carbon content	Lin, et al. (2023)
	Paecilomyces formosus	Heat stress	Rice (*Oryza sativa* L.)*subsp. Japonica*	Improve plant development biomass by increasing shoot length, fresh and dry biomass, total protein and leaf pigment level. Reduce the amount of plant stress regulators such as JA and ABA	Waqas et al. (2015)
	Paraburkholderia phytofirmans PsJN	Heat stress	Tomato (*Solanum lycopersicum* L.)	The contents of proline, malate sugars, and total amino acids increase, along with the photosynthetic efficiency, chlorophyll content, and other biochemical components	Issa et al. (2018)

(continued)

Table 2 (continued)

S. no.	Microbes	Temperature stress	Plant species	Tolerance mechanism	References
	Pezicula and Paracoccus Acidothermus and Haliangium	Heat stress	Rhododendron (*Rhododendron simsii*)	Regulate soil nitrogen, soluble sugar, and MDA contents to increase the synthesis of proteins, chlorophyll, amino acids, and other essential metabolites	Lin et al. (2022)
	Piriformospora indica, Agrobacterium rhizogenes, and Bacillus subtilis	Cold stress	Rice (*Oryza sativa* L.)	Increase biomass, photosynthetic efficiency, osmolyte accumulation, and antioxidant enzymes	Shi et al. (2024a)
	Pseudomonas brassicacearum	Heat stress	Wheat (*Triticum aestivum* L.)	Produce heat-tolerant proteins, bacteriocin, antifungal, and antibacterial activity. Increase the concentration of enzymatic antioxidants, proline, and protein	Ashraf et al. (2019)
	Pseudomonas putida	Heat stress	Wheat (*Triticum spp.*)	Enhance phytohormone, siderophore production, phosphate solubilization, and cellular metabolites (proline, sugars, starch, amino acids, protein) which reduces membrane injury and increase antioxidant enzymes activity	Ali et al. (2011)
	Pseudomonas syringae Lz4W	Cold stress	Tomato (*Solanum lycopersicum* L.)	Protects membrane, boosts antioxidants and proline supports plant growth	Subramanian et al. (2016)

(continued)

Table 2 (continued)

S. no.	Microbes	Temperature stress	Plant species	Tolerance mechanism	References
	Pseudomonas vancouverensis, *Pseudomonas frederiksbergensis*	Cold stress	Tomato (*Solanum lycopersicum* L.)	Reduce generation of ROS and membrane damage, increase the activity of antioxidants and cold tolerance genes	Subramanian et al. (2015)
	Tetrastigma hemsleyanum Diels & Gilg	Cold stress	Sanyeqing (*Tetrastigma hemsleyanum Diels & Gilg*)	Cold stress-enhanced community richness and stability enriching cold-tolerant bacteria such as *Chryseolinea*, *Lacibacterium*, *Gemmatimonas*, *Nitrospira*, *Proteobacteria*, *Acidobacteria*, and *Bacteroidetes* as well as enhance metabolite production	Li et al. (2024)
	Trichoderma harzianum	Cold stress	Tomato (*Solanum lycopersicum* L)	Promoted photosynthesis which increased the growth rate and proline concentration while electrolyte leakage, fatty acid peroxidation, and ROS levels were reduced	Ghorbanpour et al. (2018)

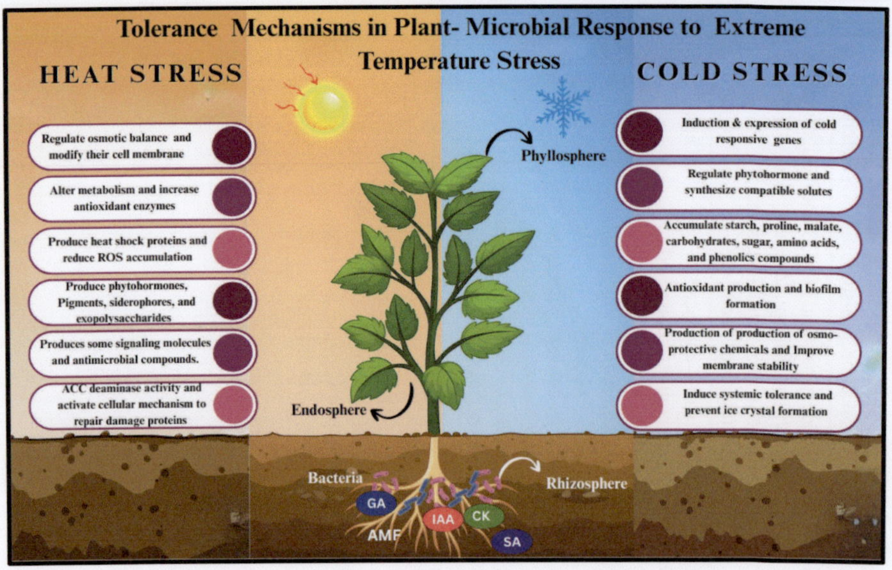

Fig. 3 Synergistic interaction of plant–microbes to mitigate extreme temperature stress (heat and cold)

lipids, proteins, RNA, and DNA, sense temperature variations and cause the activation of cellular processes to fix damaged proteins (Samtani et al., 2022; Shapiro & Cowen, 2012). Other useful microorganisms, such as Bacillus subtilis and Trichoderma spp., are capable of producing compatible solutes such as trehalose and glycerol. By shielding against the damage of cell membranes of plant cells by extreme temperatures (either freezing or excessive heat), these compounds assist plants to survive the stress (Ullah et al., 2025; Table 2, Fig. 3).

9 Consequences of Extreme Temperatures on Plant–Microbe Interactions

Among all environmental factors, temperature influences not only plant physiological processes but also the composition and behaviour of the microbes associated with them. This can disrupt the plant's natural immunity, leaving the plant more prone to disease (Choi & Lee, 2024).

9.1 *Reduced Symbiotic Efficiency*

Both extremely high and low temperatures negatively affect all growth parameters such as shoot and root development, nodule biomass and formation, and nitrogen level in shoots (Bansal et al., 2014; Table 1, Fig. 2). Low temperature decreases cell motility, delays nodule formation, inhibits metabolic activity, impairs their nitrogen-fixing capacity, and reduces symbiotic association with plants (Janczarek et al., 2024). Elevated temperature disrupts root-hair formation, delays nodulation, and lowers nitrogenase activity, which in turn reduces nodule numbers and accelerates nodule senescence (Zhang et al., 2024; Table 1). Thus, extreme temperature limits overall plant growth by inhibiting photosynthesis and carbon allocation, limiting sugar supply to roots, and ultimately impairing nitrogen fixation (Abd-Alla et al., 2023; Table 1).

9.2 *Disruption of Microbial Community Composition and Diversity*

Extreme temperature directly or indirectly affects the soil microbial community and root endophytic microbial community by affecting soil moisture, nutrient availability, vegetation growth, etc. (Zhao et al., 2024). Some researchers reported that warming reduces AMF diversity (Xu et al., 2022). High temperature also suppresses root development, thereby reducing colonization by AMF (Qiu et al., 2021). It alters their community composition, resulting in a shift from a mixed community, consequently reducing N2 fixation rates in peat moss (Carrell et al., 2019), whereas cold stress inhibits the growth and yield of leguminous plants. On the other hand, incorporation of *Rhizobium is* effective at reducing cold-induced damage (Irshad et al, 2021). Thus, extreme temperatures cause a reduction in microbial diversity and connectivity, thereby limiting nutrient availability and disrupting beneficial interactions that support native plant growth (Yu et al., 2024).

9.3 *Impaired Plant Growth and Productivity*

High temperatures, be it heat or cold, influence the growth of plants by causing physiological, anatomical, and morphological alterations. These involve nutrient absorption change, root morphology, cell division, photosynthesis, water movement, and phytohormonal cue (Sidhu et al., 2024). It also depletes and inactivates proteins, thereby affecting the synthesis of new proteins and carbohydrates. Additionally, it changes the carbon assimilation and membrane structure that eventually leads to the impairment of the plant development and decreased productivity (Sher et al., 2024).

9.4 *Increased Susceptibility to Pathogens and Loss of Disease Resistance*

Higher or lower temperature tends to make the plants more vulnerable to disease and alter the virulence characteristics of pathogens so that they become more efficiently able to infect the host (Cohen & Leach, 2020). This may result in the disintegration of genes that confer dependency of plants. In turn, severe temperature changes the flora of plants, reduces the immune system, and enhances the disease. In addition, the conditions can lead to previously unreported infections because they promote the growth and development of pathogenic microbials and more virulent strains (Trivedi et al., 2022).

9.5 *Altered Microbial Function: Metabolism and Enzyme Activities*

Extreme temperature stress may interfere with the structure, the activity of enzymes, and the metabolic routes of microbial communities, thus modifying the plant microbiome activity. High temperature stress can denature the key enzymes and disrupts the production of biomolecules. Conversely, low temperatures reduce the metabolic activity of the plant cells (Mitra et al., 2021). Therefore, drastic fluctuations intemperature can inhibit the activity of the enzymes such as nitrite reductase, and alkaline phosphatase reducing nutrient availability and consequently suppressing plant growth (Wang et al., 2023).

10 Resilience and Adaptation Strategies

Another form of stress that plants have adopted to respond to extreme temperatures is by attracting the presence of positive plant growth-promoting microorganisms. To do it, plants change their rhizosphere chemistry and emit specific root exudates which could be defined as a chemical signal, a phenomenon commonly referred to as the cry of help strategy (Rizaludin et al., 2021; Fig. 3). This chemical blend can contain sugars, amino acids, organic acids, phenolics volatile organic compounds, and secondary metabolites that is selectively used to recruit useful microbes to aid in alleviating stress. Despite such adaptive signalling being dependent on the intricate metabolic changes in the plant, the regulatory networks that regulate this action, particularly in response to changing or interacting stress, remain to be comprehended in detail. According to Tiziani et al. (2022), the presence of heat stress causes changes in the profile of root exudates of maize, with an increase in 1-aminocycloprolane-1-carboxylate (ACC), histidine, rhamnetin, and quercetin derivatives compounds, which correlates with changes in the microbial community. Equally, Liu et al. (2025) have discovered that heat-stressed *Panax notoginseng* generates high concentrations of nucleotide of ATP, GTP, UTP, and CTP. The purine and pyrimidine ones seem to specifically favour desirable microbes, which create enhanced tolerance of plants to heat. A significant role in plant association is played by bacterial taxa such as *Acetobacter, Actinobacteria, Arthrobacter, Azoarcus, Azospirillum, Azotobacter, Bacillus, Beijerinckia, Paraburkholderia, Enterobacter, Herbaspirillum, Klebsiella, Methylotrophic Nitrospira, Paenibacill*us, and *Pseudomonas* and by fungi such as *Arbuscular mycorrhiza, Aspergillus, Paecilomyces.* Such microbes enhance the overall performance of the plants in plants like Beet root, Black Gram, Chinese ginseng, Common bean, Cyclamen, Grapevines, Maize, Mugwort, Perennial ryegrass, Rhododendron, Rice, Sorghum, Soybean, Sunflower, Tomato, and Wheat (Table 2).

Phyto-microbiomes tolerate plant stress via indirect and direct mechanisms. These contain synthesis of antibacterial compounds like chloramphenicol, streptomycin, and antifungal compounds like chitinase, enhancing resistance to disease and pests (Ashraf et al., 2019; van der Voort et al., 2016). Microbes prevent photosynthetic apparatus damage, improve soil quality, and enhance photosynthesis by reducing lipid peroxidation (malondialdehyde, MDA), electrolyte leakage, reduce reactive oxygen species (ROS) concentration (Caradonia et al., 2019; Dastogeer et al., 2022; Ghorbanpour et al., 2018; Harman et al., 2021; Li et al., 2021; Lin et al., 2022; Mathur & Jajoo, 2020; Mathur et al., 2021; Shaffique et al., 2023; Zhu et al., 2010). Rhizosphere and phyllosphere microbes control the synthesis of phytohormones such as indole-3-acetic acid (IAA), gibberellic acid (GA), salicylic acid, and organic acids while reducing plant stress hormones like abscisic acid (ABA) and jasmonic acid (JA) (Khan et al., 2020a & 2020b; Mukhtar et al., 2020). These plant hormones increase the chlorophyll content, shoot-root growth, and higher biomass, and regulate leaf stomata to counteract radiation and excessive heat (Ismail et al., 2018; Schillaci et al., 2021; Gamit et al., 2024; Meza et al., 2025; Villano et al., 2025). To protect from cold stress, microbes enhance antioxidant activity, carbon fixation and promote starch, proline, malate, carbohydrates, sugar, amino acids, and phenolic compound accumulation (Abdel Latef & Chaoxing, 2011; Bruno et al., 2020; Issa et al., 2018; Kakar et al., 2016; Lin et al., 2022; Maya & Matsubara, 2013; Sarkar et al., 2018; Subramanian et al., 2015, 2016; Theocharis et al., 2012; Tiziani et al., 2022; Waqas et al., 2015) (Fig. 3, Table 2). Su et al. (2015) suggested that microbes support to strengthen the cell wall to prevent freezing damage and collapse by enhancing pigment synthesis and activating the cold stress signalling pathway. Furthermore, microbe produced siderophores under stress, which help defend against pathogens by binding essential iron and thereby limiting the growth of pathogenic microbes (Ali et al., 2011).

During heat stress, plants produce heat shock proteins and stress-responsive genes, while breaking down the enzyme ACC into ammonia and α-ketobutyrate to reduce ethylene levels and form biofilm as a protective layer (Abd El-Daim et al., 2014; Chouhan et al., 2023; Mukhtar et al., 2020; Shekhawat et al., 2021). Additionally, presence of microbes in plant increases the accumulation of γ-aminobutyric acid

(GABA) which help to maintain the carbon–nitrogen balance, the metabolism of amino acids and carbohydrates, and the regulation of plant development (Abd El-Daim et al., 2019). Besides these, microbes also solubilize essential minerals, biologically fix nitrogen, and change organic matter into accessible forms to improve the supply of nutrients (Lin et al., 2023; Yadav et al., 2015). They also enhance the metabolism of nucleotide, which leads to the enrichment of purine and pyrimidine metabolites in root exudates and enhanced resistance to osmotic and oxidative stress by upregulation of osmolyte (Jha & Mohamed, 2023; Liu et al., 2025). Mitigation of heat stress involves the synthesis of plant beneficial metabolites and enhances the phytoremediation capacity of plants (Bruno et al., 2020). Moreover, Cabral et al. (2016) suggested that under stress, beneficial plant microbes help plants stay resilient by sending nutrients to where they are needed most, filling seeds properly, and storing energy efficiently.

11 Conclusion

Drastic temperatures due to the changed environmental conditions have a tremendous effect on plant growth, microbial communities in the soil, and interaction of plants with microbes. The heat and cold stress impact the plant physiology, biochemical, and molecular processes. These are nutrient uptake, photosynthesis, membrane stability, hormonal signalling, and protein synthesis which result in retarded growth and productivity and heightened vulnerability to pathogens. At the same time, soil microbiota is vulnerable to changes in temperature levels, and it can change the size and density of the microbial population and cause cell lysis. It interferes with nutrient cycling, reduces enzyme activity, leads to dehydration and leakage of electrolytes, and eventually reduces the positive symbiosis. The microbiota is beneficial and can be tolerated to stress by rhizosphere, phytohormones, heat shock proteins, and phytohormones. They produce antioxidants, exopolysaccharides, and osmolytes and relieve oxidative stress, regulate plant metabolism, and activate stress responsive genes. Therefore, there is a potential in plant–microbe interactions to maintain plant productivity through ecological and biotechnological resource in a highly dynamic climate.

12 Future Perspectives

Well, as we are aware, to reduce the extremum temperature stress, both plants and microbes exhibit physiological, morphological, metabolic, and genetic alterations, yet there is a need to further research and understand the mechanism underlying thermotolerance. Recognition, separation, and characterization of thermotolerant microbial strains are significant to the design of bioinoculants to be used in different climatic regions. These characteristics of thermotolerant or genetically modified microbial communities including phytohormone regulation, enhanced antioxidant activity, biofilm formation, and exopolysaccharides production can improve the stability and effectiveness of plant responses to temperature stress conditions. Additionally, more sophisticated tools of molecular profiling in conjunction with genetic engineering tools such as the CRISPR can be used to develop heat-resistant plants in the future. The discovery of some important functional genes with the help of genome sequencing or molecular breeding, genetic manipulation, and omics biology has contributed to the capacity of plants to endure stress in dominating environmental conditions. Combining soil-enhancing activities, like rotating crop, green manure crop, and minimal cultivation would generate a favourable environment to promote plant growth-promoting microorganisms to multiply. Even though it is known that plant microbiota mitigates environmental and biological stress, there is still much to learn, to create cost-effective, shelf-stable bioformulations and find a promising direction to be pursued further.

References

Abd El-Daim, I. A., Bejai, S., & Meijer, J. (2014). Improved heat stress tolerance of wheat seedlings by bacterial seed treatment. *Plant and Soil, 379*(1), 337–350. https://doi.org/10.1007/s11104-014-2063-3

Abd El-Daim, I. A., Bejai, S., & Meijer, J. (2019). *Bacillus velezensis* 5113 induced metabolic and molecular reprogramming during abiotic stress tolerance in wheat. *Scientific Reports, 9*(1), 16282. https://doi.org/10.1038/s41598-019-52567-x

Abd-Alla, M. H., Al-Amri, S. M., & El-Enany, A.-W.E. (2023). Enhancing rhizobium–legume symbiosis and reducing nitrogen fertilizer use are potential options for mitigating climate change. *Agriculture, 13*(11), 2092. https://doi.org/10.3390/agriculture13112092

Abdel Latef, A. A. H., & Chaoxing, H. (2011). Arbuscular mycorrhizal influence on growth, photosynthetic pigments, osmotic adjustment, and oxidative stress in tomato plants subjected to low-temperature stress. *Acta Physiologiae Plantarum, 33*, 1217–1225. https://doi.org/10.1007/s11738-010-0650-3

Adam, S., & Murthy, S. D. S. (2013). Effect of cold stress on photosynthesis of plants and possible protection mechanisms. In: R. Gaur, P. Sharma (Eds.), *Approaches to plant stress and their management* (pp. 219–226). New Delhi: Springer India. https://doi.org/10.1007/978-81-322-1620-9_12

Adegbeye, M. J., Reddy, P. R. K., Obaisi, A. I., Elghandour, M. M. M. Y., Oyebamiji, K. J., Salem, A. Z. M., Morakinyo-Fasipe, O. T., Cipriano-Salazar, M., & Camacho-Díaz, L. M. (2020). Sustainable agriculture options for production, greenhouse gasses and pollution alleviation, and nutrient recycling in emerging and transitional nations-An overview. *Journal of Cleaner Production, 242*, Article 118319. https://doi.org/10.1016/j.jclepro.2019.118319

Adeleke, B. S., & Babalola, O. O. (2020). The endosphere microbial communities a great promise in agriculture. *International Microbiology, 24*, 1–17. https://doi.org/10.1007/s10123-020-00140-2

Adeleke, B. S., & Babalola, O. O. (2021). The plant endosphere-hidden treasures: A review of fungal endophytes. *Biotechnology and Genetic Engineering Reviews, 37*(2), 154–177. https://doi.org/10.1080/02648725.2021.1991714

Adeleke, B. S., & Babalola, O. O. (2022). Roles of plant endosphere microbes in agriculture-A review. *Journal of Plant Growth Regulation, 41*, 1411–1428. https://doi.org/10.1007/s00344-021-10406-2

Adhikari, L., Baral, R., Paudel, D., Min, D., Makaju, S., Poudel, H., Acharya, J., & Missaoui, A. (2022). Cold stress in plants: Strategies to improve cold tolerance in forage species. *Plant Stress, 4*, Article 100081. https://doi.org/10.1016/j.stress.2022.100081

Ahmad, M., Imtiaz, M., Shoib Nawaz, M., Mubeen, F., & Imran, A. (2022). What did we learn from current progress in heat stress tolerance in plants? Can microbes be a solution? *Frontiers in Plant Science, 13*, Article 794782.

Ali, S. Z., Sandhya, V., Grover, M., Linga, V. R., & Bandi, V. (2011). Effect of inoculation with a thermotolerant plant growth promoting *Pseudomonas putida* strain AKMP7 on growth of wheat (*Triticum* spp.) under heat stress. *Journal of Plant Interactions, 6*(4), 239–246. https://doi.org/10.1080/17429145.2010.545147

Arora, S., Murmu, G., Mukherjee, K., Saha, S., & Maity, D. (2022). A comprehensive overview of nanotechnology in sustainable agriculture. *Journal of Biotechnology, 355*, 21–41. https://doi.org/10.1016/j.jbiotec.2022.06.007

Ashraf, A., Bano, A., & Ali, S. A. (2019). Characterisation of plant growth-promoting rhizobacteria from rhizosphere soil of heat-stressed and unstressed wheat and their use as bio-inoculant. *Plant Biology, 21*, 762–769.https://doi.org/10.1111/plb.12972

Awasthi, A., Pattnayak, K. C., Tandon, A., Sarkar, A., & Chakraborty, M. (2023). Implications of climate change on surface temperature in North Indian states: Evidence from CMIP6 model ensembles. *Frontiers in Environmental Science, 11*, 1264757. https://doi.org/10.3389/fenvs.2023.1264757

Awasthi, A., Pattnayak, K. C., & Tiwari, P. R. (2025). Application of artificial intelligence-supported process-based climate models to understand the atmosphere/weather patterns and their prediction. *Frontiers in Environmental Science, 12*, 1543951. https://doi.org/10.3389/fenvs.2024.1543951

Babalola, O. O., & Adedayo, A. A. (2023). Endosphere microbial communities and plant nutrient acquisition toward sustainable agriculture. *Emerging Topics in Life Sciences, 7*(2), 207–217. https://doi.org/10.1042/ETLS20230069

Bansal, M., Kukreja, K., Suneja, S., & Dudeja, S. S. (2014). Symbiotic effectivity of high temperature tolerant mungbean (*Vigna radiata*) rhizobia under different temperature conditions. *International Journal of Current Microbiology and Applied Sciences, 3*(12), 807–821.

Bashir, I., War, A. F., Rafiq, I., Reshi, Z. A., Rashid, I., & Shouche, Y. S. (2022). Phyllosphere microbiome: Diversity and functions. *Microbiological Research, 254*, Article 126888. https://doi.org/10.1016/j.micres.2021.126888

Beirinckx, S., Viaene, T., Haegeman, A., Debode, J., Amery, F., Vandenabeele, S., Nelissen, H., Inzé, D., Tito, R., Raes, J., De Tender, C., & Goormachtig, S. (2020). Tapping into the maize root microbiome to identify bacteria that promote growth under chilling conditions. *Microbiome, 8*(1), 54. https://doi.org/10.1186/s40168-020-00833-w

Bonartsev, A. P., Zharkova, I. I., Yakovlev, S. G., Myshkina, V. L., Mahina, T. K., & Voinova, V. V. (2016). Biosynthesis of poly(3-hydroxybutyrate) copolymers by *Azotobacter chroococcum* 7B: A precursor feeding strategy. *Preparative Biochemistry & Biotechnology, 47*(2), 173–184. https://doi.org/10.1080/10826068.2016.1188317

Brito-Lopez, C., van der Wielen, N., Barbosa, M., & Karlova, R. (2025). Plant growth–promoting microbes and microalgae-based biostimulants: Sustainable strategy for agriculture and abiotic stress resilience. *Philosophical Transactions B, 380*(1927), 20240251. https://doi.org/10.1098/rstb.2024.0251

Bruno, L. B., Karthik, C., Ma, Y., Kadirvelu, K., Freitas, H., & Rajkumar, M. (2020). Amelioration of chromium and heat stresses in *Sorghum bicolor* by Cr6+reducing-thermotolerant plant growth promoting bacteria. *Chemosphere, 244*, Article 125521. https://doi.org/10.1016/j.chemosphere.2019.125521

Buttar, Z. A., Wu, S. N., Arnao, M. B., Wang, C., Ullah, I., & Wang, C. (2020). Melatonin suppressed the heat stress-induced damage in wheat seedlings by modulating the antioxidant machinery. *Plants, 9*(7), 809. https://doi.org/10.3390/plants9070809

Cabral, C., Ravnskov, S., Tringovska, I., & Wollenweber, B. (2016). Arbuscular mycorrhizal fungi modify nutrient allocation and composition in wheat (*Triticum aestivum L.*) subjected to heat-stress. *Plant and Soil, 408*, 385–399. https://doi.org/10.1007/s11104-016-2942-x

Caradonia, F., Francia, E., Morcia, C., Ghizzoni, R., Moulin, L., Terzi, V., & Ronga, D. (2019). Arbuscular mycorrhizal fungi and plant growth promoting rhizobacteria avoid processing tomato leaf damage during chilling stress. *Agronomy, 9*(6), 299. https://doi.org/10.3390/agronomy9060299

Carrell, A. A., Kolton, M., Glass, J. B., Pelletier, D. A., Warren, M. J., Kostka, J. E., Iversen, C. M., Hanson, P. J., & Weston, D. J. (2019). Experimental warming alters the community composition, diversity, and N2-fixation activity of peat moss (*Sphagnum fallax*) microbiomes. *Global Change Biology, 25*(9), 2993–3004.https://doi.org/10.1111/gcb.14715

Chen, Q., Song, Y., An, Y., Lu, Y., & Zhong, G. (2024). Soil microorganisms: Their role in enhancing crop nutrition and health. *Diversity, 16*(12), 734.

Cheng, Y. T., Zhang, L., & He, S. Y. (2019). Plant-microbe interactions facing environmental challenge. *Cell Host & Microbe, 26*(2), 183–192. https://doi.org/10.1016/j.chom.2019.07.009

Choi, S., & Lee, H. (2024). Interplay between plants and microbial communities: insights from holobionts and environmental interactions. *Phyton-International Journal of Experimental Botany, 93*(10), 2519–2534. https://doi.org/10.32604/phyton.2024.058012

Chouhan, D., Mathur, P., Choudhuri, C. (2023). Plant-microbe interaction and their role in mitigation of heat stress. In: P. Mathur, R. Kapoor, S. Roy (Eds.), *Microbial symbionts and plant health: Trends and applications for changing climate*. Rhizosphere Biology. Springer, Singapore. https://doi.org/10.1007/978-981-99-0030-5_6

Cohen, S. P., & Leach, J. E. (2020). High temperature-induced plant disease susceptibility: More than the sum of its parts. *Current Opinion in Plant Biology, 56*, 235–241. https://doi.org/10.1016/j.pbi.2020.02.008

Compant, S., Cambon, M. C., Vacher, C., Mitter, B., Samad, A., & Sessitsch, A. (2021). The plant endosphere world–bacterial life within plants. *Environmental Microbiology, 23*(4), 1812–1829. https://doi.org/10.1111/1462-2920.15240

Dastogeer, K. M. G., Zahan, M. I., Rhaman, M. S., Sarker, M. S. A., & Chakraborty, A. (2022). Microbe-mediated thermotolerance in plants and pertinent mechanisms- A meta-Analysis and review. *Frontiers in Microbiology, 13*, Article 833566. https://doi.org/10.3389/fmicb.2022.833566

Ding, Y., & Yang, S. (2022). Surviving and thriving: How plants perceive and respond to temperature stress. *Developmental Cell, 57*(8), 947–958. https://doi.org/10.1016/j.devcel.2022.03.010

Dong, C.-J., Wang, L.-L., Li, Q., & Shang, Q.-M. (2019). Bacterial communities in the rhizosphere, phyllosphere and endosphere of tomato plants. *PLoS ONE, 14*(11), Article e0223847. https://doi.org/10.1371/journal.pone.0223847

Dos Santos, T. B., Ribas, A. F., de Souza, S. G. H., Budzinski, I. G. F., & Domingues, D. S. (2022). Physiological responses to drought, salinity, and heat stress in plants: A review. *Stresses, 2*(1), 113–135. https://doi.org/10.3390/stresses2010009

Eggersdorfer, M., & Wyss, A. (2018). Carotenoids in human nutrition and health. *Archives of Biochemistry and Biophysics, 652*, 18–26. https://doi.org/10.1016/j.abb.2018.06.001

Erturk, Y., Ercisli, S., Haznedar, A., & Cakmakci, R. (2010). Effects of plant growth promoting rhizobacteria (PGPR) on rooting and root growth of kiwifruit (*Actinidia deliciosa*) stem cuttings. *Biological Research, 43*(1), 91–98. https://doi.org/10.4067/S0716-97602010000100011

Expósito, C. D. V., López, J. Á., Liu, J., Bao, N., Liang, J., & Zhang, J. (2022). Development of a cold-active microbial compound biofertilizer on the improvement for rice (*Oryza sativa* L.) tolerance at low temperatures. *Rhizosphere, 24*, 100586. https://doi.org/10.1016/j.rhisph.2022.100586

Fadiji, A. E., Yadav, A. N., Santoyo, G., & Babalola, O. O. (2023). Understanding the plant-microbe interactions in environments exposed to abiotic stresses: An overview. *Microbiological Research, 271*, Article 127368. https://doi.org/10.1016/j.micres.2023.127368

Gamit, H. A., & Amaresan, N. (2024). Inoculation of methylotrophic bacteria ameliorate summer heat stress and enhance the Black Gram (*Vigna mungo* L.) growth, physiology and antioxidants properties. *Journal of Plant Growth Regulation, 43*, 2077–2089. https://doi.org/10.1007/s00344-024-11251-9

Ghorbanpour, A., Salimi, A., Ghanbary, M. A. T., Pirdashti, H., & Dehestani, A. (2018). The effect of *Trichoderma harzianum* in mitigating low temperature stress in tomato (*Solanum lycopersicum* L.) plants. *Scientia Horticulturae, 230*, 134–141. https://doi.org/10.1016/j.scienta.2017.11.028

Hacquard, S., Wang, E., Slater, H., & Martin, F. (2022). Impact of global change on the plant microbiome. *New Phytologist, 234*, 1907–1909.https://doi.org/10.1111/nph.18187

Harman, G., Khadka, R., Doni, F., & Uphoff, N. (2021). Benefits to plant health and productivity from enhancing plant microbial symbionts. *Frontiers in Plant Science, 11*, Article 610065. https://doi.org/10.3389/fpls.2020.610065

Hasanuzzaman, M., Nahar, K., Alam, M. M., Roychowdhury, R., & Fujita, M. (2013). Physiological, biochemical, and molecular mechanisms of heat stress tolerance in plants. *International Journal of Molecular Sciences, 14*(5), 9643–9684. https://doi.org/10.3390/ijms14059643

Hassani, M. A., Durán, P., & Hacquard, S. (2018). Microbial interactions within the plant holobiont. *Microbiome, 6*(1), 58. https://doi.org/10.1186/s40168-018-0445-0

Huang, S., Zha, X., & Fu, G. (2023). Affecting factors of plant phyllosphere microbial community and their responses to climatic warming-A review. *Plants (Basel, Switzerland), 12*(16), 2891. https://doi.org/10.3390/plants12162891

Huot, B., Castroverde, C. D. M., Velásquez, A. C., Hubbard, E., Pulman, J. A., Yao, J., Childs, K. L., Tsuda, K., Montgomery, B. L., & He, S. Y. (2017). Dual impact of elevated temperature on plant defence and bacterial virulence in Arabidopsis. *Nature Communications, 8*(1), 1808. https://doi.org/10.1038/s41467-017-01674-2

Ibáñez, A., Garrido-Chamorro, S., & Barreiro, C. (2023). Microorganisms and climate change: A not so invisible effect. *Microbiology Research, 14*(3), 918–947. https://doi.org/10.3390/microbiolres14030064

Iqbal, B., Li, G., Alabbosh, K. F., Hussain, H., Khan, I., Tariq, M., Javed, Q., Naeem, M., & Naveed, A. (2023). Advancing environmental sustainability through microbial reprogramming in growth improvement, stress alleviation, and phytoremediation. *Plant Stress, 10*, Article 100283. https://doi.org/10.1016/j.stress.2023.100283

Irfan, M., Aslam, H., Maqsood, A., Tazeen, S. K., Mahmood, F., Shahid, M. (2023). Changes in plant microbiome in response to abiotic stress. In: S. Chhabra, R. Prasad, N. R. Maddela, N. Tuteja (Eds.) *Plant microbiome for plant productivity and sustainable agriculture.* Microorganisms for Sustainability, vol. 37. Springer, Singapore. https://doi.org/10.1007/978-981-19-5029-2_5

Irshad, A., Rehman, R. N. U., Kareem, H. A., Yang, P., & Hu, T. (2021). Addressing the challenge of cold stress resilience with the synergistic effect of rhizobium inoculation and exogenous melatonin application in *Medicago truncatula. Ecotoxicology and Environmental Safety, 226*, Article 112816. https://doi.org/10.1016/j.ecoenv.2021.112816

Ismail, Hamayun, M., Hussain, A., Iqbal, A., Khan, S. A., & Lee, I. J. (2018). Endophytic fungus *Aspergillus japonicus* mediates host plant growth under normal and heat stress conditions. *BioMed Research International, 2018*, 7696831. https://doi.org/10.1155/2018/7696831

Issa, A., Esmaeel, Q., Sanchez, L., Courteaux, B., Guise, J. F., Gibon, Y., Ballias, P., Clément, C., Jacquard, C., Vaillant-Gaveau, N., & Aït Barka, E. (2018). Impacts of *Paraburkholderia phytofirmans*strain PsJN on tomato (*Lycopersicon esculentum* L.) under high temperature. *Frontiers in Plant Science, 9*, 1397. https://doi.org/10.3389/fpls.2018.01397

Janczarek, M., Adamczyk, P., Gromada, A., Polakowski, C., Wengerska, K., & Bieganowski, A. (2024). Adaptation of *Rhizobium leguminosarum sv. trifolii* strains to low temperature stress in both free-living stage and during symbiosis with clover. *Science of the Total Environment, 951*, 175554. https://doi.org/10.1016/j.scitotenv.2024.175554

Jha, Y., Kulkarni, A., & Subramanian, R. B. (2021). Psychrotrophic soil microbes and their role in alleviation of cold stress in plants. In *Soil microbiomes for sustainable agriculture: Functional annotation* (pp. 267–286). Cham: Springer International Publishing. https://doi.org/10.1007/978-3-030-73507-4_10

Jha, Y., & Mohamed, H. I. (2023). Inoculation with *Lysinibacillus fusiformis*strain YJ4 and *Lysinibacillus sphaericus*strain YJ5 alleviates the effects of cold stress in maize plants. *Gesunde Pflanzen, 75*, 77–95. https://doi.org/10.1007/s10343-022-00666-7

Kakar, K. U., Ren, X. L., Nawaz, Z., Cui, Z. Q., Li, B., Xie, G. L., Hassan, M. A., Ali, E., & Sun, G. C. (2016). A consortium of rhizobacterial strains and biochemical growth elicitors improve cold and drought stress tolerance in rice (Oryza sativa L.). *Plant Biology, 18*(3), 471–483. https://doi.org/10.1111/plb.12427

Khan, M. A., Asaf, S., Khan, A. L., Jan, R., Kang, S. M., Kim, K. M., & Lee, I. J. (2020a). Extending thermotolerance to tomato seedlings by inoculation with SA1 isolate of *Bacillus cereus* and comparison with exogenous humic acid application. *PLoS ONE, 15*(4), Article e0232228. https://doi.org/10.1371/journal.pone.0232228

Khan, M. A., Asaf, S., Khan, A. L., Jan, R., Kang, S. M., Kim, K. M., & Lee, I. J. (2020b). Thermotolerance effect of plant growth-promoting *Bacillus cereus* SA1 on soybean during heat stress. *BMC Microbiology, 20*(1), 175. https://doi.org/10.1186/s12866-020-01822-7

Khan, R., He, P., He, P., Ahmed, A., Liu, Y., Wu, Y., Tang, G., Li, X., Di, Y., Munir, S. & He, Y., (2025). Temperature-driven modulation of endophyte efficacy in microbiome interactions during tomato spotted wilt virus disease suppression. *Journal of Agriculture and Food Research*, 102421. https://doi.org/10.1016/j.jafr.2025.102421

Khanzada, A., Wang, X., Malko, M. M., Wu, Y., Samo, A., & Dong, J. (2024). Response of the persistence of heat stress tolerance in winter wheat seedling to heat priming at early growth stages. *Plant Stress, 11*, Article 100323. https://doi.org/10.1016/j.stress.2023.100323

Koster, K., Aaltonen, H., Berninger, F., Heinonsalo, J., Köster, E., Ribeiro-Kumara, C., Sun, H., Leho, T., Zhou, X., & Pumpanen, J. (2021). Impacts of wildfire on soil microbiome in Boreal environments. *Current Opinion in Environmental Science & Health, 22*, Article 100258. https://doi.org/10.1016/j.coesh.2021.100258

Krishna, S., Bhutia, D. D., Chaubey, R. K., & Sudhir, I. (2024). Chapter 4 -Exploring plant microbiome: a holistic approach to sustainable agriculture. In J. A. Parray, N. Shameem, D. Egamberdieva (Eds.), *Microbiome drivers of ecosystem function* (pp. 61–77). Academic Press. https://doi.org/10.1016/B978-0-443-19121-3.00013-2.10.1016/B978-0-443-19121-3.00013-2

Kumar, A., & Dubey, A. (2020). Rhizosphere microbiome: Engineering bacterial competitiveness for enhancing crop production. *Journal of Advanced Research, 24*, 337–352. https://doi.org/10.1016/j.jare.2020.04.014

Kushwaha, P., Kashyap, P. L., & Kuppusamy, P. (2020). Microbes for cold stress resistance in plants: mechanism, opportunities, and challenges. In *Microbiological advancements for higher altitude agro-ecosystems & sustainability*, pp. 269–292. https://doi.org/10.1007/978-981-15-1902-4_14

Lee, J. H., Yun, H. S., & Kwon, C. (2012). Molecular communications between plant heat shock responses and disease resistance. *Molecules and Cells, 34*(2), 109–116. https://doi.org/10.1007/s10059-012-0121-3

Li, X., Zhao, C., Zhang, T., Wang, G., Amombo, E., Xie, Y., & Fu, J. (2021). Exogenous Aspergillus aculeatus enhances drought and heat tolerance of perennial ryegrass. *Frontiers in Microbiology, 12*, Article 593722. https://doi.org/10.3389/fmicb.2021.593722

Li, X., Ren, X., Su, Y., Zhou, X., Wang, Y., Ruan, S., Yan, J., Li, B., & Guo, K. (2024a). Differential effects of winter cold stress on soil bacterial communities, metabolites, and physicochemical properties in two varieties of *Tetrastigma hemsleyanum* Diels & Gilg in reclaimed land. *Microbiology Spectrum, 12*(4), e02425-e2523. https://doi.org/10.1128/spectrum.02425-23

Li, Y., Zhu, J., Xu, J., Zhang, X., Xie, Z., & Li, Z. (2024b). Effect of cold stress on photosynthetic physiological characteristics and molecular mechanism analysis in cold-resistant cotton (ZM36) seedlings. *Frontiers in Plant Science, 15*, 1396666. https://doi.org/10.3389/fpls.2024.1396666

Lin, W., Liu, L., Liang, J., Tang, X., Shi, J., Zhang, L., Wu, P., Lan, S., Wang, S., Zhou, Y., Chen, X., Zhao, Y., Chen, X., Wu, B., & Guo, L. (2022). Changes of endophytic microbial community in

Rhododendron simsii roots under heat stress and its correlation with leaf physiological indicators. *Frontiers in Microbiology, 13*, 1006686. https://doi.org/10.3389/fmicb.2022.1006686

Lin, W., Ye, Q., Liang, J., Tang, X., Shi, J., Liu, L., Duan, X., Li, X., Wu, P., Liu, Y., Chen, X., He, B., Guo, L., & Lan, S. (2023). Response mechanism of interaction between *Rhododendron hainanense* and microorganisms to heat stress. *Industrial Crops and Products, 199*, Article 116764. https://doi.org/10.1016/j.indcrop.2023.116764

Liu, H., Su, Y., Ye, C., Zuo, D., Wang, L., Mei, X., Deng, W., Liu, Y., Huang, H., Hao, J., Zhao, J., Wang, D., Zhang, X., Zhu, Y., Liu, J., Yang, M., & Zhu, S. (2025). Nucleotides enriched under heat stress recruit beneficial rhizomicrobes to protect plants from heat and root-rot stresses. *Microbiome, 13*(1), 160. https://doi.org/10.1186/s40168-025-02126-6

Mathur, S., & Jajoo, A. (2020). Arbuscular mycorrhizal fungi protects maize plants from high temperature stress by regulating photosystem II heterogeneity. *Industrial Crops and Products, 143*, Article 111934. https://doi.org/10.1016/j.indcrop.2019.111934

Mathur, S., Agnihotri, R., Sharma, M. P., Reddy, V. R., & Jajoo, A. (2021). Effect of high-temperature stress on plant physiological traits and mycorrhizal symbiosis in maize plants. *Journal of Fungi, 7*(10), 867. https://doi.org/10.3390/jof7100867

Maya, M. A., & Matsubara, Y. (2013). Influence of arbuscular mycorrhiza on the growth and antioxidative activity in cyclamen under heat stress. *Mycorrhiza, 23*(5), 381–390. https://doi.org/10.1007/s00572-013-0477-z

Meza, C., Valenzuela, F., Yáñez, M., Cabeza, R. A., Ramos, P., Plaza, A., Carrasco, B., Flores-Castañón, N., Mesquita-Neto, J., Arce-Johnson, P., & Banerjee, A. (2025). Auxin producing plant growth promoting bacteria enhance temperature stress tolerance in Chilean common bean landraces. *Scientific Reports, 15*(1), 30359. https://doi.org/10.1038/s41598-025-15645-x

Mishra, S., Spaccarotella, K., Gido, J., Samanta, I., & Chowdhary, G. (2023). Effects of heat stress on plant-nutrient relations: An update on nutrient uptake, transport, and assimilation. *International Journal of Molecular Sciences, 24*(21), 15670. https://doi.org/10.3390/ijms242115670

Mitra, D., Díaz Rodríguez A. M, Parra Cota F. I., Khoshru, B., Panneerselvam, P., Moradi, S., Sagarika, M. S., Anđelković, S., de los Santos-Villalobos, S., & Das Mohapatra, P. K. (2021). Amelioration of thermal stress in crops by plant growth-promoting rhizobacteria. *Physiological and Molecular Plant Pathology, 115*, 101679. https://doi.org/10.1016/j.pmpp.2021.101679

Mondal, S., Karmakar, S., Panda, D., Pramanik, K., Bose, B., & Singhal, R. K. (2023). Crucial plant processes under heat stress and tolerance through heat shock proteins. *Plant Stress., 10*, Article 100227. https://doi.org/10.1016/j.stress.2023.100227

Mukhtar, T., Rehman, S., Smith, D., Sultan, T., Seleiman, M. F., Alsadon, A. A., Amna, Ali, S., Chaudhary, H. J., Solieman, T. H. I., Ibrahim, A. A., & Saad, M. A. O. (2020). Mitigation of heat stress in *Solanum lycopersicum* L. by ACC-deaminase and exopolysaccharide producing *Bacillus cereus*: Effects on biochemical profiling. *Sustainability, 12* (6), 2159. https://doi.org/10.3390/su12062159

Mutambara, P., & Nkongolo, K. (2025). Effects of elevated temperatures on soil enzymatic activities and bacterial and fungal community composition and diversity. *Ecological Genetics and Genomics*, 100382. https://doi.org/10.1016/j.egg.2025.100382

Newton, A. C., Fitt, B. D., Atkins, S. D., Walters, D. R., & Daniell, T. J. (2010). Pathogenesis, parasitism and mutualism in the trophic space of microbe-plant interactions. *Trends in Microbiology, 18*(8), 365–373. https://doi.org/10.1016/j.tim.2010.06.002

Ortiz, A., & Sansinenea, E. (2022). The role of beneficial microorganisms in soil quality and plant health. *Sustainability, 14*(9), 5358. https://doi.org/10.3390/su14095358

Oubohssaine, M., Rabeh, K., & Hnini, M. (2025). Symbiosis vs pathogenesis in plants: Reflections and perspectives. *Microbial Pathogenesis, 200*, Article 107333. https://doi.org/10.1016/j.micpath.2025.107333

Parasar, B. J., Kashyap, S., Sharma, I., Marme, S. D., Das, P., & Agarwala, N. (2024). Microbe mediated alleviation of drought and heat stress in plants- current understanding and future prospects. *Discover Plants, 1*, 26. https://doi.org/10.1007/s44372-024-00022-1

Pramanik, B., Sar, P., Bharti, R., Gupta, R. K., Purkayastha, S., Sinha, S., Chattaraj, S., & Mitra, D. (2023). Multifactorial role of nanoparticles in alleviating environmental stresses for sustainable crop production and protection. *Plant Physiology and Biochemistry, 201*, Article 107831. https://doi.org/10.1016/j.plaphy.2023.107831

Qian, Z., He, L., & Li, F. (2024). Understanding cold stress response mechanisms in plants: An overview. *Frontiers in Plant Science, 15*, 1443317. https://doi.org/10.3389/fpls.2024.1443317

Qiu, Y., Guo, L., Xu, X., Zhang, L., Zhang, K., Chen, M., Zhao, Y., Burkey, K. O., Shew, H. D., Zobel, R. W., Zhang, Y., & Hu, S. (2021). Warming and elevated ozone induce tradeoffs between fine roots and mycorrhizal fungi and stimulate organic carbon decomposition. *Science Advances, 7*(28), eabe9256. https://doi.org/10.1126/sciadv.abe9256

Rekadwad, B. N., Li, W. J., Gonzalez, J. M., Punchappady Devasya, R., Ananthapadmanabha Bhagwath, A., Urana, R., & Parwez, K. (2023). Extremophiles: the species that evolve and survive under hostile conditions. *3 Biotech, 13*(9), 316. https://doi.org/10.1007/s13205-023-03733-6

Rizaludin, M. S., Stopnisek, N., Raaijmakers, J. M., & Garbeva, P. (2021). The chemistry of stress: Understanding the 'cry for help' of plant roots. *Metabolites, 11*(6), 357. https://doi.org/10.3390/metabo11060357

Samtani, H., Unni, G., & Khurana, P. (2022). Microbial mechanisms of heat sensing. *Indian Journal of Microbiology, 62*(2), 175–186. https://doi.org/10.1007/s12088-022-01009-w

Sarkar, J., Chakraborty, B., & Chakraborty, U. (2018). Plant growth promoting rhizobacteria protect wheat plants against temperature stress through antioxidant signalling and reducing chloroplast and membrane injury. *Journal of Plant Growth Regulation, 37*, 1396–1412. https://doi.org/10.1007/s00344-018-9789-8

Schillaci, M., Arsova, B., Walker, R., Smith, P. M., Nagel, K. A., Roessner, U., & Watt, M. (2021). Time-resolution of the shoot and root growth of the model cereal *Brachypodium* in response to inoculation with *Azospirillum* bacteria at low phosphorus and temperature. *Plant Growth Regulation, 93*(1), 149–162. https://doi.org/10.1007/s10725-020-00675-4

Schnecker, J., Spiegel, F., Li, Y., Richter, A., Sandén, T., Spiegel, H., Zechmeister-Boltenstern, S., & Fuchslueger, L. (2023). Microbial responses to soil cooling might explain increases in microbial biomass in winter. *Biogeochemistry, 164*(3), 521–535. https://doi.org/10.1007/s10533-023-01050-x

Seeger, M., Turner, R. J., & González, M. (2022). An exploration of microbial response to stressors with Prof. Claudio C. Vásquez Guzmán. *Biological Research, 55*(1), 25. https://doi.org/10.1186/s40659-022-00393-3

Shaffique, S., Alomrani, S. O., Injamum-Ul-Hoque, M., Peter, O., Imran, M., & Lee, I. J. (2023). Unlocking the potential of newly isolated phytohormone-producing bacterial strains for enhanced plant growth and stress tolerance. *Plant Stress, 10*, Article 100260. https://doi.org/10.1016/j.stress.2023.100260

Shapiro, R. S., & Cowen, L. E. (2012). Thermal control of microbial development and virulence: Molecular mechanisms of microbial temperature sensing. *Mbio, 3*(5), 10–1128. https://doi.org/10.1128/mBio.00238-12

Shekhawat, K., Saad, M. M., Sheikh, A., Mariappan, K., Al-Mahmoudi, H., Abdulhakim, F., Eida, A. A., Jalal, R., Masmoudi, K., & Hirt, H. (2021). Root endophyte induced plant thermotolerance by constitutive chromatin modification at heat stress memory gene loci. *EMBO Reports, 22*(3), e51049. https://doi.org/10.15252/embr.202051049

Sher, A., Noor, M. A., Li, H. X., Nasir, B., Manzoor, M. A., Hussain, S., Zhang, J., Riaz, M. W., & Hussain, S. (2024). Heat stress effects on legumes: Challenges, management strategies and future insights. *Plant Stress, 13*, Article 100537. https://doi.org/10.1016/j.stress.2024.100537

Shi, F., Zhu, S., Li, H., Zhang, B., Liu, J., & Song, F. (2024a). Further enhancement of cold tolerance in rice seedlings by *Piriformospora indica* collaborating with plant growth-promoting bacteria: Evidence from the antioxidant defense, osmoregulation, photosynthesis, and related genes. *Plant Stress, 14*, Article 100656. https://doi.org/10.1016/j.stress.2024.100656

Shi, Y., He, Y., Zheng, Y., Liu, X., Wang, S., Xiong, T., Wen, T., Duan, H., Liao, X., Cui, Q., & Nian, F. (2024b). Characteristics of the phyllosphere microbial community and its relationship with major aroma precursors during the tobacco maturation process. *Frontiers in Plant Science, 15*, 1346154. https://doi.org/10.3389/fpls.2024.1346154

Sidhu, A. K., Sharma, M., Bhickchand Agrawal, S., Pradip Bhavsar, P., & Samota, M. K. (2024). Nanomaterial strategies for enhancing plant resilience in the face of temperature stress. *CABI Agriculture and Bioscience, 5*(1), 60. https://doi.org/10.1186/s43170-024-00255-w

Sohrabi, R., Paasch, B. C., Liber, J. A., & He, S. Y. (2023). Phyllosphere microbiome. *Annual Review of Plant Biology, 74*(1), 539–568.https://doi.org/10.1146/annurev-arplant-102820-032704

Son, S., & Park, S. R. (2022). Climate change impedes plant immunity mechanisms. *Frontiers in Plant Science, 13*, 1032820. https://doi.org/10.3389/fpls.2022.1032820

Su, F., Jacquard, C., Villaume, S., Michel, J., Rabenoelina, F., Clément, C., Barka, E. A., Dhondt-Cordelier, S., & Vaillant-Gaveau, N. (2015). Burkholderia phytofirmans PsJN reduces impact of freezing temperatures on photosynthesis in Arabidopsis thaliana. *Frontiers in Plant Science, 6*, 810. https://doi.org/10.3389/fpls.2015.00810

Subramanian, P., Mageswari, A., Kim, K., Lee, Y., & Sa, T. (2015). Psychrotolerant endophytic *Pseudomonas* sp. strains OB155 and OS261 induced chilling resistance in Tomato plants (*Solanum lycopersicum* Mill.) by activation of their antioxidant capacity. *Molecular Plant-Microbe Interactions, 28*(10), 1073–1081. https://doi.org/10.1094/MPMI-01-15-0021-R

Subramanian, P., Kim, K., Krishnamoorthy, R., Mageswari, A., Selvakumar, G., & Sa, T. (2016). Cold stress tolerance in psychrotolerant soil bacteria and their conferred chilling resistance in tomato (Solanum lycopersicum Mill.) under low temperatures. *PloS one, 11*(8), e0161592. https://doi.org/10.1371/journal.pone.0161592

Tandon, A., Awasthi, A., Pattnayak, K. C., & Das, S. (2025). Rainfall variability and rising extremes in urbanizing himalayan foothills: A machine learning and data-driven exploration of hydro-climatic shifts in Uttarakhand, India. *Earth Systems and Environment*, 1–26. https://doi.org/10.1007/s41748-025-00881-y

Theocharis, A., Bordiec, S., Fernandez, O., Paquis, S., Dhondt-Cordelier, S., Baillieul, F., Clément, C., & Barka, E. A. (2012). *Burkholderia phytofirmans* PsJN primes *Vitis vinifera* L. and confers a better tolerance to low nonfreezing temperatures. *Molecular Plant-Microbe Interactions, 25*(2), 241–249. https://doi.org/10.1094/MPMI-05-11-0124

Tiwari, P., Bose, S. K., Park, K.-I., Dufossé, L., & Fouillaud, M. (2024). Plant-microbe interactions under the extreme habitats and their potential applications. *Microorganisms, 12*(3), 448. https://doi.org/10.3390/microorganisms12030448

Tiziani, R., Miras-Moreno, B., Malacrinò, A., Vescio, R., Lucini, L., Mimmo, T., Cesco, S., & Sorgonà, A. (2022). Drought, heat, and their combination impact the root exudation patterns and rhizosphere microbiome in maize roots. *Environmental and Experimental Botany, 203*, Article 105071. https://doi.org/10.1016/j.envexpbot.2022.105071

Tokas, D., Singh, S., Yadav, R., & Singh, A. N. (2023). Plant-microbe interactions: Role in sustainable agriculture and food security in a changing climate. In *Plant-microbe interaction-recent advances in molecular and biochemical approaches* (pp. 363–391). Academic Press. https://doi.org/10.1016/B978-0-323-91876-3.00008-7

Trivedi, P., Batista, B. D., Bazany, K. E., & Singh, B. K. (2022). Plant-microbiome interactions under a changing world: responses, consequences and perspectives. *The New phytologist, 234*(6), 1951–1959. https://doi.org/10.1111/nph.18016

Ullah, F., Ali, S., Siraj, M., Akhtar, M. S., & Zaman, W. (2025). Plant microbiomes alleviate abiotic stress-associated damage in crops and enhance climate-resilient agriculture. *Plants (Basel, Switzerland), 14*(12), 1890. https://doi.org/10.3390/plants14121890

Upadhyay, S. K., Srivastava, A. K., Rajput, V. D., Chauhan, P. K., Bhojiya, A. A., Jain, D., Chaubey, G., Dwivedi, P., Sharma, B., & Minkina, T. (2022). Root exudates: Mechanistic insight of plant growth promoting rhizobacteria for sustainable crop production. *Frontiers in Microbiology, 13*, Article 916488. https://doi.org/10.3389/fmicb.2022.916488

Valenzuela-Aragon, B., Cardinale, M., Rolli, E., Rustioni, L., & Francioli, D. (2025). The role of arbuscular mycorrhizal fungi in abiotic stress management in viticulture under climatic shifts. *Plant Stress, 16*, Article 100863. https://doi.org/10.1016/j.stress.2025.100863

Villano, F., Balestrini, R., Nerva, L., & Chitarra, W. (2025). Harnessing microbes as sun cream against high light stress. *New Phytologist, 245*(2), 450–457. https://doi.org/10.1111/nph.20206

Van der Voort, M., Kempenaar, M., van Driel, M., Raaijmakers, J. M., & Mendes, R. (2016). Impact of soil heat on reassembly of bacterial communities in the rhizosphere microbiome and plant disease suppression. *Ecology Letters, 19*(4), 375–382.https://doi.org/10.1111/ele.12567

Wang, S., Wang, R., Vyzmal, J., Hu, Y., Li, W., Wang, J., Lei, Y., Zhai, X., Zhao, X., Li, J., & Cui, L. (2023). Shifts of active microbial community structure and functions in constructed wetlands responded to the continuously decreasing temperature in winter. *Chemosphere, 335*, Article 139080. https://doi.org/10.1016/j.chemosphere.2023.139080

Wang, Q., Wu, Y., Wu, W., Lyu, L., & Li, W. (2025). Research progress on the physiological, biochemical and molecular regulatory mechanisms of fruit tree responses to high-temperature stress. *Horticultural Plant Journal, 11*(1), 1–14. https://doi.org/10.1016/j.hpj.2023.09.008

Waqas, M., Khan, A. L., Shahzad, R., Ullah, I., Khan, A. R., & Lee, I. J. (2015). Mutualistic fungal endophytes produce phytohormones and organic acids that promote japonica rice plant growth under prolonged heat stress. *Journal of Zhejiang University. Science. B, 16*(12), 1011–1018. https://doi.org/10.1631/jzus.B1500081

Wu, Z., Han, S., Zhou, H., Tuang, Z. K., Wang, Y., Jin, Y., Shi, H., & Yang, W. (2019). Cold stress activates disease resistance in Arabidopsis thaliana through a salicylic acid dependent pathway. *Plant, Cell & Environment, 42*(9), 2645–2663. https://doi.org/10.1111/pce.13579

Xu, X., Qiu, Y., Zhang, K., Yang, F., Chen, M., Luo, X., Yan, X., Wang, P., Zhang, Y., Chen, H., & Guo, H., (2022). Climate warming promotes deterministic assembly of arbuscular mycorrhizal fungal communities. *Global Change Biology, 28*(3), 1147–1161.https://doi.org/10.1111/gcb.15945

Yadav, A. N., Sachan, S. G., Verma, P., & Saxena, A. K. (2015). Prospecting cold deserts of northwestern Himalayas for microbial diversity and plant growth promoting attributes. *Journal of Bioscience and Bioengineering, 119*(6), 683–693. https://doi.org/10.1016/j.jbiosc.2014.11.006

Yin, W., Wang, Y., Liu, L., & He, J. (2019). Biofilms: The microbial "protective clothing" in extreme environments. *International Journal of Molecular Sciences, 20*(14), 3423. https://doi.org/10.3390/ijms20143423

Yu, H., Ren, G., Huang, Z., Qi, S., Zhao, B., Fan, X., Zhu, Z., Dai, Z., & Du, D. (2024). Effects of increasing temperature on bacterial community diversity in mixed stands of *Artemisia argyi* and *Solidago canadensis* in Eastern China. *Microorganisms, 12*(12), 2415. https://doi.org/10.3390/microorganisms12122415

Zayed, O., Hewedy, O. A., Abdelmoteleb, A., Ali, M., Youssef, M. S., Roumia, A. F., Seymour, D., & Yuan, Z. C. (2023). Nitrogen journey in plants: From uptake to metabolism, stress response, and microbe interaction. *Biomolecules, 13*(10), 1443. https://doi.org/10.3390/biom13101443

Zhang, Y., Ku, Y. S., Cheung, T. Y., Cheng, S. S., Xin, D., Gombeau, K., Cai, Y., Lam, H. M., & Chan, T. F. (2024). Challenges to rhizobial adaptability in a changing climate: Genetic engineering solutions for stress tolerance. *Microbiological Research, 288*, Article 127886. https://doi.org/10.1016/j.micres.2024.127886

Zhao, J., Lu, Z., Wang, L., & Jin, B. (2020). Plant responses to heat stress: Physiology, transcription, noncoding RNAs, and epigenetics. *International Journal of Molecular Sciences, 22*(1), 117. https://doi.org/10.3390/ijms22010117

Zhao, J., Xie, X., Jiang, Y., Li, J., Fu, X., Qiu, Y., Fu, X., Yao, Z., Dai, Z., Qiu, Y., & Chen, H. (2024). Effects of simulated warming on soil microbial community diversity and composition across diverse ecosystems. *Science of the Total Environment, 911*, Article 168793. https://doi.org/10.1016/j.scitotenv.2023.168793

Zhu, X., Song, F., & Xu, H. (2010). Influence of arbuscular mycorrhiza on lipid peroxidation and antioxidant enzyme activity of maize plants under temperature stress. *Mycorrhiza, 20*(5), 325–332. https://doi.org/10.1007/s00572-009-0285-7

Vulnerability and Resilience of Plant Biodiversity to Heat Stress Under Climate Change

Samjetsabam Bharati Devi, Kishor Sharma, Lovepreet Kaur, and Vallanattu James Jins

Abstract Global climate change has greatly threatened plant biodiversity and associated ecosystem services (ES), including forest carbon sequestration and carbon/nutrient cycling. The impacts of Heat waves (HW) and resulting heat stress (HS) on plants are understudied in many regions, like the Himalaya. Our study provides overview of global and Indian research on HW/HS, status of plant biodiversity (including threatened and endemic status). We have also summarized the biological and ecological effects of HW/HS on plants and the underlying processes, and their association with drought and pollution. We have synthesized the relationship of intensity and/or frequency of HW with the vulnerability and resilience of plant biodiversity to HS in different terrestrial and aquatic ecosystems. The study highlighted HS impacts across groups (e.g., C3 vs. C4, native vs. invasives), and coping mechanisms (by plants) and mitigation strategies (by policymakers and local communities) against HW/ HS to be employed in less studied mountain ecosystems. It outlines policy interventions and mitigation strategies against HW and HS for conserving ecosystems and native biodiversity, and future research perspectives to deepen our understanding of HS impacts on flora (and faunal) biodiversity, and ES provisioning.

Keywords Heat wave · Heat stress · Global climate change · Plant biodiversity · Ecosystem services · Mountain ecosystem

---

S. B. Devi (✉)
Department of Botany, School of Basic Sciences, Shri Ramasamy Memorial University Sikkim, Tadong, Gangtok, Sikkim, India
e-mail: sbharatidevi@gmail.com

K. Sharma
The Mountain Institute-India, Tadong, Gangtok, Sikkim, India

L. Kaur
Akal University, Talwandi Sabo, Bathinda, Punjab, India

V. J. Jins
Texas Commission on Environmental Quality, Austin, TX, USA

A. Awasthi et al. (eds.), *Rising Temperatures and Plant Ecosystems*,
https://doi.org/10.1007/978-3-032-29477-7_6

1 Introduction

1.1 Overview of Global Climate Change and Heat Waves

The global climate change (GCC) has elevated the rate of global warming in recent decades (IPCC, 2023). Global warming has made the episodes of Heat Waves (HW) more frequent, intense, and longer in India (Dar & Apurv, 2024; Devi et al., 2023a) and elsewhere (Barriopedro et al., 2023; IPCC, 2023; Zhang et al., 2025). The HW in terrestrial (Ruthrof et al., 2018), as well as freshwater (Wernberg et al., 2025), and marine ecosystems (Hermann et al., 2023; Palmas et al., 2025; Smale et al., 2019), destabilizing their structure and functions, endangering the native biodiversity and of different ecosystem service (ES) provisioning (Strydom et al., 2020; Plumlee et al., 2024; Wernberg et al., 2025). In recent decades, researchers' focus (Zhang et al., 2025) has also shifted to the Compound Heat Waves (CHW), which are characterised by persistent heat extremes across daytime (high $T_{\max}$ anomaly) and nighttime (high $\mathrm{T}_{\min}$ anomaly). These increased rates and intensity of HW are having catastrophic effects on the natural ecosystems and agroecosystems (and associated biodiversity and ES, productivity), and human well-being (Barriopedro et al., 2023; IPCC, 2023; Lixia et al., 2022; Ruthrof et al., 2018; Wernberg et al., 2025).

Impacts of HW on the terrestrial ecosystems become more severe when they are associated with drought (El-Madany et al., 2020; Barriopedro et al., 2023; Yin et al., 2023; Delerue et al., 2025), and such episodes are increasing globally (Awasthi et al., 2022; Mukherjee & Mishra, 2021; Zhang et al., 2025). HW also often interacts with pollution (e.g., due to microplastics and nutrient-enrichments), aggravating impacts on the plants (and dependent biota) in the terrestrial and aquatic environments (Bartosiewicze et al., 2016; Guo et al., 2023; Jia et al., 2023). A common index to identify heat extremes has not been achieved so far as they show variation in their threshold limit across the globe (Perkins et al., 2013). HW has been defined variably at both the global level and within India. At the global level, World Meteorological Organization (WMO), HW is defined as a period where: (i) "local excess heat accumulates over a sequence of unusually hot days and nights" (https://wmo.int/topics/heatwave)", or (ii) "the daily maximum temperature exceeds the normal maximum temperature by 5 °C (or 9^0F) for five or more consecutive days." According to the Indian Meteorological Department (IMD) HW are a "period of abnormally high temperatures, more than the normal maximum temperature that occurs during the summer season in the North-Western parts of India", which usually occurs during March to June, but can also extend till July. According to the IMD, specific criteria need to be met to declare an episode of HW in India (Fig. 1).

HW are caused by interacting effects of global warming (key factor globally), El Niño–Southern Oscillation (ENSO: in the tropics/Pacific), atmospheric planetary waves, high-pressure weather systems (in the rest of tropics), vegetation, and the soil moisture content (Barriopedro et al., 2023). Global warming has amplified the frequency, intensity and duration of HW, thereby leading to enhanced risks to ecosystems, ES, and human well-being, causing mortality and die-off in forests, grasslands,

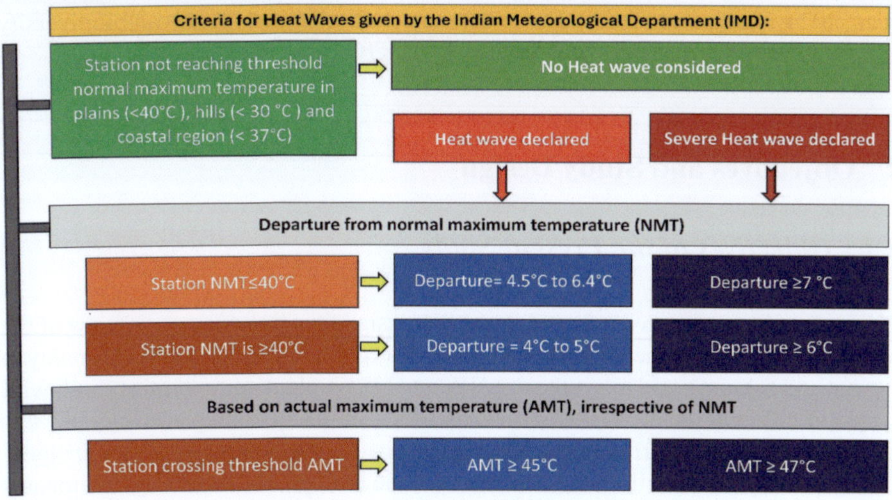

Fig. 1 Criteria for heat waves as per the Indian Meteorological Department (IMD). It is declared on the second day, if these criteria are fulfilled at ≥ 2 stations within Meteorological sub-division for ≥ 2 consecutive days. "NMT" (Normal Maximum Temperature) and "AMT" (Actual Maximum Temperature)

coral/kelp forests, human mortality, water scarcity, wildfires, losses in productivity and crop yield (https://wmo.int/topics/heatwave; Barriopedro et al., 2023; Bastos et al., 2020; Ciais et al., 2005; IPCC, 2023; Wernberg et al., 2025; Yin et al., 2023). The HW also reduces the carbon sink capacity while increasing the risk of wildfire (IPCC, 2023).

1.2 Overview of Heat Stress (HS)

Impacts of HW are often assessed based on HS or heat shock. It is defined as the "rise in temperature beyond a threshold level for a period of time sufficient to cause irreversible damage to plant growth and development" (Wahid et al., 2007), due to heat acclimation (Kan et al., 2023). HS is considered when there is a transient rise in temperature (often 10–15 °C) above the ambient temperature, but its functional complexity depends on intensity, time period, and temperature rise rate (Wahid et al., 2007). According to Bussotti and Pollastrini (2021), research in plant stress may either be agronomical (to sustain production of crops including timbers) or ecological (to preserve biodiversity and associated ES) under the global warming-induced extreme environmental conditions. The HS impacts on plants are being studied across the world (Willits et al., 2001; Zhang et al., 2010; Teskey et al., 2015; Van der Westhuizen et al., 2020; Zhao et al., 2020; Kan et al., 2023; Roach et al., 2023; Shelake et al., 2024; Wang et al., 2025). However, such studies in the Himalayan biodiversity hotspot have only been emerging in recent years (Gurung et al., 2019).

2 Objectives and Study Design

2.1 Objectives of the Present Study

The present literature review-based study was designed to provide an overview of the plant diversity across the different spatial scales (global to India, Indian Himalayan region, and Sikkim), and to synthesize HW and HS on plant communities (both wild crops) in different the ecosystems/ environments, ES provisioning and human well-being as well as summarize adaptation, and mitigation measures. The specific objectives our study are as follows: (1) Understand the biological and ecological impacts of HS on plants, such as changes in physiology, distribution range, phenology, diversity, community composition, and distribution; (2) Analyze how the intensity and/ or frequency of HW relates to plant biodiversity, including vulnerability and coping mechanisms across ecosystems; (3) Evaluate adaptation strategies by plants and mitigation efforts by policymakers and local communities against HW and HS in mountain ecosystems.

2.2 Study Design

For the present study, we referred to the main science databases (Google Scholar, PubMed, Science Direct, and Web of Science) to identify suitable research articles, monographs, scientific reports, and books. For literature survey, we used a set of keywords including: "Heat wave" or "Heatwave," "Heat stress," "Climate change," "Global warming," "Biodiversity," "Plants" or "Plant diversity," "Plant community," "Plant ecosystems," "Lake," "River," "Marine/ Sea/ Ocean," "Mountain ecosystem," "Grasslands" or "Forest" or "Temperate forest" or "Tropical forest," or "Agriculture/ Agroecosystem/ Agroforestry system", "Photosynthesis", "Physiology," "Tree mortality," "Forest die-off", "Carbon flux/carbon cycling/ carbon sequestration," "Nutrient cycling," "Greenhouse gas emissions," "Crop failure," "Crop yield," "Productivity," "Water flux," "Drought," "Pollution," "Seafood," "Plankton/ Coral/ Seagrass/ Kelp forests," "Range-shift," "Plant phenology," "Latitude," "Elevation." We limited our search to October 2025, with most of the selected literature published after the year 2000. We searched for studies on the global scale, regional scale (e.g., Asia, Himalaya), and within India, the Indian Himalayan region, and Sikkim.

3 Results and Discussions

3.1 *Global Studies on Heat Waves*

Spatio-temporal analysis of HW (including patterns, drivers, hotspots), and its impacts on different ecosystems, biodiversity, or human well-being has been conducted at the global level (Barriopedro et al., 2023; Zhang et al., 2025: for 1980–2022; Wang et al., 2024: for 1901–2020), as well as at continental or country levels, including in Africa (Engdaw et al., 2022), Europe (Knutzen et al., 2025), Australia (e.g., Chapman et al., 2025; Laz et al., 2023), the US (Bhattarai et al., 2025), Canada (Sinha et al., 2025), Brazil (Bezerra et al., 2025), China (Fan et al., 2025). A recent study (Zhang et al., 2025) has reported a high intensity of decadal rise in global CHW (3.32 °C/ decade) during 1980 to 2022, four-fold more than the reported corresponding value for daytime HW (0.73 °C/ decade) and nighttime HW (0.78 °C/ decade). The study further reported that CHW was particularly highest for the northern hemisphere high altitude (i.e., in Arctic regions: > 10 °C/ decade), and intra-annual CHW showed strong regional and differential lagged effects, especially with ENSO and Pacific Decadal Oscillation (PDO), thus has implications for the well-being of both ecosystems and humans (Zhang et al., 2025). Wang et al. (2024) reported a fivefold rise in HW risks globally during the last three decades (1991–2020) than twentieth century's first three decades (1901–1930). Barriopedro et al. (2023) also reported the three years with the highest HW magnitude for each continent, with Asia experiencing such episodes in 2010, 2021, and 2020. Global warming has increased the occurrence of the HW.

HW studies have also been conducted focusing on the aquatic ecosystems like freshwater lakes, wetlands and rivers (Hermann et al., 2023; Lyu et al., 2025; Zhang et al., 2025), reporting strong effects of HW on their ecosystem health (e.g., carbon/ water fluxes, nutrient availability, algal bloom), aquatic biodiversity (including planktons, aquatic plants), ES provisioning such as fisheries, and human well-being. Similar studies have also focussed on the marine HW impacts on the environment (e.g., acidification, carbon flux), ecological (e.g., deoxygenation, biodiversity loss, ES disruption), oyster and fisheries sustainability, and other societal aspects (Plumlee et al., 2024) in marine ecosystems (Arafeh Dalmau et al., 2025; Gomes et al., 2024; Smale et al., 2019; Starko et al., 2022; Strydom et al., 2020; Wernberg et al., 2025).

3.2 *Studies on Heat Waves in India*

Research studies on HW have also been conducted in India (covering different biogeographic zones), across longer timeframe that spans several decades, including 1950–2020 (Dar & Apurv, 2024), 1951–2013 (Devi et al., 2023a), 1951–2016 (Singh et al., 2021), 1961–2010 (Pai et al., 2013), 1961–2020 (Rajeevan et al., 2023), identifying risk zones, hotspots, or drivers. These studies in India have identified two high-risk HW zones, including Continental Zone (Northwest/Central plains) and Coastal Zone (East Coast), which showed decadal rise in HW episodes of ~2 and 1.3 days, respectively (Rajeevan et al., 2023; Singh et al., 2021). A new emerging Risk Zone has been identified in the Deccan Peninsula (South & Central parts), experiencing increased HW episodes in the recent decades (Singh et al., 2021), while the Himalayan and northeast regions are categorized as low-to-moderate-risk zones but still remain vulnerable due to low thermal acclimation (Devi et al., 2023a). At a finer scale, Azhar et al. (2017) assessed the heat vulnerability index (HVI) for 640 districts of India, and identified > 100 most vulnerable districts, including two categories of very high risk (10 districts: 2SD HVI), and high risk (97 districts: 2–1 SD HVI), which were mostly in the central India. Devi et al. (2023a) have projected that from 2050 onwards, HW in India will rapidly increase in terms of all its parameters (days, events, and duration), including in the coastal, plains and hilly regions.

Singh et al. (2021) analysed spatio-temporal shifts in HW and severe heat wave (SHW) events in India from 1951 to 2016, focusing on two seasons, i.e., March–May (pre-monsoon), June–July (early summer monsoon), and identified hotspots (northwestern, central, and south-central regions), and southward expansion of SHW. The study also reported a rise in the episodes of HW (~12/ decade) and SHW (~5/ decade) since 1980, associated with mortality in Orissa and Andhra Pradesh states.

Studies have linked ENSO to the HW in India (Dar & Apurv, 2024; Rajeevan et al., 2023), like elsewhere in the world (Barriopedro et al., 2023; Zhang et al., 2025). Rajeevan et al. (2023) reported more HW days in India during the years when it experienced El Niño (than the La Niña), but in the east coast region, HW remained similarly active or got enhanced during the years of La Niña. A recent study in India (Dar & Apurv, 2024) reported that pre-monsoon HW originated predominantly from the northwest region, which had the country's highest spatial coverage, which in turn followed two trajectories moving in directions of northeast (with higher intensity, but lower coverage and duration) and southeast (with larger area and longer duration). The study further found strong linkages with ENSO (to HW moving southeast) and larger anomalies in radiative flux (HW moving northeast), for the two major HW of India.

Nori-Sarma et al. (2019) reported 18.1% more human mortality rate during HW days than days without HW in the north-western region in India (a HW hotspot). From aquatic ecosystems, Sarkar et al. (2022) reported frequent summer episodes of Inland heat waves (IHWs) in the Lower Ganga Basin, causing severe water temperature rises (up to 42.8 and 43.5 °C, respectively, in Bhomra and Ganga River stretches). The study also reported IHWs-linked impacts on its physico-chemical parameters (including dissolved oxygen, nutrients like phosphate and nitrate, pH, etc.), shifts in dominance of phytoplankton (from blue-green algae to diatom), disrupting the food web and threatening the sustainability of regional fisheries.

3.3 Overview of Plant Biodiversity

Globally, high plant biodiversity is represented by high species richness across the angiosperms (Qian et al., 2023), gymnosperms (Earle, 2025), pteridophytes (Rothfels et al., 2025), bryophytes (Lueth & Reski, 2023), algae (Guiry & Guiry, 2025), and lichens (BSI, 2025a, 2025b; Lücking et al., 2017) (Table 1). India, being one of the 17 mega biodiversity countries, covers parts of four (out of 36) global biodiversity hotspots (Himalaya, Western Ghats & Sri Lanka, Indo-Burma and Sundaland), and is also one of the eight Vavilovian centres of crop plants' origin and diversity (Mittermeier et al., 2011; CBD, 2025). It is bestowed with very rich flora (Table 2; BSI, 2025a, 2025b), representing high percentage of global plant diversity (Earle, 2025; Guiry & Guiry, 2025; Lueth & Reski, 2023; Qian et al., 2023; Rothfels et al., 2025) across the angiosperms (6.25%), gymnosperms (7.50%), pteridophytes (10.72%), bryophytes (12.67%), algae (17.86%), and lichens (15.70%).

The Indian Himalayan region (IHR) is a biodiverse region for plants, representing a large percentage of Indian flora across angiosperms (~39%), gymnosperms (~61%), pteridophytes (~58%), bryophytes (~69%), algae (~68%), and lichens (~53%) (Nayaka & Ngangom, 2024; Singh & Pusalkar, 2020; Table 1). Despite covering only 0.22% the geographical area of India, the rich plant diversity of Sikkim state representing high percentage of the country's flora across gymnosperms and pteridophytes (~36% each), Angiosperms (~27%), lichens (~22%), and bryophytes (~13%), (Table 1; Kholia, 2010; Maity & Maiti, 2024; Nayaka & Ngangom, 2024; Singh et al., 2018; Srivastava, 2020). However, algae remain under-documented in the Sikkim Himalaya (Suseela & Toppo, 2007; Bhakta & Adhikary, 2014).

In India, plants shows high level of endemism (Table 2) across angiosperms (19.46%), gymnosperms (14.46%), pteridophytes (22.74%), bryophytes (28.44%), macro-algae (200 species of total 865 species: 23.12%), and lichens (17.74%), with most of these endemics concentrated in the Western Ghats region, and the IHR (Singh

Table 1 Plants' biodiversity under major groups globally, in the Indian Himalayan Region and Sikkim

Plant groups	Globally	Indian Himalayan region	Sikkim
Angiosperms	3,54,000[i]	8,700[vii]	5937[x]
Gymnosperms	1107[ii]	51[vii]	30[xi]
Pteridophytes	12,300[iii]	766[vii]	480 [xii]
Bryophytes	22,200[iv]	1,955[vii]	355[xiii]
Algae	50,600[v]	6158[viii]	125[xiv–xv]
Lichens	19387[vi]	1,614[ix]	657[ix]

N.B.: Literatures referred are [i]: Qian et al. (2023); [ii]: Earle (2025); [iii]: Rothfels et al. (2025); [iv]: Lueth & Reski (2023); [v]: Guiry & Guiry (2025); [vi]: Lücking et al. (2017); [vii]: Singh & Pusalkar (2020); [viii]: Singh and Singh (2024); [ix]: Nayaka & Ngangom (2024); [x]: Maity & Maiti (2024); [xi]: Srivastava (2020); [xii]: Kholia (2010); [xiii]: Singh et al. (2018); [xiv]: Suseela & Toppo (2007); [xv]: Bhakta & Adhikary (2014).

Table 2 Plant diversity in India, along with the number of endemics (and % of total diversity per group) and IUCN Red Listed species across the flora group

Plant groups	Total species [i]	Species (and %) of Endemics [ii-iv]	IUCN Red Listed threatened species [v]
Angiosperms	22,108	4,303 (19.46%)	416
Gymnosperms	83	12 (14.46%)	12
Pteridophytes	1,319	300 (22.74%)	2
Bryophytes	2,813	800 (28.44%)	7
Algae	9,035 (>865 macro-algae)	200 (only macro-algae)	NA
Lichens	3,044	540 (17.74%)	NA

N.B.: Studies referred are [i]: BSI (2025a); [ii]: Singh et al. (2015); [iii]: Ministry of Environment, Forest & Climate Change et al. (2025); [iv]: BSI (2025b); [v]: Kumar et al. (2022).

et al., 2015; BSI, 2025b; Ministry of Environment, Forest & Climate Change et al., 2025). In the country, vascular plants exhibit high species endemism (32%-40%), particularly in the Western Ghats and Eastern Himalaya (CBD, 2025).

Out of 437 IUCN-listed threatened plant species reported from India, a maximum of 416 species are Angiosperms, whereas Gymnosperms, Pteridophytes, and Bryophytes are represented by 12, 2, and 7 species, respectively (Table 2; Kumar et al., 2022). In India, the number of threatened plant species categorized either as Critically Endangered, Endangered, or Vulnerable by the IUCN Red List is significantly higher than for other faunal taxa in the country (CBD, 2025). Recently, a multi-agency collaborative vision document was released for national level Red List Assessment of Indian Flora (and Fauna) during 2025–2030, which aims to assess the threatened status of the country's thousands of plant species which are not yet evaluated based on the criteria of the IUCN Red List (Ministry of Environment, Forest & Climate Change et al., 2025). This is expected to aid in identifying more threatened species of plants, especially from pteridophytes, bryophytes, algae, and lichens, which still remain under-evaluated.

3.4 Heat Wave-Induced Heat Stress on Plants: Indicators, Responses and Adaptations

Plants are considered important and widely used indicators to assess ecosystem health (Garg et al., 2024; Schulze et al., 2004), GCC (Rumpff et al., 2010), land use change (Quétier et al., 2007), and pollution (Bayouli et al., 2021). Plants also serve as indicators of resistance against HW because of their strong and diverse responses to HS, affecting reproduction, development, photosynthesis, molecular, physiological, and enzymatic processes (Table 3).

Table 3 Different indicators, responses and adaptations to heat stress reported in different plants

Indicators of heat stress	Wild/ crop plants studied (reference)
Reproductive failure in plants (microspores degeneration and tapetum hypertrophy, male infertility, impaired sucrose metabolism)	*Hordeum vulgare* (Oshino et al., 2007), *Arabidopsis* (Deng et al., 2016), *Cicer arietinum* (Kaushal et al., 2013)
Poor seed set linked to ↓growth of pollen tube, ↓viability, ↓stigma receptivity, ↓fertilization	Diverse plant species (Swetha et al., 2025)
Disruption of plant growth and development due to heat acclimation	Diverse plants (Li et al., 2021a)
Crop failure due to absence of flowers or fruits production	*Coffea arabica* (Marias et al., 2017)
↓Stomatal conductance (gs)	*Coffea arabica* (Marias et al., 2017), *Cicer arietinum* (Devi et al., 2023b)
↑ Respiration rate—↑ stored carbohydrates use, imbalance of energy production and demand	Diverse plant species (Swetha et al., 2025)
↓ Water content in leaf	*Cicer arietinum* (Devi et al., 2023b)
↓Transpiration and ↓ water and nutrients uptake by plants	*Capsicum annuum* (Rajametov et al., 2021), diverse plants (Swetha et al., 2025)
Change in structure of chloroplast and thylakoid membrane	*Larrea divaricata* (Schreiber & Armond, 1978), *Vicia faba* (Gounaris et al., 1984), *Oryza sativa* (Vani et al., 2001), *Arabidopsis thaliana* (Zhang et al., 2010)
Change in Chlorophyll (FV/FM)	*Larrea divaricata* (Schreiber & Armond, 1978), *Solanum lycopersicum* (Willits & Peet, 2001), *Coffea arabica* (Marias et al., 2017); *Gossypium hirsutum* (Van der Westhuizen et al., 2020), *Cicer arietinum* (Devi et al., 2023b), *Pennisetum glaucum* and *Vigna unguiculata* (Kumar et al., 2025)
↓Chlorophyll content	Syrian barley landrace (Havaux & Tardy, 1999), Winter wheat (Ristic et al., 2007), Rhododendron simsii (Lin et al., 2022), *Cicer arietinum* (Devi et al., 2023b)
↓rate of Photosynthesis	Grapevine (Wang et al., 2010), diverse plants (Sangwan et al., 2022)
Imbalance in Ca2 +	Diverse food crop (Sangwan et al., 2022)
↓Photosynthesis process (RuBisCO, sucrose phosphate synthase, and sucrose activities)	*Cicer arietinum* (Devi et al., 2023b)
Inactivation of RuBisCO because of chloroplast ultrastructure destruction	Diverse plants (Scafaro et al., 2023; Swetha et al., 2025)
↓Photosystem II (PS II) ↓electron transport and ↓ATP synthesis efficiencies	Diverse plants (Scafaro et al., 2023; Swetha et al., 2025)
↑Malondialdehyde (MDA) content	*Cenchrus ciliaris* (Tommasino et al., 2012), *Rhododendron simsii* (Lin et al., 2022)

(continued)

Table 3 (continued)

Indicators of heat stress	Wild/ crop plants studied (reference)
↑Reactive oxygen species (ROS)	*Rhododendron simsii* (Lin et al., 2022), Diverse food crop (Sangwan et al., 2022; Swetha et al., 2025)
↓Total protein content	Strawberry (Gulen & Eris, 2004)
Denaturation of proteins	Diverse plants (Sangwan et al., 2022; Swetha et al., 2025)
↑Lipid peroxidation	Diverse food crop (Sangwan et al., 2022; Swetha et al., 2025)
↓Foliar non-structural carbohydrates	*Coffea arabica* (Marias et al., 2017)
↓Leaf sucrose production	*Cicer arietinum* (Devi et al., 2023b)
↓Seed yield and nutrient content	*Cicer arietinum* (Devi et al., 2023b)
↑Soluble sugar	*Rhododendron simsii* (Lin et al., 2022)
↑Proline accumulation	*Gossypium hirsutum* (De Ronde et al., 2000), Wild sorghum (Gosavi et al., 2014), *Capsicum annuum* (Rajametov et al., 2021), *Rhododendron simsii* (Lin et al., 2022)
↑Activity ofδ-1-pyrroline-5-carboxylate synthetase (P5CS)	Wild sorghum (Gosavi et al., 2014), Transgenic tobacco (Sellamuthu et al., 2024)
↑Ascorbate Peroxidase (APX) activity	*Rhododendron simsii* (Lin et al., 2022), *Primula minima* (Roach et al., 2023)
↑Peroxidase (POD) activity	Strawberry (Gulen & Eris, 2004), wild sorghum (Gosavi et al., 2014), *Rhododendron simsii* (Lin et al., 2022)
↑Glutathione reductase (GR) activity	*Primula minima* (Roach et al., 2023)
↑Glucose-6-phosphate dehydrogenase (G6PDH) activity	*Primula minima* (Roach et al., 2023)
↑Activity of Superoxide Dismutase (SOD)	Wild sorghum (Gosavi et al., 2014)
↑Activity of Catalase (CAT) enzymes	Wild sorghum (Gosavi et al., 2014), Rhododendron simsii (Lin et al., 2022), *Primula minima* (Roach et al., 2023)
Plant hormone: Salicylic acid (SA)	*Triticum aestivum* (Kumar et al., 2015), diverse crops (Sangwan et al., 2022) and other plants (Li et al., 2021a)
Level of other plant hormones: abscisic acid (ABA), auxin, brassinosteroids (BRs), cytokinin (CK), ethylene (ET), jasmonate (JA)	Diverse plants (Li et al., 2021a)
↑Heat Shock Proteins (HSPs) production	Diverse food crop (Sangwan et al., 2022), *Agave sisalana* (Batcho et al., 2021)
↑Small Heat Shock Proteins (sHSPs)	*Primula forrestii* (Zhang et al., 2013), *Primula sikkimensis* (Gurung et al., 2019)
Change in endophytic microbiota community	*Rhododendron simsii* (Lin et al., 2022)

HS interferes with the growth and reproduction, and physiology of plants (Table 3, Fig. 2), including both wild plants and food crops (Deng et al., 2016; Oshino et al., 2007). It is well documented that a mild temperature rise enhances plants' growth and development via thermomorphogenesis (Casal and Subramanian, 2019), but this effect varies in different plant species. If HS precedes the mild temperature, the plants undergo accumulation of ROS as well as lead to cell death or cell damage (Zhao et al., 2020), and ultimately disrupt growth and development in plants through heat acclimation (Kan et al., 2023; Li et al., 2021a). HS can even cause reproductive failure due to male infertility through degeneration of microspores and tapetum hypertrophy (Deng et al., 2016; Oshino et al., 2007), and by altered sucrose metabolism (Kaushal et al., 2013). HS treatment of leaves (for 90 min.) resulted in declined levels of non-structural carbohydrates in foliage, gs, and FV/FM, and subsequent crop failure (absence of flowers, fruits) in *Coffea arabica* (Marias et al., 2017).

Plants can sustain a high photosynthesis rate within the optimal temperature range (20–30 °C), but can also do well even till 35 °C (Teskey et al., 2015). The elevated leaf temperature above the threshold value co-limits photosynthesis in terrestrial plants through increased stomatal opening (Urban et al., 2017), decoupling of conductance of stomata (gs), deactivation of RuBisCO, and disruption in rates of chloroplast electron transport (Scafaro et al., 2023; Swetha et al., 2025). HS in plants is caused by a decline in photosynthesis and/or a rise in respiration rate due to damaged or declined efficiency of leaf metabolism (Feeley et al., 2020). Photosystem II (PSII) can experience permanent damage at very high temperatures of ~40–60 °C, i.e., maximum photosynthetic thermal tolerance (T50), resulting in reduced rates of electron transport and ATP synthesis, failure of photosynthesis, and ultimately leaf death (Baker, 2008; Feeley et al., 2020; Scafaro et al., 2023; Swetha et al., 2025). In terms of

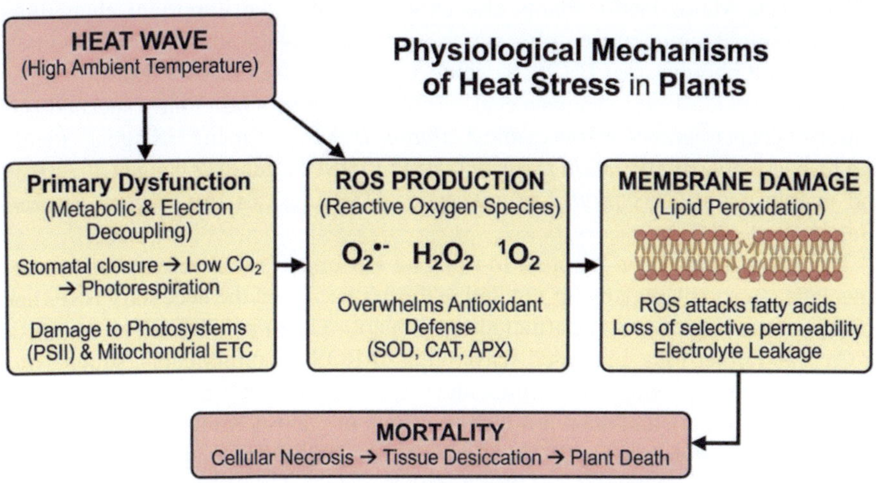

Fig. 2 Physiological mechanisms of heat stress in plants, illustrating the pathway from heat wave to ROS production to membrane damage to mortality

photosynthesis efficiency, HS alters chloroplast structure and thylakoid membrane (Schreiber & Armond, 1978; Gounaris et al., 1984; Vani et al., 2001; Zhang et al., 2010; Swetha et al., 2025), chlorophyll loss (Devi et al., 2023b; Havaux & Tardy, 1999; Ristic et al., 2007), change in chlorophyll fluorescence (FV/FM) (Marias et al., 2017; Schreiber & Armond, 1978; Van der Westhuizen et al., 2020; Willits & Peet, 2001), and transpiration rate (Rajametov et al., 2021; Roach et al., 2023; Swetha et al., 2025). A study in chickpea (a C3 plant) reported that HS can damage the membrane, change gs, and decrease water content, reduce chlorophyll content and FV/FM, and ultimately cause a decline in yield and nutrient content of seeds (Devi et al., 2023b). HS also interferes with protein content (Gulen & Eris, 2004) and its denaturation (Sangwan et al., 2022).

Plants undertake different adaptation strategies against HS through various mechanisms, including protecting reproductive systems (especially in males) and photosynthesis apparatus, creating imbalance in Ca^{2+}, producing heat shock factors (HSFs), HSPs, sHSPs, and through SA pathway (and other hormonal changes), antioxidant activity, and proline production (Table 3). Plant responds to HS by creating an imbalance in Ca2 + (Kan et al., 2023; Sangwan et al., 2022) by elevated influx of Ca^{2+} into cytosol, which binds to specialised sensor proteins like Calmodulins (CaMs) and Calmodulin-like proteins (CMLs) (La Verde et al., 2018). Through signal transductions, these activated sensors help activate transcription factors e.g., heat shock factors (HSF), which ultimately aid plants to up-regulate heat shock proteins (HSPs) productions (Batcho et al., 2021; Kan et al., 2023; Sangwan et al., 2022), small heat shock proteins (sHSPs) as protective molecules (Gurung et al., 2019; Zhang et al., 2013).

HSPs act as chaperones (Kan et al., 2023) and are classified (according to their molecular weights) under five families as HSP20, HSP60/40, HSP70, HSP90, and HSP100 (De Maio, 1999). Plants also express sHSPs, the important chaperones adapted to rapid temperature shifts, which bind to the thylakoid membrane and help in protecting photosystem II (PS II) along with the entire electron transport chain (Heckathorn et al., 1998; Sangwan et al., 2022). Molecular studies have succeeded in conferring heat tolerance in transgenic *Arabidopsis thaliana*, using HSP genes identified respectively from *Primula forrestii* (sHSP: PfHSP17 gene) (Zhang et al., 2013), and in transgenic *Gossypium hirsutum* using HSP70 genes from *Agave sisalana* (Batcho et al., 2021).

The SA has also been reported to mitigate HS impacts in diverse plant species (specially in crop plants), by up-regulating the expression of the necessary HSFs and HSPs, functioning of enzymes (including with antioxidant properties), homeostasis of Ca^{2+} and other ions, increased scavenging of ROS by antioxidants, suppressing MDA levels and peroxidation of lipids, and by protecting photosystem II (to maintain photosynthesis rate), and the reproductive system of plants (Kumar et al., 2015; Lin et al., 2022; Sangwan et al., 2022; Swetha et al., 2025). Thus, applications of SA have important role in mitigating HS-induced damage and loss of yield in agricultural crops (Sangwan et al., 2022).

Plant also has intrinsic mechanisms to counter the rise in damaging agents like Reactive Oxygen Species (ROS) like hydrogen peroxide (H_2O_2), superoxide radical

($O_2^{\bullet -}$), hydroxyl radical (•OH) associated with HS (Zhao et al., 2020; Sangwan et al., 2022; Kan et al., 2023; Swetha et al., 2025) and Malondialdehyde (MDA) contents (Lin et al., 2022; Tommasino et al., 2012). Plants also employ oxidative stress management by increasing the production of antioxidant enzymes including Peroxidase (POD), Ascorbate Peroxidase (APX), Catalase (CAT), superoxide dismutase (SOD), Glutathione reductase (GR), Glucose-6-phosphate dehydrogenase (G6PDH), and Peroxidase (POD) (Kan et al., 2023; Roach et al., 2023) as well as osmotic adjuster like δ-1-pyrroline-5-carboxylate synthetase (P5CS) which in turn is critical for synthesis of proline (Gosavi et al., 2014; Sellamuthu et al., 2024). Plants also adopt accumulation of Osmo-protectant by increasing Proline and sugar production to help stabilize cellular structure by maintaining cell turgor and water balance (De Ronde et al., 2000; Gosavi et al., 2014; Rajametov et al., 2021; Kan et al., 2023). In a review, Zhao et al. (2020), also reported that plants respond to HS through integrated physiological (membrane stability, ROS scavenging, water balance), transcriptional (HSFs, HSPs), noncoding RNA (miRNAs, SiRNAs, regulating genes), and epigenetic (DNA methylation, histone modifications) mechanisms to maintain homeostasis, with recent focus on crosstalk between these layers for tolerance and stress memory to better cope with future HS events.

Lin et al. (2022) reported a change in community composition of endophytic microbiota *in Rhododendron simsii*, with dominance of *Pezicula and Paracoccus* for medium HS, but that of *Acidothermus and Haliangium* during high HS (Lin et al., 2022). The study also found association of the representative endophyte with HS markers, positive correlation of *Pezicula* (of medium HS) with chlorophyll a, proline, sugar and MLA contents in leaf, but negative correlation with soil content of ammonium N in soil (-ve), as well as positive correlation of *Haliangium* (of high HS) with contents of soluble sugar and MLA, thus signifying the adaptive response of the plants against HS.

3.5 *Heat Stress-Driven Plant Mortality and Die-Offs*

At the levels of individual and population, HW (and resulting HS) also affects plant populations and community composition. Plants undergo avoidance or tolerance physiological adaptations under slight increases in temperature conditions beyond their optimal, minimum and maximum range. However, during prolonged HS conditions, the mortality rate increases with a rise in the rates of evaporation and transpiration, accompanied by low rainfall, drastically shortening tolerance time for the plants (Andregg et al., 2013; Hammond et al., 2022). The rise in drought conditions (often associated with HS) along with a decrease in precipitation is also reported to cause a decline in ecological productivity (Bastos et al., 2020; Yin et al., 2023; Delerue et al., 2025).

The HW-driven HS has caused large-scale tree mortality (Allen et al., 2010, 2015; Anderegg et al., 2013, 2016) as well as forest die-off (Marchin et al., 2022; Matusick et al., 2018; Park Williams et al., 2013), through wildfires (IPCC, 2023; Park

Williams et al., 2013; Rajeevan et al., 2023), and pests/ diseases (Freer-Smith et al., 2017). Although HS-induced tree mortality and die-offs are reported from the tropics (McDowell et al., 2018), neotropics (Aleixo et al., 2019), and temperate (Sterck et al., 2024) regions, but the impacts are highest in the Mediterranean/ northern temperate forests than their counterparts in tropical regions (Mantgem and Stephenson, 2007; O'sullivan et al., 2017; Zhu et al., 2018; Hauck et al., 2025). However, within the temperate forests, tree mortality rate is also higher in the case of shade-loving broad-leaved trees than the light-tolerant species (Kunert & Hajek, 2022). This mortality also varies among plant species depending on their functional traits (Aleixo et al., 2019; Margalef-Marrase et al., 2020; Sterck et al., 2024; Vitali et al., 2017), P50 levels (Vitali et al., 2017), and composition of successional stages (Caspersen, 2004). Such mortality also alters tree density and spatial arrangement in the ecosystems (Anderegg et al., 2013; Zhang et al., 2017).

The decline in forest longevity because of global warming is reported by long-term studies in tropical (Locosselli et al., 2020) and boreal forests (Chen & Luo, 2015; Searle & Chen, 2018). These changes will disrupt gross primary productivity (GPP) in the ecosystems (Ciais et al., 2005; Stephenson et al., 2011), water/ carbon fluxes (Anderegg et al, 2016), cause decline in standing aboveground biomass (Chen & Luo, 2015), carbon sequestration (Shi, 2024), provisioning of ES including the cultural services derived from forests (the protected areas), in mountain regions like the Himalaya (ul Hassan et al., 2021; Kattel, 2022).

Temperature rise associated with HW also disrupts the window of natural regeneration by significantly reducing seed lifespan and viability (Tiebel et al., 2024). The study by Tiebel et al. (2024) also found that when compared to 15–25 °C (representative of past and present temperature), and low moisture condition, seeds stored at 40 °C (for three months) in high moisture condition showed significant decline in germination due to complete loss of seed lifespan at 42 days for pioneer tree species (e.g., *Betula pendula*), 1 week for initial colonizers (e.g., *Populus tremula*), and up to 28 days for seed tolerant species (like *Picea abies*). A modelling study in China on stand-level tree mortality (Ding et al., 2023) has reported significant effects of variables related to soil (humus layer thickness), and stand (age of stand, tree height, avg. DBH, crown cover, and various biodiversity indices), that have implications for future predictions of forests and their management planning. Tree mortality is generally attributed to past episodes of drought (Kunert & Hajek, 2022; Matusick et al., 2018), but it may get worse with continued temperature rise (Allen et al., 2010, 2015; Park Williams et al., 2013). HW accelerates the rate of mortality and extinction risk for plants in forests since an increase in temperatures affects many physiological mechanisms, causing water transport failure, declined soil moisture, carbon starvation, and increased vulnerabilities to biotic stressors like attacks by pathogens and pests (Aleixo et al., 2019; Hartmann et al., 2022; López et al., 2022; Park Williams et al., 2013; Sterck et al., 2024). The resulting tree mortality ultimately leads to forest dieback and changes in forest ecosystems with forest transitioning to its simpler structure (e.g., decline in basal area, canopy coverage, above-ground biomass), composition (e.g., encroachment by savanna, shrubs, invasive species), and declined productivity (da Silva & Rossatto, 2022; French et al., 2017; Godoy et al., 2011; Park Williams et al., 2013; San-José et al., 2021; Stephenson et al., 2011).

3.6 Heat Wave-Induced Stress Impacts on Plant Community Composition

Studies have reported strong correlation of T50 [the critical temperature at which plant experiences 50% decline in Photosystem II (PSII) efficiency] with eco-physiological traits of plant leaves with relationship being negative for specific leaf area (SLA) (da Silva & Rossatto, 2022), but positive for potential evapotranspiration (PET) (Godoy et al., 2011), and leaf thickness (da Silva & Rossatto, 2022), presence of Kranz anatomy (French et al., 2017; Kumar et al., 2025), explaining the heat tolerance and projected community composition shifts in the tropical and temperate/ mediterranean region. Global warming and HW-linked HS can also alter community composition across ecosystems. A study in a neotropical region (47.65 °C) revealed that tree leaves of savanna (T50 = 49.36 °C) had greater values of heat tolerance than plants from semi-deciduous forest (T50 = 47.65 °C) (da Silva & Rossatto, 2022). The plant community composition in forests and grasslands are also projected to shift, leading to replacement of semi-deciduous forest (by savanna species) (da Silva & Rossatto, 2022), native trees/ herbs (by their invasive close-relatives) (Godoy et al., 2011), native shrubs (by exotic shrubs and grasses) (French et al., 2017), C3 grasses (by C4 species) (French et al., 2017), and C3 fodder (by C4 counterpart) plant(Kumar et al., 2025), due to higher T50 values and plasticity of the latter groups. Global warming and HS impacts on plants and their adaptation strategies also differ between plant forms (e.g., C3 and C4 plants).

A comparative study (Kumar et al. (2025), has revealed negative impact of HS on FV/FM in cowpea (a C3 plant), causing yield loss of green fodder (range: -3.83 to -18.56%), whereas opposite effects was observed for pearl millet (a C4 plant), showing yield gain (range: 9.44% to 25.02%). Thus, increased level of HW and global warming may favour the C4 plants (with Kranz anatomy), over their C3 (without Kranz anatomy) counterparts causing yield gain/ success for C4 grasses (e.g., *Imperata cylindrica*), crops (such as maize, sugarcane, and sorghum) and fodder (e.g., cowpea), but yield loss/ decline in case of C3 grasses (e.g., *Ammophila arenaria*), crops (like oats, rice, soybeans, and wheat) and fodder (e.g., pearl millet) plants (Ainsworth & Long, 2005; French et al., 2017; Kumar et al., 2025), which will have implications for agriculture and cattle rearing.

3.7 *Latitudinal and Elevational Patterns of Plants' Response to Heat Stress*

Plant species' T50 values (an indicator of HW/HS impacts) is an important factor shaping plant distribution patterns along the latitudinal and elevational gradients. The T50 in plants varies among different species (da Silva & Rossatto, 2022; Feeley et al., 2020). Studies have reported its declining trend along the latitudinal gradients (e.g., tolerance of tropical rainforest plants > boreal/ temperate forest plants) (O'sullivan et al., 2017; Zhu et al., 2018). Recent studies in neotropics (Feeley et al., 2020; O'sullivan et al., 2017) and in temperate regions (Kong et al, 2024), have also reported similar declining plants' tolerance patterns along the altitudinal gradients (i.e., T50 for low-altitude plants > T50 at colder high-altitude plants), validating the "hot–cold trade-off hypothesis" (Feeley et al., 2020). However, many tropical plant species will be facing greater threats from global warming-driven HS due to their reduced thermal safety margins (T50: adiabatic lapse rate ranging from − 0.4 oC/km to − 5.7oC/km), hence will require more attention in the future (Feeley et al., 2020).

A recent study on woody plants (including 400 species) in wet-evergreen forests Western Ghats (Page and Shankar, 2020) reported monotonic decline in species richness as well as turnover across the latitudinal gradient (8–20° N) that support Rapoport's Rule (↑ range size from lower to higher latitudes), and climatic stability hypothesis (greater tolerance towards temperature as well as rainfall seasonality by species at higher latitude). GCC studies in the mountain ecosystems of Himalaya (Agnihotri et al., 2017; Chitale et al., 2014; Kumar & Khanduri, 2024; Manish & Pandit, 2018; Telwala et al., 2013), Western Ghats (Chitale et al., 2014; Page and Shankar, 2020), Indo-Burma (Chitale et al., 2014), Andes in the neotropics (Feeley, 2012; Feeley et al., 2011, 2020; Lutz et al., 2013), and temperate regions in southern California, USA (Kelly & Goulden, 2008) and Norway, Europe (Felde et al., 2012), have reported native plant' distribution range expansion/ contraction, shifts in community composition, species range, and treelines/ timberline.

Chitale et al. (2014) reported GCC-driven future distribution shifts of endemic plants in the Himalaya and Indo-Burma (northern & north-eastern directions), and Western Ghats (in the directions of southern and south-western) showing shifts towards cooler climates. The study also projected significant shifts in distribution range (reduction: ↓; expansion: ↑) of the endemic plants by 2050 (↓23.99% and ↑7.70%), and 2080 (↓41.34% and ↑ 24.10%), with greater values of range reduction than the expansion in the three studied biodiversity hotspots of India.

Agnihotri et al. (2017) reported warming-induced shifts in species range Himalayan plants (~55 m/decade), and altered treelines (e.g., *Abies spectabilis, Betula utilis*), increasing richness in sub-alpine/alpine zones but threatening endemics, which was attributed to warming/heat response by plant's physiology (e.g., stomatal density, carbon isotopes, metabolites), impacting growth, reproduction, water use, and carbon storage. The distributional changes in the plant communities (Agnihotri et al., 2017; Chitale et al., 2014; Fordham et al., 2012; Manish & Pandit, 2018; Telwala et al., 2013), and phenology change (Kumar et al., 2017) can increase the extinction risks for endemic/ range-restricted species, particularly in the montane forests and grasslands.

Studies in tropical and Mediterranean oceans have also found strong evidence on changes in the latitudinal distribution pattern of marine ecosystems/ biomes in response to marine HW and HS. In marine biomes/ecosystems, HW/ HS has caused species' contraction/expansion of distributional range, and poleward range-shift in Seagrass, Kelp forests, and Corals, due to the strong effects of latitude (Arafeh Dalmau et al., 2025; Wernberg et al., 2025). These shifts in plant's distribution range and composition along the elevational, latitudinal and other environmental gradients will also impact associated fauna and food web in both terrestrial (Sharma et al., 2020; Shelake et al., 2024; Devi et al., 2025) and aquatic (Sarkar et al., 2022; Gomes et al., 2024) ecosystems.

3.8 *Effects of Heat Waves and Heat Stress in Various Ecosystems*

There has been rapid rise in the HW frequency and severity at both global and regional scales driven largely by the man-made GCC (Allen et al, 2010; Awasthi et al., 2022; Hartmann et al., 2022; Pattnayak et al., 2023). HW also impacts structures and functions of the forests (Kong et al., 2024), grasslands (Huang et al., 2023; Wang et al., 2025), agroecosystems (Su et al., 2021), freshwater ecosystems of lakes/ rivers/ wetlands (Sarkar et al., 2022; Hermann et al., 2023; Lyu et al., 2025), and marine ecosystems in seas and oceans (Palmas et al., 2025; Wernberg et al., 2025).

In the terrestrial ecosystems (Figures 2, 3 and 4), HW disrupts carbon and water fluxes in not only forests (Graf Pannatier et al., 2012; Tatarinov et al., 2016; Van Gorsel et al., 2016), but also in grasslands (Li et al., 2021b; Qu et al., 2018), and agroecosystems (Zanotelli et al., 2022). As a result, there is elevated HS to the plants (and dependent animals) due to a reduction in soil moisture, an increase in wildlife risks, pest infestation, diseases, and rates of evaporation (Rajeevan et al., 2023). The resulting decline in GPP due to altered carbon and water dynamics, and crop yield, ultimately affects the well-being of plants (biodiversity, composition), food chains/ webs, ecosystems, and humans (Bastos et al., 2020; Huang et al., 2023; IPCC, 2023; Kan et al., 2023; Ortiz-Bobea et al., 2021; Rajeevan et al., 2023; Schnabel et al., 2025; Sentis et al., 2013). HW also affects the stability and functional attributes of the forest ecosystem, increasing the wildfire risk and reducing its capacity for carbon sink (IPCC, 2023; Park Williams et al., 2013).

In the agroecosystems, HS is projected to reduce 21% of agricultural productivity globally (Ortiz-Bobea et al., 2021). The average temperature globally is predicted to rise by 1.5–2° Celsius by 2030 and 2060, respectively (IPCC, 2023). Yields of main food crops are projected to decline (soybean: 3.1%; rice: 3.2%; wheat: 6%; and maize: 7.4%), with each 1 °C rise in world temperature (Zhao et al., 2017), which will jeopardize the food security of the ever-increasing global human population. HS is

Fig. 3 India's representative deciduous forest in Sikkim, Eastern Himalaya (**A**); evergreen tropical evergreen forest in Western Ghats, Agasthyamalai Biosphere Reserve, Kerala (**B**); temperate evergreen forest in Barsey Rhododendron Sanctuary, Sikkim (**C**); and conifer forests in Singalila National Park, Darjeeling, West Bengal (**D**)

very crucial to understand and analyze as it affects the production of the crops as well as the food security globally (Kan et al., 2023), and thus agricultural sustainability at the regional and global levels.

The inland HW negatively impacts freshwater ecosystems, including rivers, lakes, and wetlands (Fig. 5A, B) by altering their physico-chemistry, nutrient availability, hydro-biology, causing deoxygenation, shifts in community composition (blue-green algae to diatoms), threatening native biodiversity (Planktons and higher aquatic biota such as fishes), food web, and ES provisioning, including production of fisheries (Sarkar et al., 2022; Hermann et al., 2023; Lyu et al., 2025; Zhang et al., 2025). HW also affects the overall health of marine ecosystems (Brauko et al., 2020; Wernberg et al., 2025), altering carbon balance (changing into a carbon source from a carbon sink), nutrient cycling regime (Bartosiewicze et al., 2016; Palmas et al., 2025), causing greenhouse gas emissions (Audet et al., 2017; Bartosiewicze et al., 2016), and acidification in oceans (Wernberg et al., 2025).

In the estuarine and marine ecosystems (Fig. 5C, D, HW- linked decline in dissolved oxygen (DO) and algal bloom (Bartosiewicze et al., 2016; Brauko et al., 2020) has cascading impacts on its biodiversity (including Planktons, Seagrass, Kelps, Corals, and associated biota), and community composition (Lyu et al., 2025;

Fig. 4 Representative alpine grasslands (above treeline) in Uttarakhand (**A**), organic large cardamom-based agroecosystems in Sikkim (**B**), traditional agroecosystems with maize (a C4 crops) in Darjeeling Hills (**C**), and conventional rice (a C3 crop) cultivation systems of Northern Brahmaputra plains of Assam (**D**) in India

Smale et al., 2019; Wernberg et al., 2025). These drastic changes in the marine ecosystems may ultimately lead to an altered food web (Gomes et al., 2024), increased risks for mass mortality and species extinction in the Seagrass, Corals, and Kelp forests (Arafeh Dalmau et al., 2025; Starko et al., 2022; Wernberg et al., 2025), disruptions of various ES (Smale et al., 2019) including cycling of nutrients like sedimentary organic matter (Palmas et al., 2025), and sustainability of oysters and fisheries (Cheung & Frölicher, 2020; Plumlee et al., 2024). A recent study on the foundation species has also reported a strong and significant correlation of a rise in marine HW days to elevated levels of coral bleaching, decline in density of seagrass, and biomass of kelp forests (Smale et al., 2019).

It is also observed that the HW are not isolated environmental factors, but they are often associated with other abiotic stresses like droughts (Bastos et al., 2020; Bezerra et al., 2025), and pollutions (Guo et al., 2023; Jia et al., 2023; Sharma et al., 2025). The compound heat and drought stress also amplify the disruption of carbon fluxes, and/ or ecosystem collapse in different terrestrial ecosystems as reported in temperate forests (Graf Pannatier et al., 2012), semi-arid ecosystems (El-Madany et al., 2020), and metalliferous ecosystems (Delerue et al., 2025). They also reduce the GPP of forests (Bastos et al., 2020; Bezerra et al., 2025; Ciais et al., 2005; Laz et al., 2023;

Fig. 5 The freshwater ecosystem of Khecheopalri Sacred Lake, a Ramsar designated high-altitude wetland surrounded by the broad-leaved evergreen forests in Sikkim, Eastern Himalaya, India (**A**); Periyar River surrounded by riparian forests in southern Western Ghats Kerala, India (**B**); Egret Island, a Brackish wetland in the Gulf coast of Texas, USA (**C**); and Arabian Sea representing marine ecosystem as seen from the coast of Goa, India (**D**)

Sinha et al., 2025), and socio-ecosystems (Yin et al., 2023). They even accelerate tree die-off (Hammond et al., 2022; Knutzen et al., 2025), and tree mortality in forests (Allen et al., 2015; Anderegg et al., 2013; Park Williams et al., 2013; Sterck et al., 2024; Zhang et al., 2017) and grasslands (Li et al., 2021b; Van Peer et al., 2004).

Pollution, especially resulting from microplastics and nutrient enrichments also interacts with HWs to exacerbate the negative impacts of HS on plants in both terrestrial and aquatic environments (Sarkar et al., 2022; da Fonseca et al., 2025). Microplastics can cause 7–12% reduction in photosynthesis rate in plants, with more adverse impacts in terrestrial crops (↓6–18%), followed by algae in freshwater ecosystems (↓4–14%), and plants in marine ecosystems (↓2–12%), which translates to decline in annual production of crops (↓109.73 -360.87 MT) and seafoods (↓1.05—24.33 MT), and oxygen production crucial for the survival of heterotrophic organisms including humans (Zhu et al., 2025). Interaction of microplastics with HW aggravate HS impacts on plants, by impacting the water uptake, soil properties, biogeochemical cycles (carbon & nitrogen), nutrient allocation and loss (e.g., Nitrogen), elevating oxidative stress due to increased reactive oxygen species (ROS) production, and declined chlorophyll and photosynthesis rate (Guo et al., 2023; Jia et al., 2023; da Fonseca et al., 2025). On the other hand, HWs increase the microplastic leaching rate (Guo et al., 2023), further complicating the HS in plants, and the closely associated faunal taxa like butterflies (Sharma et al., 2025). Hence, the combined effects of microplastic pollution and HWs can cause a decline in plant's GPP and crop yield (Guo et al., 2023; Jia et al., 2023; da Fonseca et al., 2025). In aquatic environments such as freshwater and marine ecosystems, microplastic pollution disrupts the cooling effects of oceans as a heat sink and increases atmospheric heating, and these feedback mechanisms drastically impact phytoplankton, GPP, and carbon and nitrogen cycling (da Fonseca et al., 2025).

Prediction of the functional and dynamic aspects of these ecosystems (both terrestrial and aquatic) under the influence of GCC is pertinent to better understand the HW impacts (and its interactions with abiotic disturbances like drought and pollution). Protection of these ecosystems is critical for the survival of human society as they provide a huge array of ES, both provisioning and regulating services (Freer-Smith et al., 2017; Kattel, 2022; Cheung & Frölicher, 2020; Plumlee et al., 2024). Phytoplankton in marine ecosystems are responsible for producing 50%-80% Oxygen globally, but are greatly threatened by global warming, HWs, and their interacting effects with the microplastic and nutrient pollution (Bartosiewicz et al., 2016; da Fonseca et al., 2025; one study). The resilience of function and structure of the forest ecosystems is also linked to the provision of cultural services, and their role as a major terrestrial carbon sink since they sequester 25% of the anthropogenic carbon emissions, helping to regulate the global climate (Kattel, 2022; Pan et al., 2011).

3.9 *Potential of Plant Communities to Mitigate Heat Wave Impacts*

High plant biodiversity and community composition can play crucial role in mitigating the adverse HW impacts by providing resistance and resilience against HS in forests (Liang et al., 2016; Schnabel et al., 2025), grasslands (Huang et al., 2023; Isbell et al., 2015a), and agroecosystems (Su et al., 2021), creating a buffered microclimate. Diverse plant communities in forests elevate canopy density and structural complexity, thereby creating a buffering microclimate against HS (Schnabel et al., 2025). They help increase levels of shading and leaf area index (LAI), thereby allowing soil temperature to stabilise, thus improving sequestration and storage of soil organic carbon, cycling of nutrients, and productivity (Isbell et al., 2015b; Chen et al., 2018; Zhu et al., 2025; Huang et al., 2023; Schnabel et al., 2025). Plant community with high diversity having varying root depths (especially in agroecosystems and grasslands) allows for efficient management of water (and nutrients) and improved photosynthesis and productivity, transpiration cooling, even during water scarcity (Isbell et al., 2015a). High plant diversity and asynchrony also help in community stability in forests and grasslands (Isbell et al., 2015a; Liang et al., 2016; Zhang et al., 2018). The stable plant community and associated food chains/web will improve the diversity and composition of beneficial wildlife responsible for providing various ES such as pollination, pest control, seed dispersal, scavenging, etc., thereby enhancing ecosystem-level productivity and mitigation potential against HS (Isbell et al., 2015a; Sentis et al., 2013). Along with focusing on improving overall plant biodiversity, there is a need to protect dominant and keystone species of the different ecosystems (Marjakangas et al., 2023; Wang et al., 2025), and include light-demanding trees having less sensitivity to HS in the forest stands (Kunert & Hajek, 2022).

Studies have also found that conservation agriculture can aid to mitigate the GCC impacts on crop productivity and yields, but it varies across regions (yield gain: tropical regions < temperate regions), and crops (increase in yield of maize over tropics) under future global warming scenarios (Bisht et al., 2016; Su et al., 2021). The traditional and organic agriculture (Fig. 4B, C) can also help to mitigate HS since they improve the diversity of wildlife and beneficial organism in soil (like arbuscular mycorrhizal fungi and rhizobacteria), insects/ wildlife crucial for ES (e.g., pollination and pest control), and ES (carbon sequestration, improved water use, as well as regulations of climate, pests, and soil health) in the agroecosystems (and nearby natural ecosystems) (Sandhu et al., 2010; Winqvist et al., 2012; Njeru et al., 2018; Sharma et al., 2020; López-Felices et al., 2022), ensuring better crop yields. Hence, there is a need to conserve human-modified ecosystems such as organic and traditional agroforestry systems due to their potential to conserve biodiversity and ES (Sharma et al., 2020; Winqvist et al., 2012), particularly against HS. Apart from taking steps to mitigate HW and HS impacts on plants, there is also need to regulate microplastic pollution, nutrient pollution (due inorganic fertilisers/ pesticides), air and soil pollution to increase resilience of forests, grasslands and agroecosystems against GCC impacts and to protect their flora (and fauna), and continued provisioning of different associated ES (da Fonseca et al., 2025; Sharma et al., 2025).

For the aquatic environments, mitigation measures should be taken against both inland and marine HWs (and other GCC triggers, such as pollution). Inland HWs can be mitigated by adopting and preserving riparian vegetation shade along the banks of lakes, rivers, and wetlands (Trimmel et al., 2018). Conserving wetland (with healthy aquatic plants and ecosystem health) can also help reduce 1 to 3 °C of surface temperature, apart from providing cooling effects to nearby areas (Liu et al., 2025), thus helping to adapt to the increasing temperature in the future. Mangrove forests can also help adapt against HWs as they are reported to help marine biota including corals (Johnson et al., 2021). There is a need to protect the plants in freshwater and marine ecosystems from pollution, particularly due to microplastics and nutrient enrichments to preserve their diverse flora (e.g., phytoplanktons, kelps, seagrass), other biota (e.g., zooplanktons, corals), and ES provisioning (Sarkar et al., 2022; Guo et al., 2023; Jia et al., 2023; da Fonseca et al., 2025).

4 Conclusions

In the present study, we highlighted the trends and drivers of HW, HS, and their impacts on plant communities across various ecosystems. The rich plant biodiversity in India (38,402 species representing 8.36% of global flora across Angiosperms, Gymnosperms, Pteridophytes, Bryophytes, Algae, and Lichens) shows significantly high diversity in the IHR (especially in Sikkim and Western Ghats. India's diverse endemic (6155 species), and IUCN Red List threatened (437 species) flora, with high vulnerability to global warming and heat stress, requires urgent conservation measures.

The continued global warming and HW/HS-linked compositional change in plants from forests, grasslands, and aquatic ecosystems will favour savanna (against semi-deciduous forest), invasive species (against native), C4 species with Kranz anatomy (against C3 species), diatoms (against blue-green algae) which will have threaten the stability of these ecosystems, causing loss of biodiversity, ES, GPP and crop yield. The flora in both terrestrial and aquatic ecosystems have shown response to HW/HS, adapting through shifts in geographical range (contraction, expansion), phenology, community composition across the elevational and latitudinal gradients, with high vulnerability seen for endemic and threatened plant species in forests, grasslands, as well as inland and marine ecosystems, (planktons, seagrass, Kelps, and corals).

High plant diversity (dominated by native and endemic species) can help mitigate HW/HS, and other abiotic stresses like drought and pollution by providing resilience and buffering effects in forests, grasslands, agroecosystems, and aquatic ecosystems. Deepening research on the impacts of HW and HS on plant communities (both wild and crops) will aid in sustaining the resilience and productivity of both natural ecosystems (e.g., forests, grasslands, lakes) and agroecosystems in future GCC scenarios.

In the future, long-term spatio-temporal studies are needed to better understand vulnerability and resilience of flora to HW/HS (e.g., plant mortality, shifts in species' range, and community compositions), especially in the understudied regions like the Himalaya. Researchers can make use of artificial intelligence modelling, GCC prediction modelling, innovative high-end omics and genetic engineering research (with plants), stress-tolerant plant varieties and transgenics, to mitigate native plants against invasives under HS. Our study underscores the need for integrated research strategies to enhance thermotolerance in plants to preserve plant biodiversity as well as to ensure global food security. The world is undergoing continued environmental change and a food insecurity crisis, which requires an urgent need for comprehensive research and policy interventions to better understand plants' vulnerability to HS across ecosystems. This study posits a future research opportunity in understanding the acclimation and adaptation patterns of plants along the elevational, latitudinal, and temperature gradients encompassing different biomes and ecosystems of the world.

Acknowledgements For providing research facilities and support, we convey our gratitude to the Shri Ramasamy Memorial University Sikkim, The Mountain Institute-India, Akal University, and

Texas Commission on Environmental Quality. We also thank Dr. Ghanashyam Sharma, Dr. Prakash Chettri, Dr. Saurabh Anand, and Shri. Binod Kumar Sharma for their various supports. We are thankful to the editorial team of the book for the opportunity, patience, and help in completing this manuscript.

References

Agnihotri, P., Husain, T., Shirke, P. A., Sidhu, O. P., Singh, H., Dixit, V., & Nautiyal, C. S. (2017). Climate change-driven shifts in elevation and ecophysiological traits of Himalayan plants during the past century. *Current Science, 112*(3), 595–601. https://doi.org/10.18520/cs/v112/i03/595-601

Ainsworth, E. A., & Long, S. P. (2005). What have we learned from 15 years of free-air CO2 enrichment (FACE)? A meta-analytic review of the responses of photosynthesis, canopy properties and plant production to rising CO2. *New Phytologist, 165*(2), 351–372. https://doi.org/10.1111/j.1469-8137.2004.01224.x

Aleixo, I., Norris, D., Hemerik, L., Barbosa, A., Prata, E., Costa, F., & Poorter, L. (2019). Amazonian rainforest tree mortality driven by climate and functional traits. *Nature Climate Change, 9*(5), 384–388. https://doi.org/10.1038/s41558-019-0458-0

Allen, C. D., Macalady, A. K., Chenchouni, H., Bachelet, D., McDowell, N., Vennetier, M., & Cobb, N. (2010). A global overview of drought and heat-induced tree mortality reveals emerging climate change risks for forests. *Forest Ecology and Management, 259*, 660–684. https://doi.org/10.1016/j.foreco.2009.09.001

Allen, C. D., Breshears, D. D., & McDowell, N. G. (2015). On underestimation of global vulnerability to tree mortality and forest die-off from hotter drought in the Anthropocene. *Ecosphere, 6*(8), 1–55. https://doi.org/10.1890/ES15-00203.1

Anderegg, W. R., Kane, J. M., & Anderegg, L. D. (2013). Consequences of widespread tree mortality triggered by drought and temperature stress. *Nature Climate Change, 3*(1), 30–36. https://doi.org/10.1038/nclimate1635

Anderegg, W. R., Martinez-Vilalta, J., Cailleret, M., Camarero, J. J., Ewers, B. E., Galbraith, D., & Trotsiuk, V. (2016). When a tree dies in the forest: Scaling climate-driven tree mortality to ecosystem water and carbon fluxes. *Ecosystems, 19*(6), 1133–1147. https://doi.org/10.1007/s10021-016-9982-1

Arafeh Dalmau, N., Schoeman, D. S., Montano-Moctezuma, G., Torres-Moye, G., Cavanaugh, K. C., Munguia-Vega, A., & Micheli, F. (2025). Marine heatwaves drive range contraction and alternative states of kelp forests at their warm limit. bioRxiv, 2025-10. https://doi.org/10.1101/2025.10.17.682914

Audet, J., Neif, E. M., Cao, Y., Hoffmann, C. C., Lauridsen, T. L., Larsen, S. E., & Davidson, T. A. (2017). Heat-wave effects on greenhouse gas emissions from shallow lake mesocosms. *Freshwater Biology, 62*(7), 1130–1142. https://doi.org/10.1111/fwb.12930

Awasthi, A., Vishwakarma, K., & Pattnayak, K. C. (2022). Retrospection of heatwave and heat index. *Theoretical and Applied Climatology, 147*(1), 589–604. https://doi.org/10.1007/s00704-021-03854-z

Azhar, G., Saha, S., Ganguly, P., Mavalankar, D., & Madrigano, J. (2017). Heat wave vulnerability mapping for India. *International Journal of Environmental Research and Public Health, 14*(4), 357. https://doi.org/10.3390/ijerph14040357

Baker, N. R. (2008). Chlorophyll fluorescence: A probe of photosynthesis in vivo. *Annual Review of Plant Biology, 59*, 89–113. https://doi.org/10.1146/annurev.arplant.59.032607.092759

Barriopedro, D., García-Herrera, R., Ordóñez, C., Miralles, D. G., & Salcedo-Sanz, S. (2023). Heat waves: Physical understanding and scientific challenges. *Reviews of Geophysics, 61*(2), e2022RG000780. https://doi.org/10.1029/2022RG000780

Bartosiewicz, M., Laurion, I., Clayer, F., & Maranger, R. (2016). Heat-wave effects on oxygen, nutrients, and phytoplankton can alter global warming potential of gases emitted from a small shallow lake. *Environmental Science & Technology, 50*(12), 6267–6275. https://doi.org/10.1021/acs.est.5b06312

Bastos, A., Ciais, P., Friedlingstein, P., Sitch, S., Pongratz, J., Fan, L., & Zaehle, S. (2020). Direct and seasonal legacy effects of the 2018 heat wave and drought on European ecosystem productivity. *Science Advances, 6*(24), eaba2724. https://doi.org/10.1126/sciadv.aba2724

Batcho, A. A., Sarwar, M. B., Rashid, B., Hassan, S., & Husnain, T. (2021). Heat shock protein gene identified from *Agave sisalana* (As HSP70) confers heat stress tolerance in transgenic cotton (*Gossypium hirsutum*). *Theoretical and Experimental Plant Physiology, 33*(2), 141–156. https://doi.org/10.1007/s40626-021-00200-6

Bauweraerts, I., Wertin, T. M., Ameye, M., McGuire, M. A., Teskey, R. O., & Steppe, K. (2013). The effect of heat waves, elevated [CO 2] and low soil water availability on northern red oak (*Quercus rubra* L.) seedlings. *Global Change Biology, 19*(2), 517–528. https://doi.org/10.1111/gcb.12044

Bayouli, I. T., Bayouli, H. T., Dell'Oca, A., Meers, E., & Sun, J. (2021). Ecological indicators and bioindicator plant species for biomonitoring industrial pollution: Eco-based environmental assessment. *Ecological Indicators, 125*, Article 107508. https://doi.org/10.1016/j.ecolind.2021.107508

Bezerra, K. F., Gomes, H. B., Baltaci, H., Ray, P., Nascimento, J. P., Herdies, D. L., & Gomes, H. B. (2025). Simultaneous Drought and Heatwave Events During Austral Summer in Northeast Brazil. *International Journal of Climatology, 45*(11), Article e8920. https://doi.org/10.1002/joc.8920

Bhakta, S., & Adhikary, S. P. (2014). Algal Diversity in the streams and waterfalls of eastern and north-eastern region of India. *Nelumbo, 56,* 1–47. https://doi.org/10.20324/nelumbo/v56/2014/86674

Bhattarai, S., Bokati, L., Sharma, S., & Talchabhadel, R. (2025). Understanding spatiotemporal variation of heatwave projections across US cities. *Scientific Reports, 15*(1), 10643. https://doi.org/10.1038/s41598-025-95097-5

Bisht, J. K, Meena, V. S, Mishra, P. K. & Pattanayak, A. (Eds.) (2016). *Conservation Agriculture: An Approach to Combat Climate Change in Indian Himalaya*, Springer. https://doi.org/10.1007/978-981-10-2558-7

Brauko, K. M., Cabral, A., Costa, N. V., Hayden, J., Dias, C. E., Leite, E. S., & Horta, P. A. (2020). Marine heatwaves, sewage and eutrophication combine to trigger deoxygenation and biodiversity loss: A SW Atlantic case study. *Frontiers in Marine Science, 7*, Article 590258. https://doi.org/10.3389/fmars.2020.590258

BSI (2025a). Plant Discoveries. Botanical Survey of India. https://bsi.gov.in/page/en/national-wild-life-action-plan (Accessed on 1 November 2025).

BSI (2025b). *Lichens*. Botanical Survey of India. https://bsi.gov.in/page/en/lichens (Accessed on 1 November 2025).

Bussotti, F., & Pollastrini, M. (2021). Revisiting the concept of stress in forest trees at the time of global change and issues for stress monitoring. *Plant Stress, 2*, Article 100013. https://doi.org/10.1016/j.stress.2021.100013

Casal, J. J., & Balasubramanian, S. (2019). Thermomorphogenesis. *Annual Review of Plant Biology, 70*(1), 321–346. https://doi.org/10.1146/annurev-arplant-050718-095919

Caspersen, J. P. (2004). Variation in stand mortality related to successional composition. *Forest Ecology and Management, 200*, 149–160. https://doi.org/10.1016/j.foreco.2004.06.015

Chapman, C. C., Sloyan, B. M., Moore, T. S., Reilly, J. A., & Matear, R. J. (2025). Marine heatwaves in the East Australian current modulated by mesoscale eddies. *Journal of Geophysical Research: Oceans, 130*(5), e2024JC021395. https://doi.org/10.1029/2024JC021395

Chen, H. Y., & Luo, Y. (2015). Net aboveground biomass declines of four major forest types with forest ageing and climate change in western Canada's boreal forests. *Global Change Biology, 21*(10), 3675–3684. https://doi.org/10.1111/gcb.12994

Chen, S., Wang, W., Xu, W., Wang, Y., Wan, H., Chen, D., & Bai, Y. (2018). Plant diversity enhances productivity and soil carbon storage. *Proceedings of the National Academy of Sciences, 115*(16), 4027–4032. https://doi.org/10.1073/pnas.1700298114

Cheung, W. W., & Frölicher, T. L. (2020). Marine heatwaves exacerbate climate change impacts for fisheries in the northeast Pacific. *Scientific Reports, 10*(1), 6678. https://doi.org/10.1038/s41598-020-63650-z

Chitale, V. S., Behera, M. D., & Roy, P. S. (2014). Future of Endemic Flora of Biodiversity Hotspots in India. *PLoS ONE, 9*(12), Article e115264. https://doi.org/10.1371/journal.pone.0115264

Ciais, P., Reichstein, M., Viovy, N., Granier, A., Ogée, J., Allard, V., & Valentini, R. (2005). Europe-wide reduction in primary productivity caused by the heat and drought in 2003. *Nature, 437*(7058), 529–533. https://doi.org/10.1038/nature03972

Convention on Biological Diversity (CBD) (2025). India—Country Profile. Retrieved [08/11/2025], from https://www.cbd.int/countries/profile/?country=in.

da Silva, B. H. P., & Rossatto, D. R. (2022). Leaves of neotropical savanna tree species are more heat-tolerant than leaves of semi-deciduous forest species. *Theoretical and Experimental Plant Physiology, 34*(2), 227–237. https://doi.org/10.1007/s40626-022-00244-2

Dar, J. A., & Apurv, T. (2024). Spatiotemporal characteristics and physical drivers of heatwaves in India. *Geophysical Research Letters, 51*, e2024GL109785. https://doi.org/10.1029/2024GL109785

Delerue, F., Randé, H., Nemer, D., Laperche, V., Sappin-Didier, V., Cormont, L., & Michalet, R. (2025). An extreme drought and heatwave event led to collapse of facilitation by metallophyte species in metalliferous ecosystems. *Plant and Soil,* 1–20. https://doi.org/10.1007/s11104-025-07611-3

Deng, Y., Srivastava, R., Quilichini, T. D., Dong, H., Bao, Y., Horner, H. T., & Howell, S. H. (2016). IRE 1, a component of the unfolded protein response signaling pathway, protects pollen development in *Arabidopsis* from heat stress. *The Plant Journal, 88*(2), 193–204. https://doi.org/10.1111/tpj.13239

Devi, R., Gouda, K. C., & Lenka, S. (2023a). Analysis of heat wave over different physiographical regions in India. *Theoretical and Applied Climatology, 154*(3), 1343–1356. https://doi.org/10.1007/s00704-023-04639-2

Devi, P., Awasthi, R., Jha, U., Sharma, K. D., Prasad, P. V., Siddique, K. H., & Nayyar, H. (2023b). Understanding the effect of heat stress during seed filling on nutritional composition and seed yield in chickpea (*Cicer arietinum* L.). *Scientific Reports, 13*, 15450 https://doi.org/10.1038/s41598-023-42586-0

Devi, S. B., Sharma, T., Rai, R. K., Irungbam, J. S., Manish, K., Sharma, K. (2025). Initial Colonization of Red Pierrot *Talicada nyseus nyseus* (GuérinMéneville, 1843) (Lepidoptera: Lycaenidae: Polyommatinae) in Sikkim, Eastern Himalaya, India. *Species, 26*, e7s1763. https://doi.org/10.54905/disssi.v26i77.e7s1763

Ding, Z., Ji, B., Yao, H., Cheng, X., Yu, S., Sun, X., & Shi, Y. (2023). An analysis of the factors affecting forest mortality and research on forecasting models in Southern China: A case study in Zhejiang Province. *Forests, 14*(11), 2199. https://doi.org/10.3390/f14112199

Earle, C. J. (Ed.). (2025, October 29). *The Gymnosperm Database.* Retrieved 10 November 2025 https://www.conifers.org/zz/gymnosperms.php

El-Madany, T. S., Carrara, A., Martín, M. P., Moreno, G., Kolle, O., Pacheco-Labrador, J., & Migliavacca, M. (2020). Drought and heatwave impacts on semi-arid ecosystems' carbon fluxes along a precipitation gradient. *Philosophical Transactions of the Royal Society B, 375*(1810), 20190519. https://doi.org/10.1098/rstb.2019.0519

Engdaw, M. M., Ballinger, A. P., Hegerl, G. C., & Steiner, A. K. (2022). Changes in temperature and heat waves over Africa using observational and reanalysis data sets. *International Journal of Climatology, 42*(2), 1165–1180. https://doi.org/10.1002/joc.7295

Fan, J., Yao, F., & Zhang, J. (2025). Characteristics of spatial and temporal variations of summer compound heat wave events in the context of abrupt change of summer maximum temperature in recent 70 years. *Stochastic Environmental Research and Risk Assessment, 39*, 2449–2460. https://doi.org/10.1007/s00477-025-02978-2

Feeley, K. J. (2012). Distributional migrations, expansions, and contractions of tropical plant species as revealed in dated herbarium records. *Global Change Biology, 18*, 1335–1341. https://doi.org/10.1111/j.1365-2486.2011.02602.x

Feeley, K. J., Silman, M. R., Bush, M. B., Farfan, W., Cabrera, K. G., Malhi, Y., & Saatchi, S. (2011). Upslope migration of Andean trees. *Journal of Biogeography, 38*(4), 783–791. https://doi.org/10.1111/j.1365-2699.2010.02444.x

Feeley, K., Martinez-Villa, J., Perez, T., Silva Duque, A., Triviño Gonzalez, D., & Duque, A. (2020). The thermal tolerances, distributions, and performances of tropical montane tree species. *Frontiers in Forests and Global Change, 3*, 25. https://doi.org/10.3389/ffgc.2020.00025

Felde, V. A., Kapfer, J., & Grytnes, J. A. (2012). Upward shift in elevational plant species ranges in Sikkilsdalen, central Norway. *Ecography, 35*(10), 922–932. http://hdl.handle.net/1956/5185

da Fonseca, E. M., & Gaylarde, C. C. (2025). Climate change and microplastics: a two-way interaction. *Emerging Contaminants and Environmental Health*, *4*(3), 15. https://doi.org/10.20517/eceh.2025.09

Fordham, D. A., Resit Akçakaya, H., Araújo, M. B., Elith, J., Keith, D. A., Pearson, R., & Brook, B. W. (2012). Plant extinction risk under climate change: Are forecast range shifts alone a good indicator of species vulnerability to global warming? *Global Change Biology, 18*(4), 1357–1371. https://doi.org/10.1111/j.1365-2486.2011.02614.x

Freer-Smith, P. H., & Webber, J. F. (2017). Tree pests and diseases: The threat to biodiversity and the delivery of ecosystem services. *Biodiversity and Conservation, 26*(13), 3167–3181. https://doi.org/10.1007/s10531-015-1019-0

French, K., Robinson, S. A., & Lia, J. (2017). Thermotolerance capacities of native and exotic coastal plants will lead to changes in species composition under increased heat waves. *Conservation Physiology*, *5*(1), cox029. https://doi.org/10.1093/conphys/cox029

Garg, T., Kumar, A., Joshi, A., Awasthi, A., Rana, A., Kumar, V., & Kaur, S. (2024). The role of phytohormones in reducing the arsenic-induced stress in plants. *South African Journal of Botany, 168*, 296–315. https://doi.org/10.1016/j.sajb.2024.03.020

Godoy, O., de Lemos-Filho, J. P., & Valladares, F. (2011). Invasive species can handle higher leaf temperature under water stress than Mediterranean natives. *Environmental and Experimental Botany, 71*(2), 207–214. https://doi.org/10.1016/j.envexpbot.2010.12.001

Gomes, D. G., Ruzicka, J. J., Crozier, L. G., Huff, D. D., Brodeur, R. D., & Stewart, J. D. (2024). Marine heatwaves disrupt ecosystem structure and function via altered food webs and energy flux. *Nature Communications, 15*(1), 1988. https://doi.org/10.1038/s41467-024-46263-2

Gosavi, G. U., Jadhav, A. S., Kale, A. A., Gadakh, S. R., Pawar, B. D., & Chimote, V. P. (2014). Effect of heat stress on proline, chlorophyll content, heat shock proteins and antioxidant enzyme activity in sorghum (*Sorghum bicolor*) at seedlings stage. Indian *Journal of Biotechnology*, *13*, 356–363. http://nopr.niscpr.res.in/handle/123456789/29616

Gounaris, K., Brain, A. R. R., Quinn, P. J., & Williams, W. P. (1984). Structural reorganisation of chloroplast thylakoid membranes in response to heat-stress. *Biochimica et Biophysica Acta (BBA)-Bioenergetics, 766*(1), 198–208.

Graf Pannatier, E., Dobbertin, M., Heim, A., Schmitt, M., Thimonier, A., Waldner, P., & Frey, B. (2012). Response of carbon fluxes to the 2003 heat wave and drought in three mature forests in Switzerland. *Biogeochemistry, 107*(1), 295–317. https://doi.org/10.1007/s10533-010-9554-y

Guiry, M.D. & Guiry, G.M. 2025. AlgaeBase. World-wide electronic publication, University of Galway. https://www.algaebase.org (Accessed on 8 November 2025).

Gulen, H., & Eris, A. (2004). Effect of heat stress on peroxidase activity and total protein content in strawberry plants. *Plant Science, 166*(3), 739–744. https://doi.org/10.1016/j.plantsci.2003.11.014

Guo, S., Mu, L., Sun, S., Hou, X., Yao, M., & Hu, X. (2023). Concurrence of microplastics and heat waves reduces rice yields and disturbs the agroecosystem nitrogen cycle. *Journal of Hazardous Materials, 452*, Article 131340. https://doi.org/10.1016/j.jhazmat.2023.131340

Gurung, P. D., Upadhyay, A. K., Bhardwaj, P. K., Sowdhamini, R., & Ramakrishnan, U. (2019). Transcriptome analysis reveals plasticity in gene regulation due to environmental cues in *Primula sikkimensis*, a high altitude plant species. *BMC Genomics, 20*(1), 989. https://doi.org/10.1186/s12864-019-6354-1

Hammond, W. M., Williams, A. P., Abatzoglou, J. T., Adams, H. D., Klein, T., López, R., & Allen, C. D. (2022). Global field observations of tree die-off reveal hotter-drought fingerprint for Earth's forests. *Nature Communications, 13*(1), 1761. https://doi.org/10.1038/s41467-022-29289-2

Hartmann, H., Bastos, A., Das, A. J., Esquivel-Muelbert, A., Hammond, W. M., Martínez-Vilalta, J., & Allen, C. D. (2022). Climate change risks to global forest health: Emergence of unexpected events of elevated tree mortality worldwide. *Annual Review of Plant Biology, 73*(1), 673–702. https://doi.org/10.1146/annurev-arplant-102820-012804

Hauck, M., Schneider, T., Bahlinger, S., Fischbach, J., Oswald, G., Csapek, G., & Dulamsuren, C. (2025). Heat tolerance of temperate tree species from Central Europe. *Forest Ecology and Management, 580*, Article 122541. https://doi.org/10.1016/j.foreco.2025.122541

Havaux, M., & Tardy, F. (1999). Loss of chlorophyll with limited reduction of photosynthesis as an adaptive response of Syrian barley landraces to high-light and heat stress. *Functional Plant Biology, 26*(6), 569–578. https://doi.org/10.1071/PP99046

Heckathorn, S. A., Downs, C. A., Sharkey, T. D., & Coleman, J. S. (1998). The small, methionine-rich chloroplast heat-shock protein protects photosystem II electron transport during heat stress. *Plant Physiology, 116*, 439–444. https://doi.org/10.1104/pp.116.1.439

Hermann, M., Peeters, E. T., & Van den Brink, P. J. (2023). Heatwaves, elevated temperatures, and a pesticide cause interactive effects on multi-trophic levels of a freshwater ecosystem. *Environmental Pollution, 327*, Article 121498. https://doi.org/10.1016/j.envpol.2023.121498

Hoover, D. L., Knapp, A. K., & Smith, M. D. (2014). Resistance and resilience of a grassland ecosystem to climate extremes. *Ecology, 95*(9), 2646–2656. https://doi.org/10.1890/13-2186.1

Huang, Y., Stein, G., Kolle, O., Kübler, K., Schulze, E. D., Dong, H., & Eisenhauer, N. (2023). Plant diversity stabilizes soil temperature. *bioRxiv,* 2023-03. https://doi.org/10.1101/2023.03.13.532451

IPCC. (2023). Climate Change 2023: Synthesis Report. *Contribution of Working Groups I, II and III to the Sixth Assessment Report of the Intergovernmental Panel on Climate Change* [Core Writing Team, H. Lee and J. Romero (Eds.)]. IPCC, Geneva, Switzerland. Geneva, Switzerland. https://doi.org/10.59327/IPCC/AR6-9789291691647.

Isbell, F., Craven, D., Connolly, J., Loreau, M., Schmid, B., Beierkuhnlein, C., & Eisenhauer, N. (2015a). Biodiversity increases the resistance of ecosystem productivity to climate extremes. *Nature, 526*(7574), 574–577. https://doi.org/10.1038/nature15374

Isbell, F., Tilman, D., Polasky, S., & Loreau, M. (2015b). The biodiversity-dependent ecosystem service debt. *Ecology Letters, 18*, 119–134. https://doi.org/10.1111/ele.12393

Jia, L., Liu, L., Zhang, Y., Fu, W., Liu, X., Wang, Q., & Huang, L. (2023). Microplastic stress in plants: Effects on plant growth and their remediations. *Frontiers in Plant Science, 14*, 1226484. https://doi.org/10.3389/fpls.2023.1226484

Johnson, J. V., Dick, J. T., & Pincheira-Donoso, D. (2021). Mangrove forests mitigate coral bleaching under thermal stress from climate change. *bioRxiv,* 2021-06. https://doi.org/10.1101/2021.06.04.447049

Kan, Y., Mu, X. R., Gao, J., Lin, H. X., & Lin, Y. (2023). The molecular basis of heat stress responses in plants. *Molecular Plant, 16*(10), 1612–1634. https://doi.org/10.1016/j.molp.2023.09.013

Kattel, G. R. (2022). Climate warming in the Himalayas threatens biodiversity, ecosystem functioning and ecosystem services in the 21st century: Is there a better solution? *Biodiversity and Conservation, 31*(8), 2017–2044. https://doi.org/10.1007/s10531-022-02417-6

Kaushal, N., Awasthi, R., Gupta, K., Gaur, P., Siddique, K. H., & Nayyar, H. (2013). Heat-stress-induced reproductive failures in chickpea (Cicer arietinum) are associated with impaired sucrose metabolism in leaves and anthers. *Functional Plant Biology, 40*(12), 1334–1349. https://doi.org/10.1071/FP13082

Kelly, A. E., & Goulden, M. L. (2008). Rapid shifts in plant distribution with recent climate change. *Proceedings of the National Academy of Sciences, 105*(33), 11823–11826. https://doi.org/10.1073/pnas.080289110

Kholia, B. S. (2010). *Ferns and Fern-allies of Sikkim, A Pictorial Handbook Part-I*. Botanical Survey of India, Sikkim Himalayan Regional Center, Gangtok.

Knutzen, F., Averbeck, P., Barrasso, C., Bouwer, L. M., Gardiner, B., Grünzweig, J. M., & Gliksman, D. (2025). Impacts on and damage to European forests from the 2018–2022 heat and drought events. *Natural Hazards and Earth System Sciences, 25*(1), 77–117. https://doi.org/10.5194/nhess-25-77-2025

Kong, X., Zhang, S., Wang, A., Liu, Y., Wang, K., Zhao, X., & Xi, B. (2024). Trees in cooler regions are more vulnerable to thermal stress: Evidence from temperate poplar plantations in Northern China during the 2022 heatwaves. *Agricultural and Forest Meteorology, 356*, Article 110181. https://doi.org/10.1016/j.agrformet.2024.110181

Kumar, S., & Khanduri, V. P. (2024). Impact of climate change on the Himalayan alpine treeline vegetation. *Heliyon, 10*(23), Article e40797. https://doi.org/10.1016/j.heliyon.2024.e40797

Kumar, R. R., Sharma, S. K., Goswami, S., Verma, P., Singh, K., Dixit, N., & Rai, R. D. (2015). Salicylic acid alleviates the heat stress-induced oxidative damage of starch biosynthesis pathway by modulating the expression of heat-stable genes and proteins in wheat (*Triticum aestivum*). *Acta Physiologiae Plantarum, 37*(8), 143. https://doi.org/10.1007/s11738-015-1899-3

Kumar, V., Sandhu, S. S., Kaur, S., Walia, S. S., & Gill, K. K. (2025). Impact of heat stress on growth and fodder yield of C3 (Cowpea) and C4 (Pearl Millet) Crops: Insights from field and modelling experiments. *Journal of Agronomy and Crop Science, 211*(3), Article e70063. https://doi.org/10.1111/jac.70063

Kumar, S., Chopra, N., & Al-Tawaha, A. R. M. S. (2017). Phenological variations in tree and shrub species of a banj oak (*Quercus leucotrichophora* A. Camus) dominated forest in Kumaun, Central Himalaya. *Advances in Environmental Biology, 11*(5), 35–44. https://www.aensiweb.net/AENSIWEB/aeb/aeb/2017/May/35-43.pdf

Kumar, P., Panwar, G. S., & Singh, S. K. (2022). A brief analysis of IUCN red listed threatened plants of India. *Indian Forester, 147*(12), 1155–1171. https://doi.org/10.36808/if/2021/v147i12/166613

Kunert, N., & Hajek, P. (2022). Shade-tolerant temperate broad-leaved trees are more sensitive to thermal stress than light-demanding species during a moderate heatwave. *Trees, Forests and People, 9*, Article 100282. https://doi.org/10.1016/j.tfp.2022.100282

La Verde, V., Dominici, P., & Astegno, A. (2018). Towards understanding plant calcium signaling through calmodulin-like proteins: A biochemical and structural perspective. *International Journal of Molecular Sciences, 19*(5), 1331. https://doi.org/10.3390/ijms19051331

Laz, O. U., Rahman, A., & Ouarda, T. B. (2023). Compound heatwave and drought hotspots and their trends in Southeast Australia. *Natural Hazards, 119*(1), 357–386. https://doi.org/10.1007/s11069-023-06115-6

Li, N., Euring, D., Cha, J. Y., Lin, Z., Lu, M., Huang, L. J., & Kim, W. Y. (2021a). Plant hormone-mediated regulation of heat tolerance in response to global climate change. *Frontiers in Plant Science, 11*, Article 627969. https://doi.org/10.3389/fpls.2020.627969

Li, L., Zheng, Z., Biederman, J. A., Qian, R., Ran, Q., Zhang, B., & Wang, Y. (2021b). Drought and heat wave impacts on grassland carbon cycling across hierarchical levels. *Plant, Cell & Environment, 44*(7), 2402–2413. https://doi.org/10.1111/pce.13767

Liang, J., Crowther, T. W., Picard, N., Wiser, S., Zhou, M., Alberti, G., & Reich, P. B. (2016). Positive biodiversity-productivity relationship predominant in global forests. *Science, 354*(6309), aaf8957. https://doi.org/10.1126/science.aaf8957

Lin, W., Liu, L., Liang, J., Tang, X., Shi, J., Zhang, L., & Guo, L. (2022). Changes of endophytic microbial community in *Rhododendron simsii* roots under heat stress and its correlation with leaf physiological indicators. *Frontiers in Microbiology, 13*, 1006686. https://doi.org/10.3389/fmicb.2022.1006686

Liu, C., Hiraga, Y., Mbugua, J., Hokson, J. A., & Kazama, S. (2025). Wetland loss intensifies regional heat and stress: Insights from the 2023 Southeast Asia heatwave. *Preprint, Research Square*. https://doi.org/10.21203/rs.3.rs-6812001/v1

Lixia, Y. U. A. N., Ruilin, M. E. N. G., Jiali, L. I., Lifeng, L. I. N., Xiaojun, X. U., Jianpeng, X. I. A. O., & Tao, L. I. U. (2022). Relationship between heatwave and years of life lost associated with stroke in Guangdong Province: Based on Bayesian spatio-temporal model. *Journal of Environmental and Occupational Medicine, 39*(3), 268–274. https://doi.org/10.11836/JEOM21458

Locosselli, G. M., Brienen, R. J., Leite, M. D. S., Gloor, M., Krottenthaler, S., Oliveira, A. A. D., & Buckeridge, M. (2020). Global tree-ring analysis reveals rapid decrease in tropical tree longevity with temperature. *Proceedings of the National Academy of Sciences, 117*(52), 33358–33364. https://doi.org/10.1073/pnas.2003873117

López, R., Ramírez-Valiente, J. A., & Pita, P. (2022). How plants cope with heatwaves in a drier environment. *Flora, 295*, Article 152148. https://doi.org/10.1016/j.flora.2022.152148

López-Felices, B., Aznar-Sánchez, J. A., Velasco-Muñoz, J. F., & Mesa-Vázquez, E. (2022). Installation of hedgerows around greenhouses to encourage biological pest control: Farmers' perspectives from Southeast Spain. *Journal of Environmental Management, 323*, Article 116210. https://doi.org/10.1016/j.jenvman.2022.116210

Lücking, R., Hodkinson, B. P., & Leavitt, S. D. (2017). The 2016 classification of lichenized fungi in the Ascomycota and Basidiomycota-Approaching one thousand genera. *The Bryologist, 119*(4), 361–416. https://doi.org/10.1639/0007-2745-119.4.361

Lueth, V. M., & Reski, R. (2023). Mosses. *Current Biology, 33*, R1163–R1185. https://doi.org/10.1016/j.cub.2023.09.042

Lutz, D. A., Powell, R. L., & Silman, M. R. (2013). Four decades of Andean timberline migration and implications for biodiversity loss with climate change. *PLoS ONE, 8*(9), Article e74496. https://doi.org/10.1371/journal.pone.0074496

Lyu, J., Shi, Y., Liu, T., Xu, X., Liu, S., Yang, G., & Gao, J. (2025). Extreme drought-heatwave events threaten the biodiversity and stability of aquatic plankton communities in the Yangtze River ecosystems. *Communications Earth & Environment, 6*(1), 171. https://doi.org/10.1038/s43247-025-02143-1

De Maio, A. (1999). Heat shock proteins: facts, thoughts, and dreams. *Shock (Augusta, Ga.)*. 11, 1–12. https://doi.org/10.1097/00024382-199901000-00001

Maity, D., and Maiti, G.G. (2024). *Flora of Sikkim: A Comprehensive Account of Angiosperms with Identification Keys Volume I-III (Set)*. First Edition. Dehradun: M/s Bishen Singh Mahendra Pal Singh.

Manish, K., & Pandit, M. K. (2018). Geophysical upheavals and evolutionary diversification of plant species in the Himalaya. *PeerJ, 6*, Article e5919. https://doi.org/10.7717/peerj.5919

Marchin, R. M., Esperon-Rodriguez, M., Tjoelker, M. G., & Ellsworth, D. S. (2022). Crown dieback and mortality of urban trees linked to heatwaves during extreme drought. *Science of the Total Environment, 850*, Article 157915. https://doi.org/10.1016/j.scitotenv.2022.157915

Margalef-Marrase, J., Pérez-Navarro, M. Á., & Lloret, F. (2020). Relationship between heatwave induced forest die-off and climatic suitability in multiple tree species. *Global Change Biology, 26*, 3134–3146. https://doi.org/10.1111/gcb.15042

Marias, D. E., Meinzer, F. C., & Still, C. (2017). Impacts of leaf age and heat stress duration on photosynthetic gas exchange and foliar nonstructural carbohydrates in *Coffea arabica*. *Ecology and Evolution, 7*(4), 1297–1310. https://doi.org/10.1002/ece3.2681

Marjakangas, E. L., Santangeli, A., Kujala, H., Mammola, S., & Lehikoinen, A. (2023). Identifying 'climate keystone species' as a tool for conserving ecological communities under climate change. *Diversity and Distributions, 29*(11), 1341–1354. https://doi.org/10.1111/ddi.13764

Matusick, G., Ruthrof, K. X., Kala, J., Brouwers, N. C., Breshears, D. D., & Hardy, G. E. S. J. (2018). Chronic historical drought legacy exacerbates tree mortality and crown dieback during acute heatwave-compounded drought. *Environmental Research Letters, 13*(9), Article 095002. https://doi.org/10.1088/1748-9326/aad8cb

McDowell, N., Allen, C. D., Anderson-Teixeira, K., Brando, P., Brienen, R., Chambers, J., & Xu, X. (2018). Drivers and mechanisms of tree mortality in moist tropical forests. *New Phytologist, 219*(3), 851–869. https://doi.org/10.1111/nph.15027

Ministry of Environment, Forest & Climate Change (MoEFCC), Botanical Survey of India (BSI), Zoological Survey of India (ZSI), & IUCN India. (2025). *National Red List Assessment of Indian Flora and Fauna: Vision 2025–2030*. https://bsi.gov.in/uploads/documents/publications/e-publications/hindi/IUCN_Vision_2025-2030-_published.pdf.

Mittermeier RA, Turner WR, Larsen FW, Brooks TM, Gascon C (2011) Global biodiversity conservation: The critical role of hotspots. In F. Zachos & J. Habel (Eds.), *Biodiversity Hotspots*. Springer, Berlin, Heidelberg. https://doi.org/10.1007/978-3-642-20992-5_1

Mukherjee, S., & Mishra, A. K. (2021). Increase in compound drought and heatwaves in a warming world. *Geophysical Research Letters, 48*(1), e2020GL090617. https://doi.org/10.1029/2020GL090617

Nayaka, S., & Ngangom, R. (2024). Current status of lichen diversity in the Himalayan biodiversity hotspot. In T. Pullaiah (Ed.), *Biodiversity Hotspot of the Himalaya* (pp. 67–90). CRC Press, Taylor & Francis. https://doi.org/10.1201/9781003455202

Njeru, E. M. (2018). Exploiting diversity to promote arbuscular mycorrhizal symbiosis and crop productivity in organic farming systems. *AIMS Agriculture and Food, 3*(3), 280–294.

Nori-Sarma, A., Anderson, G. B., Rajiva, A., ShahAzhar, G., Gupta, P., Pednekar, M. S., & Bell, M. L. (2019). The impact of heat waves on mortality in Northwest India. *Environmental Research, 176*, Article 108546. https://doi.org/10.1016/j.envres.2019.108546

Ortiz-Bobea, A., Ault, T. R., Carrillo, C. M., Chambers, R. G., & Lobell, D. B. (2021). Anthropogenic climate change has slowed global agricultural productivity growth. *Nature Climate Change, 11*(4), 306–312. https://doi.org/10.1038/s41558-021-01000-1

Oshino, T., Abiko, M., Saito, R., Ichiishi, E., Endo, M., Kawagishi-Kobayashi, M., & Higashitani, A. (2007). Premature progression of anther early developmental programs accompanied by comprehensive alterations in transcription during high-temperature injury in barley plants. *Molecular Genetics and Genomics, 278*(1), 31–42. https://doi.org/10.1007/s00438-007-0229-x

O'sullivan, O. S., Heskel, M. A., Reich, P. B., Tjoelker, M. G., Weerasinghe, L. K., Penillard, A., & Atkin, O. K. (2017). Thermal limits of leaf metabolism across biomes. *Global Change Biology, 23*(1), 209–223. https://doi.org/10.1111/gcb.13477

Page, N. V., & Shanker, K. (2020). Climatic stability drives latitudinal trends in range size and richness of woody plants in the Western Ghats, India. *Plos ONE, 15*(7), Article e0235733. https://doi.org/10.1371/journal.pone.0235733

Pai, D. S., Nair, S., & Ramanathan, A. N. (2013). Long term climatology and trends of heat waves over India during the recent 50 years (1961–2010). *Mausam, 64*(4), 585–604. https://doi.org/10.54302/mausam.v64i4.742

Palmas, F., Cariccia, F., Pasquini, V., Cabiddu, S., Addis, P., & Pusceddu, A. (2025). Marine heatwaves effects on quantity, composition, and turnover of sedimentary organic matter in a Mediterranean coastal lagoon: A benthocosm study. *Estuarine, Coastal and Shelf Science, 318*, Article 109252. https://doi.org/10.1016/j.ecss.2025.109252

Pan, Y., Birdsey, R. A., Fang, J., Houghton, R., Kauppi, P. E., Kurz, W. A., & Hayes, D. (2011). A large and persistent carbon sink in the world's forests. *Science, 333*(6045), 988–993. https://doi.org/10.1126/science.120160

Park Williams, A., Allen, C. D., Macalady, A. K., Griffin, D., Woodhouse, C. A., Meko, D. M., & McDowell, N. G. (2013). Temperature as a potent driver of regional forest drought stress and tree mortality. *Nature Climate Change, 3*(3), 292–297. https://doi.org/10.1038/nclimate1693

Pattnayak, K. C., Awasthi, A., Sharma, K., & Pattnayak, B. B. (2023). Fate of rainfall over the North Indian states in the 1.5 and 2 C warming scenarios. *Earth and Space Science, 10*(2), e2022EA002671. https://doi.org/10.1029/2022EA002671

Perkins, S. E., & Alexander, L. V. (2013). On the measurement of heat waves. *Journal of Climate, 26*(13), 4500–4517. https://doi.org/10.1175/JCLI-D-12-00383.1

Plumlee, J. D., Powers, S. P., Kimbro, D. L., Lehrter, J. C., Herrmann, J., & Mareska, J. (2024). Increasing duration of heatwaves poses a threat to oyster sustainability in the Gulf of Mexico. *Ecological Indicators, 162*, Article 112015. https://doi.org/10.1016/j.ecolind.2024.112015

Qian, H., Zhang, J., & Jiang, M. (2023). Global patterns of taxonomic and phylogenetic diversity of flowering plants: Biodiversity hotspots and coldspots. *Plant Diversity, 45*(3), 265–271. https://doi.org/10.1016/j.pld.2023.01.009

Qu, L., Chen, J., Dong, G., & Shao, C. (2018). Heavy mowing enhances the effects of heat waves on grassland carbon and water fluxes. *Science of the Total Environment, 627*, 561–570. https://doi.org/10.1016/j.scitotenv.2018.01.287

Quétier, F., Thébault, A., & Lavorel, S. (2007). Plant traits in a state and transition framework as markers of ecosystem response to land-use change. *Ecological Monographs, 77*(1), 33–52. https://doi.org/10.1890/06-0054

Rajametov, S. N., Yang, E. Y., Cho, M. C., Chae, S. Y., Jeong, H. B., & Chae, W. B. (2021). Heat-tolerant hot pepper exhibits constant photosynthesis via increased transpiration rate, high proline content and fast recovery in heat stress condition. *Scientific Reports, 11*(1), 14328. https://doi.org/10.1038/s41598-021-93697-5

Rajeevan, M., Rohini. P, Smitha Anil Nair, A. N., Tirkey, S., Goswami, T., Kumar, N. (2023). *Meteorological Monograph: Heat and Cold Waves in India Processes and Predictability* (Met. Monograph No. MoES/IMD/Synoptic Met/01(2023)/28). Meteorological Department, Ministry of Earth Sciences India, Government of India.

Ristic, Z., Bukovnik, U., & Prasad, P. V. (2007). Correlation between heat stability of thylakoid membranes and loss of chlorophyll in winter wheat under heat stress. *Crop Science, 47*(5), 2067–2073. https://doi.org/10.2135/cropsci2006.10.0674

Roach, T., Neuner, G., Kranner, I., & Buchner, O. (2023). Heat acclimation under drought stress induces antioxidant enzyme activity in the alpine plant *Primula minima. Antioxidants, 12*(5), 1093. https://doi.org/10.3390/antiox12051093

De Ronde, J. A., Van Der Mescht, A., & Steyn, H. S. F. (2000). Proline accumulation in response to drought and heat stress in cotton. *African Crop Science Journal, 8*(1), 85–91. https://doi.org/10.4314/acsj.v8i1.27718

Rothfels, C. J., Lee, J., Sundue, M. A., Smith, A. R., Kasameyer, A., Gross, J., & Looy, C. V. (2025). The PteridoPortal: A publicly accessible collection of over three million records of extant and extinct pteridophytes. *Applications in Plant Sciences, 13*(2), Article e70003. https://doi.org/10.1002/aps3.70003

Rumpff, L., Coates, F., & Morgan, J. W. (2010). Biological indicators of climate change: Evidence from long-term flowering records of plants along the Victorian coast, Australia. *Australian Journal of Botany, 58*(6), 428–439. https://doi.org/10.1071/BT10053

Ruthrof, K. X., Breshears, D. D., Fontaine, J. B., Froend, R. H., Matusick, G., Kala, J., & Hardy, G. E. S. J. (2018). Subcontinental heat wave triggers terrestrial and marine, multi-taxa responses. *Scientific Reports, 8*(1), 13094. https://doi.org/10.1038/s41598-018-31236-5

Sandhu, H. S., Wratten, S. D., & Cullen, R. (2010). Organic agriculture and ecosystem services. *Environmental Science & Policy, 13*(1), 1–7. https://doi.org/10.1016/j.envsci.2009.11.002

Sangwan, S., Shameem, N., Yashveer, S., Tanwar, H., Parray, J. A., Jatav, H. S., & Poczai, P. (2022). Role of salicylic acid in combating heat stress in plants: Insights into modulation of vital processes. *Frontiers in Bioscience-Landmark, 27*(11), 310. https://doi.org/10.31083/j.fbl2711310

San-José, M., Werden, L., Peterson, C. J., Oviedo-Brenes, F., & Zahawi, R. A. (2021). Large tree mortality leads to major aboveground biomass decline in a tropical forest reserve. *Oecologia, 197*(3), 795–806. https://doi.org/10.1007/s00442-021-05048-w

Sarkar, U. K., Paul, T. T., Puthiyottil, M., Karnatak, G., Das, V. G., Lianthuamluaia, & Das, B. K. (2022). Inland heat waves (IHWs) and associated impacts on hydro-biology of aquatic ecosystems in lower Ganga basin, India. *International Journal of Biometeorology, 66*(1), 235–245. https://doi.org/10.1007/s00484-021-02210-0

Scafaro, A. P., Posch, B. C., Evans, J. R., Farquhar, G. D., & Atkin, O. K. (2023). Rubisco deactivation and chloroplast electron transport rates co-limit photosynthesis above optimal leaf temperature in terrestrial plants. *Nature Communications, 14*(1), 2820. https://doi.org/10.1038/s41467-023-38496-4

Schnabel, F., Beugnon, R., Yang, B., Richter, R., Eisenhauer, N., Huang, Y., & Bruelheide, H. (2025). Tree diversity increases forest temperature buffering via enhancing canopy density and structural diversity. *Ecology Letters, 28*(3), Article e70096. https://doi.org/10.1111/ele.70096

Schreiber, U., & Armond, P. A. (1978). Heat-induced changes of chlorophyll fluorescence in isolated chloroplasts and related heat-damage at the pigment level. *Biochimica et Biophysica Acta (BBA)-Bioenergetics, 502*(1), 138–151. https://doi.org/10.1016/0005-2728(78)90138-X

Schulze, C. H., Waltert, M., Kessler, P. J., Pitopang, R., Veddeler, D., Mühlenberg, M., & Tscharntke, T. (2004). Biodiversity indicator groups of tropical land-use systems: Comparing plants, birds, and insects. *Ecological Applications, 14*(5), 1321–1333. https://doi.org/10.1890/02-5409

Searle, E. B., & Chen, H. Y. H. (2018). Temporal declines in tree longevity associated with faster lifetime growth rates in boreal forests. *Environment Research Letters, 13*, Article 125003. https://doi.org/10.1088/1748-9326/aaea9e

Sellamuthu, G., Tarafdar, A., Jasrotia, R. S., Chaudhary, M., Vishwakarma, H., & Padaria, J. C. (2024). Introgression of Δ 1-pyrroline-5-carboxylate synthetase (PgP5CS) confers enhanced resistance to abiotic stresses in transgenic tobacco. *Transgenic Research, 33*(3), 131–147. https://doi.org/10.1007/s11248-024-00385-x

Sentis, A., Hemptinne, J. L., & Brodeur, J. (2013). Effects of simulated heat waves on an experimental plant–herbivore–predator food chain. *Global Change Biology, 19*(3), 833–842. https://doi.org/10.1111/gcb.12094

Sharma, K., Acharya, B. K., Sharma, G., Valente, D., Pasimeni, M. R., Petrosillo, I., & Selvan, T. (2020). Land use effect on butterfly alpha and beta diversity in the Eastern Himalaya, India. *Ecological Indicators, 110*, Article 105605. https://doi.org/10.1016/j.ecolind.2019

Sharma, K., Devi, S. B., & Kumar, A. (2025). Microplastics and toxic leachate from littered cigarette butts threaten the environment, biodiversity, and human well-being. In B. Singh & S. K. Upadhyay (Eds.), *Microplastics: Origins, Risks, and Mitigation* (pp. 299–316). Elsevier. https://doi.org/10.1016/B978-0-443-29804-2.00012-3

Shelake, R. M., Wagh, S. G., Patil, A. M., Červený, J., Waghunde, R. R., Kim, J. -Y., & Suzuki, N. (2024). Heat stress and plant–biotic interactions: Advances and perspectives. *Plants, 13*(15), https://doi.org/10.3390/plants13152022

Shi, H. (2024). Impacts of extreme weather events on terrestrial carbon sequestration revealed by weather stations in the Northern Hemisphere. arXiv preprint, arXiv:2405.16207. https://doi.org/10.48550/arXiv.2405.16207

Singh, S., Mall, R. K., & Singh, N. (2021). Changing spatio-temporal trends of heat wave and severe heat wave events over India: An emerging health hazard. *International Journal of Climatology, 41*, E1831–E1845. https://doi.org/10.1002/joc.6814

Singh, D. K., & Pusalkar, P. K. (2020). Floristic diversity of the Indian Himalaya. In G. H. Dar &A. A. Khuroo (Eds.), *Biodiversity of the Himalaya: Jammu and Kashmir State* (pp.93–126). https://doi.org/10.1007/978-981-32-9174-4_5

Singh, Y., & Singh, G. (2024). Algal diversity of the Western and Central Himalaya. In Y. Singh & G. Singh (Eds.), *Biodiversity Hotspot of the Himalaya* (pp. 67–90). CRC Press, Taylor & Francis. https://doi.org/10.1201/9781003455202

Singh P., Karthigeyan, K., Lakshminarasimhan, P., Dash, S. S. (2015). *Endemic Vascular Plants of India* (p. 355). Botanical Survey of India, Kolkata.

Singh, D. K., Singh, S. K., Singh, D., and Saha, P. (2018). *Bryophytes of India*. Botanical Survey of India. https://bsi.gov.in/uploads/userfiles/file/ENVIS/PAMPLETS/Pamphlet_Bryophytes%20of%20India.pdf.

Sinha, D., Rajulapati, C. R., & Crawford, A. (2025). Concurrent heatwaves and droughts in Canada: Spatio-temporal changes, climate drivers, and persistence properties. *Earth's Future, 13*(10), e2025EF006104. https://doi.org/10.1029/2025EF006104

Smale, D. A., Wernberg, T., Oliver, E. C., Thomsen, M., Harvey, B. P., Straub, S. C., & Moore, P. J. (2019). Marine heatwaves threaten global biodiversity and the provision of ecosystem services. *Nature Climate Change, 9*(4), 306–312. https://doi.org/10.1038/s41558-019-0412-1

Srivastava, R. C., (2020). *Gymnosperms in Sikkim* (p. 108). Independently published.

Starko, S., Neufeld, C. J., Gendall, L., Timmer, B., Campbell, L., Yakimishyn, J., & Baum, J. K. (2022). Microclimate predicts kelp forest extinction in the face of direct and indirect marine heatwave effects. *Ecological Applications, 32*(7), Article e2673. https://doi.org/10.1002/eap.2673

Stephenson, N. L., Van Mantgem, P. J., Bunn, A. G., Bruner, H., Harmon, M. E., O'Connell, K. B., & Franklin, J. F. (2011). Causes and implications of the correlation between forest productivity and tree mortality rates. *Ecological Monographs, 81*(4), 527–555. https://doi.org/10.1890/10-1077.1

Sterck, F. J., Song, Y., & Poorter, L. (2024). Drought-and heat-induced mortality of conifer trees is explained by leaf and growth legacies. *Science Advances, 10*(15), eadl4800. https://doi.org/10.1126/sciadv.adl4800

Strydom, S., Murray, K., Wilson, S., Huntley, B., Rule, M., Heithaus, M., & Zdunic, K. (2020). Too hot to handle: Unprecedented seagrass death driven by marine heatwave in a World Heritage Area. *Global Change Biology, 26*(6), 3525–3538. https://doi.org/10.1111/gcb.15065

Su, Y., Gabrielle, B., & Makowski, D. (2021). The impact of climate change on the productivity of conservation agriculture. *Nature Climate Change, 11*, 628–633. https://doi.org/10.1038/s41558-021-01075-w

Suseela, M. R., & Toppo, K. (2007). Contribution to the desmid flora of Sikkim Himalayas, India. *Bulletin of the National Museum of Nature and Science, Series B, 33*(3–4), 105–114.

Swetha, M. S., Rajavel, M., Senthil, A., Djanaguiraman, M., Anitha, K., Karthikeyan, R., & Suresh, R. (2025). High temperature stress-Physiological mechanism in crop plants. Plant Science Today. 12(sp*3*), 01–13.https://doi.org/10.14719/pst.8598

Tatarinov, F., Rotenberg, E., Maseyk, K., Ogée, J., Klein, T., & Yakir, D. (2016). Resilience to seasonal heat wave episodes in a Mediterranean pine forest. *New Phytologist, 210*(2), 485–496. https://doi.org/10.1111/nph.13791

Telwala, Y., Brook, B. W., Manish, K., & Pandit, M. K. (2013). Climate-induced elevational range shifts and increase in plant species richness in a Himalayan biodiversity epicentre. *PLoS ONE, 8*(2), Article e57103. https://doi.org/10.1371/journal.pone.0057103

Teskey, R., Wertin, T., Bauweraerts, I., Ameye, M., McGuire, M. A., & Steppe, K. (2015). Responses of tree species to heat waves and extreme heat events. *Plant, Cell & Environment, 38*(9), 1699–1712. https://doi.org/10.1111/pce.12417

Tiebel, K., Dahlmann, J., & Karge, A. (2024). Global warming could shorten the seed lifespan of pioneer tree species and thus natural regeneration window of damaged areas. *European Journal of Forest Research, 143*(2), 437–450. https://doi.org/10.1007/s10342-023-01633-1

Tommasino, E., Griffa, S., Grunberg, K., Ribotta, A., Lopez Colomba, E., Carloni, E., & Luna, C. M. (2012). Malondialdehyde content as a potential biochemical indicator of tolerant *Cenchrus ciliaris* L. genotypes under heat stress treatment. *Grass and Forage Science, 67*(3), 456–459. https://doi.org/10.1111/j.1365-2494.2012.00851.x

Trimmel, H., Weihs, P., Leidinger, D., Formayer, H., Kalny, G., & Melcher, A. (2018). Can riparian vegetation shade mitigate the expected rise in stream temperatures due to climate change during heat waves in a human-impacted pre-alpine river? *Hydrology and Earth System Sciences, 22*(1), 437–461. https://doi.org/10.5194/hess-22-437-2018

ul Hassan, W., Nayak, M. A., & Lyngwa, R. V. (2021). Recent changes in heatwaves and maximum temperatures over a complex terrain in the Himalayas. *Science of the Total Environment, 794*, 148706. https://doi.org/10.1016/j.scitotenv.2021.148706

Urban, J., Ingwers, M. W., McGuire, M. A., & Teskey, R. O. (2017). Increase in leaf temperature opens stomata and decouples net photosynthesis from stomatal conductance in *Pinus taeda* and *Populus deltoides* x *nigra. Journal of Experimental Botany, 68*(7), 1757–1767. https://doi.org/10.1093/jxb/erx052

Van Gorsel, E., Wolf, S., Cleverly, J., Isaac, P., Haverd, V., Ewenz, C., & Silberstein, R. (2016). Carbon uptake and water use in woodlands and forests in southern Australia during an extreme heat wave event in the "Angry Summer" of 2012/2013. *Biogeosciences, 13*(21), 5947–5964. https://doi.org/10.5194/bg-13-5947-2016

Van Peer, L., Nijs, I., Reheul, D., & De Cauwer, B. (2004). Species richness and susceptibility to heat and drought extremes in synthesized grassland ecosystems: Compositional versus physiological effects. *Functional Ecology, 18*(6), 769–778. https://doi.org/10.1111/j.0269-8463.2004.00901.x

Vani, B., Saradhi, P. P., & Mohanty, P. (2001). Alteration in chloroplast structure and thylakoid membrane composition due to in vivo heat treatment of rice seedlings: Correlation with the functional changes. *Journal of Plant Physiology, 158*(5), 583–592. https://doi.org/10.1078/0176-1617-00260

Vitali, V., Büntgen, U., & Bauhus, J. (2017). Silver fir and Douglas fir are more tolerant to extreme droughts than Norway spruce in south-western Germany. *Global Change Biology, 23*, 5108–5119. https://doi.org/10.1111/gcb.13774

Wahid, A., Gelani, S., Ashraf, M., & Foolad, M. R. (2007). Heat tolerance in plants: An overview. *Environmental and Experimental Botany, 61*(3), 199–223. https://doi.org/10.1016/j.envexpbot.2007.05.011

Wang, L. J., Fan, L., Loescher, W., Duan, W., Liu, G. J., Cheng, J. S., & Li, S. H. (2010). Salicylic acid alleviates decreases in photosynthesis under heat stress and accelerates recovery in grapevine leaves. *BMC Plant Biology, 10*(1), 34. https://doi.org/10.1186/1471-2229-10-34

Wang, Z., Dong, X., De Boeck, H. J., Dong, G., Jiang, S., Chen, J., & Qu, L. P. (2025). Dominant species rather than plant biodiversity shape grassland resistance and recovery in response to heatwaves and mowing. *Journal of Ecology, 113*(10), 2935–2952. https://doi.org/10.1111/1365-2745.70138

Wang, C., Li, Z., Chen, Y., Li, Y., Ouyang, L., Zhu, J., & Li, H. (2024). Changes in global heatwave risk and its drivers over one century. *Earth's Future, 12*(10), e2024EF004430. https://doi.org/10.1029/2024EF004430

Wernberg, T., Thomsen, M. S., Burrows, M. T., Filbee-Dexter, K., Hobday, A. J., Holbrook, N. J., & Smith, K. (2025). Marine heatwaves as hot spots of climate change and impacts on biodiversity and ecosystem services. *Nature Reviews Biodiversity, 1*(7), 461–479. https://doi.org/10.1038/s44358-025-00058-5

Van der Westhuizen, M. M., Oosterhuis, D. M., Berner, J. M., & Boogaers, N. (2020). Chlorophyll a fluorescence as an indicator of heat stress in cotton (*Gossypium hirsutum* L.). *South African Journal of Plant and Soil, 37*(2), 116–119. https://doi.org/10.1080/02571862.2019.1665721

Willits, D. H., & Peet, M. M. (2001). Measurement of chlorophyll fluorescence as a heat stress indicator in tomato: Laboratory and greenhouse comparisons. *Journal of the American Society for Horticultural Science, 126*(2), 188–194. https://doi.org/10.21273/JASHS.126.2.188

Winqvist, C., Ahnström, J., & Bengtsson, J. (2012). Effects of organic farming on biodiversity and ecosystem services: Taking landscape complexity into account. *Annals of the New York Academy of Sciences, 1249*(1), 191–203. https://doi.org/10.1111/j.1749-6632.2011.06413.x

Yin, J., Gentine, P., Slater, L., Gu, L., Pokhrel, Y., Hanasaki, N., & Schlenker, W. (2023). Future socio-ecosystem productivity threatened by compound drought–heatwave events. *Nature Sustainability, 6*(3), 259–272. https://doi.org/10.1038/s41893-022-01024-1

Zanotelli, D., Montagnani, L., Andreotti, C., & Tagliavini, M. (2022). Water and carbon fluxes in an apple orchard during heat waves. *European Journal of Agronomy, 134*, Article 126460. https://doi.org/10.1016/j.eja.2022.126460

Zhang, R., Wise, R. R., Struck, K. R., & Sharkey, T. D. (2010). Moderate heat stress of *Arabidopsis thaliana* leaves causes chloroplast swelling and plastoglobule formation. *Photosynthesis Research, 105*(2), 123–134. https://doi.org/10.1007/s11120-010-9572-6

Zhang, Q., Shao, M. A., Jia, X., & Wei, X. (2017). Relationship of climatic and forest factors to drought-and heat-induced tree mortality. *PLoS ONE, 12*(1), Article e0169770. https://doi.org/10.1371/journal.pone.0169770

Zhang, Y., Loreau, M., He, N., Wang, J., Pan, Q., Bai, Y., & Han, X. (2018). Climate variability decreases species richness and community stability in a temperate grassland. *Oecologia, 188*(1), 183–192. https://doi.org/10.1007/s00442-018-4208-1

Zhang, K., Li, J. B., Ng, M. K. P., Guo, Z. F., Tai, A. P., Liu, S. W., & Wu, J. (2025). Global prevalence of compound heatwaves from 1980 to 2022. *Advances in Climate Change Research, 16*(3), 565–575. https://doi.org/10.1016/j.accre.2025.04.010

Zhang, L., Gao, Y., Pan, H., Hu, W., & Zhang, Q. (2013). Cloning and characterisation of a *Primula* heat shock protein gene, PfHSP17. 1, which confers heat, salt and drought tolerance in transgenic *Arabidopsis thaliana. Acta Physiologiae Plantarum, 35*(11), 3191–3200. https://doi.org/10.1007/s11738-013-1354-2

Zhao, C., Liu, B., Piao, S., Wang, X., Lobell, D. B., Huang, Y., & Asseng, S. (2017). Temperature increase reduces global yields of major crops in four independent estimates. *Proceedings of the National Academy of Science USA, 114*, 9326–9331. https://doi.org/10.1073/pnas.1701762114

Zhao, J., Lu, Z., Wang, L., & Jin, B. (2020). Plant responses to heat stress: Physiology, transcription, noncoding RNAs, and epigenetics. *International Journal of Molecular Sciences, 22*(1), 117. https://doi.org/10.3390/ijms22010117

Zhu, Y., Queenborough, S. A., Condit, R., Hubbell, S. P., Ma, K. P., & Comita, L. S. (2018). Density-dependent survival varies with species life-history strategy in a tropical forest. *Ecology Letters, 21*(4), 506–515. https://doi.org/10.1111/ele.12915

Zhu, R., Zhang, Z., Zhang, N., Zhong, H., Zhou, F., Zhang, X., & Xing, B. (2025). A global estimate of multiecosystem photosynthesis losses under microplastic pollution. *Proceedings of the National Academy of Sciences, 122*(11), Article e2423957122. https://doi.org/10.1073/pnas.2423957122

Impact of Rising Temperature on Plant Ecosystem

A. Waheed Deshmukh, Pradeep Kumar Ramteke, Ajit P. Rathod, and Payal Bhautik

Abstract Climate change, particularly the continuous increase in global temperatures, is profoundly influencing plant ecosystems across the world. The gradual warming of the planet has altered the timing of seasonal events, such as leaf emergence and flowering, and shifted the geographical distribution of many species. Rising temperatures affect plant physiological functions, including photosynthesis, respiration, and water use efficiency, which in turn influence growth rate and crop productivity. Extended heat exposure often leads to water stress, reduced biomass, and lower yields. Changes in temperature regimes are also disrupting pollination cycles and seed dispersal patterns, resulting in ecological imbalance within plant–pollinator networks. Many species are being displaced from their native habitats, moving toward higher elevations or latitudes in search of favorable conditions, although such migrations may not guarantee survival. Increased evapotranspiration further depletes soil moisture, intensifying drought conditions in semi-arid regions. Consequently, plants become more susceptible to pests, diseases, and invasive species, posing a serious threat to global food security. These changes cascade through ecosystems, affecting herbivores, pollinators, and soil microorganisms that depend on stable plant communities. This chapter examines the physiological and genetic responses of plants to heat stress and highlights adaptation strategies such as the development of heat-tolerant crop varieties, conservation of plant genetic resources, and restoration of ecosystem functions to enhance agricultural resilience under changing climatic conditions.

Keywords Heat stress · Climate change · Plant ecosystem · Phenology shift · Adaptive strategies

A. Waheed Deshmukh (✉) · P. K. Ramteke
Department of Chemical Engineering, Shri Guru Gobind Singhji Institute of Engineering and Technology, Nanded, India
e-mail: awasdk@gmail.com

A. P. Rathod
Department of Chemical Engineering, Visvesvaraya National Institute of Technology (VNIT), Nagpur, India

P. Bhautik
Department of Chemical Engineering, Priyadarshini College of Engineering, Nagpur, India

A. Awasthi et al. (eds.), *Rising Temperatures and Plant Ecosystems*,
https://doi.org/10.1007/978-3-032-29477-7_7

1 Introduction

Global warming is a rise in the average temperature of the atmosphere and oceans of the earth. Greenhouse gases, particularly carbon dioxide is the prime cause of this warming (Raizada et al., 2022). Climate change can also be caused due to natural and manmade factors such as, volcanic activities, changes in natural carbon cycle, solar variations, forest fires, oceanic processes, and burning of fossil fuels, industrialization, deforestation and land cleaning, excessive usage of chemicals in agricultural practices. The consequences of these changes are severe, such as melting of the ice caps, which impacts on agriculture and eventually on the livelihood of human beings (Awasthi et al., 2022). To deal with climate change, requires a clear strategy to curb greenhouse gas emission and concrete measures to reduce emissions (Chakraborty et al., 2025). It is well known fact that increase in greenhouse gases increases the temperature on the earth and temperature of earth further impacts structural, molecular, and ecological processes that are essential to life on Earth. The temperature may dictate the growth rate in plants, the quantity of leaves and flowers, among others, which contribute to producing a healthy plant and above all a healthy crop (Gates, 1980). It is also the determinant of the flow of energy in the organism and influences the growth of all other organisms. Temperature affects survival mechanisms in insects, especially at extreme cold temperatures. Insects can survive by freezing or avoiding it through physiological adaptations, including cryoprotectants and quick freezing of the insect body. The temperature also determines the distribution, abundance, and invisibility of insects (Bale, 2002). Knowledge of temperature impacts is crucial in forecasting species reactions to climate change as well as controlling agricultural systems.

Figure 1 presents the interconnected climate systems. The global climate system consists of different elements such as the atmosphere, hydrosphere, cryosphere, lithosphere, and biosphere, which interact dynamically with each other. This model highlights the cascading effects of the temperature changes on biosphere response, especially the response of plants.

Different types of ecosystems that are dominated by plants are globally significant to ecological stability and biodiversity. Besides the indispensable ecosystem functions such as pollination and carbon sequestration, plants are important in maintaining and reinstating regulated processes and stability in ecosystems. Without them, homeostasis would not be possible as they are the primary producers of land ecosystems, which are the backbone of all terrestrial ecosystems (Shahdin & Siddiqa, 2025). In order to be effective, the ecosystem needs to be diverse and varied to provide these services (Hoque et al., 2025). Nevertheless, man has been putting pressure on these processes by deforestation and climate change. Still, resilience to ecosystems faces an increasing risk because of the habitats fragmentation and inappropriate management, which results in the destabilization of plant diversity that is crucial to ecosystem health and land stability (Sarfraz et al., 2022). Agriculture is now being revolutionized using the recent developments in plant science such as the use of genomics and

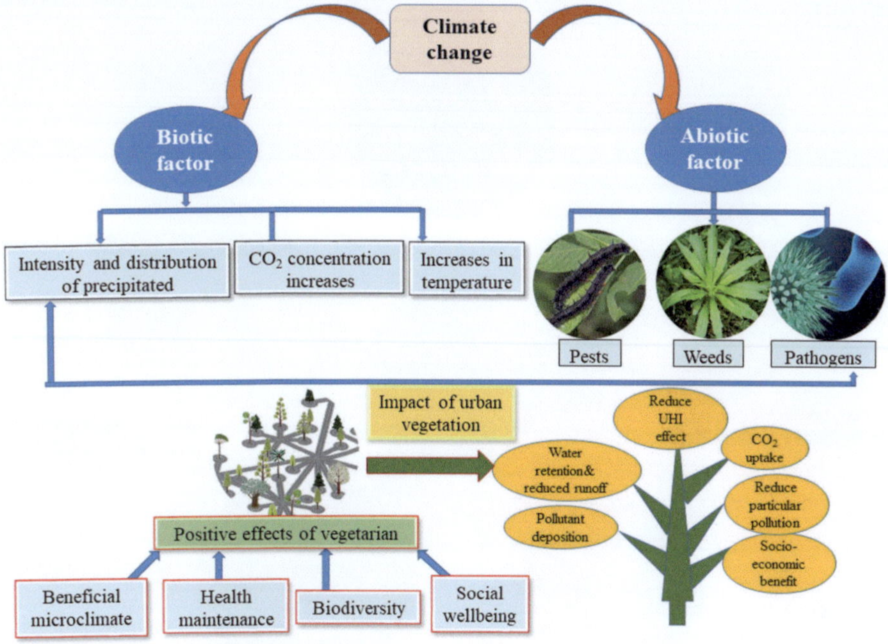

Fig. 1 Global climate systems of the components

the optimization of photosynthesis to tackle global challenges. Figure 2 indicates the contribution of the plant ecosystem to biodiversity in the world.

This chapter discusses the complex effects of climate change-induced temperature variation within biosphere reserves, focusing on plant and their associated ecosystem to large-scale shifts in species distribution and migration timing as well as changes in ecosystem composition and functioning. Table 1 shows the significance of the plant ecosystem in global biodiversity and ecological balance.

In these respects, entire biomes can undergo significant changes, and the fundamental processes of the ecosystem, such as carbon storage, nutrient cycling, and other functions, may undergo profound alteration.

2 Global Trends in Rising Temperature

Global surface temperature has shown a consistent warming trend since the 1970s, with recent years setting new heat records (Rahmstorf et al., 2017). The IPCC AR6 report indicates a global temperature rise of about +1.1 °C with 2024 estimated at 1 °C above pre-industrial levels, aiming to limit warming below 1.5 °C. The increase in global temperature is primarily driven by rising tropical ocean temperature, which enhance the hydrologic cycle (Graham, 1995). Even though there is some short-term

Fig. 2 Plant ecosystem in global biodiversity

Table 1 Significance of plant ecosystem in global biodiversity and ecological balance (Barber, 2008; Prajapati et al., 2023; Shih, 2015)

Aspect	Significance
Biodiversity support	Provide habitat and food for a wide range of organisms, maintaining species diversity
Oxygen production	Produce oxygen through photosynthesis, which is essential for life on earth
Carbon sequestration	Absorb CO_2 and help mitigate climate change
Soil fertility	Add organic matter, prevent erosion, and maintain nutrient cycles
Water regulation	Regulate water cycles through transpiration and maintain groundwater recharge
Pollination and reproduction	Support pollinators and enable the reproduction of flowering plants
Climate regulation	Influence the local and global climate by controlling temperature and humidity
Ecosystem services	Provide food, medicine, timber, and raw material for humans and wildlife

variability, the overall tread is still statistically significant and surpasses measurement and analysis uncertainties (Brohan et al., 2006). In contrast to previous warm period, such as the one around 1940, the warming seen in recent decades is more widely distributed rather than concentrated at high northern latitudes. Recent reports have indicated that there are vast regional variations of the global warming patterns.

Despite an increase in global temperature, regional differences in the speed of warming and cooling are appearing in certain areas, and cooling variables atmospheric processes are in an intricate interplay that determines these regional differences (Tamarin-Brodsky et al., 2020). Depending on the various emission scenarios and climate models, the global warming varies. Temperature increases are more likely to be experienced in all regions and seasons, but minor decrease is likely to be experienced in places like Australia, Southern Africa, and Mediterranean (Ruoateenoja et al., 2003). Northern and southern hemispheres are quite different, and this impact on the local ecosystem, weather conditions, and the risk of extreme temperatures occurrence (Shahdin et al., 2025; Cheshmehzangi, 2020). It is possible to estimate the relationship between CO_2 emission and the increase in temperature using historical data and project the occurrence of subsequent warming under different conditions (Donald, 2024).

Figure 3 demonstrates the data on the comparison of global mean temperature changes reported by major agencies where each of the datasets is standardized to the 1860–2020 average presenting the consistent picture of the recent climate change. Studies have always shown a tight linkage between the rise in temperature around the globe and the emission of greenhouse gases. Research reports on carbon dioxide (CO_2) and methane (CH_4) as the major causes of increase in temperature. To alleviate the risks of the associates, there is an urgent need to curtail the release of greenhouse gases, especially CO_2 and CH_4. It is recommended to stay at 370.665–405.79 ppm of CO_2 concentration and 1831.49–1879.59 ppb of methane concentration every year (Wu et al., 2023). It has been found that the effects of emissions on temperature abnormalities over the last several decades have been unchanging, and the likelihood of extreme values in both the records of emissions and temperature is high (Agligardi et al., 2019).

3 Phenological Shift and Physiological Response of Plants to Climate Change

Climate change is altering the phenology of plants and pollinators, potentially disrupting their interaction. However, the effects of these shifts are complex and species-specific, with some species advancing their activity, some delaying, and others remaining unchanged (Sola et al., 2014). Experimental studies have shown that plant species with historically advanced flowering received fewer visits (Rafferty &

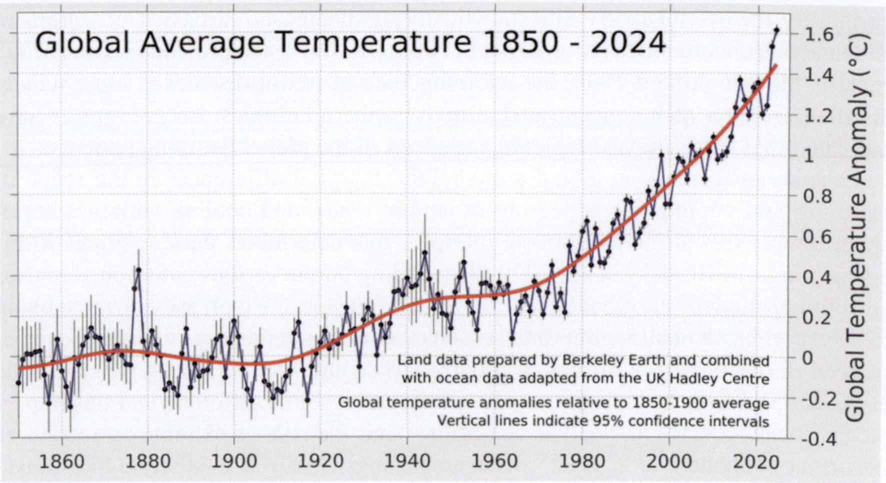

Fig. 3 Comparison of global mean temperature change among major reporting agencies (*Source* https://berkeleyearth.org/global-temperature-report-for-2024/)

Ives, 2011). Traits associated with mutualism may impact the likelihood of phenological mismatches, with habituated, short-term, seasonal, and non-symbiotic interactions suspected to be more susceptible (Rafferty et al., 2015). It is suggested that to gain a clearer understanding of the implications of shifts in phenology, researchers measure interaction rates and fitness consequences, focus on individual variation along an environmental gradient, and take an integrated life cycle perspective (Fisogni et al., 2025). Table 2 presents the phenological changes and physiological reactions observed in the plant.

At issue are recent studies of how plant respond to varied environmental stresses, as well as some level of elevated CO_2 (eCO_2). While there are some other factors that can dictate its effects, eCO_2 mainly relates to photosynthesis and water-use efficiency (Xu et al., 2015). To promote sustainable food production, a better understanding of the different levels of plant stress response from the molecular level to the whole plant level is needed because plants can have complex physiological and biochemical adaptations to combined drought and heat stress, including regulating

Table 2 Phenological shifts and physiological response of plants

Phenological shifts	Physiological response
Earlier flowering, leaf unfolding, and fruiting due to rising temperature	Altered photosynthesis rates under elevated CO_2 levels
Changes in seed dispersal timing and growing season length	Modification to water-use efficiency and stomatal conductance
Potential mismatch between plant and pollinators, affecting reproduction	There is increased respiration and metabolic activity under heat stress

carbon metabolism, antioxidant level, and many more (Haghpanah et al., 2024). Although we will review and promote plant-based production, the combination of disparate omics methodologies with system biology could allow for the development of climate-resilient yield while also understanding the actual mechanics of adaptation (Ahuja et al., 2010). New ecosystems and ecological communities are being formed as a result of the global redistribution of plant and animal species brought about by climate change (Sugden, 2017). The functioning of the ecosystem, human well-being, and the dynamics of climate change are all impacted by these changes, which upset important species interactions. They also affect the availability of natural resources, which has serious environmental repercussions. Using ecologically based management techniques can improve farming while lessening the impact off-site (Matson et al., 1997; Schirpke et al., 2024). Furthermore, ecological interaction may be disrupted by climate change even before discernible changes in distribution take place, which could lead to extinction cascades. For example, plants under drought-induced water stress produce fewer flowers and fruits, which consequently reduces animal foraging opportunities (Fonturbel et al., 2021).

4 Climate Change and Its Impact on Plant Distribution and Biodiversity

In this chapter, recent research has found major changes in species ranges in response to climate change. A meta-analysis indicated that species are moving to higher elevation at 11.0 m per decade and to higher latitude at 16.9 km per decade and both rates were significantly faster than what has been reported previously (Chem et al., 2011). The reaction varies by species optional elevation were steadily rising, averaging 29 m every ten years in western Europe (Lenoir et al., 2008), other observed more complex patterns. In western north America, plat species were equally, likely to shift upward or downward, with regional variation influenced by the interaction between temperature changes and seasonal precipitation patterns. In western north America, plant species were equally likely to shift upward or downward with regional variation influenced by the interaction between temperature changes and seasonal precipitation patterns (Harsh & Hille 2016).

Figure 4 presents the effects of climate change on plant distribution and biodiversity. It highlights shifts in species range, biodiversity loss, and ecosystem disruption. The main threats to the global biodiversity and ecosystem stability include habitat loss, fragments, and degradation (Banks-Leite et al., 2020). The processes break up prolonged habitats into small and secluded units leading to the death of species and the change in ecosystem functionality (Wu & Liu, 2014). Changes in habitat have an impact that depends on the situation, and are shaped by the evolutionary histories of species and biogeography. Although biodiversity loss has the potential to influence ecosystem functioning the relation is highly complicated and can be saturated as little as low to moderate level of diversity is lost (Gonzzalez et al., 2009).

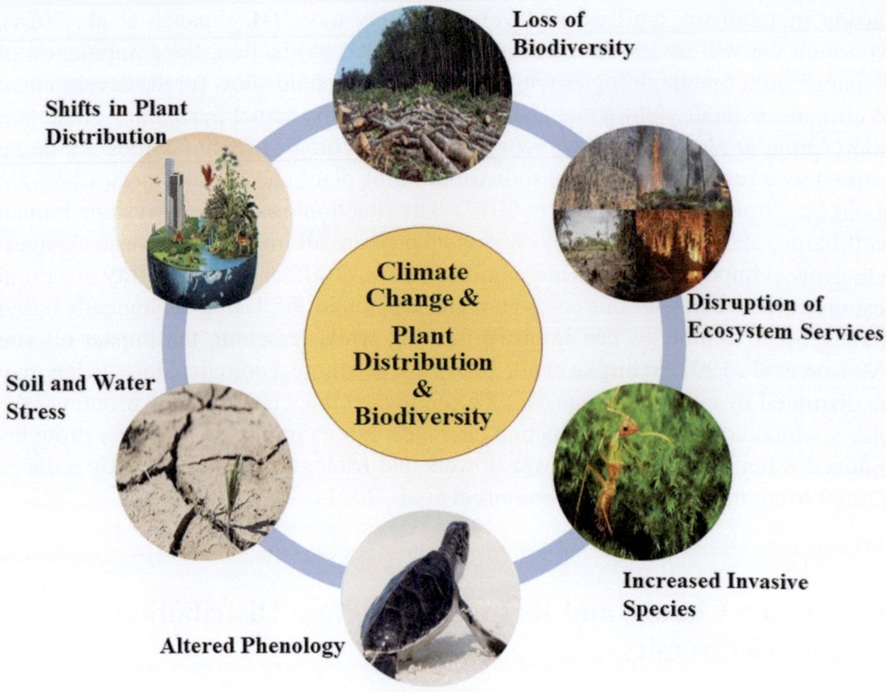

Fig. 4 Effect of climate change on biodiversity and plant distribution

Its research has highlighted the significance of habitat restoration, ceasing habitat loss and investing research on habitat fragmentation because these measures are essential in enhancing conservation biology and landscape ecology (Wilson et al., 2016). Ecological interaction, especially pollination and seed disturbance, and global warming (Memmott et al., 2007). In addition, invasive species may alter ancient ecological relations that involve the interference of the mutualism of pollination and dispersal between plants and insects (Traveset & Richardson, 2006). The scale of seed disperser loss risk to the plant communities and ecosystem processes is determined by the characteristics of the plants and the susceptibility of the disperser species (Rogers et al., 2021).

However, increased resource availability in disturbed habitats may mitigate some of these disruptions. For instance, greater sunlight exposure can benefit shade-intolerant plants, potentially preserving key ecological interactions by maintaining or enhancing floral and fruit resources for pollinators and seed dispersers (Fonturbel et al., 2017).

5 Ecosystem-Level Impact

Climate change and habitat fragmentation pose significant threats to plant–pollinator mutualistic relationships through multiple mechanisms. Global warming can cause phonological disruption, though plant–pollinator relationships are particularly vulnerable to disruption (Aizen et al., 2012). Fragmented landscapes reduce pollinator visits and nesting sites, leading to pollen limitation and decreased seed production that can drive plant populations toward extinction spirals. However, some pollination systems demonstrate surprising resilience with pollen flow occurring across kilometers between fragments (Harris & Johnson, 2004). Figure 5 shows the ecosystem-level impacts, which refer to the changes that occur in entire ecosystems due to natural or human-induced factors.

Research consistently demonstrates that warmer temperature increases pest pressure and disease dynamics across agricultural and forest systems. Climate warming alters the relative growth rates of crops and insect pests, with projections suggesting increased pest pressure across much of the American Midwest, potentially requiring greater pesticide use (Taylor et al., 2018). Empirical studies show that higher temperature increases disease transmission and outbreak intensity in host–pathogen systems, not through changes in mean transmission rates but by reducing variability in disease risk across individuals (Elderd & Reilly, 2014). Temperature and humidity are key epidemic drivers, affecting pathogen survival, multiplication rates, and disease

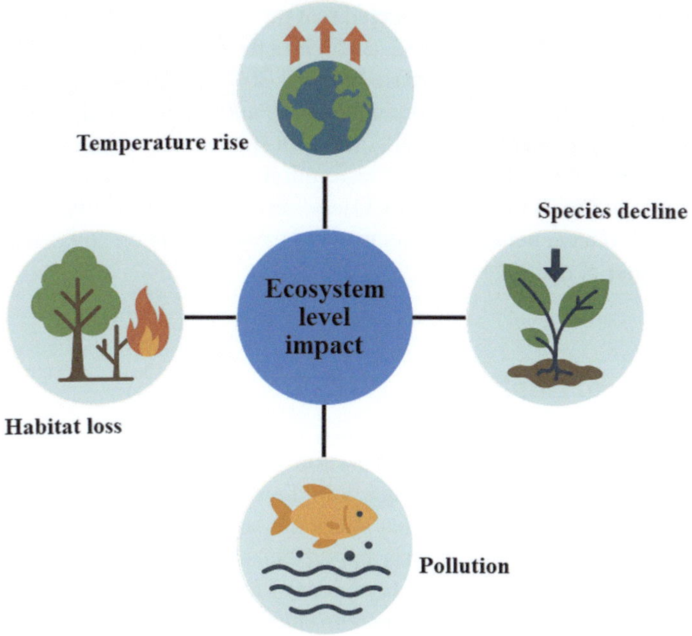

Fig. 5 Ecosystem-level impact

cycle components, including infection processes and inoculum dispersal (Legreve & Duveiller, 2010). The dynamics of forest pest are especially susceptible and according to the models, the entire behavior of insect outbreak will increase as the climate warms, due to the fact that pest species features predispose them to climate chaos (Jesse et al., 2003).

Conversion and disturbance of forests have a big impact on the biomass loss of potential to capture carbon. On agricultural conversion in the Amazon, above-ground biomass and carbon stocks can decline by 95% or greater than in unaltered rainforests with 87% of the biomass losses in Brazil due to agricultural conversion (Denich et al., 2019). The same can be said about timber harvesting, which decreases the potential of carbon absorption by 81% in the natural forest since it is 28.37-ton CO_2 eq ha^{-1} before trees are cut and 4.44-ton CO_2 eq ha^{-1} after the process (Yuniawati & Tampulolon, 2021). There is also an impact of fire regimes on carbon sequestration, yet they can act as carbon sinks (Williams et al., 2004). The potential of natural forests toward carbon storage is significant with 112.19 MgC/ha of vegetation in the tropical dry forests and this indicates the significance of natural forest ecosystem protection in the process of carbon sequestration (Varghese, 2011).

Grasslands constitute an important carbon deposit in the world with about one-third of the terrestrial carbon stock and up to 30% of all soils in the world contain organic carbon. Nevertheless, the potential of these ecosystems in carbon sequestration is greatly challenged and on a downward trend. Climate change has a tremendous influence on the carbon storage in the grasslands because it modifies the carbon contribution of plants and the carbon transformation in microbes (Bai & Cotufo, 2022). In Qinghai Province, it is projected that the potential of the grassland carbon sequestration will be reduced by 20.82–25.55% in the future under the climate conditions that grassland cover will decrease by 28.5–30.71%. The state of drought may affect the level of soil organic carbon in the rangelands substantially and exceed the benefits of the management (Liu et al., 2011). In spite of these issues, both solution grazing management and biodiversity restoration would sequester 148–699 megatons CO_2e/y and 2.3–7.3 billion-ton CO_2e/y, respectively, globally. Light grazing has been found to have beneficial effects over heavy grazing on soil organic carbon storage (Spangler et al., 2011).

6 Agricultural Implications

Heat stress significantly threatens stable crop yield, with the impact exacerbated by global warming. At +3 °C warming both direct yield reductions and decreased agricultural labor capacity due to heat stress are equally important, particularly in exposed regions such as Southeast Asia and sub-Saharan Africa, where labor capacity could drop by 30–50%, leading to higher prices of food and global maize yield losses (Lima et al., 2021). Extreme heat during reproductive stages can double global maize yield losses and reduce projected gains in spring wheat and soybean, with regional disparities and lower-income countries most affected, substantial climate mitigation could avoid over 80% of these losses (Deryng et al, 2014). Accurate assessment of heat stress impact requires considering canopy temperature, which can be much higher than standard air temperature, leading to underestimation of yield declines if ignored (Siebert et al., 2014). Heat stress impairs photosynthesis, damages cellular structures, and reduces reproductive success, but breeding transgenic approaches and thermos protectants can enhance crop thermos tolerance (Arachchige et al., 2024). Recent discussion has highlighted significant challenges and strategies related to grain quality, including reduction in protein and micronutrient concentration. Global nutrition is at risk due to climate changes, particularly the increased CO_2 levels, which have been shown to lower grain protein, iron, and zinc concentration (Miner et al., 2024). The introduction of dwarfing genes into wheat has resulted in significant declines in grain, zinc, iron, magnesium, and manganese, with the decline differing among genotypes (Velu et al., 2017). While it is possible to develop high zinc genotypes, bio-fortification initiative in bread wheat has documents low protein and gluten contents in many of these types. Conversely, they may also have higher content of other important nutrient, such as proline and carotenoids (Kaur et al., 2020). Breeding finger grain quality and micronutrient content is supported by genomic and phenotyping advancement, facilitating food and nutritional security (Pandey et al., 2023). As a result of climate change, farmers are experiencing deceased productivity due to unpredictable rainfall, water stress, extreme weather events, and new pest infestation (Pattnayak et al., 2023; Tandon et al., 2025a; Verma et al., 2025). Low yield, outbreaks of the disease and pests and flooring damage caused yield reduction raise production constraints. Farmers also are not aware of valuable practice such as crop mapping or intercropping (Tuble & Lacuesta, 2024). Food security is impacted by rising temperature and erratic rainfall, particularly for smallholder farmers. By using indigenous techniques, conserving water improving drainage and utilizing fewer chemical, farmers have adjusted. Pest management can improve climate resilience by improving soil health and water retention. Although larger farms and a range of water source results in higher adoption rates, practice uncertainly continues to be the biggest obstacle to the implementation of climate-smart practice (Taulu, 2025).

7 Adaptation and Mitigation Strategies

7.1 Development of Heat-Tolerant Varieties

To combat the effects of climate change, plant breeding and genetic engineering are essential for creating crop varieties that can withstand high temperatures. High temperatures significantly reduce crop productivity, so tolerant cultivars are essential for sustainable farming. Due to the low levels of genetic diversity and complexities in heat tolerance traits, producing such varieties remains difficult (Driedonks et al., 2016). We can utilize marker-assisted selection (MAS) to create lines that possess the genetic basis of heat tolerance. Furthermore, conventional breeding focuses on

Table 3 Adaptation and mitigation strategies (Del Rio, 2012)

Aspect	Adaptation strategies	Mitigation strategies
Objective	Reduce vulnerability and strengthen resilience	Reduce greenhouse gas (GHG) emissions
Approach	Adjust to climate impacts	Prevent or reduce the factors contributing to climate change
Examples	Developing drought-resistant crops, rainwater harvesting and early warning systems	Renewable energy adoption, afforestation, and carbon capture
Time frame	Short to medium term	Long-term, with gradual benefits
Focus area	Local/regional level	Global/national level
Outcome	Coping with climate variability	Slowing or reversing climate change

phenotypic traits for heat tolerance, such as increased cuticle thickness, reduced leaf area, and increased rooting depth. Genetic engineering will support these efforts by introducing genes associated with resistance, thereby increasing the resilience of plants, as seen in the case of heat shock proteins (HSPs) (Agalya et al., 2025). The use of breeding lines based on pollen sterility and spikelet sterility at spike emergence has advanced research identifying heat-tolerant breeding lines. Several screening programs have demonstrated success in identifying heat-tolerant rice breeding lines that can yield, e.g., N22 and IDSA 77. In summary, a multidisciplinary method containing robust heat tolerance (Thangapandian et al., 2023). Table 3 provides adaptation and mitigation strategies for different parameters.

7.2 Climate-Smart Agricultural Practices

Crop rotation, mulching, and effective irrigation are examples of climate-smart agricultural (CSA) techniques that greatly improve agricultural sustainability and productivity while tackling the effects of climate change. Crop rotation has proven to be beneficial, increasing maize productivity by 5–10% and wheat yield by 30–45% (Regmi & Paudel, 2024). Also (Salihi & Himatkhwah, 2024), water-efficient irrigation technologies improve water use effectiveness by 9–68%. However, problems with land tenure, a lack of funds, and inexperience are obstacles to broad adoption. Crop rotation, precision irrigation, intercropping, and legume-based cover crops are highlighted as successful CSA techniques in European studies. These practices support the three pillars of CSA by lowering greenhouse gas emissions, boosting food security, and enhancing systems resilience (Bhattacharyya et al., 2020). In a rainfed system in Uganda, the study evaluated the extent to which CSA techniques, such as mulching half-moon pits and permanent planting basins, enhanced soil moisture storage, maize yields, and water use efficiency. The outcome showed that these parameters had a significant enhancement in comparison with the conventional approach, which indicated a higher resilience to maize production in the sub-humid areas (Zizinga et al., 2021). A Zambian study that investigated the effect of CSA practices on maize yield established that intercropping legumes with a lot of yield and a low yield resilience and the use of climate factors to determine the adoption of modern input (Asrlam et al., 2015). Other CSA methods, including crop shifting, conservation tillage, agroforestry, residue management, and intercropping, are also significant to assist farmers in coping with climate-stressors, particularly temperature fluctuations and unpredictable rainfall patterns (Kumari et al., 2019; Tandon et al., 2025b). The evidence demonstrates that CSA practices like half-moon pits, mulch, and permanent planting basins in sub-humid Uganda have a much better water use efficiency in storing moisture and yielding more maize plants and therefore increasing resilience in regions with erratic precipitation (Zizinga et al., 2022).

7.3 Ecosystem Restoration and Conservation

Forest restoration and reforestation project has become a key tool of dealing with the issue of deforestation, loss of biodiversity, and climate changes in the world. Such initiatives span different strategies such as natural regeneration afforestation and reforestation efforts that have a high degree of associated socio-economic value such as carbon capture and biodiversity protection (Prati & Pomero, 2022). To restore forest functions and stability, biodiversity—ecosystem functioning viewpoint is needed in effective forest restoration, which implies species diversity instead of monoculture plantations (Aerts & Honnay, 2011). Conservation efforts could be focused on biodiversity hotspots, where 44% of vascular plant species and 35% of vertebrate species are limited to just 25 areas covering only 1.4% of the earth's land surface. This hotspot approach enables efficient resource allocation for maximum conservation impact (Myers et al., 2000).

7.4 Policy and Governance Measure

International climate agreement and policy governance have undergone significant evolution particularly after COP15 in 2009, when the paradigm shifted from top-down to bottom-top initiative. Current policy approaches including sectoral measures under pledge-and review regimes, carbon pricing mechanisms and indirect mitigation measures such as the liberalization of trade in climate-friendly products (Yamaguchi, 2012). Research in this filed has expanded rapidly since the implementation of the Kyoto Protocol in 2005, with environmental studies, economics, and public administration emerging as the dominant disciplines. Future research trends emphasize green technology innovation local climate governance and the strengthening of political awareness (Chukui et al., 2021). A central challenge lies in integrating sustainable development policies and measure (SD-PANs) into international framework, which requires agreed guidelines, pledge processes, tracking mechanisms and progress assessment procedure (Baumert, 2005). Contemporary governance systems also face institutional inertia in addressing climate change, as powerful groups benefiting from existing arrangements often resist policy changes that alter production and consumption practices in the energy, agricultural, and transportation sectors. (Meadowcroft, 2009).

National adaptation plans for climate change are increasingly recognized as essential policy and governance measures, with many countries developing strategic frameworks and detailed plans to address climate risks (Mullan et al., 2013). Effective adaptation requires robust institutional capacity and one of the best indicators of a successful national adaptation strategy is excellent governance in their absence other determinants of adaptive capacity have limited impact (Berrang Ford et al., 2014). Policy framework emphasizes the importance of mainstreaming adaptation into government operation, establishing coordination mechanisms across sectors and level of government and attractive stakeholders throughout the revision process (Lim et al., 2004). However, challenges remain in overcoming information gaps, securing financing, and measuring adaptation outcomes. Institutional and policy processes, including cross-sectional integration, vertical coordination, and policy learning, are critical for embedding adaptation into governance systems (Dovers & Hezri, 2010).

8 Future and Limitations of the Impact of Rising Temperature on Plant Ecosystems

Plant ecosystems are seriously threatened by rising temperatures brought on by climate change through a number of interrelated pathways. From seed germination through growth, development, flowering, and fruiting, temperature increases have a direct impact on plant physiological processes, which ultimately lower crop yields (Khan et al., 2013). With changes in leaf emergence and senescence impacting temperature regulation, soil moisture, and atmospheric dynamics, climate modifies plant phenology, or the timing of life cycle events, offering strong evidence of ecosystem-wide effects (Rai, 2015). The biggest threat to agro-ecosystems and forest, especially in arid areas, is the confluence of rising temperature with increased drought and forests especially in arid areas is the confluence of rising temperature with increased drought and salinity stress. These stresses can harm photosynthetic machinery irreversibly and reduce carbon acquisition by increasing stomatal resistance (Loreto & Centritto, 2008). Plant diversity has already been shown to be impacted by climate change through changes in species distribution, new plant assemblages, and endemic species extinction (Shekhar Seth, 2015). Future research requires long-term climate data, modern tools like remote sensing, and a comprehensive understanding of plant responses to project atmosphere conditions.

9 Conclusion

Plant ecosystems are under a lot of stress due to rising global temperature, which sets off a domino effect that affects wildlife, the environment and even local economies. Temperature changes have caused plants and animal to relocate interfered with breeding and blooming schedules, and even changed the color and texture of entire landscapes. These disruptions affect oil and gas operations, nutrient cycles and the delicate balance of biodiversity, increasing ecosystems susceptibility to invasive plants that creep also riverbanks, swarms of pests, and quickly spreading pathogens. Farmers are feeling the pressure as increased heat stress is causing crops to wilt, yields to decrease, seeds to lose their vigor, and soils to try into dust all pressing challenges to global food systems. Smallholder farmers in developing countries and other marginalized communities with few resources for adaptation are left more vulnerable to food scarcity, conflict over water or land, and negative economic impacts. This chapter underscores the inherent relationship between climate change and the people it impacts by connecting the effect of warming temperatures on the whole ecosystem to the subsequent effect on the local economy and farming. There is scant research that looks obliquely at the relationship between climate change in the environment, economy, and living system. As example, drought will have an impact on crop yield, but warming will have an impact on energy costs. One instance of how climate adaptation strategies continue to overlook a significant amount of the knowledge found in local and indigenous knowledge systems, making solutions less inclusive, is the construction of a seawall without consulting the fishermen who are familiar with the tides. In order to weave green leaves, soil and human stories into a single fabric, future research must draw from a variety of disciplines, including ecological monitoring, innovative farming techniques, and socioeconomic modeling. With strong governance and well-defined policies to support them like fences that protect a young forest as it grows, we must test climate-smart farming, sustainable land use, and ecosystem restoration on a much larger scale. These directions will close knowledge gaps and lay out practical steps like restoring a mangrove shoreline for strengthening resilience in ecosystems and communities facing an unpredictable climate ahead.

References

Aerts, R., & Honnay, O. (2011). Forest restoration, biodiversity and ecosystem functioning. *BMC Ecology, 11*(1), 29. https://doi.org/10.1186/1472-6785-11-29

Agalya, P., Vijai Selvaraj, K. S., Beaulah, A., Ramalingam, J., Bharathi, A., & Gurusamy, K. (2025). Genetic frontiers in tomato breeding: Overcoming heat stress for sustainable yield. *Plant Science Today*. https://doi.org/10.14719/pst.7710

Agliardi, E., Alexopoulos, T., & Cech, C. (2019). On the relationship between GHGs and global temperature anomalies: Multi-level rolling analysis and copula calibration. *Environmental and Resource Economics, 72*(1), 109–133. https://doi.org/10.1007/s10640-018-0259-3

Ahuja, I., de Vos, R. C. H., Bones, A. M., & Hall, R. D. (2010). Plant molecular stress responses face climate change. *Trends in Plant Science, 15*(12), 664–674. https://doi.org/10.1016/j.tplants.2010.08.002

Aizen, M. A., Sabatino, M., & Tylianakis, J. M. (2012). Specialization and rarity predict nonrandom loss of interactions from mutualist networks. *Science, 335*(6075), 1486–1489. https://doi.org/10.1126/science.1215320

Arachchige, S. M., et al. (2024). Confronting heat stress in crops amid global warming: Impacts, defense mechanisms, and strategies for enhancing thermotolerance. *Crop Breeding, Genetics and Genomics, 6*(4). https://doi.org/10.20900/cbgg20240011

Arslan, A., McCarthy, N., Lipper, L., Asfaw, S., Cattaneo, A., & Kokwe, M. (2015). Climate smart agriculture? Assessing the adaptation implications in Zambia. *Journal of Agricultural Economics, 66*(3), 753–780. https://doi.org/10.1111/1477-9552.12107

Awasthi, A., Dewan, R., & Singh, G. (2022). Glacier melting: Drastic future. *Advances in Behavioral Based Safety: Proceedings of HSFEA, 2020*, 255–266. https://doi.org/10.1007/978-981-16-8270-4_19

Bai, Y., & Cotrufo, M. F. (2022). Grassland soil carbon sequestration: Current understanding, challenges, and solutions. *Science, 377*(6606), 603–608. https://doi.org/10.1126/science.abo2380

Bale, J. S. (2002). Insects and low temperatures: From molecular biology to distributions and abundance. *Philosophical Transactions of the Royal Society of London. Series B: Biological Sciences, 357*(1423), 849–862. https://doi.org/10.1098/rstb.2002.1074

Banks-Leite, C., Ewers, R. M., Folkard-Tapp, H., & Fraser, A. (2020). Countering the effects of habitat loss, fragmentation, and degradation through habitat restoration. *One Earth, 3*(6), 672–676. https://doi.org/10.1016/j.oneear.2020.11.016

Barber, J. (2008). Photosynthetic generation of oxygen. *Philosophical Transactions of the Royal Society B: Biological Sciences, 363*(1504), 2665–2674. https://doi.org/10.1098/rstb.2008.0047

Baumert, K. A. (2005). *Sustainable development policies and measures and international climate agreements.*

Berrang-Ford, L., Ford, J. D., Lesnikowski, A., Poutiainen, C., Barrera, M., & Heymann, S. J. (2014). What drives national adaptation? A global assessment. *Climatic Change,* 124(1–2), 441–450. https://doi.org/10.1007/s10584-014-1078-3

Bhattacharyya, P., Pathak, H., & Pal, S. (2020). *Crop management for climate-smart agriculture* (pp. 85–111). https://doi.org/10.1007/978-981-15-9132-7_7

Brohan, P., Kennedy, J. J., Harris, I., Tett, S. F. B., & Jones, P. D. (2006). Uncertainty estimates in regional and global observed temperature changes: A new data set from 1850. *Journal of Geophysical Research: Atmospheres, 111*(D12). https://doi.org/10.1029/2005JD006548

Chakraborty, M., Singh, R., Awasthi, A., Kumar, V., Kumar, A., & Kaur, S. (2025). Assessment of greenhouse gas emissions in delhi landfills: Analyzing factors affecting emissions and environmental implications. *Journal of Material Cycles and Waste Management, 27*(1), 330–342. https://doi.org/10.1007/s10163-024-02114-2

Chen, I.-C., Hill, J. K., Ohlemüller, R., Roy, D. B., & Thomas, C. D. (2011). Rapid range shifts of species associated with high levels of climate warming. *Science, 333*(6045), 1024–1026. https://doi.org/10.1126/science.1206432

Cheshmehzangi, A. (2020). The analysis of global warming patterns from 1970s to 2010s. *Atmospheric and Climate Sciences, 10*(03), 392–404. https://doi.org/10.4236/acs.2020.103022

Chunkui, Z., Peishan, T., & Jingyuan, X. (2021). Progress and prospects of research on climate policy and governance. *Journal of Advances in Health, 3*(2), 135–156. https://doi.org/10.3724/SP.J.2096-5141.2021.0008

Del Rio, C. R. (2012). The Role of ecosystems in building climate change resilience and reducing greenhouse gases. In *Integrating ecology and poverty reduction* (pp. 331–352). Springer New York. https://doi.org/10.1007/978-1-4419-0633-5_20

De Lima, C. Z., Buzan, J. R., Moore, F. C., Baldos, U. L. C., Huber, M., & Hertel, T. W. (2021). Heat stress on agricultural workers exacerbates crop impacts of climate change. *Environmental Research Letters, 16*(4), Article 044020. https://doi.org/10.1088/1748-9326/abeb9f

Denich, M., Kanashiro, M., & Vlek, P. L. G. (2019). The potential and dynamics of carbon sequestration in traditional and modified fallow systems of the Eastern Amazon Region, Brazil. In *Global climate change and tropical ecosystems* (pp. 213–229). CRC Press. https://doi.org/10.1201/9780203753187-11

Deryng, D., Conway, D., Ramankutty, N., Price, J., & Warren, R. (2014). Global crop yield response to extreme heat stress under multiple climate change futures. *Environmental Research Letters, 9*(3), Article 034011. https://doi.org/10.1088/1748-9326/9/3/034011

Donald, R. (2024). Estimate of temperature rise in the 21st century for various scenarios. *IgMin Research, 2*(7), 564–569. https://doi.org/10.61927/igmin218

Dovers, S. R., & Hezri, A. A. (2010). Institutions and policy processes: The means to the ends of adaptation. *Wires Climate Change, 1*(2), 212–231. https://doi.org/10.1002/wcc.29

Driedonks, N., Rieu, I., & Vriezen, W. H. (2016). Breeding for plant heat tolerance at vegetative and reproductive stages. *Plant Reproduction, 29*(1–2), 67–79. https://doi.org/10.1007/s00497-016-0275-9

Elderd, B. D., & Reilly, J. R. (2014). Warmer temperatures increase disease transmission and outbreak intensity in a host–pathogen system. *Journal of Animal Ecology, 83*(4), 838–849. https://doi.org/10.1111/1365-2656.12180

Fisogni, A., de Manincor, N., Kaminskaia, E., & Rafferty, N. E. (2025). Complexities of phenological shifts for plant-pollinator interactions and ways forward. *Integrative and Comparative Biology*. https://doi.org/10.1093/icb/icaf034

Fontúrbel, F. E., Nespolo, R. F., Amico, G. C., & Watson, D. M. (2021). Climate change can disrupt ecological interactions in mysterious ways: Using ecological generalists to forecast community-wide effects. *Climate Change Ecology, 2*, Article 100044. https://doi.org/10.1016/j.ecochg.2021.100044

Fontúrbel, F. E., Salazar, D. A., & Medel, R. (2017). Increased resource availability prevents the disruption of key ecological interactions in disturbed habitats. *Ecosphere, 8*(4). https://doi.org/10.1002/ecs2.1768

Gates, D. M. (1980). *Temperature and organisms* (pp. 527–569). https://doi.org/10.1007/978-1-4612-6024-0_15

Gonzalez, A., Mouquet, N., & Loreau, M. (2009). Biodiversity as spatial insurance: the effects of habitat fragmentation and dispersal on ecosystem functioning. In *Biodiversity, ecosystem functioning, and human wellbeing* (pp. 134–146). Oxford: Oxford University Press. https://doi.org/10.1093/acprof:oso/9780199547951.003.0010

Graham, N. E. (1995). Simulation of recent global temperature trends. *Science, 267*(5198), 666–671. https://doi.org/10.1126/science.267.5198.666

Haghpanah, M., et al. (2024). Drought tolerance in plants: Physiological and molecular responses. *Plants (Basel)*. https://doi.org/10.3390/plants13212962

Harris, L. F., & Johnson, S. D. (2004). The consequences of habitat fragmentation for plant-pollinator mutualisms. *International Journal of Tropical Insect Science, 24*(1), 29–43. https://doi.org/10.1079/IJT20049

Harsch, M. A., & HilleRisLambers, J. (2016). Climate warming and seasonal precipitation change interact to limit species distribution shifts across Western North America. *PLoS ONE, 11*(7), Article e0159184. https://doi.org/10.1371/journal.pone.0159184

Hoque, M. B., Oyshi, T. H., Hannan, M. A., Haque, P., Rahman, M. M., & Jalil, M. A. (2025). Plant science: Pioneering sustainable solutions for global challenges. *Probe—Plant & Animal Sciences, 7*(1). https://doi.org/10.18686/ppas2266

Kaur, N., Kaur, H., & Mavi, G. S. (2020). Assessment of nutritional and quality traits in biofortified bread wheat genotypes. *Food Chemistry, 302*, Article 125342. https://doi.org/10.1016/j.foodchem.2019.125342

Khan, M. I. R., Asgher, M., & Khan, N. A. (2013). Rising temperature in the changing environment: A serious threat to plants. *Climate Change and Environmental Sustainability, 1*(1), 25. https://doi.org/10.5958/j.2320-6411.1.1.004

Kumari, S., Singh, T. P., & Prasad, S. (2019). Climate smart agriculture and climate change. *International Journal of Current Microbiology and Applied Sciences, 8*(03), 1112–1137. https://doi.org/10.20546/ijcmas.2019.803.134

Legrève, A., & Duveiller, E. (2010). Preventing potential disease and pest epidemics under a changing climate. In *Climate change and crop production* (pp. 50–70). CABI. https://doi.org/10.1079/9781845936334.0050

Lenoir, J., Gégout, J. C., Marquet, P. A., de Ruffray, P., & Brisse, H. (2008). A significant upward shift in plant species optimum elevation during the 20th century. *Science, 320*(5884), 1768–1771. https://doi.org/10.1126/science.1156831

Lim, B., Spanger-Siegfried, E., Burton, I., Malone, E., & Huq, S. (2004). *Adaptation policy frameworks for climate change: Developing strategies, policies and measures.*.

Liu, B., Hou, F., & Du, Y. (2011). Research of potential carbon sequestration for grassland in Qinghai province. In *2011 International Conference on Multimedia Technology* (pp. 6499–6505). https://doi.org/10.1109/ICMT.2011.6002694

Logan, J. A., & Jacques Régnière, J. A. P. (2003). Assessing the impacts of global warming on forest pest dynamics. *Frontiers in Ecology and the Environment*. https://doi.org/10.1890/1540-9295(2003)001[0130:ATIOGW]2.0.CO;2

Loreto, F., & Centritto, M. (2008). Leaf carbon assimilation in a water-limited world. *Plant Biosystems—An International Journal Dealing with All Aspects of Plant Biology, 142*(1), 154–161. https://doi.org/10.1080/11263500701872937

Matson, P. A., Parton, W. J., Power, A. G., & Swift, M. J. (1997). Agricultural intensification and ecosystem properties. *Science, 277*(5325), 504–509. https://doi.org/10.1126/science.277.5325.504

Meadowcroft, J. (2009). Climate change governance. *World Bank*. https://doi.org/10.1596/1813-9450-4941

Memmott, J., Craze, P. G., Waser, N. M., & Price, M. V. (2007). Global warming and the disruption of plant–pollinator interactions. *Ecology Letters, 10*(8), 710–717. https://doi.org/10.1111/j.1461-0248.2007.01061.x

Miner, G. L., Stewart, C. E., Delgado, J. A., Ippolito, J. A., Mason, R. E., Haley, S. D., Guttieri, M. J., Ainsworth, E. A., McGrath, J. M., & Beebout, S. J. (2024). Global change impacts on mineral nutritional quality of cereal grains: Coordinated datasets and analyses to advance a systems-based understanding. *Field Crops Research, 310*, Article 109338. https://doi.org/10.1016/j.fcr.2024.109338

Mullan, M., et al. (2013). *National adaptation planning: Lessons from OECD countries* (OECD Environment Working Papers No. 54). Paris: OECD Publishing. https://doi.org/10.1787/5k483jpfpsq1-en.

Myers, N., Mittermeier, R. A., Mittermeier, C. G., da Fonseca, G. A. B., & Kent, J. (2000). Biodiversity hotspots for conservation priorities. *Nature, 403*(6772), 853–858. https://doi.org/10.1038/35002501

Pandey, M. K., Mir, R. R., & Sreenivasulu, N. (2023). The plant genome special section: Grain quality and nutritional genomics for breeding next-generation crops. *The Plant Genome, 16*(4). https://doi.org/10.1002/tpg2.20396

Pattnayak, K. C., Awasthi, A., Sharma, K., & Pattnayak, B. B. (2023). Fate of rainfall over the North Indian states in the 1.5 and 2 C warming scenarios. *Earth and Space Science, 10*(2), e2022EA002671. https://doi.org/10.1029/2022EA002671

Prajapati, S., Choudhary, S., Kumar, V., Dayal, P., Srivastava, R., Gairola, A., & Borate, R. (2023). *Carbon Sequestration: A Key Strategy for Climate Change Mitigation Towards a Sustainable Future, 1*, 1–14. https://doi.org/10.18782/2583-4770.128

Prati, R., & Pomero, A. (2022). Forest restoration and reforestation initiatives. *International Journal of Humanities and Education Research, 4*(1), 07–10. https://doi.org/10.33545/26649799.2022.v4.i1a.65

Rafferty, N. E., & Ives, A. R. (2011). Effects of experimental shifts in flowering phenology on plant-pollinator interactions. *Ecology Letters, 14*(1), 69–74. https://doi.org/10.1111/j.1461-0248.2010.01557.x

Rafferty, N. E., CaraDonna, P. J., & Bronstein, J. L. (2015). Phenological shifts and the fate of mutualisms. *Oikos, 124*(1), 14–21. https://doi.org/10.1111/oik.01523

Rahmstorf, S., Foster, G., & Cahill, N. (2017). Global temperature evolution: Recent trends and some pitfalls. *Environmental Research Letters, 12*(5), Article 054001. https://doi.org/10.1088/1748-9326/aa6825

Rai, P. K. (2015). A concise review on multifaceted impacts of climate change on plant phenology. *Environmental Skeptics and Critics, 4*(4), 106–115.

Raizada, A., Sharma, S., & Srivastava, N. (2022). Climate change—An overview. *Journal of Mountain Research, 17*(2). https://doi.org/10.51220/jmr.v17i2.23

Regmi, S., & Paudel, B. (2024). Climate-smart agriculture: A review of sustainability, resilience, and food security. *Archives of Agriculture and Environmental Science, 9*(4), 832–839. https://doi.org/10.26832/24566632.2024.0904028

Rogers, H. S., Donoso, I., Traveset, A., & Fricke, E. C. (2021). Cascading impacts of seed disperser loss on plant communities and ecosystems. *Annual Review of Ecology, Evolution, and Systematics, 52*(1), 641–666. https://doi.org/10.1146/annurev-ecolsys-012221-111742

Ruosteenoja, K., Carter, T. R., Jylhä, K., & Tuomenvirta, H. (2003). Future climate in world regions: An intercomparison of model-based projections for the new IPCC emissions scenarios. *The Finnish Environment*.

Salihi, M. S., & Himatkhwah, R. (2024). Impacts of climate-smart agriculture on crop production: A review. *Journal of Natural Science Review, 2*(4), 84–94. https://doi.org/10.62810/jnsr.v2i4.99

Sarfraz, M., Ullah, H., Iqbal, R., Rehman, Z. U., Sajid, S., & Ullah, M. (2022). Goals, definitions, principles, and issues in biodiversity reporting: A critical review. *Journal of Plant and Environment, 4*(1), 65–75. https://doi.org/10.33687/jpe.004.01.3766

Schirpke, U., et al. (2024). Effects of climate-related environmental changes on non-material benefits from human-nature interactions: A literature review. *Ecosystem Services*. https://doi.org/10.1016/j.ecoser.2024.101650

Shahdin, R., & Siddiqa, A. (2025). Ecosystem architects: the vital role of plants in nature's balance. *BioScientific Review, 7*(2), 54–72. https://doi.org/10.32350/bsr.72.05

Shekhar Seth, C. (2015). A review on effects of climate change on plants and ecosystems and certain approaches for plant response studies under climate change scenario with specific focus on FACE. *Journal of Food and Nutritional Disorders, 04*(01). https://doi.org/10.4172/2324-9323.1000165

Shih, P. M. (2015). Photosynthesis and early Earth. *Current Biology, 25*(19), R855–R859. https://doi.org/10.1016/j.cub.2015.04.046

Siebert, S., Ewert, F., Eyshi Rezaei, E., Kage, H., & Graß, R. (2014). Impact of heat stress on crop yield—On the importance of considering canopy temperature. *Environmental Research Letters, 9*(4), Article 044012. https://doi.org/10.1088/1748-9326/9/4/044012

Solga, M. J., Harmon, J. P., & Ganguli, A. C. (2014). Timing is everything: An overview of phenological changes to plants and their pollinators. *Natural Areas Journal, 34*(2), 227–234. https://doi.org/10.3375/043.034.0213

Spangler, L., Vance, G., Schuman, G., & Derner, J. (2011). Rangeland sequestration potential. *Assessment*. https://doi.org/10.2172/1037875

Sugden, A. M. (2017). Consequences of shifting species distributions. *Science, 355*(6332), 1386–1388. https://doi.org/10.1126/science.355.6332.1386-j

Tamarin-Brodsky, T., Hodges, K., Hoskins, B. J., & Shepherd, T. G. (2020). Changes in Northern Hemisphere temperature variability shaped by regional warming patterns. *Nature Geoscience, 13*(6), 414–421. https://doi.org/10.1038/s41561-020-0576-3

Tandon, A., Awasthi, A., & Pattnayak, K. C. (2025a). Efficacy of machine learning in simulating precipitation and its extremes over the capital cities in North Indian states. *Scientific Reports, 15*(1), 10345. https://doi.org/10.1038/s41598-024-84360-w

Tandon, A., Awasthi, A., & Pattnayak, K. C. (2025b). Extreme precipitation projections in Asian Subregions regions: Study on CMIP6 development, geographic patterns, and research gaps. *Physics and Chemistry of the Earth, Parts a/b/c, 104148.* https://doi.org/10.1016/j.pce.2025.104148

Taulu, S. (2025). Climate resilience through conservation farming: A review of adaptive practices among smallholder farmers. *International Journal of Science and Research Archive, 15*(3), 1026–1039. https://doi.org/10.30574/ijsra.2025.15.3.1817

Taylor, R., Herms, D., Cardina, J., & Moore, R. (2018). Climate change and pest management: Unanticipated consequences of trophic dislocation. *Agronomy, 8*(1), 7. https://doi.org/10.3390/agronomy8010007

Thangapandian, R., Jayaraj, T., Redoña, E., & Sheryl, S. (2023). Development of heat tolerant breeding lines and varieties in rice. *Indian Journal of Plant Genetic Resources, 23*(01), 1–3.

Traveset, A., & Richardson, D. (2006). Biological invasions as disruptors of plant reproductive mutualisms. *Trends in Ecology & Evolution, 21*(4), 208–216. https://doi.org/10.1016/j.tree.2006.01.006

Tuble, R., & Lacuesta, L. (2024). Challenges and adaptation practices of farmers on climate change. *Journal of Interdisciplinary Perspectives, 2*(7). https://doi.org/10.69569/jip.2024.0091

Varghese, A.O., & Thawale, P. (2011). *Carbon sequestration potential in above ground biomass of natural reserve forest of CENTRAL India.*

Velu, G., Singh, R. P., Huerta, J., & Guzmán, C. (2017). Genetic impact of Rht dwarfing genes on grain micronutrients concentration in wheat. *Field Crops Research, 214*, 373–377. https://doi.org/10.1016/j.fcr.2017.09.030

Verma, K. K., Song, X., Kumari, A., Jagadesh, M., Singh, S. K., Bhatt, R., Singh, M., Seth, C. S., & Li, Y. (2025). Climate change adaptation: Challenges for agricultural sustainability. *Plant, Cell & Environment, 48*(4), 2522–2533. https://doi.org/10.1111/pce.15078

Williams, R. J., Hutley, L. B., Cook, G. D., Russell-Smith, J., Edwards, A., & Chen, X. (2004). Viewpoint: Assessing the carbon sequestration potential of mesic savannas in the Northern Territory, Australia: Approaches, uncertainties and potential impacts of fire. *Functional Plant Biology, 31*(5), 415. https://doi.org/10.1071/FP03215

Wilson, M. C., Chen, X.-Y., Corlett, R. T., Didham, R. K., Ding, P., Holt, R. D., Holyoak, M., Hu, G., Hughes, A. C., Jiang, L., Laurance, W. F., Liu, J., Pimm, S. L., Robinson, S. K., Russo, S. E., Si, X., Wilcove, D. S., Wu, J., & Yu, M. (2016). Habitat fragmentation and biodiversity conservation: Key findings and future challenges. *Landscape Ecology, 31*(2), 219–227. https://doi.org/10.1007/s10980-015-0312-3

Wu, J., & Liu, Z. M. (2014). Effect of habitat fragmentation on biodiversity: A review. *Chinese Journal of Ecology, 33*(7), 1946–1952.

Wu, X., et al. (2023). Research on the relationship between global temperature rise and annual average concentrations of major greenhouse gas emissions. *Environment, Resource and Ecology Journal, 7*(6). https://doi.org/10.23977/erej.2023.070605

Xu, Z., Jiang, Y., & Zhou, G. (2015). Response and adaptation of photosynthesis, respiration, and antioxidant systems to elevated CO_2 with environmental stress in plants. *Frontiers in Plant Science, 6*. https://doi.org/10.3389/fpls.2015.00701

Yamaguchi, M. (2012). *Policies and measures* (pp. 129–159). https://doi.org/10.1007/978-1-4471-4228-7_7

Yuniawati, & Tampubolon, R. M. (2021). A decrease in carbon absorption potential due to timber harvesting in natural forest. *Jurnal Sylva Lestari, 9*(3), 389–399. https://doi.org/10.23960/jsl.v9i3.516

Zizinga, A., Mwanjalolo, J.-G.M., Tietjen, B., Bedadi, B., Gabiri, G., & Luswata, K. C. (2021). Climate smart agriculture practices in rainfed systems in Uganda: Effectivity and results. *SSRN Electronic Journal*. https://doi.org/10.2139/ssrn.3937099

Zizinga, A., Mwanjalolo, J.-G.M., Tietjen, B., Bedadi, B., Gabiri, G., & Luswata, K. C. (2022). Impacts of climate smart agriculture practices on soil water conservation and maize productivity in rainfed cropping systems of Uganda. *Frontiers in Sustainable Food Systems, 6*. https://doi.org/10.3389/fsufs.2022.889830

Effects of Temperature Rise on Growth and Ecosystem Interactions of Himalayan Medicinal Plant Tulsi (Ocimum Sanctum): A Case Study from Uttarakhand

Deeksha and **Ranjit Kumar Mukherji**

Abstract The Indian Himalayan region, which is a biodiversity hot spot, is becoming more susceptible to climate change, and both the increasing temperature and the regular heatwaves are becoming a major ecological problem. The paper has chosen Tulsi (Ocimum sanctum), which is a medicinally and culturally significant plant in Ayurveda and contemporary pharmaceuticals to analyze the impact of climatic variations on its growth and ecological relationship in the Almora district, Uttarakhand. The study adopts a case study method, whereby it incorporates long-term climatic data (2000–2024) with field measurements and experimental simulations with +2 and +4 °C conditions to understand how Tulsi can change in terms of physiology and phenology. Field plots 60 of them were examined. The findings indicate that high temperatures have a severe impact on reducing the plant height, leaf biomass, and flowering period. Pollinator activity also decreases significantly, soil moisture decreases, and there is a decrease in the activity of microorganisms, which means stress on the ecosystem on a global scale. Those scientific observations were also supplemented by the interviews of local communities, which stated that in the recent years there were earlier flowering and less potency in medicinal use. The paper has generalized implications to the sustainability of medicinal plants and pharmaceutical supply chain that relies on the Himalayan flora. The suggested adaptive mechanisms to reduce these effects include shade-net farming, mixed farming, and community-based seed banks as part of the overall proposal like the National AYUSH Mission and Medicinal Plants Board. Future studies are to study long-term molecular adaptation processes and participatory conservation theories in order to increase the resilience of Himalayan medicinal resources during the condition of warming.

Keywords Climate change · Case study · Himalayan ecosystem · Medicinal plants · Rising temperature · Tulsi

Deeksha (✉) · R. K. Mukherji
Department of Commerce, Graphic Era deemed to be University, Dehradun, UK, India
e-mail: deeksha6498@gmail.com

A. Awasthi et al. (eds.), *Rising Temperatures and Plant Ecosystems*,
https://doi.org/10.1007/978-3-032-29477-7_8

1 Introduction

The rhythms of nature are changing under the pressure of the climate change in high the Indian Himalayas, which is the water tower of Asia. Increasing at an above-average rate, the temperatures in Uttarakhand are destabilizing ecosystems, changing flowering cycle, rainfall pattern, and posing a threat to the biodiversity that communities have relied on for long (IPCC, 2023; Negi et al., 2021; Tandon et al., 2023, 2025). Temperatures and heatwaves are starting to change the landscape, endangering both traditional medicine and modern medicines with a significant threat to medicinal plants (Bhutiyani et al., 2020).

Tulsi (Ocimum sanctum), the so-called Queen of Herbs is one of such plants, which has a healing and heritage value. Its leaves are well-known in Ayurveda as non-inflammatory and adaptogenic agents; they contain a large number of bio-active substances, including eugenol, and have been used in remedies and pharmaceutical products globally (Kumar & Goel, 2022). Outside the field of medicine, Tulsi has a very strong cultural significance, which is cultivated in the family yards as a sign of good health and holiness.

But the Himalayan home of Tulsi is unsteady. The biodiversity hot spot of Uttarakhand is characterized by more than 1,800 medicinal species (Samant & Palni, 2020), and it experiences the impacts of warming that endanger pollination, flowering, and phytochemical quality. These changes have economic and ecological implications to the farmers and pharmaceutical industries.

This chapter discusses the impacts of increasing temperature on the growth, physiology, and ecological relationships of Tulsi in Uttarakhand and emphasize adaptive responses and policy alternatives to maintain the medicinal biodiversity of a warming Himalaya.

Research Objectives

1. Examine temperature and heatwave changes in the medicinal plant ecosystems in Uttarakhand.
2. Test how increasing temperatures will influence the growth and phenology of Tulsi (Ocimum sanctum).
3. Examine ecosystem interactions—such as soil situations, pollinator behavior, and companion flora-in warmed conditions.
4. Provide adaptive policies on sustainable production and conservation applicable to the pharmaceutical industry.

The chapter addresses the effects of increasing tempos on the growth, physiology, and ecological interactions among Tulsi in the Himalayan region of Uttarakhand and the adaptive practices and policy directions that can be undertaken to maintain medicine biodiversity in a warmer Himalaya.

2 Literature Review

2.1 International Researches on Temperature Increase and Pharmaceutical Plants

Global warming, especially the increase in the global temperature (Awasthi et al., 2023), has turned out to be a major factor affecting the plant life of the world. The average global temperature has increased by approximately 1.1 °C during the industrial era and it is expected that the average temperature will increase by 1.5–2 °C by 2050 (IPCC, 2023). This warming trend has critical implications on plant physiology, growth cycles, and secondary metabolite production, which are crucial to the medicinal value of most species (Sharma & Reddy, 2021).

Even the slight change of the micro-climatic conditions can cause imbalance in the events of life cycle of medicinal plants, such as, germination, flowering, and fruiting (Singh et al., 2022). As an example, increased temperatures have been found to induce flowering in Withania somnifera (Ashwagandha) and Mentha arvensis (peppermint), although it also slows down growth and increase biomass to harvest (Pinto et al., 2020).

Prolonged heat stress in tropical and subtropical areas causes a reduction in the yield of essential oil and changes in phytochemical profiles, which eventually impact the quality of medicines and marketability of these crops (Kumar et al., 2021). In addition to specific species, the factors of climate change affect more complicated ecosystem relationships, such as pollination, soil microbial organisms, and inter-species rivalry (Rao et al., 2021).

In the recent global studies, it is stressed that to sustain the medicinal plant diversity in changing climatic conditions there is a need to use integrated ecological monitoring and adaptive management (Bhatia & Singh, 2023).

2.2 Growth, Phenology, and Phytochemical Response Studies of Tulsi

Tulsi (Ocimum sanctum/O. tenuiflorum) is an important Ayurvedic herb because it is anti-fungal, antibacterial, and adaptogenic (Kumar & Goel, 2022). The plant can survive in tropical and subtropical conditions although its physiological behavior is significantly altered in response to changes in temperature. According to laboratory and field research, high temperatures increase phenological progression resulting in premature flowering and low leaf biomass and essential oil content (Choudhary et al., 2021).

Temperature has a strong influence on the secondary metabolite profile of Tulsi specifically eugenol and ursolic acid. Its best biosynthesis is at 20–30 °C, and higher temperatures (above 35 °C) lead to oxidative damage and ineffective phytochemical production (Verma et al., 2020). The same tendencies in phenology have been observed in central and northern India, where the vegetative period is becoming shorter and all plants are becoming less vigorous due to warming (Singh & Bhatnagar, 2023).

Tulsi is ecologically dependent on pollinators, helpful soil microbes, and other related plants. Changes in temperature may interfere with these relationships, such as the lack of pollinators caused by heat stress decreases seed set, and changed soil microbial activity means the nutrients are not absorbed (Rathi et al., 2022). It is interesting to note that there is a research gap on populations of Tulsi in high-altitude Himalayas and this chapter attempts to fill this gap.

2.3 Biodiversity and Ecosystem Services Affected by Climate in the Himalaya

Himalayan region also is one of the great hotspots of biodiversity in the world with more than 10,000 plant species, among them 1800 species utilized in the Ayurveda, Unani and modern medicine are found (Samant & Palni, 2020). Climate change is however becoming a threat to this diversity. According to the latest studies, the Indian Himalayan Region (IHR) is warming at an almost 2 times higher rate compared to the rest of the world, and Uttarakhand and Himachal Pradesh experience severe temperature and rainfall pattern changes (Bhutiyani et al., 2020; Negi et al., 2021; Tandon et al., 2025).

Increase in temperature in the Himalayan ecosystems has several implications:

a. The higher the species moves to a cooler environment, the higher the altitudinal change in plant distribution (Telwala et al., 2013).
b. Disturbed phenology, where flowering and fruiting are advanced or delayed (change in pollination and seed dispersal) (Rawat et al., 2021).
c. Loss of ecosystem services, such as soil stability, water retention, and carbon sequestration, which are vital in the livelihoods of the area and the global climate system (Singh & Thadani, 2022).

Events of habitat degradation and extreme climate conditions are also changing the phytogeography of medicinal plants which are already under strain due to the pressure toward collection to use in pharmaceuticals. In Uttarakhand, local farmers and traditional healers report that the growth phases and the strength of medicinal herbs have changed evidently within the last twenty years (Joshi et al., 2022). These observations underscore the fact that there is an urgent need to conduct empirical research on the connection between temperature increase and ecology of medicinal flora in the Himalayan setting.

2.4 *Identified Research Gaps*

The current body of literature provides definite connections between the increase of temperature and medicinal plant physiology on a worldwide scale. However:

i. There is little literature on Tulsi in high-altitude Himalayan ecosystem.
ii. There is a paucity of studies that relate temperature-mediated phenological variation to interactions within an ecosystem (e.g., pollinators, soil microorganisms).
iii. Uttarakhand has a dearth of case studies involving the observation of local communities being combined with scientific climate evidence.

The chapter aims at filling these gaps as much as possible using a narrowed down case study of Tulsi in Uttarakhand in conjunction with climatic trends, growth responses, and observations at the ecosystem level.

3 Methodology

The research took place in Almora district, Uttarakhand, in the Himalayas, which is especially prone to climate changes because of its altitude (1,600–1,800 m) and sensitive ecosystems. The chosen site was to ensure that Tulsi (Ocimum sanctum) is a common garden plant and grows in home farms and small farms, and thus a good location to be studied to determine the effects of climate on the growth and sustainability of medicinal plants (Samant & Palni, 2020). According to climatic records, the central Himalayas is warming at 0.3–0.4 °C per 10 years, and heatwaves and changes in rainfall are redefining the ecosystems (Negi et al., 2021). The examination of Tulsi in this case therefore gives ideas on the local expression of global climate change in local agricultural and medicinal systems.

3.1 *Data Sources*

Indian Meteorological station in Almora was used to obtain temperature and rainfall data (2000–2024) to determine warming patterns and heatwave occurrences in the long term. Field observations were also supported with installation of on-site data loggers in the Tulsi plots to record microclimatic variations in order to bring the field observations to larger climate trends.

3.2 *Plant Growth Parameters*

Five community gardens and farms were chosen and Tulsi was monitored by growing in natural field conditions. Important parameters such as plant height, leaf biomass, flowering initiation, and seed set were recorded after every two weeks March to October 2023, a time of the year when Tulsi thrives well. These observations aided in attributing temperature increase to physiological and reproductive alteration in the plant.

3.3 Experimental Design

A controlled experiment was carried out to provide a simulation of future warming to supplement field monitoring. Polyhouses with passive heating materials were used to expose plants to ambient, +2 and +4 °C conditions. The design was such that a direct comparison could be made between the effects of incremental warming on Tulsi growth and reproduction, which would be useful in the same regard as climate change forecasts.

3.3.1 Observational Study

Field-based observational study design was used to study the population behavior of naturally growing plants of Ocimum tenuiflorum (Tulsi) on the community garden to observe the population behavior of this plant for three conditions, i.e., under ambient or normal temperature conditions. Phenological data were systematically recorded in terms of germination, vegetative growth, flowering, and seed setting. To determine natural variability, data were gathered in several locations that reflected different micro-climatic conditions, including shaded area and the open field. These observations were used as a baseline in determining growth and phenologic pattern under the current environmental conditions.

3.3.2 Experimental Setup

An experimental factorial location-year design was adopted, applied to controlled conditions in polyhouses to recreate high temperature conditions. Tulsi plants were subjected to three temperature conditions namely ambient temperature (+0), +2, and +4 °C as compared to ambient temperature. Passive heating was also realized by applying translucent shading and reflective mulch. To determine the impact of temperature increase on plant development, growth parameters and phenology, responses were taken in an organized manner compared with the control treatments.

3.4 Experimental Simulation Results

Polyhouse experiments showed that there were apparent differences in the growth and reproduction of plants subjected to high temperature. Exposure of Tulsi plants to +2 and + 4 °C demonstrated:

i. **Smaller Plant Height**—Under ambient conditions plants were on average 55 cm, whereas at high temperatures, it was about 42 cm.
ii. **Fewer Numbers of Leaves**—The numbers of leaves per plant reduce to about 78 leaf numbers instead of 120 leaf numbers.
iii. **Reduced Biomass**—With a fall of about 180 g down to 118 g per plant.
iv. **Earlier Flowering**—Flowering about one week before ambient conditions.
v. **Reduced Seed Set**—Deteriorating massively to 60% in elevated temperatures under ambient conditions to 85%.

The above results show that an increase in temperature affects vegetative growth and reproductive success adversely resulting in lower yields and therapeutic potential of Tulsi.

3.5 Statistical Analysis

All the data were analyzed statistically in order to identify the impact of temperature treatment on growth and phenological parameters of Tulsi. One-way Analysis of Variance (ANOVA) was used to compare the mean of the three temperature treatments (ambient, +2, and +4 °C). In cases where the major differences were noted, post hoc pairwise comparisons were done with the help of Tukey honest significant difference (HSD) test. All analyses were conducted at a level of significance of $p < 0.05$. The data will be expressed as mean SD. IBM SPSS Statistics version 26 (IBM Corp., Armonk, NY, USA) was used to do all the statistical analyses.

3.6 Pollinator Observation

The activity of pollinators was observed during the maximum flowering phase when ambient and high temperature conditions were applied to Tulsi plants. The observations were conducted within the period (**08:00 and 11:00 h**), the most active pollination period. The sites were observed by 20-min sessions, and the sessions were repeated three times a day, and three days apart in the flowering season.

Three plants were randomly selected at every observation point and using direct visual counts in addition to hand tally counters, the number and type of visits by the pollinators (e.g., bees, flies, butterflies) were counted. The observer bias was prevented as all the observations were made by the same trained observer in accordance with standardized procedures. Simultaneously, such environmental parameters as temperature and light intensity were observed.

The rate of pollinator visitation was estimated as number of visits to the plant per 20 min session and compared between treatments and microclimatic conditions to determine how increase in temperature influenced the pollination activity.

3.7 *Ecosystem Interactions*

Since Tulsi is not in isolation, interactions between the ecosystems were also examined:

3.7.1 Pollinators

Animal (bee and butterfly) visits were observed at various times of the day to learn how warming changes the relationships between plants and pollinators, and such relations are essential in seed production (Rawat et al., 2021).

3.7.2 Soil Health

The sample of soil was measured at the beginning and ending of summer to determine microbial activity and respiration (Rathi et al., 2022), which connected the process of warming with the work of below-ground ecosystems.

3.7.3 Water Stress

Plant-soil interactions the soil moisture (through TDR) and the water content on the leaf surface were taken to measure the indirect influence of increasing temperatures.

3.7.4 Ethnobotanical Understanding

Community-level findings were gathered through semi-structured interviews with farmers, herbal practitioners, on the evolving growth and potency of Tulsi. Climatic and field data was used to cross-check these views and combine scientific and local knowledge (Joshi et al., 2022).

Table 1 Effects of elevated temperature on growth and reproductive traits of Tulsi

Parameter Ambient	Ambient (0 °C)	+2 °C	+4 °C
Height of plant (cm)	55.3 ± 3.2	48.7 ± 2.8	42.1 ± 2.5
Leaves	120 ± 8	95 ± 6	78 ± 5
Leaf biomass (g FW)	180 ± 12	150 ± 10	118 ± 9
Flowering time (days)	32 ± 3	28 ± 2	24 ± 2
Seed set (%)	85 ± 5	72 ± 4	60 ± 3

Table 2 Soil moisture and microbial activity

Temperature treatment	Soil moisture (%)	Microbial activity (CO_2 μmol/g soil/h)
Ambient	25 ± 2	3.5 ± 0.2
+2 °C	20 ± 1	2.8 ± 0.2
+4 °C	15 ± 1	2.0 ± 0.1

3.8 Linking Methods to Findings

Table 1 presented a steady decrease in the height of the plant, the number of leaves, the leaf biomass, and the percentage of seed set with increasing temperature. Even though flowering time was lower under high temperature, the total reproductive output was lower. This implies that heat stress has a negative impact on vegetative and reproductive characteristics of Tulsi.

Table 2 shows that soil moisture and microbial respiration (CO_2 evolution) were lower when the temperature was high. The weakened microbial activity indicates the lack of biological functioning of the soil in the conditions of heat stress. This will decrease the level of nutrients available to plants.

The experimental results are summarized in Tables 1 and 2. Tulsi grown under +2 and +4 °C temperatures exhibited lower growth, biomass, flowering, and seed set, and soil moisture and microbial activity were also less. These findings show that the increased temperature not only subjected the individual plants to stress but also undermined the processes of the supporting ecosystems, eventually posing medicinal value and productivity in Himalayan settings.

4 Result and Discussion

4.1 Temperature Trends and Anomalies

The 24-year (2000–2024) temperature data of Almora reveals an annual warming rate of 0.3–0.4 °C, and there were numerous pre-monsoon heat waves (April–May) during the years 2022–2023 (Negi et al., 2021). This kind of warming is necessary

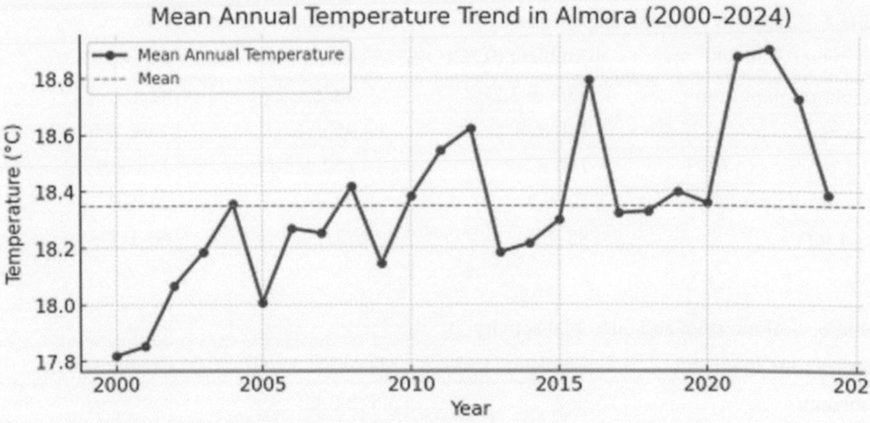

Fig. 1 Mean annual temperature trend in Almora (2000–2024)

in the case of Tulsi (Ocimum sanctum) that prefers 20–30 °C but temperatures that are excessively high, early in the season, would interfere with its development and multiplication (IMD, 2023; Kumar & Goel, 2022; Singh et al., 2021).

Figure 1 Mean Annual Temperature Trend in Almora (2000–2024) indicates the variation in the annual average temperature of Almora during the 24-year interval. The value shows a slowly rising general tendency in average annual temperature, implying the factor of the current climate change on the territory. This increase in temperature can have some effects on the local ecosystem, farmland, and biodiversity.

4.2 Tulsi Growth and Phenological Effects

4.2.1 Growth Rate and Biomass

High temperatures caused significant decrease in the plant height and the leaf biomass where the significant reduction was observed at +4 °C. The elevation of temperature increased the rate of metabolism and decreased vegetative growth, which decreased the yield (Table 3).

Table 3 indicates that the height of the plant, the biomass of the leaves, and the duration of flowering reduced with increase in temperature. Due to high temperature,

Table 3 Growth parameters of Tulsi under different temperature treatments

Parameter	Ambient (0 °C)	+2 °C	+4 °C
Plant height (cm)	55.3 ± 3.2	48.7 ± 2.8	42.1 ± 2.5
Leaf biomass (g FW)	180 ± 12	150 ± 10	118 ± 9
Flowering duration (days)	32 ± 3	28 ± 2	24 ± 2

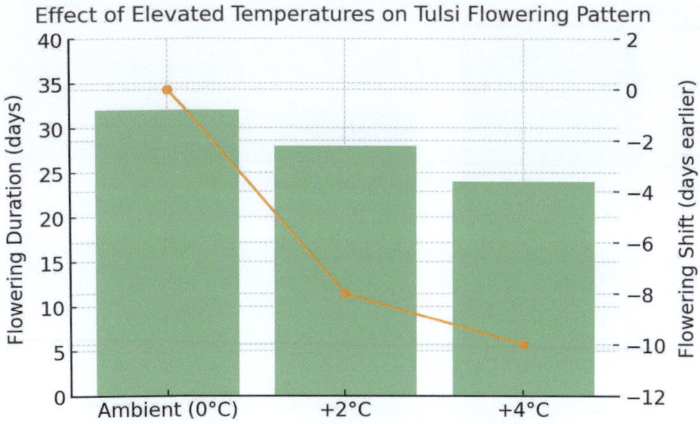

Fig. 2 Effect of elevated temperatures on Tulsi flowering pattern

the growth cycle was reduced and the final biomass production was low. These data show that temperature stress adversely affects the performance of Tulsi in growth.

4.2.2 Flowering Pattern

Higher temperatures facilitated flowering by 8–10 days and reduced flowering time by 25% a result also found in ethnobotanical surveys where farmers indicated earlier flowering and rapid drying than ten years earlier.

Figure 2 Effect of Elevated Temperatures on Tulsi Flowering Pattern shows how the level of temperature affects flowering of the plant. This shows that an increase in temperature is more likely to change the flowering period and flowering intensity resulting in decreased or earlier flowering. This is an indication that an increase in temperature is likely to impact the reproductive cycle and the overall productivity of Tulsi plants.

4.3 Ecosystem Interactions

4.3.1 Pollinator Activity

At +2 and +4, the number of pollinator visits decreased by 30 and 50%, respectively, and this had a remarkable impact on seed set and genetic diversity (Rawat et al., 2021).

Figure 3 Pollinator Visits under Temperature Treatments demonstrate that varying levels of temperature have an impact on the rate of pollinator visits on plants. The figure reveals that moderate temperature conditions are conducive to increasing the pollinator activity, whereas high temperatures are likely to decrease the number of

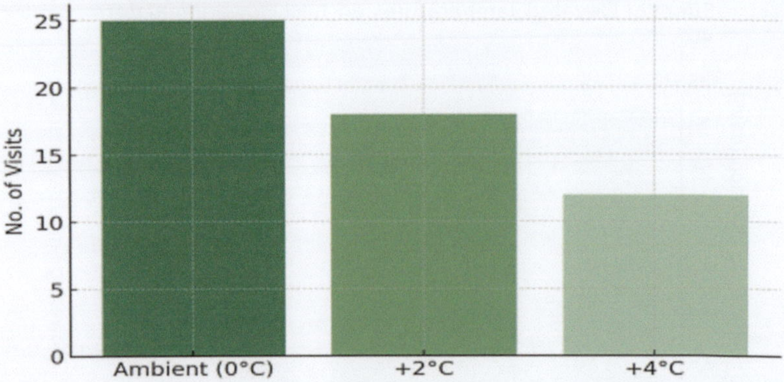

Fig. 3 Pollinator visits under temperature treatments

pollinator visits. It indicates that the increase in temperatures can have a detrimental impact on the interactions between plants and pollinators as well as the general pollination efficiency.

4.3.2 Moisture and Soil Microorganisms

The moisture level in the soil was lower in high temperature conditions (decreased by 25% at +4 °C), and the microbiota activity was decreased in the soil, especially in the fixation and reduction of nitrogen-fixing bacteria and mycorrhizal fungi, which are crucial in nutrient uptake (Rathi et al., 2022). Such developments can have an indirect impact on productivity and resilience of Tulsi.

Table 4 indicates that soil moisture and proportional microbial activity decrease gradually as the temperature level increases. Greater temperatures were correlated with reduced positive microbial performance and nutrient absorption. This points to the adverse effect of warming on the soil–plant interactions.

4.4 Medicinal and Pharma Supply Chain Implications

The diminished biomass and the changed phenology pose a risk to both household and pharmaceutical production chains of Tulsi-based products, in particular, immunity and anti-inflammatory preparations (Kumar & Goel, 2022). The adaptive instruments to protect productivity and quality include cultivation in lower altitudes, polyhouse cultivation, community-based conservation, and training farmers.

Table 4 Soil moisture and microbial activity of Tulsi under temperature treatments

Temperature treatment	Soil moisture (%)	Microbial activity (Relative)	Key observations
Ambient (0 °C)	25	High	Optimal soil and microbial activity; healthy Tulsi growth
+2 °C	20	Moderate	Decline in soil moisture; slight reduction in nitrogen-fixing bacteria
+4 °C	15	Low	Significant drop in soil moisture; reduced mycorrhizal fungi and nutrient uptake

4.5 Discussion

The present study proved that increasing temperature directly influences the growth, flowering, and related ecological interactions of Tulsi in Uttarakhand. Contrasting the generalized discussions on the climate change, these findings are based on Tulsi as the case study, where heat stress causes decreased biomass, reduced duration of reproductive stages, reduced pollinator activity, and reduced health of microorganisms in the soil. Combined, these stressors decrease the number and possible quality of the Tulsi to be used locally and at pharmaceutical industry.

Significantly, taking advantage of integrating field data and farmer perceptions, this study highlights the dual value of Tulsi as both a medication crop that lies at the core of the household traditions and a raw material in commercial supply chains. The results highlight the acuteness of creating climate-resilience cultivation measures that are specific to Tulsi so that it could be ecologically preservative as well as livelihood-insuring among Himalayan communities.

5 Policy and Conservation Implications

The results of the present research highlight the timeliness of the intervention of multi-levels to protect such medicinal plants as Tulsi against the growing climate change risks in the Himalayan region. The alterations in growth forms, phenology, and ecosystem relationships brought about by changes in temperatures do not only endanger the local diversity but threaten the health of the communities living in rural areas, as well as the part of the supply chain of the pharmaceutical industry that relies on the local diversity.

5.1 Communicative-Based Medicinal Plant Management

Tulsi has been grown as a home and community garden crop as a combination of ecology and culture by local communities in Uttarakhand. Amplification of these community-based systems can be one of the portals of conservation. For instance:

a. Training and awareness session to sensitize the farmers in matters of effects of climate on medicine plants.
b. Establishment of self-help groups or cooperative in sustainable harvesting and value addition.
c. Integration of traditional ecological knowledge (TEK) into the scientific systems of conservation, which not only provides conservation of biodiversity but also of cultural heritage (Joshi et al., 2022).

Experiences in other Himalayan states show that community-based initiatives are effective since participatory conservation strategies resulted in a surge in the survival of plants and income earnings (Samant & Palni, 2020).

5.2 Adaptive Strategies of Cultivation

Considering the anticipated increase in temperature, the adaptive agronomic practices are necessary:

a. The polyhouses and shade nets may protect plants against the heat stress with no cost in phytochemical quality.
b. Companion planting (e.g., mint, lemongrass) enhances the health of the soil and lowers the number of pests.
c. Setting up the seed bank managed by communities will make the presence of climate-adapted Tulsi varieties available to be cultivated in the future (Negi et al., 2021).

Adaptive practice not only stabilizes production levels but also increases the resilience of farmers in reaction to uncertainty in the market as well as climate changes.

5.3 Integration with National and State Medicinal Plant Programs

There are a number of current medicinal plant conservation projects in India and they include:

a. The organization enables promotion of cultivation of preferred medicinal species.

b. The National Medicinal Plants Board (NMPB). This organization provides ex-situ and in-situ conservation projects.
c. In Uttarakhand, State Medicinal Plant Boards (SMPB) are developed that benefit the local farmers and the herbal cooperatives.

Inculcating the Tulsi conservation plans to these programs can bring in funds, research infrastructure, and policy incentives. Further, incorporating climate adaptation in these schemes would make medicinal plant ecosystems long lasting.

5.4 Bridging Science, Policy, and Traditional Ecological Knowledge

The gap that this paper reveals in policy is that though there are already medicinal plant conservation programs, there are few programs with adaptive strategies of plan changes in response to climate changes. The gap between them needs to be bridged by:

a. Constant monitoring of the temperature effects to major types of species such as Tulsi.
b. Integration into official databases of community observations (phenological changes, yield patterns).
c. Formulation of climate-smart herbal farming protocols that both the government and the pharmaceutical companies can promote.

It is possible to develop a resilient and comprehensive approach to conservation of the Himalayan medicinal plants by linking scientific facts with local wisdom and policies.

6 Conclusion

This chapter has shown that high temperatures in the Almora region of the Indian Himalayas are affecting the growth and ecological relationship of Tulsi (Ocimum sanctum). The evaluation of 20 years of climatic records and field observations revealed a decrease in the biomass of leaf, a shortened flowering duration, a change in pollinator activity and a decrease in the moisture level and microbial activity of soil. Although these changes can be seen as moderate in terms of their effects per year, when they are accumulated, their influences are very risky to the Tulsi planting and the life-dependent on it.

The observations of earlier flowering and a faster rate of drying of Tulsi by local farmers are in accordance with the scientific studies, and supports the importance of combining empirical information with traditional knowledge in this regard. Adaptive strategies that are region-specific which should be based on ecological strategy, policy strategy, and community-based strategy are necessary to maintain the resilience of Tulsi in Almora.

Impact statement: To preserve Tulsi in the warming environment, it is necessary to combine science, indigenous knowledge, and regional-level adaptations.

7 Future Scope and Limitations

The present research is restricted to the Almora district and the 20-year climatic data may not reflect the major Himalayan trends. They did not involve biochemical and molecular studies, especially of secondary metabolites such as eugenol, and touched briefly on socio-economic issues of adaptive practices.

Such multi-scalar, interdisciplinary studies such as molecular, biochemical, and ecological should be adapted in future research. Cross-zone studies of the Himalayas can help to understand region-specific weakness. There is a need to conduct integrated ecosystem research on Tulsi, pollinators, soil microbiota as well as companion flora in response to warming. Interconnection of scientific research and national conservation plans and pharmaceutical distribution lines can inform climate-resilient actions, including breeding Tulsi varieties that are not heat resistant so that ecological, as well as socio-economic viability is ensured.

References

Awasthi, A., Pattnayak, K. C., Tandon, A., Sarkar, A., & Chakraborty, M. (2023). Implications of climate change on surface temperature in North Indian states: Evidence from CMIP6 model ensembles. *Frontiers in Environmental Science, 11*, 1264757. https://doi.org/10.3389/fenvs.2023.1264757

Bhatia, S., & Singh, A. (2023). Climate change and medicinal plant diversity: A global review. *Environmental Research, 215*, Article 114358.

Bhutiyani, M. R., et al. (2020). Climate variability and its impact on Himalayan ecosystems. *Environmental Monitoring and Assessment, 192*(5), 301. https://doi.org/10.1007/s10661-020-8228-2

Choudhary, R., et al. (2021). Temperature stress and phenological responses of *Ocimum sanctum*. *Journal of Plant Physiology, 257*, Article 153349.

Intergovernmental Panel on Climate Change. (2023). In H. Lee & J. Romero (Eds.), *Climate change 2023: Synthesis report. Contribution of Working Groups I, II and III to the sixth assessment report of the intergovernmental panel on climate change*. IPCC. https://doi.org/10.59327/IPCC/AR6-9789291691647

IMD World Competitiveness Center. (2023). *IMD world competitiveness yearbook 2023*. International Institute for Management Development. https://www.imd.org/wcc/world-competitiveness-center

Joshi, H., et al. (2022). Ethnobotanical knowledge and climate change perceptions in Uttarakhand Himalaya. *Journal of Ethnobiology and Ethnomedicine, 18*(1), 22.

Kumar, R., Gupta, L., & Ahmed, S. (2021). Exploring green HRM practices in Indian SMEs. *International Journal of Sustainable Business, 8*(2), 75–89.

Kumar, S., & Goel, R. (2022). Phytochemical and pharmacological properties of *Ocimum sanctum*: A review. *Journal of Ethnopharmacology, 283*, Article 114125. https://doi.org/10.1016/j.jep.2021.114125

Negi, G. C. S., et al. (2021). Rising temperatures and biodiversity in Indian Himalaya: A review. *Current Science, 120*(3), 423–430.

Pinto, S. R., et al. (2020). Heat stress and growth responses of medicinal plants under climate change scenarios. *Plant Physiology Reports, 25*(2), 157–165.

Samant, S. S., & Palni, L. M. S. (2020). Diversity, distribution, and conservation status of medicinal plants in Indian Himalayan Region. *Biodiversity and Conservation, 29*(4), 1347–1365. https://doi.org/10.1007/s10531-019-01890-0

Sharma, X., & Reddy, Y. (2021). *Green human resource practices and organizational sustainability: A conceptual exploration.*

Singh, P., Rawat, G. S., & Tiwari, R. (2021). Recent temperature trends and their ecological implications in the Central Himalaya, Uttarakhand. *Journal of Mountain Science, 18*(9), 2305–2317. https://doi.org/10.1007/s11629-021-6792-5

Singh, V., & Thadani, R. (2022). Human resource development and environmental sustainability in service industries. *Journal of Service and Management Research, 7*(4), 150–165.

Singh, A., Singh, B., & Singh, C. (2022). Sustainable human resource management: Trends and future directions. *Journal of Management Perspectives, 20*(4), 45–52.

Singh, H., & Bhatnagar, P. (2023). Leadership and green HRM: Moderating influences on employee professionalism. *Journal of Corporate Sustainability, 13*(1), 88–104.

Tandon, A., Awasthi, A., & Pattnayak, K. C. (2023, November). Comparison of different machine learning methods on precipitation dataset for Uttarakhand. In *2023 2nd International Conference on Ambient Intelligence in Health Care (ICAIHC)* (pp. 1–6). IEEE. https://doi.org/10.1109/ICAIHC59020.2023.10431402

Tandon, A., Awasthi, A., Pattnayak, K. C., & Das, S. (2025). Rainfall variability and rising extremes in urbanizing himalayan foothills: A machine learning and data-driven exploration of hydro-climatic shifts in Uttarakhand, India. *Earth Systems and Environment, 1–26*. https://doi.org/10.1007/s41748-025-00881-y

Telwala, D., Desai, R., & Mehta, P. (2013). *Sustainability management practices in Indian industries. Sustainable Agriculture, Forest and Environmental Management* (pp. 50–64).

Verma, K., Singh, J., & Lal, R. (2020). Role of human resource practices in sustainability initiatives. *International Journal of Business Sustainability, 5*(1), 29–40.

Rao, P., Singh, M., & Patel, T. (2021). Linking HRM with environmental performance: Evidence from manufacturing firms. *Global Journal of Human Resource Management, 9*(3), 124–137.

Rathi, R., et al. (2022). Soil microbial responses to warming in subalpine ecosystems. *Plant and Soil, 471*(1–2), 235–250.

Rawat, B., et al. (2021). Climate-driven shifts in pollinator behavior in Himalayan ecosystems. *Journal of Mountain Science, 18*(6), 1463–1474.

Heat Stress Impacts on the Himalayan Forest Ecosystems: Present Status and Future Perspectives

Kumar Manish, Neelu Anand Jha, Samjetsabam Bharati Devi, Suratna Sur Shan Sher Sherpa, Prakash Chhetri, and Kishor Sharma

Abstract The Himalayan forests provide many ecosystem functions to about 18% of the global population, including ensuring water security, agricultural benefits, carbon sequestration, groundwater recharge, medicinal benefits, etc. The longevity of the Himalayan forest ecosystems is also critical for the survival of the seven Himalayan nations, because they form the headwater regions of all the major rivers of Asia. In recent years, however, the Himalayan forests have been suffering from multiple natural and anthropogenic threats. While the effects of different threats are well researched in the Himalayan forests, studies documenting the effects of heatwave-induced stress (heat stress) are rare. The present review aims to fill this literature gap by extrapolating global heat stress mechanisms to Himalayan vegetation zones to predict vulnerabilities. The review study uses different scientific literature databases to derive conclusions on the effect of heat stress in the Himalayan forest ecosystems.

Keywords Forests · Ecosystem functions · Heat stress · Himalaya · Productivity · Range shift

K. Manish · N. A. Jha
Jindal School of Environment and Sustainability, O.P. Jindal Global University, Sonipat, Haryana, India

Centre for Biodiversity and Conservation, O.P. Jindal Global University, Sonipat, Haryana, India

S. B. Devi
Department of Botany, School of Basic Sciences, Shri Ramasamy Memorial University Sikkim, Tadong, Gangtok, Sikkim, India

S. S. S. S. Sherpa
West Bengal Forest Department, New C.I.T. Building, Kolkata, West Bengal, India

P. Chhetri
GB Pant National Institute of Himalayan Environment, Sikkim Regional Centre, Pangthang, Gangtok, Sikkim, India

K. Sharma (✉)
The Mountain Institute-India, Tadong, Gangtok, Sikkim, India
e-mail: kisarmaudl@gmail.com

A. Awasthi et al. (eds.), *Rising Temperatures and Plant Ecosystems*,
https://doi.org/10.1007/978-3-032-29477-7_9

1 Introduction

The Himalayan Mountain range, extending across seven countries (Afghanistan, Pakistan, India, Nepal, China, Bhutan, and Myanmar), is one of the youngest and tallest mountain chains globally (Manish & Pandit, 2018). It is home to over 52 million people (Apollo, 2017) and is highly critical for sustaining the food and water security, and livelihoods of approximately1.5 billion people in the south and south-east Asia, especially for the poor and marginalized human societies (Kattel, 2022; KC et al., 2022; Shrestha et al., 2012). It has high coverage of protected areas, forest cover and endemic plant species (BSI, 2025; Chettri et al., 2008; Nayaka & Ngangom, 2024; Pandit, 2017; Manish, 2019; Singh & Pusalkar, 2020).

In the absence of the Himalayan forests, food and water security of all seven nations will be severely compromised (Dhimalet al., 2021; Shrestha et al., 2012; Singh et al., 2010). Therefore, saving the Himalayan forests from natural and anthropogenic threats becomes critical, due to its great significance for conservation of biodiversity, and provisioning different ecosystems functioning and services (Kattel, 2022; Pandit, 2017). However, to conserve and save forests, it is necessary to identify the different threats first. While the effects of various threats, such as climate change, deforestation, hydropower development, unplanned urbanization, tourism, etc. on the Himalayan forests are well documented (Banerjee et al., 2019; Kattel, 2022; Krishnan et al., 2019; Pandit, 2017; Pandit et al., 2007; Pandit & Grumbine, 2012; Shrestha et al., 2012; Singh & Kumar, 2022; Telwala et al., 2013; Xu et al., 2009), the impacts of heat-related stress are relatively not well documented (Gurung et al., 2019; Lamsal et al., 2017). The effects of heat stress and heat waves on the different ecosystems are well documented in Europe and Australia (Bastos et al., 2020; Granier et al., 2007; Salomón et al. 2022; Van Gorsel et al., 2016), but those from the Asian region are largely lacking, especially in the global biodiversity hotspots like the Himalaya (but see Gurung et al., 2019; Lamsal et al., 2017; Mohan & Rajeevan, 2017; ul Hassan et al., 2021; Yuan et al., 2016). The Himalayan forests and the plants (and animals) harboured within them are at greater threats from anthropogenic disturbance (Pandit, 2017) and different stresses including heat stress (Gurung et al., 2019). Studies have documented warming-induced shifts in species' range (upward shifts, contraction/expansion), and treeline (Kumar & Khanduri, 2024; Telwala et al., 2013). The present review aims to fill this literature gap by documenting the effects of heat stress on the Himalayan forests, ranging from alterations in carbon sequestration potential, ecosystem functioning, biodiversity losses, hydrological regimes, tree growth declines, microbial changes, phenological changes, forest productivity, ecosystem homogenization and species range shifts. It underlines why the responses in the Himalayan forests might be different from the effects observed in the European and Australian forests. This contribution also suggests some policy measures that can be adopted at the institutional level to conserve the highly threatened Himalayan Forest ecosystems.

2 Methodology

We used four standard databases, namely Science Direct, Web of Science, Google Scholar, and PubMed to shortlist peer-reviewed research articles, literature reviews and book chapters with the following keywords: "Himalaya" AND "Heat stress," "Himalaya" AND "Heat waves," "Himalaya" AND "Water stress" and "Himalaya" AND "Water scarcity" AND "Forest fire," "Carbon sequestration/carbon cycling," "Nutrient cycling," "Productivity," "Range shifts/Range contraction/expansion," "Treeline shift," "Plant phenology," "Endemic species," "Plant invasion." All peer-reviewed research articles, literature review articles and book chapters in the English language published from 1986 up to 15 November 2025 (as available on the databases) were considered in this study. We limited our search to Himalayan plants and Himalayan forests and excluded zoological records, conference proceedings, data papers, book reviews, and newspaper articles. Surprisingly, there were only a few reports (4) that have attempted to document the effect of heat stress and heat waves on the structural composition and functioning of the Himalayan forests.

3 Results and Discussions

3.1 Effects of Heat Stress

The effects of heat stress on the biological communities can vary at different levels. At the ecosystem level, heat stress and heat wave-related events can lead to a decrease in tree growth, ecosystem productivity, carbon uptake and storage, increased evapotranspiration levels, and disrupt water flux, cause forest damage and die-off (Anderegg et al, 2016; Brzosteket al., 2014; Ciais et al., 2005; Elizabeth & Arain, 2024; Knutzen et al., 2025; Margalef-Marrase et al., 2020; Tiwari et al., 2021; Van Gorsel et al., 2016; Yuan et al., 2016). For example, Ciais et al. (2005) showed a 30% reduction in gross primary productivity (GPP) and a reversal of the carbon sink during the heatwave event in Europe.

At the individual plant level, heat stress can lead to stomatal closure, declined photosynthesis, increased cavitation, damage to thylakoid membranes and Rubisco enzyme, shorten seed viability and regeneration, delay in seed germination, early or delayed leaf emergence, disrupted root systems and reproduction, mortality of both individual plants and seedlings, etc. (Breshears et al., 2021; Deng et al., 2016; Devi et al., 2023; Li et al., 2021; Luo et al., 2020; Margalef-Marrase et al., 2020; Mishra et al., 2023; Pandey et al., 2021; Scafaro et al., 2023; Stangler et al., 2017; Tiebel et al., 2014; Zhang et al., 2010). At the soil level, heat stress can lead to decreased soil moisture levels, changes in soil nutrients, decreased litter decomposition rates, etc. (Mishra et al., 2023; Nasong et al., 2025; Pan et al., 2021; Van Gorsel et al., 2016). The most lethal effect of heat stress is plant mortality and forest dieback, as observed in Europe and Australia (Allen, 2009; Breshearset al., 2021; Margalef-Marrase et al.,

2020; Van Gorsel et al., 2016). Literature indicates that heat stress commonly leads to water stress and water scarcity, and in combination, they lead to mortality and effects such as changes in seed germination time, soil nutrient availability, and declined photosynthesis, etc. (Allen, 2009; Breshears et al., 2021; Mishra et al., 2023). In the Himalaya, climate-induced heat and drought stress have been reported to cause the widespread decline of *Quercus leucotrichophora* (Banj oak) forests (Das et al., 2021). However, the above-generalised effects of heat stress and heat wave-related events can vary in the Himalayan region, as there is a distinct elevational and latitudinal gradation. Thus, it is important to first understand the structure and zonation of the Himalayan forests along the elevational and latitudinal gradients and then visualise the different impacts of heat stress within these ecosystems.

3.2 The Himalayan Forests

The Himalayan Forest's composition and biodiversity are structured by a dominant moisture gradient. The primary driver is the stark precipitation difference between Eastern and Western Himalaya (Burbank et al., 2012; Singh et al., 2017; Tandon et al., 2025a). This directly shapes forest community structure. The Eastern Himalaya, receiving intense monsoon rainfall, supports highly biodiverse and structurally complex tropical evergreen and broad-leaved forests (Pandit et al., 2014; Rana et al., 2019; Tandon et al., 2025b). In contrast, the Western Himalaya, influenced more by winter snowfall, hosts drier forests and seasonal cold, dominated by drought-tolerant conifers (like pines) and deciduous species such as oaks and Sal (Singh et al., 2017; Thadani & Singh, 2025). Angiosperm diversity triples along this west-to-east gradient, driven jointly by precipitation and temperature (Rana et al., 2019). Thus, the vulnerability of these ecosystems to heat stress is not uniform. The forests of the Western Himalaya are inherently more susceptible to heatwave-induced drought and associated impacts like intensified forest fires due to their lower baseline moisture availability. The Eastern Himalayan forests, comparatively, are buffered by higher ambient humidity and precipitation.

Global warming has led to a rise in the rates of occurrence, spread, and intensity of heatwaves in the Himalayan region (Krishnan et al., 2019; Rana et al., 2025; ul Hassan et al., 2021). The heatwave has resulted in more frequent and destructive forest fires, especially in the Western Himalaya, which are more vulnerable to drought conditions, due to lower rainfall (than the Eastern Himalayan region), and increased release and dispersion of air pollutants (Negi et al., 2012; Sharma et al., 2023; Singh et al., 2017). The release of air pollutants like ozone (O_3) and their daytime exposure can even cause a decline in photosynthesis rates in plants (Negi et al., 2012). Studies have also reported that forest fire has negative consequences for carbon sequestration and sink (Khan et al., 2025), and endemic plant species (Chitale & Behera, 2019) in the Himalaya. Moreover, climate change and heatwave-induced forest fire) will also promote the encroachment by invasive/alien plant species such as *Lantana, Eupatorium*, etc. into the habitat and range of native species (Chitale & Behera,

2019; Khuroo et al., 2010; Negi et al., 2012; Sharma et al., 2023; Trivedi et al., 2025). Climate change and heatwave is expected to change boundaries of forest types (and thus species composition), phenology change, cause northward and uphill shifts of treeline, and species range in the Himalayan region (Devkota et al., 2019; Hamid et al., 2018; He et al., 2019; Kumar & Khanduri, 2024; Negi et al., 2012; Singh et al., 2010; Telwala et al., 2013). Warming and heatwave induced stress can also lead to a rise in infestation by pests and diseases in the Himalaya plants (Negi et al., 2012). The Himalayan region is projected to experience a continued rise in heat stress, which will show 15% rise by 2100 (than the current time), with Sichuan and Yunnan becoming new hotspots (Lamsal et al., 2017). In the sections below, we attempt to hypothesize the impacts of heatwaves and heat stress on the different forest types in the Himalayan region.

3.3 Deciduous and Semi-deciduous Forests

The deciduous and semi-deciduous forests (Fig. 1) in the Himalaya commonly include Sal (*Shorea robusta*), Maple (*Acer* spp.), Birches (*Betula utilis*), Khair (*Acacia catechu*), and Sissoo (*Dalbergia sissoo*) (Negi & Singh, 1992; Singh & Singh, 1987; Thadani & Singh, 2025). Monodominant stands of deciduous species such as *Shorea robusta* and various species of Oaks are more common in the western Himalaya than in the eastern regions (Thadani & Singh, 2025). Their phenology is tightly synchronized with the seasonal monsoon cycle. The deciduous tree species in the Himalaya commonly shed their leaves in the late spring season and then produce new leaves when the monsoon season starts (Negi & Singh, 1992; Singh & Singh, 1987). The deciduous species' comparatively higher leaf mass loss during the dry pre-monsoon season is compensated to some extent by higher concentration of Nitrogen (N) in their mature leaves, and a greater percentage of N mass retranslocated away from leaves during leaf abscission (Negi & Singh, 1993). This is a strategy to utilize energy and carbon efficiently in producing new leaves when maximum water availability is present (Negi & Singh, 1992). Heat stress can prematurely trigger leaf emergence, desynchronizing it from the monsoon. Thus, if there are heat waves and heat stress along with milder winter and hotter summer episodes in the Himalaya, the springtime leaf production in deciduous tree species will be hastened (Khanduri et al., 2008; Negi & Singh, 1992).The heatwaves and heat stress will also lead to water stress and an increase in the evapotranspiration rates in the deciduous species due to their shallow roots (Breil et al., 2023; Negi, 2006).This can also lead to decreased photosynthesis rates and overall productivity levels as the species tend to close their stomata to cope with water stress (Dreesen et al., 2012; Negi & Singh, 1993). All these changes will lead to a loss of synchrony between leaf emergence, flowering, herbivory, litter decomposition, seed maturation and regeneration, lower biomass, reduced seed mass, seed set, pollen viability and overall loss of ecosystem services (Khanduri et al., 2008; Negi, 2006; Negi & Singh, 1992; Negi et al., 2012; Shrestha et al., 2012) in the deciduous ecosystems in the Himalaya. The cumulative effect is

Fig. 1 Deciduous forests dominated by *Shorea robusta* and *Tectona grandis* in Sikkim (**A**, **B**), and Uttarakhand (**C**, **D**), in the Indian Himalayan Region

a breakdown in ecological synchrony, leading to reduced regeneration, biomass, and ecosystem services.

3.4 Evergreen and Semi-evergreen Forests

The evergreen and semi-evergreen forests in the Himalaya (Fig. 2) are dominated by oaks, including Banj oak (*Quercus leucotrichophora*), Moru oak (*Quercus floribunda*), Kharsu oak (*Quercus semecarpifolia*), Rianj oak (*Quercus lanuginosa*), Alder (*Alnus nepalensis*), Laurels (e.g., Cinnamomum spp.), *Lithocarpus*, *Schima*, *Castanopsis* and *Rhododendron* species (Garkoti & Singh, 1999; Negi, 2006; Singh et al., 2017; Thadani & Singh, 2025). The majority of the evergreen tree species like Oaks (Fig. 2A, B), Laurels (2 B), Alders (Fig. 2C), Rhododendrons (Fig. 2D), etc., have leaves with 1 year of life span (which fall over an extended period of time to allow uninterrupted photosynthesis) and new leaves emerge as soon as leaf drop happens (simultaneously) in the pre-monsoon season (approximately springtime) (Negi, 2006; Thadani & Singh, 2025). They are reported to have higher primary productivity (than the deciduous counterparts) due to their higher leaf mass, which stays for longer periods, providing an extended period of photosynthesis (Negi,

2006; Singh et al., 1994). Unlike the deciduous species, evergreen species also have comparatively lower loss of leaf mass during senescence and thus lower dependency on the N mass (retranslocated from leaf to stems) for new leaf growth during dry season (Negi & Singh, 1993). A study in the western Himalaya found that when compared to the deciduous forests (stands of *Betula utilis* and *Acer cappadocicum*), evergreen forests (with stands of *Rhododendron arboreum*, *Rhododendron campanulatum*) have a slower litter decomposition rate but higher initial and critical values of litter C:N ratio, and extended nutrient immobilization periods (Garkoti & Singh, 1999).

Therefore, warming-induced heat waves and heat stress can change leaf phenology in these species (Kumar & Khanduri, 2024). Species might have to wait until heat stress subsides to produce new leaves. This will ultimately lead to changes in the timing of leaf decomposition, nutrient cycling, and delayed seed production (Awasthi et al., 2022; Garkoti & Singh, 1995, 1999; Joshi & Tewari, 2009; Singh et al., 2024), similar to the reports from tropical montane wet forests elsewhere (Bothwell et al., 2014). The delayed seed production can have a serious ecological consequence. It can lead to reduced seedling establishment success and forest regeneration. In the Himalayan oaks, seed germination and seed maturation also coincide with the availability of adequate moisture availability during the monsoon rains (Negi et al., 2012;

Fig. 2 Broad-leaved evergreen forests (temperate oak forests) in Uttarakhand, Western Himalaya (**A**), temperate Oak-Laurel mixed forests (**B**) Alder forests (**C**), Rhododendron forests (**D**) in Sikkim, Eastern Himalaya, India

Singh et al., 2024). Under water stress conditions, the oak seeds have been reported to develop early, as indicated by their larger size (Singh et al., 2024). The early seed development can lead to a "carbon drain" situation where most of the carbon is used up in seed development, and sufficient carbon is not left in the individuals for leaf production (Holland et al., 2019; Singh et al., 2024). Thus, early seed maturation due to water stress and heat stress can cause disruption or likely breakdown of the natural cycle of seed germination, leading to large-scale changes in the plant population, composition and structure of forests (Negi et al., 2012; Singh et al., 2024).

The heat sensitivity of these forests is not uniform but is mediated by significant intraspecific variation. For example, populations of *Quercus floribunda* within the same region exhibit divergent responses; one population delayed leaf emergence until the monsoon under drought, while another, with better physiological adaptations to maintain water potential, exhibited normal spring phenology (Singh et al., 2000). However, there are also adaptations in oaks and similar broad-leaved evergreen tree species that may allow them to withstand heat and water stress conditions better. The broad-leaved evergreen tree species have deep root systems, thick leaves, large leaf area per unit mass, large leaf mass, the ability to delay stomatal closure till critical thresholds are reached, and the ability to maintain carbon gain even under the conditions of heat and water stress (Negi & Singh, 1992; Negi, 2006; Singh et al., 2006). All these properties ensure high stem hydraulic capacitance in broad-leaved evergreen tree species and consequently give them the competitive edge over deciduous species under conditions of heat stress and related events like water stress, higher evaporative demand, etc. (Garkoti & Singh, 1999; Singh et al., 2006). Plant invasions (e.g., *Ageratina adenophora*) into the Oak and alder forests disrupts their ecosystem health with invaded sites showing better vegetation structure (basal area, density, and Importance Value Index) as well as properties of soil (Soil moisture, dehydrogenase enzyme activity, organic and microbial biomass carbon components) than the uninvaded sites, thus reflecting the better adaptability of the invasive species than the native species (Rao et al., 2025). Because of such conflicting reports, we have reasons to believe that a generalised response of evergreen and semi-evergreen species in the Himalaya is unlikely. We are likely to see site-specific responses to the conditions of heat stress and heat wave-related events in the Himalaya as far as evergreen and semi-evergreen species are concerned. To summarize, heat stress makes Himalayan evergreen forests less able to adapt by messing with their phenology and putting stress on carbon allocation. However, their natural ability to buffer changes in physiology and the fact that there is genetic heterogeneity in the area will lead to very site-specific results, from resilience to decline.

3.5 Conifer Forests

The conifer forests in the Himalaya (Figs. 3 and 4) are dominated by Chir pine (*Pinus roxburghii*) and Kesiya pine (*Pinus kesiya*) at lower elevations and by Blue pine (*Pinus wallichiana*), Chilgoza pine (*Pinus gerardiana*), Firs (*Abies spp.*),

Hemlock (*Tsuga dumosa*), Spruce (*Picea smithiana*), Cypress (*Cupressus*), Junipers (*Juniperus*), and Deodar (*Cedrus deodara*) at inner valleys and higher elevations (Thadani & Singh, 2025). The leaves of the Himalayan evergreen conifers range from needle-like, scale-like and flattened types (Singh & Singh, 1987; Thadani & Singh, 2025). Studies specific to these Himalayan species support their resilience; for instance, *Pinus roxburghii* exhibits significant stomatal control and cavitation-resistant xylem, while *Cedrus deodara* demonstrates a highly conservative water-use strategy due to its structural adaptations (Thadani & Singh, 2025). The low surface area of these leaves enables the conifers to adopt a conservative water use strategy at all times (Thadani & Singh, 2025). Additionally, the conifers have a high degree of stomatal control, waxy cuticle, cavitation-resistant xylem, and the ability to do year-round photosynthesis (Brodribb et al., 2014; Lodge et al., 2018; Thadani & Singh, 2025).

Conifers are known to use two types of water-stress adaptation mechanisms: (i) hormonal control (abscisic acid-dependent stomate closure) found in plants experiencing continued drought conditions, and (ii) water transport channel (which can function in extreme tensions) allowing for leaf dehydration and damage resistance in plant species against water-stress (Brodribb et al., 2014). Even within the conifers, there is a marked difference among the evergreen (e.g., Spruce) and deciduous (e.g., *Larix* spp.) species in their carbon (larch had low carbon investment in new leaves

Fig. 3 Conifer forests in the Himalaya: deodar forest stands in Uttarakhand (**A**, **B**), and *Cryptomeria* forest stands in south Sikkim (**C**, **D**)

Fig. 4 Himalayan coniferous forests and conifer-broad-leaved mixed forests in Uttarakhand (**A**, **B**) and Sikkim (**C**, **D**)

due to foliage longevity), nitrogen (higher N content and capacity for photosynthesis in Larix spp.) as well as water fluxes and relations (high efficiency of water use in spruce, but greater water demand in *Larix* spp.) (Matyssek, 1986). All these adaptations make conifers highly resistant to conditions of water stress. At higher elevations in the Himalaya, there are fewer chances of heat waves and heat stress events because of persistent snowfall and short growing seasons. Once established, conifers generally are drought and heat-resistant. Therefore, it can be expected that conifers will be the ones that will be least impacted under the conditions of heat waves and heat stress events in the Himalaya.

3.6 Herbaceous Ecosystems

There is scant literature on the effect of heat waves and heat stress on the herbaceous ecosystems. The available literature suggests that the herbaceous ecosystems are the worst affected due to heat waves and heat stress due to weak stomatal control and high evapotranspiration rates (Pandey et al., 2021; Xu et al., 2019; Yang et al., 2023). The herbaceous species also have shallow root systems, due to which they cannot survive even marginal water stress (Pandey et al., 2021).However, growths of herbaceous plants grown in the oak forest and pine forests are shown to have marked differences in the above-ground parts, such as leaves (e.g., leaf area greater in pine forest for *Artemisia vulgaris*, but reduced leaf area for shade-adapted *Gerbera gossypina, Arisaema concinnum*, and *Onychium contiguum)* and below ground parts such as roots (e.g., populations in pine forest had higher secondary roots for *A. vulgaris*, greater root-shoot ratio for all species) (Mehrotra et al., 1999). Pandey et al. (2021) reported that heat stress individually (or together with water stress) enhanced stomatal conductance, transpiration rate, contents of intracellular CO_2, hydrogen peroxide and malondialdehyde, but lowered water and carboxylation efficiencies, photosynthesis rate and contents of chlorophyll in two of the Himalayan medicinal herbs namely *Valeriana jatamansi* and *Hedychium spicatum.*

Thus, commonly they suffer hydraulic failure owing to water stress as a result of heatwaves (Pandey et al., 2021). Amongst all the ecosystems, herbaceous systems have been reported to possess longer recovery time (Wang & Huang, 2004). Evapotranspiration levels are often reported to be highest in herbaceous ecosystems (Pandey et al., 2021; Tiwari et al., 2021; Yang et al., 2021). Because of low canopy cover at high elevations where herbs dominate in the Himalaya (Fig. 5), heat waves are likely to lead to more soil evaporation and low moisture contents (Müller et al., 2016). This would result in severe water stress and hydraulic failure among the herbs afterwards. We anticipate that since most of the high-elevation Himalayan herbaceous species are annuals (Figs. 5 and 6), the immediate effect of heatwaves will be mortality, while for perennials, it can cause sublethal and lethal effects such as leaf damage, damage to Rubisco, decreased photosynthesis levels, decrease on productivity, etc. (Mehrotra et al., 1999; Pandey et al., 2021; Tiwari et al., 2021). Even literature on the effects of climate change on Himalayan flora indicates that herbaceous species are most affected, especially in the higher elevation sub-alpine and alpine zones (Dimri et al., 2019; Manish et al., 2016; Telwala et al., 2013). However, a recent study in the western Himalayan alpine meadow reports that future global warming would lead increase in ecosystem respiration (due to rise in growing seasons' above-ground respiration), but enhance net uptake of carbon (driven by increased plant growth, as well as resistant respiration of soil) in both herbaceous and sedge meadows (Tiwari et al., 2021).

Native Himalayan plants in high-altitude region are also adapting to the heat stress at the genetic and molecular levels. During heat stress, plants also produce heat shock proteins and Salicylic acid as an adaptation mechanism (Batcho et al., 2021; Sangwan et al., 2022). They also upregulate expressions of small heat shock proteins, which protect the photosystem II, electron transport chains, and thus maintain photosynthesis (Gurung et al., 2019; Sangwan et al., 2022). A recent transcriptome-based study in *Primula sikkimensis* (Fig. 6B) has reported gene expression plasticity during heat stress, showing increased expression and production of Small Heat Shock Proteins (Gurung et al., 2019). Global warming and heat stress can also increase the rate of plant invasion (due to their higher adaptability to abiotic stress) in the grassland and alpine meadows, further threatening their native plant diversity and community composition (Khuroo et al., 2010; Telwala et al., 2013; Trivedi et al., 2025). Climate warming led to decline in the diversity of plants but rise in community biomass in

Fig. 5 *Rhododendron* krummholz in Sikkim (**A**), Alpine meadows of the Himalaya in East Sikkim with foraging Yaks (**B**), in Uttarakhand above treeline (**C**), and of the Trans-Himalaya in the Union Territory of Ladakh (**D**)

the grasslands of the high-altitude regions (Wangchuk et al., 2021). The decline in plant diversity, community homogenization in the alpine region, will have negative consequences for the dependent Himalayan Yaks (Fig. 5B) as well as other wild and domesticated herbivores (Sapkota et al., 2022).

In summary, herbaceous ecosystems are the most vulnerable to heat waves and heat stress in the Himalayan ecosystems. This vulnerability arises due to physiological traits, such as weak stomatal control, high evapotranspiration and shallow roots. As a result, the herbaceous ecosystems are likely to experience high mortality and reduced productivity compared to other Himalayan ecosystems.

3.7 *Shifts in Treeline, Species Range and Phenology*

A recent meta-analysis study reports climate change-driven elevational treeline shift of 2.54 m/year (range: 0.11 to >10 m/year) across the Himalaya, with trends being significantly positive in the Eastern and Central regions, but non-significant for Western parts and more clear shifts in treeline observed for *Betula utilis* and *Pinus wallichiana* (Reshi et al., 2026). Kumar and Khanduri (2024) reported upward shifts

Fig. 6 Himalayan herbaceous flowering plants *Rheum nobile* (**A**), *Primula sikkimensis* (**B**), *Caltha* sp. (**C**), and *Ligularia* sp. (**D**) in the Alpine meadows of Sikkim, Eastern Himalaya

in alpine treeline, altered regeneration and phenology in the Himalaya, with species shifting northward from a low of 0.80 m to a high of 503 m, *Abies spectabilis*, *Betula utilis* being the most sensitive species. The study also reported the highest average treeline shifts in Arunachal Pradesh (452 m), followed by Jammu and Kashmir (441 m) and Uttarakhand (411 m). Climate change is projected to cause range-shifts in plant communities, with more pronounced effects, especially for the eastern Himalayan endemics and high-altitude alpine species, leading to community composition shifts and increased species richness in ecotones (Van Peer et al., 2004; Joshi, 2015; Kharal et al., 2015; Manish et al., 2016; Telwala et al., 2013). Studies have also reported range shifts of specific tree species, including Himalayan pine (Dubey et al., 2003) and Himalayan birch (Hamid et al., 2018), as well as herbaceous alpine plant such as *Meconopsis* species (He et al., 2019), epiphytic lichens like *Lobaria pindarensis* (Devkota et al., 2019), etc.

Climate warming has also led to changes in temperature and precipitation regimes, and as a result, phenological changes in the plants. Shrestha et al. (2012) reported that during the period of 1982–2006, Himalaya experienced a significant rise in mean annual temperature (1.5 °C) and rainfall (166 mm), as well as ecosystems-level phenology change with the average values of start of growing season (SOS), and length of growing season (LOS) advancing by 4.7 days each. The continued warming and heatwave-induced stress is going to have catastrophic effects for natural grasslands and forest ecosystems in the Himalaya, as well as its flora (and the dependant fauna), pollination (due to asynchrony in the phenology of plants and their pollinators), seed dispersal, pest control, and other ecosystems services, and health of both ecosystems and humans (Dhimal et al., 2021; Pandit, 2017; Kumar et al., 2017; Sharma et al., 2021; Tiwari et al., 2021), as reported elsewhere (Barriopedro et al., 2023; Isbell et al., 2015; Sentis et al., 2013; Cremonese et al., 2017). One of the mitigation measures for protecting the Himalayan forests against the heat stress could be promoting the pine-oak mixed forest stands (which complement in their vulnerabilities to abiotic stress, together surviving million years of evolutionary time) in their natural ranges (Fig. 4B, D) due to their suitability and resilience against water stress and forest fires (Singh & Zobel, 2025).

3.8 Future Scope and Limitations

Studies have found that diverse plant communities provide wide range of functional traits (e.g., roots, shoots, leaf types and plasticity), helping to maintain high photosynthesis rate by efficiently managing water, nutrients, C and N fluxes (Bothwell et al., 2014; Isbell et al., 2015; Liang et al., 2016; Zhang et al., 2018). Thus, forests in the Eastern Himalaya with high plant diversity (BSI, 2025; Nautiyal and Manish, 2024; Nayaka & Ngangom, 2024; Devi et al., 2026; Singh & Pusalkar, 2020) and phenological asynchrony structured by monsoon-driven high precipitation and moisture regime (Pandit, 2017; Singh et al., 2017) would be better adapted to the heatwave and heat stress (Liang et al., 2016). However, given their higher numbers of endemic and threatened plant species, it will be necessary to take policy and management strategies to safeguard the natural ecosystems in the Himalaya from further disturbances like heatwave-induced forest fire, anthropogenic land use change, unplanned tourism, over-exploitation and trade of rare plants/herbs, and by adopting a sustainable development and well-planned urbanization (Banerjee et al., 2019; Nayaka & Ngangom, 2024; Pandit, 2017; Rajeevan et al., 2023; Singh & Pusalkar, 2020; Sharma et al., 2021).

We have tried to present an overall picture of the differential responses of different ecosystems in the Himalayan region, viz. deciduous, evergreen, coniferous and herbaceous systems (Table 1; Fig. 7). We argue that because of the different adaptations and anatomical and physiological characteristics of different ecosystems, there will be different responses to heat waves and heat stress in the Himalayan region. One of the major limitations of the present study is that there is not enough literature (either field-based or lab-based) on the effect of heat stress on the Himalayan ecosystems and plant communities (Gurung et al., 2019; Lamsal et al., 2017). Most of the literature is from Europe and Australia, where researchers have documented various changes, such as decreased productivity levels, decreased regeneration, longevity, plant mortality, forest die-off, changes in nutrient cycling, carbon and water fluxes, litter decomposition, effects on seed regeneration, etc. (Allen, 2009; Bastos et al.,

2020; IPCC, 2023; Knutzen et al., 2025; Locosselli et al., 2020; Margalef-Marrase et al., 2020; Tiebel et al., 2024; Van Gorsel et al., 2016).

We believe that till now, the effects of heat stress are not very evident in the Himalaya because moisture and water availability are the main limiting factors for plant growth, instead of temperature. Moisture and water availability have not been affected due to climate change so much in the Himalaya because of the monsoon and western westerlies weather systems. However, it is anticipated that heat waves and heat stress events will likely rise by 15% by 2100 (Lamsal et al., 2017). Therefore, empirical studies are needed to document the effect of heat stress on the Himalayan ecosystems focussing on three broad research directions: (1) Manipulative heat stress experiments in oak and pine forests; (2) Long term monitoring of phenology under heat events (for indicator plant groups like Oaks, Rhododendrons, Orchids, etc.), and shifts in treeline and species distribution range, and (3) Physiological stress measurements across elevations. This will aid policymakers in getting enough evidence to design pragmatic conservation actions for saving the Himalayan forests.

Table 1 Summary of the potential impacts of heatwaves and heat stress on the Himalayan ecosystems

Himalayan ecosystem type	Key impacts of heat stress & heatwaves	Resilience level
Deciduous & semi-deciduous forests	• Early leaf emergence and disrupted phenology • Increased evapotranspiration and water stress • Reduced photosynthesis and productivity • Lower seed viability and regeneration • Loss of ecosystem services	Low (Highly vulnerable)
Evergreen & semi-evergreen forests	• Altered leaf phenology and nutrient cycling • Delayed seed production • Intraspecific variation in response	Moderate (Better adapted than deciduous)
Conifer forests	• Minimal impact due to drought/heat resistance • Conservative water-use strategy	High (Least impacted)
Herbaceous ecosystems	• Severe water stress and hydraulic failure • High evapotranspiration • Immediate mortality in annuals • Sublethal effects in perennials • Longer recovery time	Very low (Worst affected)

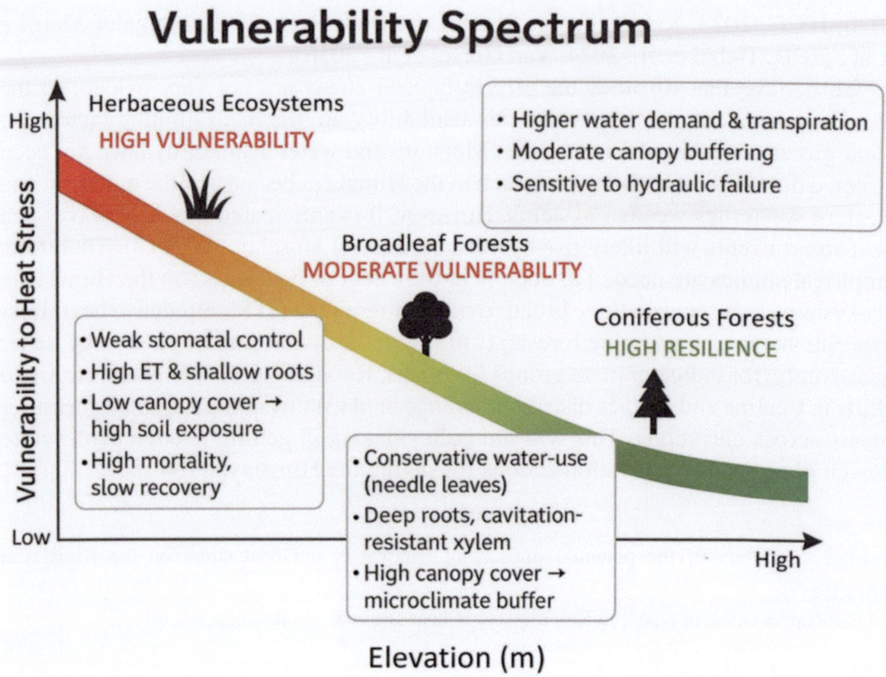

Fig. 7 Generalization of the vulnerability of major Himalayan ecosystems towards heat stress and heat wave-related events. Among all the ecosystems, herbaceous ecosystems (meadows) are likely to be most vulnerable and coniferous forests are likely to be most resilient towards heat stress

4 Conclusions

The observations in the present study suggest that the different ecosystems within the Himalaya will likely have differential responses to heat stress and heat waves in the future. Herbaceous ecosystems, followed by shrublands and deciduous forests, will suffer the most in case of heat stress and heat waves (Table 1; Fig. 7). Evergreen forests, including broad-leaved forests and conifers, will suffer the least and are likely to have the highest resilience and greatest recovery time to heat stress and heat waves. Future conservation actions should be focused on protecting the herbaceous ecosystems, especially in the high-elevation areas, as these ecosystems are likely to suffer the most because of climate-induced warming and heat stress. The high-elevation herbaceous systems also form a critical resource base for medicinal plants and are important for the livelihoods of the local people. We conclude that urgent conservation actions are needed in the high-elevation areas in the Himalaya.

Acknowledgements For providing research facilities and support, we convey our gratitude to the O.P. Jindal Global University, Shri Ramasamy Memorial University Sikkim, West Bengal Forest Department, and GB Pant National Institute of Himalayan Environment (Sikkim Regional Centre. We also thank Dr. Ghanashyam Sharma, Mr. Gaurav Rawat for their various supports. We are thankful to the editorial team of the book for the opportunity, patience, and help in completing this manuscript.

References

Allen, C. D. (2009). Climate-induced forest dieback: An escalating global phenomenon? *Unasylva, 60*(231/232), 43–49.

Anderegg, W. R., Martinez-Vilalta, J., Cailleret, M., Camarero, J. J., Ewers, B. E., Galbraith, D., et al. (2016). When a tree dies in the forest: Scaling climate-driven tree mortality to ecosystem water and carbon fluxes. *Ecosystems, 19*(6), 1133–1147. https://doi.org/10.1007/s10021-016-9982-1

Apollo, M. (2017). The population of Himalayan regions—By the numbers: Past, present and future. In R. Efe & M. Öztürk (Eds.), *Contemporary studies in environment and tourism* (pp. 145–160). Cambridge Scholars Publishing.

Awasthi, P., Bargali, K., Bargali, S. S., & Khatri, K. (2022). Nutrient return through decomposing *Coriaria nepalensis* litter in degraded hills of Kumaun Himalaya, India. *Frontiers in Forests and Global Change, 5*, 1008939. https://doi.org/10.3389/ffgc.2022.1008939

Banerjee, A., Chakraborty, P., & Bandopadhyay, R. (2019). Urgent conservation needs in the Sikkim Himalaya biodiversity hotspot. *Biodiversity, 20*(2–3), 88–97. https://doi.org/10.1080/14888386.2019.1656547

Barriopedro, D., García-Herrera, R., Ordónez, C., Miralles, D. G., & Salcedo-Sanz, S. (2023). Heat waves: Physical understanding and scientific challenges. *Reviews of Geophysics, 61*(2), e2022RG000780. https://doi.org/10.1029/2022RG000780

Bastos, A., Ciais, P., Friedlingstein, P., Sitch, S., Pongratz, J., Fan, L., et al. (2020). Direct and seasonal legacy effects of the 2018 heat wave and drought on European ecosystem productivity. *Science Advances, 6*(24), eaba2724. https://doi.org/10.1126/sciadv.aba2724

Batcho, A. A., Sarwar, M. B., Rashid, B., Hassan, S., & Husnain, T. (2021). Heat shock protein gene identified from Agave sisalana (As HSP70) confers heat stress tolerance in transgenic cotton (Gossypium hirsutum). *Theoretical and Experimental Plant Physiology, 33*(2), 141–156. https://doi.org/10.1007/s40626-021-00200-6

Bothwell, L. D., Selmants, P. C., Giardina, C. P., & Litton, C. M. (2014). Leaf litter decomposition rates increase with rising mean annual temperature in Hawaiian tropical montane wet forests. *PeerJ, 2*, e685. https://doi.org/10.7717/peerj.685

Breil, M., Weber, A., & Pinto, J. G. (2023). The potential of an increased deciduous forest fraction to mitigate the effects of heat extremes in Europe. *Biogeosciences, 20*(12), 2237–2250. https://doi.org/10.5194/bg-20-2237-2023

Breshears, D. D., Fontaine, J. B., Ruthrof, K. X., Field, J. P., Feng, X., Burger, J. R., et al. (2021). Underappreciated plant vulnerabilities to heat waves. *New Phytologist, 231*(1), 32–39. https://doi.org/10.1111/nph.17348

Brodribb, T. J., McAdam, S. A., Jordan, G. J., & Martins, S. C. (2014). Conifer species adapt to low-rainfall climates by following one of two divergent pathways. *Proceedings of the National Academy of Sciences, 111*(40), 14489–14493. https://doi.org/10.1073/pnas.1407930111

Brzostek, E. R., Dragoni, D., Schmid, H. P., Rahman, A. F., Sims, D., Wayson, C. A., et al. (2014). Chronic water stress reduces tree growth and the carbon sink of deciduous hardwood forests. *Global Change Biology, 20*(8), 2531–2539. https://doi.org/10.1111/gcb.12528

BSI. (2025). Plant discoveries. Botanical Survey of India. Retrieved November 20, 2025, from https://bsi.gov.in/page/en/national-wild-life-action-plan

Burbank, D. W., Bookhagen, B., Gabet, E. J., & Putkonen, J. (2012). Modern climate and erosion in the Himalaya. *Comptes Rendus Geoscience, 344*, 610–626. https://doi.org/10.1016/j.crte.2012.10.010

Chettri, N., Shakya, B., Thapa, R., & Sharma, E. (2008). Status of a protected area system in the Hindu Kush-Himalayas: An analysis of PA coverage. *The International Journal of Biodiversity Science and Management, 4*(3), 164–178. https://doi.org/10.3843/biodiv.4.3:4

Chitale, V., & Behera, M. D. (2019). How will forest fires impact the distribution of endemic plants in the Himalayan biodiversity hotspot? *Biodiversity and Conservation, 28*, 2259–2273. https://doi.org/10.1007/s10531-019-01733-8

Ciais, P., Reichstein, M., Viovy, N., Granier, A., Ogée, J., Allard, V., et al. (2005). Europe-wide reduction in primary productivity caused by the heat and drought in 2003. *Nature, 437*(7058), 529–533. https://doi.org/10.1038/nature03972

Cremonese, E., Filippa, G., Galvagno, M., Siniscalco, C., Oddi, L., di Cella, U. M., & Migliavacca, M. (2017). Heat wave hinders green wave: The impact of climate extreme on the phenology of a mountain grassland. *Agricultural and Forest Meteorology, 247*, 320–330. https://doi.org/10.1016/j.agrformet.2017.08.016

Das, A., Menon, T., Ratnam, J., Thadani, R., Rajashekar, G., Fararoda, R., & Shahabuddin, G. (2021). Expansion of pine into mid-elevation Himalayan oak forests: Patterns and drivers in a multiple-use landscape. *Forest Ecology and Management, 497*, 119491. https://doi.org/10.1016/j.foreco.2021.119491

Deng, Y., Srivastava, R., Quilichini, T. D., Dong, H., Bao, Y., Horner, H. T., & Howell, S. H. (2016). IRE 1, a component of the unfolded protein response signaling pathway, protects pollen development in Arabidopsis from heat stress. *The Plant Journal, 88*(2), 193–204. https://doi.org/10.1111/tpj.13239

Devi, P., Awasthi, R., Jha, U., et al. (2023). Understanding the effect of heat stress during seed filling on nutritional composition and seed yield in chickpea (*Cicer arietinum* L.). *Scientific Reports, 13*, 15450. https://doi.org/10.1038/s41598-023-42586-0

Devi, S. B., Sherpa, S. S. S. S., Manish, K., & Sharma, K. (2026). Variation of tree diversity, structure and composition in the different forest types of Eastern Himalaya, India. *Trees, Forests and People, 24*, 101217. https://doi.org/10.1016/j.tfp.2026.101217

Devkota, S., Dymytrova, L., Chaudhary, R. P., Werth, S., & Scheidegger, C. (2019). Climate change–induced range shift of the endemic epiphytic lichen *Lobaria pindarensis* in the Hindu Kush Himalayan region. *The Lichenologist, 51*, 157–173. https://doi.org/10.1017/s002428291900001x

Dhimal, M., Bhandari, D., Dhimal, M. L., Kafle, N., Pyakurel, P., Mahotra, N., et al. (2021). Impact of climate change on health and well-being of people in Hindu Kush Himalayan region: A narrative review. *Frontiers in Physiology, 12*, 651189. https://doi.org/10.3389/fphys.2021.651189

Dimri, A. P., Choudhary, A., & Kumar, D. (2019). Elevation dependent warming over Indian Himalayan region. In A.P. Dimri, B. Bookhagen, M. Stoffel & T. Yasunari (Eds.), *Himalayan weather and climate and their impact on the environment* (pp. 141–156). Springer. https://doi.org/10.1007/978-3-030-29684-1_9

Dreesen, F. E., De Boeck, H. J., Janssens, I. A., & Nijs, I. (2012). Summer heat and drought extremes trigger unexpected changes in productivity of a temperate annual/biannual plant community. *Environmental and Experimental Botany, 79*, 21–30. https://doi.org/10.1016/j.envexpbot.2012.01.005

Dubey, B., Yadav, R. R., Singh, J., & Chaturvedi, R. (2003). Upward shift of Himalayan pine in western Himalaya, India. *Current Science, 85*, 1135–1136.

Elizabeth, A. R., & Arain, M. A. (2024). Impacts of heat and drought on the dynamics of water fluxes in a temperate deciduous forest from 2012 to 2020. *Agricultural and Forest Meteorology, 344*, 109791. https://doi.org/10.1016/j.agrformet.2023.109791

Garkoti, S., & Singh, S. (1995). Variation in net primary productivity and biomass of forests in the high mountains of Central Himalaya. *Journal of Vegetation Science, 6*(1), 23–28. https://doi.org/10.2307/3236252

Garkoti, S. C., & Singh, S. P. (1999). Litter decomposition and nutrient release in central Himalayan high altitude forests. *Tropical Ecology, 40*(1), 19–26.

Granier, C., Cookson, S. J., Tardieu, F., & Muller, B. (2007). Cell cycle and environmental stresses. In Inzé, D. (Ed.), *Annual plant reviews 32: Cell cycle control and plant development* (pp. 335–355). Blackwell Publishing Ltd. https://onlinelibrary.wiley.com/doi/10.1002/9780470988923.ch15

Gurung, P. D., Upadhyay, A. K., Bhardwaj, P. K., Sowdhamini, R., & Ramakrishnan, U. (2019). Transcriptome analysis reveals plasticity in gene regulation due to environmental cues in *Primula sikkimensis*, a high altitude plant species. *BMC Genomics, 20*(1), 989. https://doi.org/10.1186/s12864-019-6354-1

Hamid, M., Khuroo, A. A., Charles, B., Ahmad, R., Singh, C. P., & Aravind, N. A. (2018). Impact of climate change on the distribution range and niche dynamics of Himalayan birch, a typical treeline species in Himalayas. *Biodiversity and Conservation, 28*, 2345–2370. https://doi.org/10.1007/s10531-018-1641-8

He, X., Burgess, K. S., Yang, X. F., Ahrends, A., Gao, L. M., & Li, D. Z. (2019). Upward elevation and northwest range shifts for alpine Meconopsis species in the Himalaya-Hengduan Mountains region. *Ecology and Evolution, 9*, 4055–4064. https://doi.org/10.1002/ece3.5034

Holland, B. L., Monk, N. A., Clayton, R. H., & Osborne, C. P. (2019). A theoretical analysis of how plant growth is limited by carbon allocation strategies and respiration. *In Silico Plants, 1*(1), diz004. https://doi.org/10.1093/insilicoplants/diz004

IPCC. (2023). In H. Lee & J. Romero (Eds.), *Climate change 2023: Synthesis report. Contribution of working groups I, II and III to the sixth assessment report of the intergovernmental panel on climate change*. IPCC. https://doi.org/10.59327/IPCC/AR6-9789291691647

Isbell, F., et al. (2015). Biodiversity increases the resistance of ecosystem productivity to climate extremes. *Nature, 526*, 574–577. https://doi.org/10.1038/nature15374

Joshi, B., & Tewari, A. (2009). Irregularity in frequency of mast seed years in *Quercus* floribunda a late successional species of Central Himalaya. *Russian Journal of Ecology, 40*(7), 482–485. https://doi.org/10.1134/s1067413609070054

Joshi, M. D. (2015). Impacts of climate change on *Abies spectabilis*: An approach integrating a species distribution model (MaxEnt) and a dynamic vegetation model (LPJ-GUESS). Lund University GEM thesis series.

Kattel, G. R. (2022). Climate warming in the Himalayas threatens biodiversity, ecosystem functioning and ecosystem services in the 21st century: Is there a better solution? *Biodiversity and Conservation, 31*(8), 2017–2044. https://doi.org/10.1007/s10531-022-02417-6

KC, K. B., Tzadok, E., & Pant, L. (2022). Himalayan ecosystem services and climate change driven agricultural frontiers: A scoping review. *Discover Sustainability, 3*(1), 35. https://doi.org/10.1007/s43621-022-00103-9

Khan, Z., Raza, M., Ali, S. A., Kumar, A., Saqib, M., Ahmad, A., & Khare, S. (2025). Predicting forest fire risk and carbon pool vulnerability in the Himalayas: A machine learning approach. *Earth Systems and Environment, 1–27*. https://doi.org/10.1007/s41748-025-00863-0

Khanduri, V. P., Sharma, C. M., & Singh, S. P. (2008). The effects of climate change on plant phenology. *The Environmentalist, 28*(2), 143–147. https://doi.org/10.1007/s10669-007-9153-1

Kharal, D. K., Meilby, H., Rayamajhi, S., Bhuju, D., & Thapa, U. K. (2015). Tree ring variability and climate response of *Abies spectabilis* along an elevation gradient in Mustang, Nepal. *Banko Janakari, 24*(1), 3–13. https://doi.org/10.3126/banko.v24i1.13473

Khuroo, A. A., Weber, E., Malik, A. H., Dar, G. H., & Reshi, Z. A. (2010). Taxonomic and biogeographic patterns in the native and alien woody flora of Kashmir Himalaya, India. *Nordic Journal of Botany, 28*, 685–696. https://doi.org/10.1111/j.1756-1051.2010.00750.x

Knutzen, F., Averbeck, P., Barrasso, C., Bouwer, L. M., Gardiner, B., Grünzweig, J. M., et al. (2025). Impacts on and damage to European forests from the 2018–2022 heat and drought events. *Natural Hazards and Earth System Sciences, 25*(1), 77–117. https://doi.org/10.5194/nhess-25-77-2025

Krishnan, R., Shrestha, A. B., Ren, G., Rajbhandari, R., Saeed, S., Sanjay, J., et al. (2019). Unravelling climate change in the Hindu Kush Himalaya: Rapid warming in the mountains and increasing extremes. In Wester, A. Mishra, A. Mukherji & A. B. Shrestha (Eds.), *The Hindu Kush Himalaya assessment: Mountains, climate change, sustainability and people* (pp. 57–97). Springer. https://doi.org/10.1007/978-3-319-92288-1_3

Kumar, S., Chopra, N., & Al-Tawaha, A. R. M. S. (2017). Phenological variations in tree and shrub species of a banj oak (*Quercus leucotrichophora* A. Camus) dominated forest in Kumaun, Central Himalaya. *Advances in Environmental Biology, 11*(5), 35–44.

Kumar, S., & Khanduri, V. P. (2024). Impact of climate change on the Himalayan alpine treeline vegetation. *Heliyon, 10*(23), e40797. https://doi.org/10.1016/j.heliyon.2024.e40797

Lamsal, P., Kumar, L., Shabani, F., & Atreya, K. (2017). The greening of the Himalayas and Tibetan Plateau under climate change. *Global and Planetary Change, 159*, 77–92. https://doi.org/10.1016/j.gloplacha.2017.09.010

Li, N., Euring, D., Cha, J. Y., Lin, Z., Lu, M., Huang, L. J., & Kim, W. Y. (2021). Plant hormone-mediated regulation of heat tolerance in response to global climate change. *Frontiers in Plant Science, 11*, 627969. https://doi.org/10.3389/fpls.2020.627969

Liang, J., Crowther, T. W., Picard, N., Wiser, S., Zhou, M., Alberti, G., et al. (2016). Positive biodiversity-productivity relationship predominant in global forests. *Science, 354*(6309), aaf8957. https://doi.org/10.1126/science.aaf8957

Locosselli, G. M., Brienen, R. J., Leite, M. D. S., Gloor, M., Krottenthaler, S., Oliveira, A. A. D., et al. (2020). Global tree-ring analysis reveals rapid decrease in tropical tree longevity with temperature. *Proceedings of the National Academy of Sciences, 117*(52), 33358–33364. https://doi.org/10.1073/pnas.2003873117

Lodge, A. G., Dickinson, M. B., & Kavanagh, K. L. (2018). Xylem heating increases vulnerability to cavitation in longleaf pine. *Environmental Research Letters, 13*(5), 055007. https://doi.org/10.1088/1748-9326/aabbe5

Luo, H., Xu, H., Chu, C., He, F., & Fang, S. (2020). High temperature can change root system architecture and intensify root interactions of plant seedlings. *Frontiers in Plant Science, 11*, 160. https://doi.org/10.3389/fpls.2020.00160

Manish, K., Telwala, Y., Nautiyal, D. C., & Pandit, M. K. (2016). Modelling the impacts of future climate change on plant communities in the Himalaya: A case study from Eastern Himalaya, India. *Modeling Earth Systems and Environment, 2*(2), 92. https://doi.org/10.1007/s40808-016-0163-1

Manish, K., & Pandit, M. K. (2018). Geophysical upheavals and evolutionary diversification of plant species in the Himalaya. *PeerJ, 6*, e5919. https://doi.org/10.7717/peerj.5919

Manish, K. (2019). Macroecological patterns and drivers of Himalayan plant species diversity and distribution through the Ages. *Frontiers in Biogeography, 11*, e38754. https://doi.org/10.21425/f5fbg38754

Margalef-Marrase, J., Pérez-Navarro, M. Á., & Lloret, F. (2020). Relationship between heatwave induced forest die-off and climatic suitability in multiple tree species. *Global Change Biology, 26*, 3134–3146. https://doi.org/10.1111/gcb.15042

Matyssek, R. (1986). Carbon, water and nitrogen relations in evergreen and deciduous conifers. *Tree physiology, 2*(1-2-3), 177–187. https://doi.org/10.1093/treephys/2.1-2-3.177

Mehrotra, P., Garkoti, S., Rawat, Y. S., & Singh, S. P. (1999). Morphological differences and habitat relationship of four common herb species of oak and pine forests of Central Himalaya. *Oecologia Montana, 8*(1–2), 15–20.

Mishra, S., Spaccarotella, K., Gido, J., Samanta, I., & Chowdhary, G. (2023). Effects of heat stress on plant-nutrient relations: An update on nutrient uptake, transport, and assimilation. *International Journal of Molecular Sciences, 24*(21), 15670. https://doi.org/10.3390/ijms242115670

Mohan, T. S., & Rajeevan, M. (2017). Past and future trends of hydroclimatic intensity over the Indian monsoon region. *Journal of Geophysical Research: Atmospheres, 122*, 896–909. https://doi.org/10.1002/2016jd025301

Müller, M., Schwab, N., Schickhoff, U., Böhner, J., & Scholten, T. (2016). Soil temperature and soil moisture patterns in a Himalayan alpine treeline ecotone. *Arctic, Antarctic, and Alpine Research, 48*(3), 501–521. https://doi.org/10.1657/aaar0016-004

Nasong, D., Zhou, S., Kornhuber, K., & Yu, B. (2025). Concurrent heat extremes in relation to global warming, high atmospheric pressure and low soil moisture in the Northern Hemisphere. *Earth's Future, 13*(1), e2024EF005256. https://doi.org/10.1029/2024EF005256

Nautiyal, D. C., & Manish, K. (2024). Anthropogenic disturbance produces divergent effects in the community structure and composition of tropical semi-evergreen forests in the Eastern Himalaya. *BioRisk, 22*, 1–15. https://doi.org/10.3897/biorisk.22.120802

Nayaka, S., & Ngangom, R. (2024). Current status of lichen diversity in the Himalayan biodiversity hotspot. In T. Pullaiah (Ed.), *Biodiversity hotspot of the Himalaya* (pp. 67–90). Apple Academic Press. https://doi.org/10.1201/9781003455202-6

Negi, G. C. S., & Singh, S. P. (1992). Leaf growth pattern in evergreen and deciduous species of the Central Himalaya, India. *International Journal of Biometeorology, 36*(4), 233–242. https://doi.org/10.1007/BF02726404

Negi, G. C. S., & Singh, S. P. (1993). Leaf nitrogen dynamics with particular reference to retranslocation in evergreen and deciduous tree species of Kumaun Himalaya, Canadian. *Journal of Forest Research, 23*(3), 349–357. https://doi.org/10.1139/x93-051

Negi, G. C. S., Samal, P. K., Kuniyal, J. C., Kothyari, B. P., Sharma, R. K., & Dhyani, P. P. (2012). Impact of climate change on the western Himalayan mountain ecosystems: An overview. *Tropical Ecology, 53*(3), 345–356.

Negi, G. C. S. (2006). Leaf and bud demography and shoot growth in evergreen and deciduous trees of central Himalaya, India. *Trees, 20*, 416–429. https://doi.org/10.1007/s00468-006-0056-4

Pan, M., Wang, T., Hu, B., Shi, P., Xu, J., & Zhang, M. (2021). Mesocosm experiments reveal global warming accelerates macrophytes litter decomposition and alters decomposition-related bacteria community structure. *Water, 13*(14), 1940. https://doi.org/10.3390/w13141940

Pandey, V., Tiwari, D. C., Dhyani, V., Bhatt, I. D., Rawal, R. S., & Nandi, S. K. (2021). Physiological and metabolic changes in two Himalayan medicinal herbs under drought, heat and combined stresses. *Physiology and Molecular Biology of Plants, 27*(7), 1523–1538. https://doi.org/10.1007/s12298-021-01027-w

Pandit, M. (2017). *Life in the Himalaya: An Ecosystem at Risk*. Harvard University Press.

Pandit, M. K., & Grumbine, R. E. (2012). Potential effects of ongoing and proposed hydropower development on terrestrial biological diversity in the Indian Himalaya. *Conservation Biology, 26*(6), 1061–1071. https://doi.org/10.1111/j.1523-1739.2012.01918.x

Pandit, M. K., Manish, K., & Koh, L. P. (2014). Dancing on the roof of the world: Ecological transformation of the Himalayan landscape. *BioScience, 64*, 980–992. https://doi.org/10.1093/biosci/biu152

Pandit, M. K., Sodhi, N. S., Koh, L. P., Bhaskar, A., & Brook, B. W. (2007). Unreported yet massive deforestation driving loss of endemic biodiversity in Indian Himalaya. *Biodiversity and Conservation, 16*(1), 153–163. https://doi.org/10.1007/s10531-006-9038-5

Rajeevan, M., Rohini. P, Smitha Anil Nair, A. N., Tirkey, S., Goswami, T., & Kumar, N. (2023). Meteorological monograph: Heat and cold waves in India processes and predictability (Met. Monograph No. MoES/IMD/Synoptic Met/01(2023)/28). Meteorological Department, Ministry of Earth Sciences India, Government of India.

Rana, S. K., Price, T. D., & Qian, H. (2019). Plant species richness across the Himalaya driven by evolutionary history and current climate. *Ecosphere, 10*(11), e02945. https://doi.org/10.1002/ecs2.2945

Rana, S., Kant, C., Meena, R. S., & Singh, S. K. (2025). Heatwave magnitude impact and projected changes over Himachal Pradesh: CMIP6 projections. *Environmental Quality Management, 34*(3), e70066. https://doi.org/10.1002/tqem.70066

Rao, R., Kumar, S., Mishra, A., & Garkoti, S. C. (2025). *Ageratina adenophora* invasion alters vegetation structure and soil properties in alder and oak forests in the central Himalayas, India. *Environmental Sustainability, 8*(1), 71–83. https://doi.org/10.1007/s42398-025-00341-3

Reshi, Z. A., Nanda, S. A., Rafiq, I. (2026). Treeline shifts in the Himalaya: A meta-analytical perspective. In S. P. Singh & R. Thadani (Eds.), *Himalayan forests in a changing world* (pp. 41–66). Springer. https://doi.org/10.1007/978-981-95-1577-6_2

Sangwan, S., Shameem, N., Yashveer, S., Tanwar, H., Parray, J. A., Jatav, H. S., et al. (2022). Role of salicylic acid in combating heat stress in plants: insights into modulation of vital processes. *Frontiers in Bioscience-Landmark, 27*(11), 310. https://doi.org/10.31083/j.fbl2711310

Sapkota, S., Acharya, K. P., Laven, R., & Acharya, N. (2022). Possible consequences of climate change on survival, productivity and reproductive performance, and welfare of Himalayan Yak (*Bos grunniens*). *Veterinary Sciences, 9*(8), 449. https://doi.org/10.3390/vetsci9080449

Scafaro, A. P., Posch, B. C., Evans, J. R., Farquhar, G. D., & Atkin, O. K. (2023). Rubisco deactivation and chloroplast electron transport rates co-limit photosynthesis above optimal leaf temperature in terrestrial plants. *Nature Communications, 14*(1), 2820. https://doi.org/10.1038/s41467-023-38496-4

Sentis, A., Hemptinne, J. L., & Brodeur, J. (2013). Effects of simulated heat waves on an experimental plant–herbivore–predator food chain. *Global Change Biology, 19*(3), 833–842. https://doi.org/10.1111/gcb.12094

Sharma, S., Khadka, N., Nepal, B., Ghimire, S. K., Luintel, N., & Hamal, K. (2021). Elevation dependency of precipitation over southern slope of central Himalaya. *Jalawaayu, 1*, 1–14. https://doi.org/10.3126/jalawaayu.v1i1.36446

Sharma, A., Srivastava, S., Mitra, D., & Singh, R. P. (2023). Spatiotemporal distribution of air pollutants during a heat wave–induced forest fire event in Uttarakhand. *Environmental Science and Pollution Research, 30*(51), 110133–110160. https://doi.org/10.1007/s11356-023-29906-7

Shrestha, U. B., Gautam, S., & Bawa, K. S. (2012). Widespread climate change in the Himalayas and associated changes in local ecosystems. *Plos One, 7*(5), e36741. https://doi.org/10.1371/journal.pone.0036741

Singh, S. P., Adhikari, B. S., & Zobel, D. B. (1994). Biomass, productivity, leaf longevity, and forest structure in the central Himalaya. *Ecological Monographs, 64*(4), 401–421. https://doi.org/10.2307/2937143

Singh, J. S., & Singh, S. P. (1987). Forest vegetation of the Himalaya. *The Botanical Review, 53*(1), 80–192. https://doi.org/10.1007/BF02858183

Singh, S. P., Tewari, A., Singh, S. K., & Pathak, G. C. (2000). Significance of phenologically asynchronous populations of the central Himalayan oaks in drought adaptation. *Current Science, 79*, 353–357.

Singh, S. P., Singh, V., & Skutsch, M. (2010). Rapid warming in the Himalayas: Ecosystem responses and development options. *Climate and Development, 2*, 221–232. https://doi.org/10.3763/cdev.2010.0048

Singh, S. P., Singh, R. D., Gumber, S. U. R. A. B. H. I., & Bhatt, S. P. A. R. S. H. (2017). Two principal precipitation regimes in Himalayas and their influence on tree distribution. *Tropical Ecology, 58*(4), 679–691.

Singh, D. K., & Pusalkar, P. K. (2020). Floristic diversity of the Indian Himalaya. In G. H. Dar & A. A. Khuroo (Eds.), *Biodiversity of the Himalaya: Jammu and Kashmir State* (pp. 93–126). https://doi.org/10.1007/978-981-32-9174-4_5

Singh, H., & Kumar, M. (2022). Climate change and its impact on Indian Himalayan forests: Current status and research needs. In S. Rani & R. Kumar (Eds.), *Climate change: Impacts, responses and sustainability in the Indian Himalaya* (pp. 223–242). Springer Climate. Springer. https://doi.org/10.1007/978-3-030-92782-0_11

Singh, N., Tewari, A., Mittal, A., Shah, S., Bisht, M., Siddiqui, S., et al. (2024). Seed ecology and seedling dynamics of western Himalayan treeline tree species. *Frontiers in Forests and Global Change, 7*, 1437954. https://doi.org/10.3389/ffgc.2024.1437954

Singh, S. P., Zobel, D. B., Garkoti, S. C., Tewari, A., & Negi, C. M. S. (2006). Patterns in water relations of central Himalayan trees. *Tropical Ecology, 47*(2), 159–182.

Singh, S. P., & Zobel, D. B. (2025). Pine-Oak type: A unique resilient global system with high potential for nature-based solutions. *Ecological Indicators, 175*, 113536. https://doi.org/10.1016/j.ecolind.2025.113536

Stangler, D. F., Hamann, A., Kahle, H. P., & Spiecker, H. (2017). A heat wave during leaf expansion severely reduces productivity and modifies seasonal growth patterns in a northern hardwood forest. *Tree Physiology, 37*(1), 47–59. https://doi.org/10.1093/treephys/tpw094

Tandon, A., Awasthi, A., & Pattnayak, K. C. (2025a). Efficacy of machine learning in simulating precipitation and its extremes over the capital cities in North Indian states. *Scientific Reports, 15*(1), 10345. https://doi.org/10.1038/s41598-024-84360-w

Tandon, A., Awasthi, A., Pattnayak, K. C., & Das, S. (2025b). Rainfall variability and rising extremes in urbanizing himalayan foothills: A machine learning and data-driven exploration of hydroclimatic shifts in Uttarakhand, India. *Earth Systems and Environment,* 1–26.

Telwala, Y., Brook, B. W., Manish, K., & Pandit, M. K. (2013). Climate-induced elevational range shifts and increase in plant species richness in a Himalayan biodiversity epicentre. *Plos One, 8*(2), e57103. https://doi.org/10.1371/journal.pone.0057103

Thadani, R., & Singh, S. P. (2025). Major forest types and associations in the Himalayas. In S. Singh & R. Thadani (Eds.), *Himalayan forests in a changing world* (pp. 95–148). Springer. https://doi.org/10.1007/978-981-95-0978-2_4

Tiebel, K., Dahlmann, J., & Karge, A. (2024). Global warming could shorten the seed lifespan of pioneer tree species and thus natural regeneration window of damaged areas. *European Journal of Forest Research, 143*(2), 437–450. https://doi.org/10.1007/s10342-023-01633-1

Tiwari, P., Bhattacharya, P., Rawat, G. S., Rai, I. D., & Talukdar, G. (2021). Experimental warming increases ecosystem respiration by increasing above-ground respiration in alpine meadows of Western Himalaya. *Scientific Reports, 11*(1), 2640. https://doi.org/10.1038/s41598-021-82065-y

Trivedi, V. L., Sati, P., & Attri, C. (2025). Non-native invasive and native range shifter plant species in Alpine Himalaya: A major threat to biodiversity of Alpine Meadows. In R. Kumar & S. Kumar (Eds.), *Biodiversity of the Himalaya: Utilization, conservation, and management of plant resources* (pp.27–47). CRC Press, Taylor & Francis. https://doi.org/10.1201/9781774919934-2

ul Hassan, W., Nayak, M. A., & Lyngwa, R. V. (2021). Recent changes in heatwaves and maximum temperatures over a complex terrain in the Himalayas. *Science of the Total Environment, 794*, 148706. https://doi.org/10.1016/j.scitotenv.2021.148706

Van Gorsel, E., Wolf, S., Cleverly, J., Isaac, P., Haverd, V., Ewenz, C.,... & Silberstein, R. (2016). Carbon uptake and water use in woodlands and forests in southern Australia during an extreme heat wave event in the "Angry Summer" of 2012/2013. *Biogeosciences, 13*(21), 5947–5964. https://doi.org/10.5194/bg-13-5947-2016

Van Peer, L., Nijs, I., Reheul, D., & De Cauwer, B. (2004). Species richness and susceptibility to heat and drought extremes in synthesized grassland ecosystems: Compositional vs physiological effects. *Functional Ecology, 769–778*. https://doi.org/10.1111/j.0269-8463.2004.00901.x

Wang, Z., & Huang, B. (2004). Physiological recovery of Kentucky bluegrass from simultaneous drought and heat stress. *Crop Science, 44*(5), 1729–1736. https://doi.org/10.2135/cropsci2004.1729

Wangchuk, K., Darabant, A., Nirola, H., Wangdi, J., & Gratzer, G. (2021). Climate warming decreases plant diversity but increases community biomass in high-altitude grasslands. *Rangeland Ecology & Management, 75*, 51–57. https://doi.org/10.1016/j.rama.2020.11.008

Xu, J., Grumbine, R. E., Shrestha, A., Eriksson, M., Yang, X., Wang, Y. U. N., & Wilkes, A. (2009). The melting Himalayas: Cascading effects of climate change on water, biodiversity, and livelihoods. *Conservation Biology, 23*(3), 520–530. https://doi.org/10.1111/j.1523-1739.2009.01237.x

Xu, J., Badola, R., Chettri, N., Chaudhary, R. P., Zomer, R., Pokhrel, B., et al. (2019). Sustaining biodiversity and ecosystem services in the Hindu Kush Himalaya. In P. Wester, A. Mishra, A. Mukherji & A. B. Shrestha (Eds.), *The Hindu Kush Himalaya assessment: Mountains, climate change, sustainability and people* (pp. 127–165). Springer. https://doi.org/10.1007/978-3-319-92288-1_5

Yang, Z., Jiang, Y., Qiu, R., Gong, X., Agathokleous, E., Hu, W., & Clothier, B. (2023). Heat stress decreased transpiration but increased evapotranspiration in gerbera. *Frontiers in Plant Science, 14*, 1119076. https://doi.org/10.3389/fpls.2023.1119076

Yuan, W., Cai, W., Chen, Y., Liu, S., Dong, W., Zhang, H., et al. (2016). Severe summer heatwave and drought strongly reduced carbon uptake in Southern China. *Scientific Reports, 6*(1), 18813. https://doi.org/10.1038/srep18813

Zhang, Y., Loreau, M., He, N., Wang, J., Pan, Q., Bai, Y., & Han, X. (2018). Climate variability decreases species richness and community stability in a temperate grassland. *Oecologia, 188*(1), 183–192. https://doi.org/10.1007/s00442-018-4208-1

Zhang, R., Wise, R. R., Struck, K. R., & Sharkey, T. D. (2010). Moderate heat stress of Arabidopsis thaliana leaves causes chloroplast swelling and plastoglobule formation. *Photosynthesis Research, 105*(2), 123–134. https://doi.org/10.1007/s11120-010-9572-6

Impact of Rising Temperature on Habitat Distribution of Wild Edible Fruit, *Malus indica* (Wall.) B.B.Liu (syn. *Docynia indica*): A Case Study in Eastern Himalaya

Bijoy Chhetri and Aditya Pradhan

Abstract The fragile Himalayan ecosystem is under threat from climate change, with rising temperatures that may significantly impact biodiversity. The Himalayas are considered a data-deficient region due to limited availability of long-term monitoring data; however, the existing data, albeit fragmented, remains inadequate for addressing conservation directives and formulating appropriate conservation planning and habitat management for the sensitive Himalayan species. Rising temperatures already impact global animal and plant populations. As the climate overheats, these effects worsen, potentially causing heat stress and death in plants due to their inability to cope with rising temperatures. In this chapter, we presented a case study on Himalayan sensitive species using a spatial distribution model to reveal effect of increasing temperature on the spatial distribution of the plant species. We identified the sensitive and vulnerable habitats of one of the under-rated wild edible fruits (*Docynia indica*) in the Eastern Himalayas and addressed the conservationist to plan suitable strategies for habitat management. This case study will serve as a model for other associate species of the mountain ecosystem to address the climate shift in the Eastern Himalayas.

Keywords Climate change · Conservation · Habitat management · Mountain ecosystem · Spatial modelling

---

B. Chhetri
Department of Zoology, School of Basic Sciences, SRM University Sikkim, Tadong, Gangtok, Sikkim, India

A. Pradhan (✉)
Department of Botany, School of Basic Sciences, SRM University Sikkim, Tadong, Gangtok, Sikkim, India
e-mail: apradhan512@gmail.com

A. Awasthi et al. (eds.), *Rising Temperatures and Plant Ecosystems*,
https://doi.org/10.1007/978-3-032-29477-7_10

1 Introduction

Environmental changes are dramatically impacting community composition in Earth's biodiversity hotspots. The fragile Himalayan ecosystems are experiencing uncertainty in temperature, rainfall, wind, and snowfall (Awasthi et al., 2023; Subba et al., 2018; Tandon et al., 2025). However, the scenarios are predominant in the Himalayas, where temperature plays a pivotal role. It is documented that global temperatures have increased by 0.74 °C in the past 100 years (IPCC, 2007). According to Stocker et al. (2013), the temperature is also projected to rise by 0.3–4.5 °C by the end of the twenty-first century. These increases have led to significant impact in climatic scenarios, which in turn have noticeable impacts on Himalayan biodiversity (Sharma et al., 2009). Himalaya is considered a data-deficient region, and existing data remains at a fragmented level. Most of the Himalayan ecosystems are under-surveyed and un-surveyed due to the geographical undulating terrain and unbearable climatic conditions that hinder exploratory missions (Chhetri & Badola, 2017). Furthermore, it makes it very difficult and challenging to make an appropriate plan for the conservation and management of the fragile Himalayan ecosystems (Chhetri et al., 2021). Nowadays, the conservationists have given emphasis to conserving charismatic species using a spatial distribution model based on machine learning and artificial intelligence tools to fulfil this gap (Chhetri & Badola, 2017; Wiens et al., 2009). Ecological modelling is more prevalent, and appropriate methodologies are used to find out the range of the species and associate their biophysical variables (Peterson, 2006) in the Himalayan region. However, most ecological models' accuracy depends on sample size; however, sometimes sample size may not be adequate to predict habitat patterns for rare and threatened plants (Chhetri & Badola, 2017). MaxEnt (Maximum Entropy modelling), along with ArcGIS toolkits, is now used globally for better understanding species distribution also using a small sample size, making it ideal for predicting species distribution and dynamic changes in environmental factors (Chhetri & Badola, 2017; Phillips et al., 2009), even in the fragile Himalayan ecosystem. In recent decades, the Himalayan landscapes have been shifting due to rapid urbanization. This shift may change land use and land cover patterns and be further boosted by climate change. Due to habitat loss, habitat fragmentation, and excessive use of natural resources worldwide, many wild edible fruits (WEFs) are vulnerable in the Himalayas, which may further neglect those economically and medicinally important WEFs in inaccessible terrains and extreme climatic conditions.

In this article, our research addresses the following objectives taking *D. indica*, one of the wild edible fruits as a model species were to (a) predict spatial habitat distribution of the species using with a small sample size of the existing data to address their conservation status; (b) address and identify the potential environmental predictors used in the model for the species; (c) predict future habitat distribution of the species; (d) quantify the dynamic changes in habitats distribution between the present and future habitat distributions of the species; (e) quantify the existing protected areas effectiveness on the both present and future climatic scenarios.

2 Materials and Methods

2.1 *Studied Species*

D. indica is a WEF under the Rosaceae family. It is distributed in the lower temperate forests of the eastern Himalayas with an altitudinal range of 1200–1800 m asl (Biswas et al., 2021). The deciduous evergreen forests are widely found in the Himalayas, stretching along the eastern part of Bhutan, Darjeeling, Nepal, and Sikkim (Sundriyal & Sundriyal, 2003). The species is found in wet temperate forests, including riverine slopes, as well as in cultivated agroecosystems in the Sikkim Himalayas (Sharma et al., 2019). Its fruit is widely used in traditional medicine by local communities. Locals prepare a fruit product called "Chuk," which is available in Sikkim Himalayan markets (Sharma et al., 2019). Associated species include *Cryptomeria japonica, Macaranga denticulate, Ficus* sp., *Rhododendron arboreum, and Schima wallichii,* among others, in the Sikkim Himalayas.

2.2 *Study Area*

We conducted extensive field studies in Sikkim, covering the three districts: East Sikkim, West Sikkim, and South Sikkim, with an altitudinal gradient ranging from 657 to 1982 m asl. (Fig. 1). The total area of Sikkim is 7,096 sq. km, with a geographical extent of 27°04′46″–28°07′48″ N and 88°00′58″–88°55′25″ E with an altitude gradient ranging from 284 to 8586 m asl. It falls under one of the 36 global biodiversity hotspot regions (Mittermeier et al., 2004). The region contains a rich diversity of flora and fauna due to its unique features, spanning from tropical to arctic zones. This variation leads to a complex vegetation structure and diverse communities, shaped by the area's remarkable physiographic diversity and climatic differences.

2.3 *Ecological Distribution Modelling*

MaxEnt (version 3.2.4) is used to delineate species habitat distribution using the occurrence record and environmental variables, with the help of ArcGIS software version 10.4 (Followed a similar approach by Chhetri et al. (2021)).

2.4 *Environmental Predictors*

We used 19 bioclimatic variables (worldclim, 1 km resolution), NDVI (24 layers, downloaded from MODIS), and land use/cover data (downloaded from DIVA-GIS).

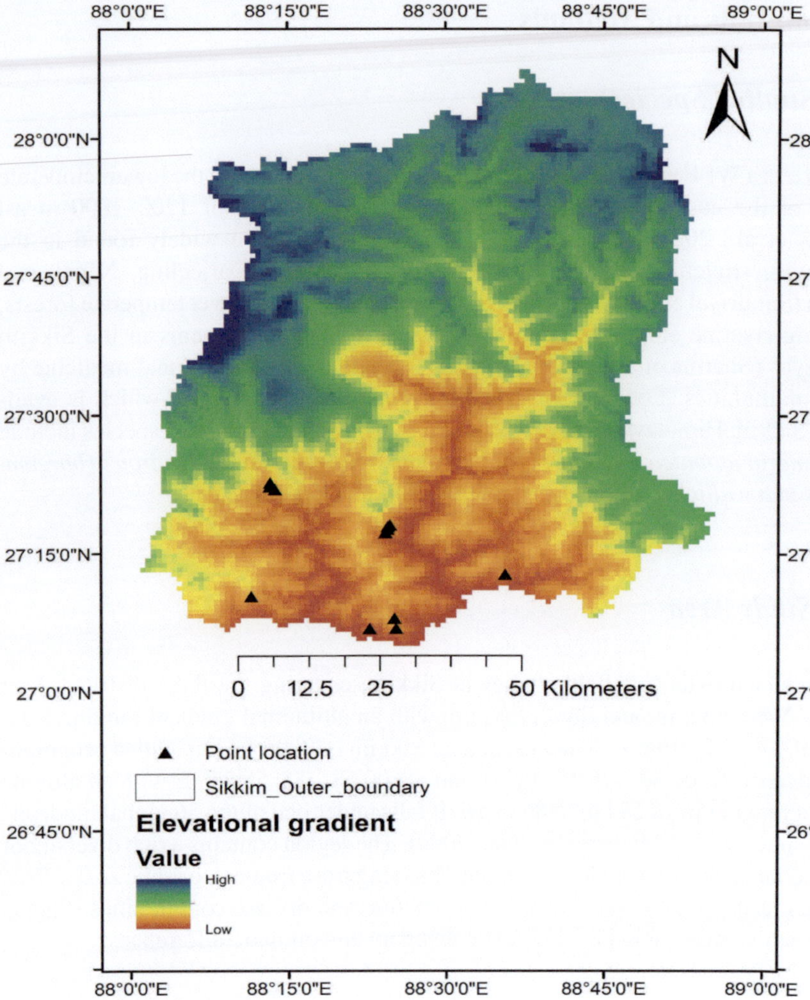

Fig. 1 The point location with the digital elevation model of the species in the eastern Himalayas

Using DEM (ASTGTM2), we generated slope, aspect and altitude using ArcGIS 10.4. The bioclimatic variables were derived from monthly temperature and rainfall data for further environmental analysis.

2.5 *Future Climatic Predictors*

We used a set of RCPs (Representative Concentration Pathways) scenarios, AR5-2050 (downloaded from www.ccafs-climatic.org/data), then selected three widely used global circulation models (GCMs) for regional climate impact studies: Miroc-ESM (Japan), HadGEM2-AO (Korea), and GFDL-CM3 (USA). These future climate scenario models were used to predict the distribution of habitats of the species. For minimizing model bias, we applied an ensemble method by taking the mean values of the three models, followed by Chhetri et al. (2021). The Representative Concentration Pathways (RCPs) deal with the possible future greenhouse gas emissions based on trajectories adopted by the IPCC. There are four RCP scenarios: +2.6, +4.5, + 6.0, and +8.5 W/m^2. These levels of radiative forcing, as estimated for 2100 relative to pre-industrial levels, are based on IPCC (2018). Greenhouse gases are considered a key driver of climate warming. The RCPs represent trajectories of intensified greenhouse gas emissions and are used to better understand the impact of rising temperatures on Himalayan species.

2.6 *Calibrating the Model*

We identified the crucial environmental predictors for the potential distribution of the species based on PCA (Principal Components Analysis), followed by Chhetri and Badola (2017)). PCA is used for quantifying mutual relationships among variables and can also be used for testing the autocorrelation between each pair of environmental variables. We excluded those variables from the final model that had their highest loading values in the principal components using the SPSS software version 21. A total of 49 environmental predictors; 14 environmental predictors were selected based on their minimum loading values, 14 were treated as independent and non-correlated predictors at a 95% confidence level. The Jackknife validation method and 10-percentile threshold rule were applied along with 20% of the data for testing and 80% for training (Chhetri & Badola, 2017; Flory et al., 2012).

2.7 *Execution of MaxEnt*

To evaluate model accuracy, goodness-of-fit was assessed used based on the AUC (area under the receiver operating characteristic curve). The AUC value ranges from 0.5 to 1, with maximum values denote better performance: values above 0.9 denote excellent predictive ability, 0.7–0.9 indicate good performance, and values near 0.5 suggest poor or uninformative predictions (Swets, 1988). The final output was categorized into five threshold habitat classes regrouped into percentages from 0–100%: very low class ($\leq$10%), low class (11–30%), moderate class (31–50%), high class (51–70%), and very high class ($\geq$70%).

2.8 Dynamic Changes in Habitat

Dynamic changes in the species' suitable habitat were assessed by comparing future and present habitat distribution, using binary maps of potential and non-potential habitat generated from probability thresholds (Jimenez-Valverde & Lobo, 2007). We applied a Python-based ArcGIS toolkit, Species Distribution Model (SDM), to quantify changes in the species' habitat in future climatic scenarios under loss, gain, and unchanged categories (Abdelaal et al., 2019).

2.9 Effectiveness of the Protected Zones

We assessed the effectiveness of the current protected areas (PAs) in the study region for the target species by overlaying the predicted current distribution map onto the boundaries of the PAs. This method allowed us to quantitatively evaluate how well the present protected zones encompass the species' suitable habitats.

3 Results

3.1 Habitat Suitability of the Species

The MaxEnt modeling results identified Bio7, Bio2, NDVI (a201313161), Bio5, and Bio12 as the most influential variables among the 14 potential environmental predictors in shaping the final habitat distribution of *D. indica* in the Sikkim Himalayas (Fig. 2 and Table 1). The accuracy of the result of the model for the species, based on the AUC value, which emerged very well (AUC training = 0.944 and AUC test = 0.926). We evaluated the area of the suitable predicted habitat of *D. indica* based on the present scenarios (Fig. 2). The total area of Sikkim is 7096 sq km. Under the very high class, the area covered is minimal (230.45 km^2) compared to the other threshold classes. However, the moderate threshold class (592.47 km^2) and high threshold class (525.55 km^2) have covered nearly equal suitable habitat areas. The predicted habitat under the low threshold class showed slightly higher area covered (665.49 km^2) than the very high and moderate threshold classes. The predicted habitat under "very high threshold class" emerged in small pockets in the undulating terrain of the Sikkim Himalayas (Fig. 3). We also quantified the suitable areas of the species based on the future scenarios (RCP-2050). Under the RCP 2050, the highest suitable area emerged for the species for RCP 8.5 (983.95 km^2), followed by RCP 6.0 (942.82 km^2). Comparatively, RCP2.6 (928.48 km^2) and RCP4.5 (928.48 km^2) have shown equal suitable areas covered for the species.

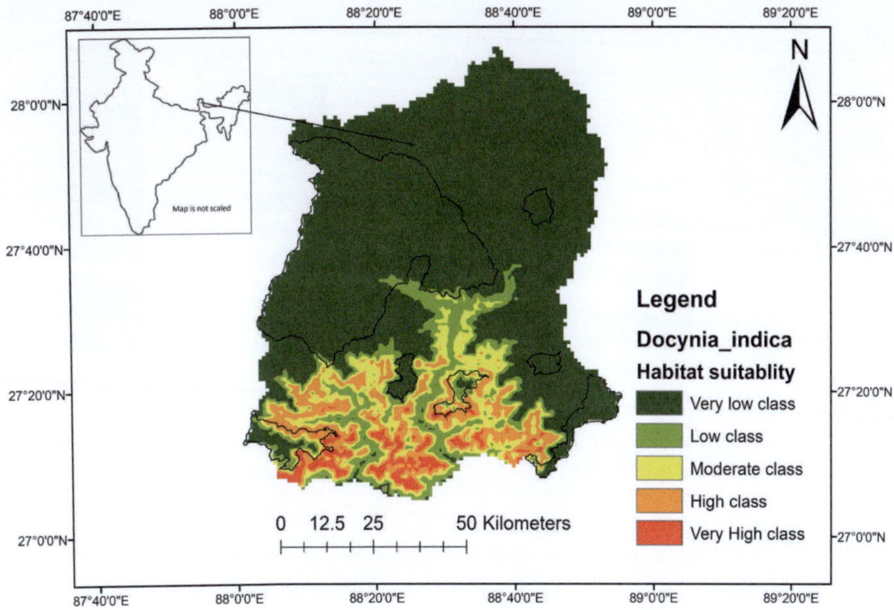

Fig. 2 Habitat suitability map of *D. indica* in Sikkim with protected areas, eastern Himalaya

Table 1 Environmental predictors' contribution to the model of *D. indica* in Sikkim, eastern Himalaya

Environmental predictors	Percent contribution	Permutation importance
bio-7	81.5	50.3
bio-2	11.7	0
NDVI (a2013161)	6	14.5
bio-5	0.7	8.4
bio-12	0.1	26.8

3.2 Dynamic Habitat Shift

Under RCP-2050 models, the maximum predicted suitable habitat gain area seems to rise by 166.38 km^2 (17.65%) for the RCP6.0 and 55.46 km^2 (5.64%) for RCP8.5, and there is no gain in the suitable habitat areas in both RCP 2.6 and RCP 4.5 (Fig. 3; Table 2). However, the predicted highest loss in the suitable habitat area is found under RCP 4.5 (433.16 km^2, 46.65%) model, followed by RCP 2.6 (396.28, 42.73%), RCP 8.5 (31.56 km^2, 3.20%), and RCP 6.0 (14.38 km^2, 1.52%). The highest predicted in the suitable habitat area is found remains unchanged under model RCP8.5 (896.93 km^2, 91.15%), followed by RCP6.0 (762.10 km^2, 80.83%), RCP2.6 (531.66 km^2, 57.26%), and RCP4.5 (495.31 km^2, 53.34%). The majority expansion of habitat emerged as a shift under different scenarios in latitudinal and higher altitudinal

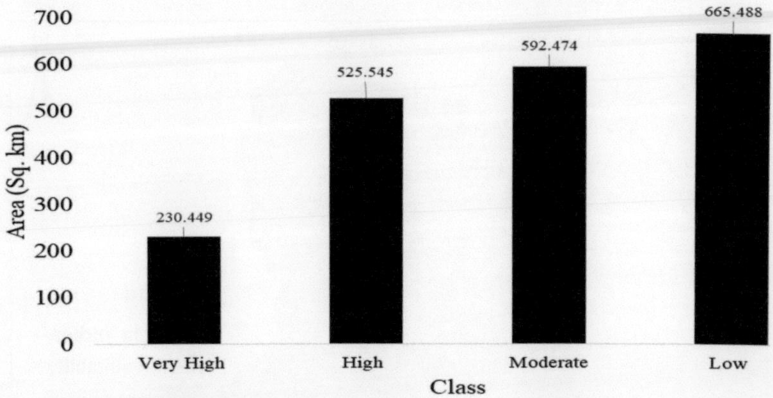

Fig. 3 Areas covered in different threshold classes of suitable habitat of *D. indica* in the present scenario of Sikkim, eastern Himalaya

Table 2 The areas dynamic change (km^2) of *D. indica* in Sikkim, eastern Himalaya

Area of habitat (km^2)						Proportion of area (%)			
Model	Future	Lost	Gained	Unchanged	Total	Lost	Gained	Unchanged	Total
RCP2.6	928.49	396.82	0	531.66	−396.83	42.73	0	57.26	−42.73
RCP4.5	928.49	433.16	0	495.31	−433.16	46.65	0	53.35	−46.65
RCP6.0	942.82	14.34	166.38	762.10	152.03	1.52	17.64	80.83	16.12
RCP8.5	983.94	31.56	55.46	896.93	23.90	3.20	5.63	91.15	2.42

regions. The highest total suitable habitat area may be reduced by RCP4.5 (433.16 km^2, 46.65%) of the current suitable area, followed by RCP2.6 (396.82 km^2, 42.74%), RCP8.5 (31.56, 3.20%), and RCP6.0 (14.34 km^2, 1.52%). The total suitable habitat area may be increased by 152.03 km^2 or 16.13 km^2 under RCP6.0 and RCP8.5 (23.90 km^2, 2.43%) and the total suitable habitat area would be decreased by 433.16 km^2 (46.65%) under RCP4.5 and RCP2.6 (396.83 km^2, 42.73%) of the present area for the species (Fig. 4). The analysis of current protected areas indicated that the majority of suitable habitat for the species lies outside these protected zones and is likely to extend further beyond protected regions under future scenarios.

4 Discussion

WEFs serve as important sources of nutrition and traditional medicine for local communities in the Himalayas (Ritika et al., 2024). Most of the area is primarily unexplored due to inaccessible terrain and extreme environmental conditions, resulting in

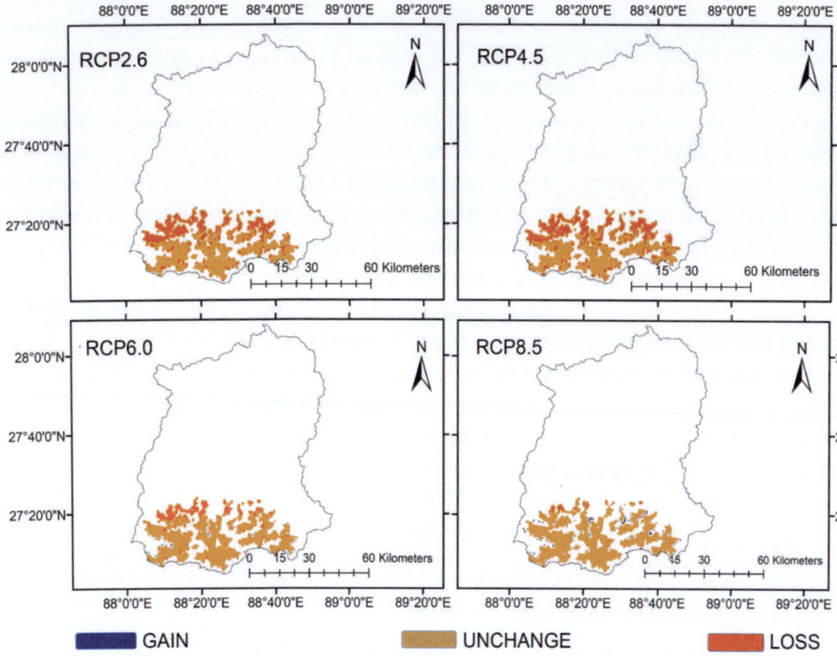

Fig. 4 Dynamic shift of habitat of *D. indica* in the future scenarios (RCPs-2050) of Sikkim, eastern Himalaya

the loss of many economically and medicinally important WEFs, without their potential medicinal and nutritional values (Barman & Mishra, 2025). Sikkim, the eastern Himalayan region, is India's most diverse and species-rich state. WEFs account for 65% of the state's wild edible plant resources, providing an alternative source of livelihood for the local inhabitants (Sundriyal & Sundriyal, 2004). Although *D. indica* can tolerate extreme weather conditions (as low as −0.5 °C), temperature and precipitation patterns changes due to climate change may impact on fruit size, weight, and overall production (Rahman et al., 2025). Studies on related species indicate potential shifts in suitable habitat resulting from climate change, which could similarly impact *D. indica* in the future (Gufi et al., 2023). Moreover, the rapid increase in temperature plays a pivotal role in the Himalayan ecosystem, causing many species to shift their habitats (Subba et al., 2018), extend their ranges (Shrestha & Bawa, 2014), or experience range contractions (Hu et al., 2015). Numerous studies have suggested that temperature is a driving force behind population declines and range shifts with the intensification of greenhouse gases. Similar results were found in the present study, where different RCP-2050 scenarios indicated that temperature changes could significantly affect *D. indica*. Specifically, the annual temperature range (Bio7) made the greatest contribution to shaping the species' habitat distribution.

The Mountain landscape is predominant by harsh climatic conditions and undulating terrain; sometimes it is challenging for the researcher to assess the whole target study properly (Chhetri et al., 2021). In this scenario, the MaxEnt model offers an effective tool for studying the Himalayan landscape to predict important areas for species distribution, including regions that have not been surveyed or are under-surveyed due to these constraints. Furthermore, the Himalayas are considered a data-deficient region; the lack of long-term monitoring sites results in insufficient systematic data, which hinders the development of effective strategies to mitigate climate change. The IPCC (2018) report recommends that maintaining benefits for human and natural ecosystems requires mitigation of the increase in global temperature up to 1.5 °C rather than 2 °C. Accelerating climate change and increasing anthropogenic pressures may further put Himalayan species at risk. Dunn et al. (2016) highlighted that existing protected areas in the Himalayan landscape are inadequate for conservation and management, and are unlikely to achieve Aichi Target 11 or the CBD targets for mountain species. While protected areas are crucial for preventing extinction, Clark et al. (2013) noted that the present protected areas in Southeast Asia may not be sufficient to prevent habitat loss for many species.

5 Conclusion

The article addresses the recent advances in species distribution modeling that could be used to identify the effect of climate change on Himalayan species, which will help to address conservationist to formulate appropriate conservation and management strategies for the mountain ecosystem. The results revealed that the current suitable habitat for *D. indica* based on the very high threshold class covers minimum area only 230.44 km^2 among all threshold classes. The environmental predictors Bio2 (mean diurnal range), NDVI, Bio5 (maximum temperature of the warmest month), Bio7 (temperature annual range), and Bio12 (annual precipitation) were the most influential environmental predictors shaping the species distribution model. The Bio7 (temperature annual range) variable plays a significant role in shaping the suitable habitat of the species, as reflected by its significant contribution to the model. Major fluctuations in temperature may affect the species within the Himalayan ecosystem. However, anthropogenic pressures and overexploitation in some areas may make the species even more vulnerable, highlighting the need for sustainable use to conserve its genetic resources. Although the species has socio-economic and cultural value, there is still limited scientific exploration on its nutritional and medicinal properties.

Acknowledgements The authors thanks to the Vice-Chancellor of SRM University Sikkim for extended the facilities and support. Also, thanks go to the official of the State Forest Department, Government of Sikkim for their help in the field study.

References

Abdelaal, M., Fois, M., Fenu, G., & Bacchetta, G. (2019). Using MaxEnt modeling to predict the potential distribution of the endemic plant Rosa arabica Crép in Egypt. *Ecological Informatics, 50*, 68–75. https://doi.org/10.1016/j.ecoinf.2019.01.003

Awasthi, A., Pattnayak, K. C., Tandon, A., Sarkar, A., & Chakraborty, M. (2023). Implications of climate change on surface temperature in North Indian states: Evidence from CMIP6 model ensembles. *Frontiers in Environmental Science, 11*, 1264757. https://doi.org/10.3389/fenvs.2023.1264757

Barman, M., & Mishra, P. (2025). Comprehensive review on ethnobotanical, nutritional and functional properties of wild edible fruits of North Eastern Himalaya, India. *Journal of Food Science Technology, 62*(8), 1407–1424. https://doi.org/10.1007/s13197-025-06351-7

Biswas, J., Barman, P., Rahman, M., & Ghosh, R. (2021). A comprehensive review on the endemic plant Docyniaindica: Its ethnobotanical, phytochemical, and ethnomedicinal perspective. *Asian Journal of Biological and Life Sciences, 10*(3), 515.

Chhetri, B. & Badola, H. K. (2017). Predicting suitable habitats for the vulnerable species, *Rhododendron niveum* Hook.f. (Ericaceae) in Eastern Himalaya. *NeBIO, 8*(3), 139–146.

Chhetri, B., Badola, H. K., & Barat, S. (2021). Modelling climate change impacts on distribution of Himalayan pheasants. *Ecological Indicators, 123*, 107368. https://doi.org/10.1016/j.ecolind.2021.107368

Clark, N. E., Boakes, E. H., McGowan, P. J. K., Mace, G. M., & Fuler, R. A. (2013). Protected areas in South Asia have not prevented habitat loss: A study using historical models of land use change. *Plos One, 8*, e65298. https://doi.org/10.1007/s13197-025-06351-7

Dunn, J. C., Buchanan, G. M., Stein, R. W., Whittinngham, M. J., & McGowan, P. J. K. (2016). Optimising different types of biodiversity coverage of protected areas with a case study using Himalayan Galliformes. *Biological Conservation, 196*, 22–30. https://doi.org/10.1016/j.biocon.2016.01.015

Flory, A. R., Kumar, S., Stohlgren, T. J., & Cryan, P. M. (2012). Environmental conditions associated with bat white nose syndrome mortality in the north-eastern United States. *Journal of Applied Ecology, 49*, 680–689. https://doi.org/10.1111/j.1365-2664.2012.02129.x

Gufi, Y., Manaye, A., Tesfamariam, B., Abrha, H., Tesfaye, M., & Hintsa, S. (2023). Modeling impacts of climate change on the geographic distribution and abundances of *Tamarindus indica* in Tigray region, Ethiopia. *Heliyon, 9*(7), e17471. https://doi.org/10.1016/j.heliyon.2023.e17471

Hu, X.-G., Jin, Y., Wang, X.-R., Mao, J.-F., Li, Y., & Yin, T. (2015). Predicting impacts of future climate change on the distribution of the Widespread Conifer Platycladus orientalis. *Plos One, 10*(7), e0132326. https://doi.org/10.1371/journal.pone.0132326

IPCC. (2007). In S. Solomon, D. Qin, M. Manning, Z. Chen, M. Marquis, K.B. Averyt, M. Tignor & H.L. Miller (Eds.), *Climate change 2007: The physical science basis. Contribution of working group I to the fourth assessment report of the intergovernmental panel on climate change*. Cambridge University Press.

IPCC. (2018). Summary for policymakers. In V. Masson-Delmotte, P. Zhai, H.-O. Pörtner, D. Roberts, J. Skea, P. R. Shukla, A. Pirani, W. Moufouma-Okia, C. Péan, R. Pidcock, S. Connors, J. B. R. Matthews, Y. Chen, X. Zhou, M. I. Gomis, E. Lonnoy, T. Maycock, M. Tignor & T. Waterfield (Eds.), *Global Warming of 1.5 °C. An IPCC special report on the impacts of global warming of 1.5 °C above pre-industrial levels and related global greenhouse gas emission pathways, in the context of strengthening the global response to the threat of climate change, sustainable development, and efforts to eradicate poverty* (pp. 3–24). Cambridge University Press. https://doi.org/10.1017/9781009157940.001

Jimenez-Valverde, A., & Lobo, J. M. (2007). Threshold criteria for conversion of probability of species presence to either-or presence-absence. *Acta Oecologica, 31*, 361–369. https://doi.org/10.1016/j.actao.2007.02.001

Mittermeier, R. A., Gils, P. R., Hoffman, M., Pilgrim, J., Brooks, T., Mittermeier, C. G., Lamoreaux, J., & Da-fonseca, G. A. B. (Eds.). (2004). *Hotspots revisited. Earth's biologically richest and most endangered terrestrial ecoregions*. CEMEX. USA.

Peterson, A. T. (2006). Uses and requirements of ecological niche models and related distributional models. *Biodiversity Informatics, 3*, 59–72.

Phillips, S. J., Dudík, M., & Schapire, R. E. (2009). MaxEnt software for species habitat modelling, version 3.3.1. http://biodiversityinformatics.amnh.org/open_source/maxent/

Rahman, M., Islam, M.A., Das, R., Hazarika, N.K., Goswami, K., & Barman, P. (2025). Enriching genomic resources of wild apple *Docynia indica (Wall.)* decne using tissue-specific transcriptome profiling. *Discover Plants,* 2, 283. https://doi.org/10.1007/s44372-025-00365-3

Ritika, Bora, B., Ismail, B. B., Garba, U., Mishra, S., Jha, A. K., Naik, B., Kumar, V., Rather, M. A., Rizwana & Preet, M. S. (2024). Himalayan fruit and circular economy: Nutraceutical potential, traditional uses, challenges and opportunities. *Food Production, Processing and Nutrition, 6*(1), 71.

Sharma, E., Chettri, N., Tse-ring, K., Shrestha, A. B., Fang, J., Mool, P., & Eriksson, M. (2009). *Climate change impacts and vulnerability in the Eastern Himalayas*. ICIMOD.

Sharma, G., Pradhan, B. K., Chettri, S., Chettri, A., & Chhetri, D. R. (2019). Indigenous knowledge and phytochemical screening of medicinal chuk from Rhus chinensis, Docynia indica and Hippophae salicifolia in Sikkim Himalaya. *Indian Journal of Traditional Knowledge, 18*, 250–260.

Shrestha, U. B., & Bawa, K. S. (2014). Impact of climate change on potential distribution of Chinese caterpillar fungus (Ophiocordyceps sinensis) in Nepal Himalaya. *Plos One, 9*(9), e106405.

Stocker, T. F., Qin, D., Plattner, G.-K., Tignor, M., Allen, S. K., Boschung, J., Nauels, A., Xia, Y., Bex, V., & Midgley, P. M. (2013). Climate change 2013. The physical science basis. In *Fifth assessment report of the Intergovernmental Panel on Climate Change*. Intergovernmental Panel on Climate Change.

Subba, B., Sen, S., Ravikanth, G., & Nobis, M. P. (2018). Direct modelling of limited migration improves projected distributions of Himalayan amphibians under climate change. *Biological Conservation, 227*, 352–360. https://doi.org/10.1016/j.biocon.2018.09.035

Sundriyal, M., & Sundriyal, R. C. (2003). Underutilized edible plants of the Sikkim Himalaya: Need for domestication. *Current Science, 85*(6), 731–736.

Sundriyal, M., & Sundriyal, R. C. (2004). Wild edible plants of the Sikkim Himalaya: Nutritive values of selected species. *Economic Botany, 58*(2), 286–299. https://doi.org/10.1663/0013-0001(2004)058[0286:WEPOTS]2.0.CO;2

Swets, J. A. (1988). Measuring the accuracy of diagnostic systems. *Science, 240*, 1285–1293.

Tandon, A., Awasthi, A., Pattnayak, K. C., & Das, S. (2025). Rainfall variability and rising extremes in urbanizing himalayan foothills: A machine learning and data-driven exploration of hydroclimatic shifts in Uttarakhand, India. *Earth Systems and Environment, 1–26*. https://doi.org/10.1007/s41748-025-00881-y

Wiens, J. A., Stralberg, D., Jongsomjit, D., Howell, C. A., & Snyder, M. A. (2009). Niches, models, and climate change: Assessing the assumptions and uncertainties. *PNAS, 106*, 9729–19736. https://doi.org/10.1073/pnas.0901639106

Evidence of Forest Fire During the Pliocene: A Case Study from the Sub-Himalaya of Arunachal Pradesh, India

Shreyasi Basak, Dipak Kumar Paruya, Oindrila Biswas, Ranita Biswas, Manjuree Karmakar, and Subir Bera

Abstract Forest fires have a substantial impact on the ecosystem, influencing the diversity of forest resources, carbon cycles, and global climate change. Palaeoecological studies help in reconstructing the natural fire history of a region, revealing the relationships between vegetation, climate, and past fire occurrences. This information, in turn, helps define the full range of natural disturbance patterns and the resilience of a biome. The key indicators of ancient forest fires preserved in the sediments include charred charcoal particles, burned leaf cuticular parts, phytoliths and wood remains. This case study documents the incidence of wildfire in the Arunachal Himalaya, attributed to rising temperature and declining precipitation during ~9.6–6.8 Ma. We isolated a substantial number of dark coloured charred cuticular remains, sometimes with attached phytoliths and charcoal remains from the middle Siwalik (Subansiri Formation, Pliocene, ~9.6–6.8 Ma) sediments of Bhalukpong area, Arunachal Pradesh, eastern Himalaya. These remains indicate the occurrence of widespread forest fires during this period. The presence of plenty of SADDLE and BILOBATE morphotypes in charred grass cuticles recovered from the studied sediments suggests the presence of Panicoideae and Chloridoideae grass subfamilies in the vegetation. These grass subfamilies usually grow in warm and dry conditions. A significant number of pollen grains of deciduous elements such as *Acer* sp., *Terminalia* sp., Betulaceae, Dipterocarpaceae, *Quercus* ap., *Abies* sp., *Pinus* sp., *Cedrus* sp. etc. were also recovered. Earlier works from the Indian Siwalik sediments inferred that prior to the Late Miocene, the vegetation was predominantly evergreen. However, a remarkable shift occurred during the Late Miocene, with vegetation pattern changing to semi-evergreen/moist deciduous with open grassland. This shift corresponds with a decrease in precipitation with an increase in seasonality. Climate Leaf Analysis Multivariate Program (CLAMP) analysis of fossilized leaves of the Siwalik sedimentary strata also indicates a decrease in winter precipitation,

S. Basak · D. K. Paruya · R. Biswas · M. Karmakar · S. Bera (✉)
Center of Advanced Studies, Department of Botany, University of Calcutta, Kolkata, India
e-mail: berasubir@yahoo.co.in

O. Biswas
Department of Botany, School of Basic Sciences, SRM University Sikkim, Gangtok, India

A. Awasthi et al. (eds.), *Rising Temperatures and Plant Ecosystems*,
https://doi.org/10.1007/978-3-032-29477-7_11

which might have led to extended dry seasons. Thus, higher temperature, extended dry periods, and high biomass content of the forest floor by excessive leaf shedding of semi-evergreen/ deciduous forest elements during the dry season might have triggered the fire.

Keywords Palaeofire · Pollen · Phytolith · Wood remains · Middle Siwalik · Arunachal Himalaya

1 Introduction

A fast chemical oxidative reaction process that produces heat, light and a variety of chemical compounds are known as fire. Fire is generally caused due to the combined effect of high-temperature heat source. In modern times, it may occur due to lightning, volcanic activity, rock spark or anthropogenic activities. The result of wildfire causes breakdown of cellulosic plant materials that leads to the formation of a range of gaseous components like ammonia, carbon dioxide and methane. Various types of fires occur that put a detrimental effect on the plant materials. For instance, surface fires, where only the forest floor gets burned, whereas grass fires/scrub fires burn at a much faster rate. In case of ground fires, the movement of fire is quite slow and they only burn thick peat layers. On the other hand, in case of crown fires, the fire spread till the trunk of the trees, producing high temperatures around 800–900 °C. Another type of ground fire that occurs is known as peat fires, where biomass burning occurs across peat-forming areas. The result of such fires leads to variable temperatures that exceed up to 900 °C (Rein, 2013). However, fire and life have shaped each other through geological time, creating characteristic patterns of fire activity known as "fire regimes." According to Morgan et al. (2001), fire regimes are a general form in which fires occur naturally in a particular ecosystem over an long period of time. Ecological processes like soil formation, nutrient recycling, wildlife habitats, carbon dynamics and regional climates are generally influenced by these fire regimes.

Study of palaeofire through the ages with the help of preserved plant traits like fire scars on tree rings, fossil charcoal, and palynomorphs in sedimentary strata, presence of pyrolytic polycyclic aromatic hydrocarbons (PAHs) and charcoalified tracheids or wood fragments in the fossil records can relate to past chronic wildfire behaviour and the vegetation involved (Bhattacharya et al., 2025). Evidence of fire known from deep time can also be related to the atmospheric oxygen content, since oxygen plays an important role in the start and spread of forest fire. Palaeofire evidences as old as Late Devonian to Early Carboniferous age from the Ohio Basin shows a rise in microscopic charcoal, which coincides with records of macroscopic charcoal found in terrestrial and near-shore environments, indicating increased wildfire activity during that time (Rimmer et al., 2004). This widespread increase in wildfires may be linked to a rise in atmospheric oxygen during that period (Brand et al., 2021). Again, charcoal concentration in Carboniferous peat is much more when compared with those in modern peats, suggesting increased fire activity during deposition (Scott et al., 2014).

Thus, the palaeoenvironmental proxies impart important perceptions into long-term vegetation–climate and fire interactions across different ecosystems and geological periods (Scott et al., 2014).

Wood or plant remains as fuel load with sufficient supply of oxygen and natural heat sources like lightning, sparks from rockfall, or volcanic activities acted as the three elements of the fire triangle and were sufficient to start wildfire on ancient Earth. As mentioned earlier, fossil charcoal remains from the Late Devonian period indicate the development of woody plant species which might have acted as a fuel load for ancient fires (Scott et al., 2014). Fossil charcoal records also indicate that forest fire incidence started taking place after the appearance of land plants about 420 million years ago of the Palaeozoic era in the Late Silurian period (Glasspool et al., 2004; Scott & Glasspool, 2006). With expectations of few age gaps, the constant presence of charcoal in the sediments indicates that fire has continuously shaped the forest ecosystems throughout the last ~400 Ma (Diessel, 2010; Glasspool & Scott, 2010; Rimmer et al., 2015). Moreover, microscopic charcoal fragments, which can be easily windblown from the place of origin to lake sediments, alluvial fan deposits, flood plain deposits and marine sediments, are indicators of large regional fire events (Falcon-Lang, 1998; Nicholas & Jones, 1992).

During the occurrence of forest fire pyrolysis takes place that leads to the conversion of wood to charcoal. However, there are various forms of plant charcoal materials like soot formed due to the combustion of vegetation along with fossil fuels like coal, oil, and gas, whereas black charcoal is the main residue formed due to the combustion of biomass materials. Charcoal is mainly inert in nature and the quantification of the presence of charcoal residues are utilised in depicting the inter-relationship between occurrence of fire and climate. Charcoal comes in varied range of sizes, such as fine charcoal (<125 μm), mesoscopic (125 μm–1 mm) and macroscopic (>1 mm) (Glasspool & Scott, 2010; Scott, 2010). Thus, charcoal acts as one of the important fire-related proxies. Recent studies have shown that non-pollen palynomorphs like phytoliths, which are generally used as a proxy for palaeovegetation and climate reconstructions, can also be used as an ancient fire proxy (Esteban et al., 2023; Meng et al., 2022). Fire causes a dull opaque discoloration of phytoliths through changes in their refractive indices (Cordova et al., 2011; Elbaum et al., 2003; Jones & Milne, 1963; Kealhofer & Penny, 1998; Parr, 2006). Direct contact with fire induces darkening or charring of phytoliths, giving them brown or black colour (Cordova et al., 2011; Piperno, 2006). Therefore, the presence and intensity of prehistoric wildfire can be inferred from the presence of fire-discoloured phytoliths, grass cuticles and other charcoalified plant remnants in the sedimentary strata.

Arunachal Pradesh, easternmost among the seven Siwalik sectors situated in the eastern Himalaya, can be broadly divided into three sedimentary strata: i. lower (Middle to Upper Miocene), which is primarily made up of mudstone; ii. middle (Pliocene), dominated by sandstone; iii. upper (Upper Pliocene to Lower Pleistocene), which is composed of conglomerates (Chakraborty et al., 2020; Joshi & Chakraborty, 2001; Joshi et al., 2003; Karunakaran & Ranga Rao, 1976; Kumar, 1997). However, there are several records of fossilized megafloral assemblages from Arunachal Pradesh's Siwalik sediments, which serve as a rich source for providing important signatures of previous climate shifts (Bera et al., 2014; Khan et al., 2014a, 2014b, 2015, 2023), palynological or other micro fossil records are rare (Jain & Dutta, 1978; Dutta, 1980; Dutta & Singh, 1980; Khan et al., 2006). The present work deals with the palynological evidence of ancient forest fire for the first time in the tropical moist deciduous forest during the middle Siwalik (Pliocene) sedimentation.

2 Methodology

2.1 *Geological Setting*

The strata, which is present along the fluvial section of Kameng River, are exposed over a stretch of about 12 km flowing in a nearly North–South direction in West Kameng District of Arunachal Pradesh. Sedimentation occurred between ~10.5 and 2.6 Ma, as is established by palaeomagnetic age (Chirouze et al., 2012).

In Arunachal Himalaya, the middle Siwalik rock system (Subansiri Formation, Pliocene) is more loosely packed, brittle, and poorly indurated than the comparatively compact and indurated rocks of the lower Siwalik. While on the other hand, the middle Siwalik rocks are more compacted and indurated than the very coarse and friable upper Siwalik rocks. The middle Siwalik rocks are made of multi-storied sandstone having calcareous concretions of varying size and shape, the texture of which is like salt and pepper; the grains are medium to coarse, intercalated with grey shale. The conglomerates and siltstone preserve ample leaf compression and impression fossils (Joshi & Chakraborty, 2001; Khan et al., 2015).

The middle Siwalik sediments in the vicinity of Bhalukpong area which a part of West Kameng district of Arunachal, are the subject of the current study. In the Bhalukpong area (27°01′09.8″ N, 92°38′11.7″ E), the sampling was done from the roadside exposures along the river Kameng from the Subansiri Formation (Fig. 1). A generalised litholog is presented in Fig. 2.

2.2 *Extraction of Pollen Grains and Phytoliths*

For palynological study, about 10 samples were macerated following standard maceration techniques of Batten (1999) with some modifications. The samples were mixed with several chemicals (sequentially with hydrochloric acid, then hydrogen fluoride, followed by nitric acid, and finally with potassium hydroxide) according to the rock types to release the organic content from the rock and to concentrate the palynomorphs. The palynomorphs were fixed on slides using polyvinyl alcohol and mounted with canada balsam. The palynomorphs were analysed i.e., studied and

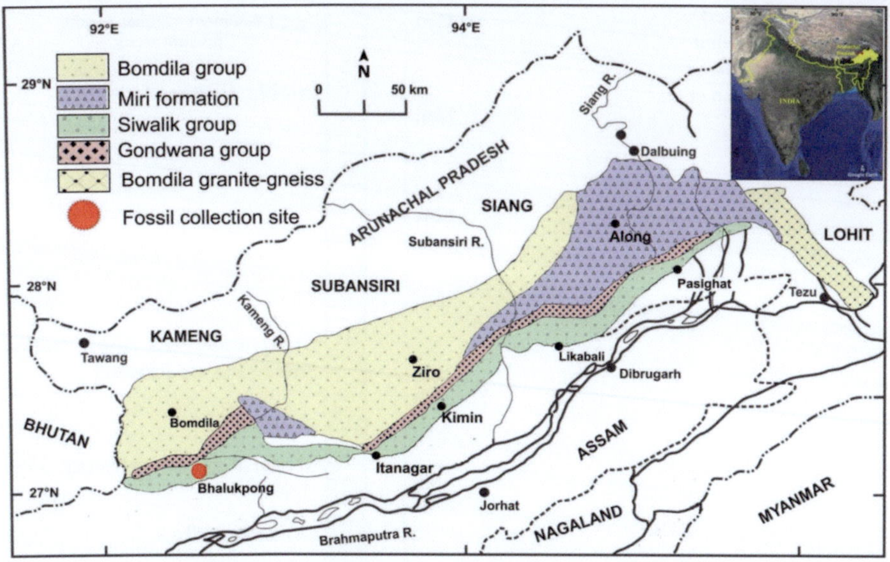

Fig. 1 The red circle in the geological map of the Arunachal Himalayan foothills indicates the location of fossils. Adapted after Singh and Tripathi (1990)

photographed under the light microscope (Zeiss Axioskop 40). All the slide specimens are kept in the Herbarium cum Museum of the Department of Botany, University of Calcutta, Kolkata, India (CUH).

Phytoliths were macerated following the procedure of Pearsall (2000) with small alterations. For each sample, 5–7 gm of sediment was crushed and then oxidised with a mixture of hydrochloric acid and nitric acid mixed in a 1:1 ratio. The mixture was left overnight, and then the excess acid was washed by repeated centrifugation with distilled water at 2500 rpm. Subsequently, the acid-free samples were dried and to remove any trace of organic matters, they were further treated with hydrogen peroxide. The treated samples were cleaned with distilled water till hydrogen peroxide was completely removed and then dried. The dried samples were subjected to the heavy liquid flotation method using a mixture of potassium iodide and cadmium iodide with 2.3 specific gravity. The supernatant thus obtained was centrifuged several times with distilled water and the precipitates containing phytolith morphotypes were studied. For phytolith identification, counting and photography, slides were prepared using polyvinyl alcohol and Dibutylphthalate Polystyrene Xylene. Light microscopic images of isolated plant remain are taken in a Zeiss Axioskop 40 at 40x magnification.

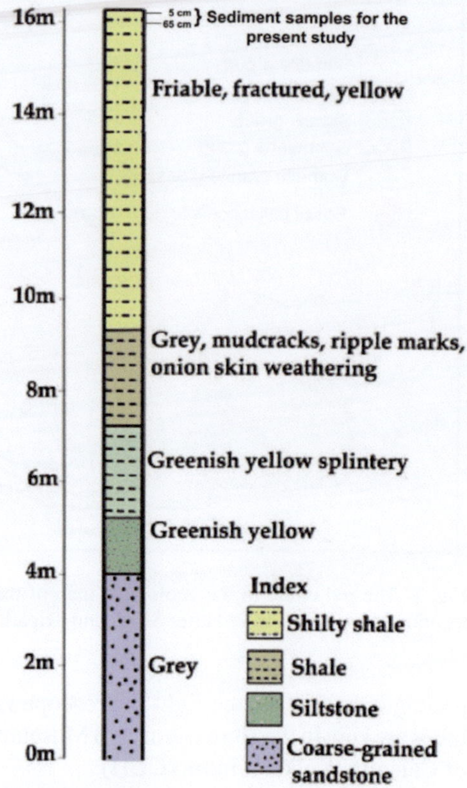

Fig. 2 A generalised litho-log of Arunachal Himalayan foothills. Modified after Khan et al. (2009)

3 Results and Discussions

3.1 Plant Microfossils as Fire Proxies

A total of 12 pollen types belonging to 9 angiosperm and one gymnosperm family, 4 pteridophytic spore types, along with 13 grass and 5 non-grass phytolith morphotypes, fungal remains, wood fragments, leaf cuticles and charcoal fragments were recovered from the sediments. Among the palynomorphs, pollen grains of *Acer* sp., *Terminalia* sp., Betulaceae, Dipterocarpaceae, *Quercus* sp., *Abies* sp., *Pinus* sp., *Cedrus* sp., and the common deciduous forest elements are recovered from throughout the sediments.

A significant amount of both grass and non-grass phytolith morphotypes are recovered (Figs. 3, 4 and 5) from these sediment samples. Several burnt phytolith morphs as well as charred grass cuticle and wood fragments, are recovered from the studied sediments along with the unburnt phytolith morphs. Phytolith evidences indicate presence of different grass subfamilies during the middle Siwalik depositions. BILOBATE, CROSS types indicate Panicoideae grasses whereas, SADDLE types indicate the presence of Chloridoideae grasses during sedimentation (Biswas et al., 2016, 2025; Karmakar et al., 2024; Neumann et al., 2017). Chloridoideae grasses have C_4 photosynthetic pathway, while most of the Panicoideae grasses also follow C_4 pathway (Pardo & Vanburen, 2021). C_4 grasses/plants are common in locations with high wildfire risks/hazards as suggested previously by Ratnam et al. (2011), Tix and Charvat (2005), Ripley et al. (2015). Other recovered grass phytoliths are BILOBATE SCOOPED, RONDEL, TOWER, BULLIFORM FLABELLATE, BLOCKY and ELONGATE ENTIRE. Table 1 provides detailed climatic indications of the different grass phytolith morphotypes.

Non-grass phytoliths include POLYGONAL ACHENE of Cyperaceae, SPHEROID ECHINATE of Arecaceae, OPAQUE PERFORATED plate of Asteraceae and SPHEROID PSILATE, IRREGULAR FACETATE types recovered from other woody plants (An & Xie, 2022). Burnt remains of SPHEROID ECHINATE type indicate plants of Arecaceae as a fuel during ancient fire (Fig. 3).

Several studies demonstrated that grass subfamily diagnostic phytolith morphotypes act as an excellent proxy in interpreting palaeoclimate, since different grass subfamilies have different ecological preferences (Aleman et al., 2014; Barboni & Bremond, 2009; Biswas et al., 2016). Again, wildfires are directly proportional to biomass availability (Whitlock et al., 2010). Now, C_4 grasses have high productivity rate and can generate large biomass or fuel loads required to support fire events (Ripley et al., 2015). The recovered burnt BILOBATES and SADDLES, thus, act as a clue proving C_4 grass as the fuel load for ancient fire. Therefore, burnt or charcoalified phytoliths and wood fragments point to the fact that ancient fire events occurred during middle Siwalik sedimentation in the Arunachal Himalayas. Moreover, during the middle Siwalik, the temperature of Arunachal Pradesh was comparatively warm and dry as indicated by the recovery of BILOBATES and SADDLES in significant amounts from the sedimentary deposits.

3.2 *Causes and Consequences of Fire in the Middle Siwalik Forest*

Thus, plant microfossil evidences indicate presence of deciduous and semi-evergreen elements with different grass sub-families during the middle Siwalik forests in the Arunachal Himalaya. Khan et al. (2023) reported an increase of deciduous forest elements during the middle Siwalik sedimentation, which continued till upper Siwalik time shifting the wet evergreen lower Siwalik Forest towards a deciduous one. Notably, at present a similar forest with both wet evergreen and deciduous elements is found in the area. However, this change in vegetation can be attributed to climate change throughout the Siwalik sedimentation, indicating an increase in rainfall seasonality (Khan et al., 2023).

The previous studies have shown that the Indian Siwalik houses a significant number of plant (leaf, wood, spore-pollen, etc.) and animal fossils (vertebrates

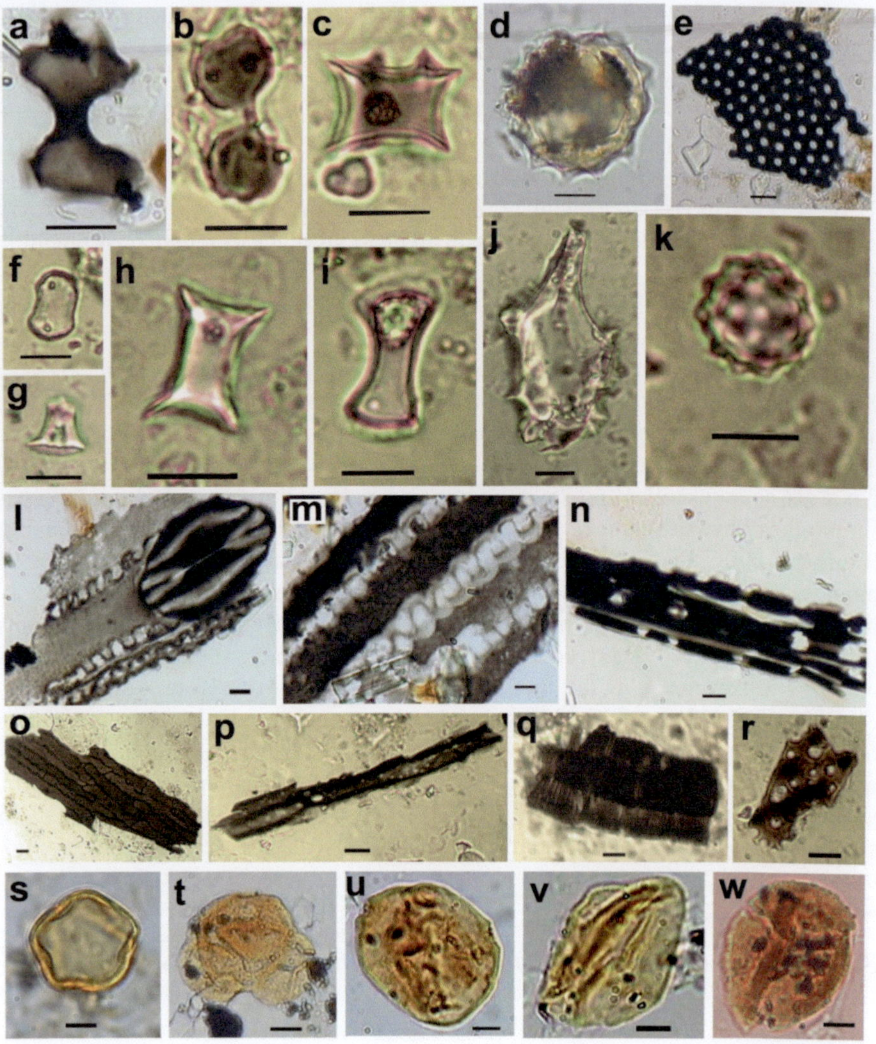

Fig. 3 Light photomicrographs of recovered pollens, phytoliths and other plant remains, **a**, **b** burnt BILOBATES, **c** burnt RONDEL, **d** burnt SPHEROID ECHINATE, **e** OPAQUE PERFORATED, **f** SADDLE, **g** RONDEL, **h** SADDLE COLLAPSED, **i** SADDLE TALL, **j** BULLIFORM FLABELLATE, **k** SPHEROID ECHINATE, **l**, **n** burnt grass cuticle, **o**–**r** burnt wood fragments, **s** Terminalia sp., **t** Pinus sp., **u** Betula sp., **v** Fagaceae, **w** Dipterocarpaceae (Scale: 10 μm)

including mammals and invertebrates such as ostracods, molluscs, etc.) (Gupta, 1976; Munthe et al., 1983; West et al., 1978). However, there are a few studies attempted so far focusing on both macro and micro plant remains, such as spore/pollen, wood and leaves from the Indian part of the Siwalik strata (Banerjee, 1968; Mathur, 1984; Prasad, 1993). According to Banerjee (1968), the subtropical coastal

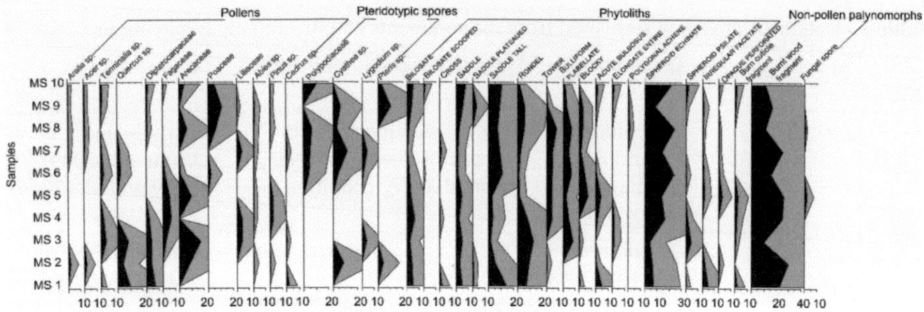

Fig. 4 Frequency distribution of the recovered polymorphs and non-pollen polymorphs from the samples under study

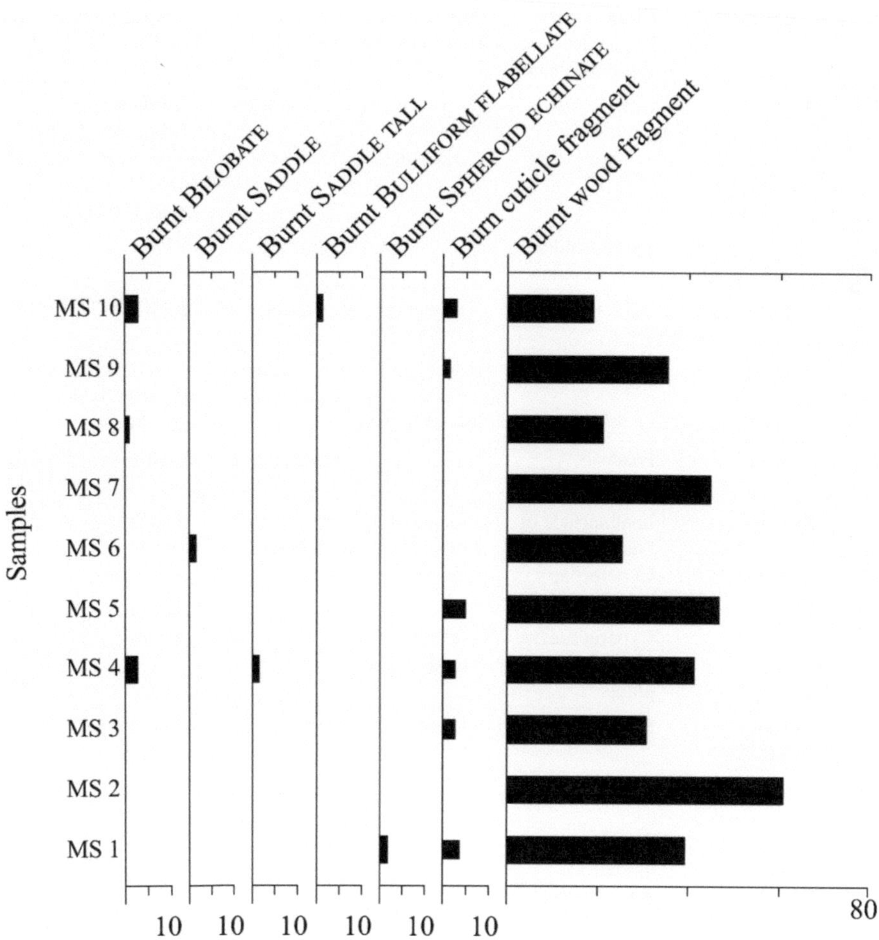

Fig. 5 Graphical representation of the isolated burnt phytoliths, grass cuticle and wood fragments

Table 1 Details of phytolith recovered from the sediments and its climatic implications

Sl. No.	Morphotypes	Taxonomical affinity	Environmental implication	References
1	BILOBATE	Panicoideae, Chloridoideae, Bambusoideae	Associated mostly with C_4 grasses growing in warm, humid or dry condition	Barboni and Bremond (2009), Biswas et al. (2016)
2	BILOBATE SCOOPED	Oryzoideae	Often associated with rice family suggests moist or waterlogged conditions or presence of water availability for plants like *Oryza*	Bourel and Novello (2019)
3	CROSS	Panicoideae, Chloridoideae, Oryzoideae, Bambusoideae	Associated mostly with C_4 grasses growing in warm, humid or dry condition	Twiss et al. (1969), Biswas et al. (2016)
4	SADDLE	Chloridoideae	Associated with C_4 grasses growing in dry to arid condition	Barboni and Bremond (2009), Neumann et al. (2017), Karmakar et al. (2024)
5	SADDLE PLATEAUED	Bambusoideae, Arundinoideae	Associated with grasses growing in warm condition	Biswas et al. (2016)
6	SADDLE TALL	Bambusoideae	Associated with grasses growing in warm condition	Biswas et al. (2024)
7	RONDEL	Pooideae	Associated with C_3 grasses growing in cool and moist environments	Twiss et al. (1969), Barboni et al. (2007)
8	TOWER	Poaceae	Indicative of grassland vegetation	Biswas et al. (2016)
9	BULLIFORM FLABELLATE	Leaf surface of Poaceae and Cyperaceae	Indicative of high evapotranspiration due to high temperature	Bremond et al. (2005), Biswas et al. (2016)
10	BLOCKY	Poaceae, Cyperaceae, Woody dicot plants	Linked with warm evergreen-deciduous broad-leaved forest	Barboni et al. (2007)
11	ACUTE BULBOSUS	Poaceae	Indicative of grassland vegetation	Strömberg (2004), Barboni et al. (2007), Biswas et al. (2024)
12	ELONGATE ENTIRE	Poaceae epidermis, sub-epidermal tissue and inflorescence	Indicative of grassland vegetation	Biswas et al. (2016)

(continued)

Table 1 (continued)

Sl. No.	Morphotypes	Taxonomical affinity	Environmental implication	References
13	POLYGONAL ACHENE	Cyperaceae	Associated with sedges; suggests wetland or riparian habitats	Piperno (2006), Barboni et al. (2007)
14	SPHEROID ECHINATE	Arecaceae	Associated with palms; indicative of warm, humid/ tropical environments	Piperno (2006), Barboni et al. (2007)
15	SPHEROID PSILTAE	Tropical woody plants	Associated with warm board leaf woody trees growing in tropical to sub-tropical forest	Piperno (2006), Barboni et al. (2007)
16	IRREGULAR FACETATE	Tropical woody plants	Associated with warm board leaf woody trees growing in tropical to sub-tropical forest	Piperno (2006), Barboni et al. (2007)
17	OPAQUE PERFORATED	Asteraceae	Cosmopolitan family, the phytolith recovered indicate mesophytic condition, but some members might show xerophytic adaptation	Wang et al. (2022)

depositional environment of the lower Siwalik got transformed into a temperate and more inland one during the middle Siwalik in Punjab, i.e., the western part of the Siwalik strata. An increase of grassland from lower to middle Siwalik in the north and north-western region was suggested by Mathur (1984) based on palynological analyses. Furthermore, Awasthi (1992) observed that the floristic composition of the overall Siwalik Forest changed from tropical evergreen in the lower Siwalik to mixed moist deciduous type during middle Siwalik. Thus, it is quite apparent that there was a significant shift in vegetation pattern during the Siwalik sedimentation and expansion of grassland and deciduous elements are also evident throughout the Indian Siwalik sectors through time from west to east, as we also have observed in its easternmost part, i.e., Arunachal Siwalik sector. A similar trend of expansion of grassland is also reported by Hoorn et al. (2000) in central Nepal Himalaya during the lower part of the middle Siwalik (~8 to ~6.5 Ma), which further got established as dominant vegetation during the middle-upper Siwalik (~6.5 to <2 Ma) in the area. Moreover, on the basis of carbon isotopic study, Awasthi and Prasad (1990) have proved that the grasslands were actually of C_4 type, a similar observation based on the recovered phytolith morphotypes in this work.

Consequently, it can be concluded that the dried grasses and fallen leaves of deciduous trees might have served as fuel for the wildfire incidence. In Southeast Asia, including India, the late Neogene period is considered as a crucial time for speciation and evolution of flowering plants (Khan et al., 2023). In the middle Siwalik (Pliocene) of the Arunachal Himalaya, forest fire might have helped shape the forest ecosystem.

4 Limitations

The Siwalik sedimentary strata were formed during the Middle Miocene to Lower Pleistocene, when alluvial detritus from the rising Himalayas was accumulated into a long, narrow foredeep.

The Siwalik sediments were then folded, faulted and upheaved during the last stage of the Himalaya's rise, or the Pleistocene-recent, before a continuous mountain range of relatively low elevation (1000–1200 m asl) was formed. Thus, being tectonically active, there is a chance of scarcity and deformation/corrosion of the plant remains, especially the pollen grains (Banerjee, 1968; Dutta & Singh, 1980; Hoorn, 2000). As a result, proper identification of the palynomorphs, which is the prerequisite in such a study, becomes a bit challenging.

5 Conclusion and Future Scope

Palaeofire evidences help in understanding and interpreting the palaeoecological impact of fire on the evolving plant communities throughout Earth's long evolutionary history. It can be said that the expansion of ancient forests and events of ancient forest fires are directly related. Microscopic plant remains, or commonly microfossils like pollen grains and microcharcoals, act as excellent proxies for ancient fire activity. Charcoal is made up of carbon and is formed during the incomplete burning of plant materials (MacDonald et al., 1991). Because of its chemical nature, charcoal remains preserved in sediments and acts as effective direct evidence of palaeo-fire in reconstruction studies (Gosling et al., 2019; MacDonald et al., 1991; Unkelbach et al., 2025). Pollen grains, on the other hand, act as the direct proof for standing vegetation during sedimentation and can also infer palaeo-climate (Chauhan et al., 2015; Bali et al., 2017; Quamar & Bera, 2017). Palynological records along with charcoal from sediments indicate which type of vegetation was destroyed by past fire events (Gosling et al., 2019; Swain, 1978).

The following can be concluded from the present work.

1. The recovery of burnt plant remains indicate occurrence of a regional wildfire in the Arunachal Himalaya during the middle Siwalik (Pliocene).
2. The palynological evidence also points towards the prevalence of a deciduous forest, and the leaf litter and dried grasses might have fuelled the forest fire.

Wildfire signatures based on a multiproxy approach from all three Siwalik strata, i.e., lower, middle and upper, need to be studied from Arunachal Himalaya to find out the history of forest fire in the entire Siwalik sedimentation and impact of forest fire events in shaping the past vegetation through time in this part of eastern Himalaya.

Acknowledgements The authors. acknowledge the UGC-CAS VII, Department of Botany, University of Calcutta, DST-FIST, and DBT-BUILDER for the necessary facilities.

References

Aleman, J. C., Canal-Subitani, S., Favier, C., & Bremond, L. (2014). Influence of the local environment on lacustrine sedimentary phytolith records. *Palaeogeography, Palaeoclimatology, Palaeoecology, 414*, 273–283.

An, X., & Xie, B. (2022). Phytoliths from woody plants: A review. *Diversity, 14*(5), 339. https://doi.org/10.3390/d14050339

Awasthi, N. (1992). Changing patterns of vegetation through Siwalik succession. *Palaeobotanist, 40*, 312–327.

Awasthi, N. & Prasad, M. (1990). Siwalik plant fossils from Surai Khola area, western Nepal. In Jain KP & Tiwari RS (eds)-Proc' Symp. Vistas in Indian palaeobotany. *Palaeobotanist, 3*, 298–318.

Bali, R., Chauhan, M. S., Mishra, A. K., Ali, S. N., Tomar, A., Khan, I., et al. (2017). Vegetation and climate change in the temperate-subalpine belt of Himachal Pradesh since 6300 cal. yrs. BP, inferred from pollen evidence of Triloknath palaeolake. *Quaternary International, 444*, 11–23.

Banerjee, D. (1968). Siwalik microflora from Punjab, India. *Review of Palaeobotany and Palynology, 6*, 171–176.

Barboni, D., Bremond, L., & Bonnefille, R. (2007). Comparative study of modern phytolith assemblages from inter-tropical Africa. *Palaeogeography, Palaeoclimatology, Palaeoecology, 246*(2–4), 454–471. https://doi.org/10.1016/j.palaeo.2006.10.012

Barboni, D., & Bremond, L. (2009). Phytoliths of East African grasses: An assessment of their environmental and taxonomic significance based on floristic data. *Review of Palaeobotany and Palynology, 158*(1–2), 29–41.

Batten, D. J. (1999). Small palynomorphs. In T. P. Jones & N. P. Rowe (Eds.), *Fossil plants and spores: Modern techniques.*

Bera, S., Gupta, S., Khan, M. A., De, A., & Mukhopadhyay, R. (2014). First megafossil evidence of Cyatheaceous tree fern from the Indian Cenozoic. *Journal of Earth System Science, 123*(6), 1433–1438.

Bhattacharya, S., Summons, R. E., Murthy, S., Foster, C., Husain, F., & Ankit, Y. (2025). Impact of wildfires on Gondwanan flora during the Permian-Triassic transition. *Geochemical Perspective Letters, 36*, 35–41. https://doi.org/10.7185/geochemlet.2532

Biswas, O., Ghosh, R., Paruya, D. K., Thapa, K. K., & Bera, S. (2016). Can grass phytoliths and indices be relied on during vegetation and climate interpretations in the eastern Himalayas? Studies from Darjeeling and Arunachal Pradesh, India. *Quaternary Science Reviews, 134*, 114–132. https://doi.org/10.1016/j.quascirev.2016.01.003

Biswas, O., Naskar, M., Saikia, K., Datta, B. K., Chowdhury, A. K., Paruya, D. K., Karmakar, M., Biswas, R., Ghosh, R., & Bera, S. (2024). Phytoliths in bamboos from eastern and north-eastern India: Implications in distinguishing different ecoclimatic conditions and in deciphering Late Holocene climate variability. *The Holocene [Sage Journal], 34*(11). https://doi.org/10.1177/095968362412663

Biswas, R., Karmakar, M., Naskar, M., Biswas, O., Kabeer, K. A. A., & Bera, S. (2025). Phytolith spectra of the monotypic grass *Danthonidium gammiei* (Bhide) Hubb. in Hook.(Danthonioideae) from Northern Western Ghats, India: Taxonomic and environmental significance of its distinctive Bilobates. *Review of Palaeobotany and Palynology, 332*, 105235.

Bourel, B.A. & Novello, A. (2019). Bilobate phytolith size matters for taxonomical and ecological identification of Chad grasses: A case study on 15 species. *Review of Palaeobotany and Palynology, 275*, 104–114.

Brand, U., Davis, A. M., Shaver, K. K., Blamey, N. J., Heizler, M., & Lécuyer, C. (2021). Atmospheric oxygen of the Paleozoic. *Earth-Science Reviews, 216*, Article 103560.

Bremond, L., Alexandre, A., Peyron, O., & Guiot, J. (2005). Grass water stress estimated from phytoliths in West Africa. *Journal of Biogeography, 32*(2), 311–327.

Chakraborty, T., Taral, S., More, S., & Bera, S. (2020). Cenozoic Himalayan foreland basin: An overview and regional perspective of the evolving sedimentary succession. In N. Gupta & S. Tandon (Eds.), *Geodynamics of the Indian Plate*. Springer Geology. Springer.

Chauhan, M. S., Pokharia, A. K., & Srivastava, R. K. (2015). Late Quaternary vegetation history, climatic variability and human activity in the Central Ganga Plain, deduced by pollen proxy records from Karela Jheel, India. *Quaternary International, 371*, 144–156.

Chirouze, F., Dupont-Nivet, G., Huyghe, P., van der Beek, P., Chakraborty, T., Bernet, M., & Erens, V. (2012). Magnetostratigraphy of the Neogene Siwalik Group in the far eastern Himalaya: Kameng section, Arunachal Pradesh, India. *Journal of Asian Earth Sciences, 44*, 117–135.

Cordova, C. E., Johnson, W. C., Mandel, R. D., & Palmer, M. W. (2011). Late quaternary environmental change inferred from phytoliths and other soil-related proxies: Case studies from the central and southern Great Plains, USA. *Catena, 85*(2), 87–108.

Diessel, C. F. K. (2010). The stratigraphic distribution of inertinite. *International Journal of Coal Geology, 81*(4), 251–268.

Dutta, S.K. (1980). Palynostraigraphy of the sedimentary formations of the Arunachal Pradesh, 2. Palynology of the Siwalik equivalent rocks of Kameng district. *Geophytology, 10*(1), 5–13.

Dutta, S. K., & Singh, H. P. (1980). Palynostratigraphy of sedimentary formations in Arunachal Pradesh. Palynology of Siwalik Rocks of the lesser Himalayas, Kameng Dirtrict, Arunachal Pradesh. In *Proceedings of the IV International Palynological Conference Lucknow* (1976–77) (Vol. 2, pp. 617–626).

Elbaum, R., Weiner, S., Albert, R. M., & Elbaum, M. (2003). Detection of burning of plant materials in the archaeological record by changes in the refractive indices of siliceous phytoliths. *Journal of Archaeological Science, 30*, 217–226.

Esteban, I., Stratford, D., Sievers, C., de La Peña, P., Mauran, G., Backwell, L., d'Errico, F., & Wadley, L. (2023). Plants, people and fire: Phytolith and FTIR analyses of the post-Howiesons Poort occupations at Border Cave (KwaZulu-Natal, South Africa). *Quaternary Science Reviews, 300*, Article 107898.

Falcon-Lang, H. J. (1998). The impact of wildfire on an Early Carboniferous coastal system, North Mayo, Ireland. *Palaeogeography, Palaeoclimatology, Palaeoecology, 139*, 121–138.

Glasspool, I. J., Edwards, D., & Axe, L. (2004). Charcoal in the Silurian as evidence for the earliest wildfire. *Geology, 32*, 381–383.

Glasspool, I. J., & Scott, A. C. (2010). Phanerozoic concentrations of atmospheric oxygen reconstructed from sedimentary charcoal. *Nature Geoscience, 3*, 627–630.

Gosling, W. D., Cornelissen, H. L., & McMichael, C. N. H. (2019). Reconstructing past fire temperatures from ancient charcoal material. *Palaeogeography, Palaeoclimatology, Palaeoecology, 520*, 128–137.

Gupta, V. J. (1976). *Indian cenozoic stratigraphy*. Hindustan Publishing Corporation.

Hoorn, C., Ohja, T., & Quade, J. (2000). Palynological evidence for vegetation development and climate change in the sub-Himalayan zone (Neogene, central Nepal). *Palaeogeography, Palaeoclimatology, Palaeoecology, 163*, 133–161.

Jain, K. P., & Dutta, S. K. (1978). Lower Tertiary dinoflagellates, spores, pollen grains from Siang district, Arunachal Pradesh. *Journal of the Palaeontological Society of India, 21–22*, 106–111.

Jones, L. H. P., & Milne, A. A. (1963). Studies of silica in the oat plant. *Plant and Soil, 18*(2), 207–220.

Joshi, A., & Chakraborty, P. P. (2001). Systematic geological mapping in parts of East and West Kameng districts, Arunachal Pradesh. Geological Survey of India Unpublished progress report for F.S., 1999–2000.

Joshi, A., Tewari, R., Mehrotra, R. C., Chakraborty, P. P., & De, A. (2003). Plant remains from the Upper Siwalik sediments of West Kameng District, Arunachal Pradesh. *Journal of the Geological Society of India, 61*, 319–324.

Karmakar, M., Biswas, R., Mukherjee, B., Biswas, O., Naskar, M., Paruya, D. K., Ghosh, R., & Bera, S. (2024). Chloridoideae grass phytoliths can distinguish two different rainfall zones: A case study from northern parts of Western Ghats, India. *Quaternary International, 681*, 11–23.

Karunakaran, C., & Ranga Rao, A. (1976). Status of exploration for hydrocarbons in the Himalayan region-contributions to stratigraphy and structure. Himalayan Geology Seminar, New Delhi. Geological Survey of India. *Miscellaneous Publication, 41*(5) Sec III, 1–66.

Kealhofer, L., & Penny, D. (1998). A combined pollen and phytolith record for fourteen thousand years of vegetation change in northeastern Thailand. *Review of Palaeobotany and Palynology, 103*(1–2), 83–93.

Khan, M. A., Gupta, S., Paruya, D. K., De, A., De, B., & Bera, S. (2006). Palynoassemblage from the upper Siwalik sediments of Papumpare district, Arunachal Pradesh with remarks on palaeoenvironment. *Journal of the Botanical Society of Bengal, 60*(1), 44–49.

Khan, M., De, B. & Bera, S. (2009). Leaf-impressions of Calophyllum L. from the Middle Siwalik sediments of Arunachal Sub-Himalaya, India. *Pleione. 3,* 101–106.

Khan, M. A., Spicer, R. A., Bera, S., Ghosh, R., Yang, J., & Spicer, T. E. V. (2014a). Miocene to Pleistocene floras and climate of the Eastern Himalayan Siwaliks, and new palaeoelevation estimates for the Namling-Oiyug Basin, Tibet. *Global Planetary Change, 113*, 1–10.

Khan, M. A., Spicer, T. E. V., Spicer, R. A., & Bera, S. (2014b). Occurrence of *Gynocardia odorata* Robert Brown (Achariaceae, formerly Flacourtiaceae) from the Plio-Pleistocene sediments of Arunachal Pradesh, northeast India and its palaeoclimatic and phytogeographic significance. *Review of Palaeobotany and Palynology, 211*, 1–9.

Khan, M. A., Bera, S., Ghosh, R., Spicer, R. A., & Spicer, T. E. V. (2015). Leaf cuticular morphology of some angiosperm taxa from the Siwalik sediments (middle Miocene to lower Pleistocene) of Arunachal Pradesh, eastern Himalaya: Systematic and palaeoclimatic implications. *Review of Palaeobotany and Palynology, 214*, 9–26.

Khan, M. A., Mahato, S., Spicer, R. A., Spicer, T. E. V., Ali, A., Hazra, T., & Bera, S. (2023). Siwalik plant megafossil diversity in the Eastern Himalayas: A Review. *Plant Diversity, 43*(3), 243–264.

Kumar, G. (1997). *Geology of the Arunachal Pradesh.* Geological Society of India.

MacDonald, G. M., Larsen, C. P., Szeicz, J. M., & Moser, K. A. (1991). The reconstruction of boreal forest fire history from lake sediments: A comparison of charcoal, pollen, sedimentological, and geochemical indices. *Quaternary Science Reviews, 10*(1), 53–71.

Mathur, Y. K. (1984). Cenozoic palynofossils vegetation ecology and climate of the north and northwestern sub-Himalayan region, India. In R. O. White (Ed.), *The evolution of the East Asian environment*(Vol. 2, pp. 504–549). Centre of Asian Studies.

Meng, M., Jie, D., Gao, G., Gao, T., Xu, S., Lian, Y., et al. (2022). Characteristics of burned phytolith from representative plants in Northeast China and implications for paleo-fire reconstruction. *Review of Palaeobotany and Palynology, 300*, Article 104628.

Morgan, P., Hardy, C. C., Swetnam, T. W., Rollins, M. G., & Long, D. G. (2001). Mapping fire regimes across time and space: Understanding coarse and fine-scale fire patterns. *International Journal of Wildland Fire, 10*, 329–342.

Munthe, J., Dongol, B., Hutchinson, J. H., Kean, W. F., Munthe, K., & West, R. M. (1983). New fossil discoveries from the Miocene of Nepal include a hominoid. *Nature, 303*, 331–333.

Neumann, K., Fahmy, A. G., Müller-Scheeßel, N., & Schmidt, M. (2017). Taxonomic, ecological and palaeoecological significance of leaf phytoliths in West African grasses. *Quaternary International, 434*, 15–32.

Nichols, G. J., & Jones, T. P. (1992). Fusain in Carboniferous shallow marine sediments, Donegal, Ireland: The sedimentological effects of wildfire. *Sedimentology, 39*, 487–502.

Pardo, J., & Vanburen, R. (2021). Evolutionary innovations driving abiotic stress tolerance in C_4 grasses and cereals. *The Plant Cell, 33*(11), 3391–3401. https://doi.org/10.1093/plcell/koab205

Parr, J. F. (2006). Effect of fire on phytolith coloration. *Geoarchaeology: An International Journal, 21*(2), 171–185.

Pearsall, D. (2000). Palaeoethnobotany–A handbook of procedures, 2nd (edn). *Academic Press, New York*, pp. 695.

Piperno, D. R. (2006). Phytoliths: A comprehensive guide for archaeologists and paleoecologists. Rowman Altamira.

Prasad, M. (1993). Siwalik (Middle Miocene) woods from Kalagarh area in the Himalayan foothills and their bearing on palaeoclimate and phytogeography. *Review of Palaeobotany and Palynology, 76*, 49–82.

Quamar, M. F., & Bera, S. K. (2017). Pollen records related to vegetation and climate change from northern Chhattisgarh, central India during the late Quaternary. *Palynology, 41*(1), 17–30.

Ratnam, J., Bond, W. J., Fensham, R. J., Hoffmann, W. A., Archibald, S., Lehmann, C. E., Anderson, M. T., Higgins, S. I., & Sankaran, M. (2011). When is a 'forest' a savanna, and why does it matter? *Global Ecology and Biogeography, 20*(5), 653–660.

Rein, G. (2013). Smouldering fires and natural fuels. In M. Belcher Claire (Ed.), *Fire phenomena in the earth system* (pp. 15–34). Wiley.

Rimmer, S. M., Thompson, J. A., Goodnight, S. A. & Robl, T. L. (2004) Multiple controls on the preservation of organic matter in Devonian-Mississippian marine black shales: Geochemical and petrographic evidence. *Palaeogeography, Palaeoclimatology, Palaeoecology, 215*(1–2), 125–154. http://doi.org/10.1016/S0031-0182(04)00466-3

Rimmer, S. M., Hawkins, S. J., Scott, A. C., & Cressler, W. L. (2015). The rise of fire: Fossil charcoal in late Devonian marine shales as an indicator of expanding terrestrial ecosystems, fire, and atmospheric change. *American Journal of Science, 315*(8), 713–733.

Ripley, B., Visser, V., Christin, P. A., Archibald, S., Martin, T., & Osborne, C. (2015). Fire ecology of C_3 and C_4 grasses depends on evolutionary history and frequency of burning but not photosynthetic type. *Ecology, 96*(10), 2679–2691.

Scott, A. C., & Glasspool, I. J. (2006). The diversification of Palaeozoic fire systems and fluctuations in atmospheric oxygen concentration. *Proceedings of the National Academy of Sciences, 103*(29), 10861–10865.

Scott, A. C. (2010). Charcoal recognition, taphonomy and uses in palaeoenvironmental analysis. *Palaeogeography, Palaeoclimatology, Palaeoecology, 291*, 11–39.

Scott, A. C., Bowman, D. M. J. S., Bond, W. J., Pyne, S. J., & Alexander, M. E. (2014). *Fire on earth: An introduction* (1st ed., pp. 1–120). Wiley Blackwell.

Singh, T. & Tripathi, S.K.M. (1990) Siwalik sediments of Arunachal Himalaya: Palynology, Palaeoecology and Palaeogeography. *Palaeobotanist, 38*, 325–332.

Strömberg, C. A. (2004). Using phytolith assemblages to reconstruct the origin and spread of grass-dominated habitats in the great plains of North America during the late Eocene to early Miocene. *Palaeogeography, Palaeoclimatology, Palaeoecology, 207*(3–4), 239–275.

Swain, A. M. (1978). Environmental changes during the past 2000 years in north-central Wisconsin: Analysis of pollen, charcoal, and seeds from varved lake sediments. *Quaternary Research, 10*(1), 55–68.

Tix, D., & Charvat, I. (2005). Aboveground biomass removal by burning and raking increases diversity in a reconstructed prairie. *Restoration Ecology, 13*(1), 20–28.

Twiss, P. C., Suess, E., & Smith, R. M. (1969). Morphological classification of grass phytoliths. *Soil Science Society of America Journal, 33*(1), 109–115.

Unkelbach, J., Behling, H., & Schmidt, C. (2025). Reconstruction of vegetation, climate, fire and human activity at an Iron Age oasis site in Lizq, northern Central Oman, beginning in 400 cal yr BCE. *Palaeogeography, Palaeoclimatology, Palaeoecology, 113091.*

Wang, M., Yang, Q., Yang, W., Shi, L., Zhang, Y., Zhou, Z., Zhang, W., & Zheng, H. (2022). Surface phytolith and pollen assemblages of a low-latitude subtropical region in Southwest China and their implications for vegetation and climate. *Frontiers in Plant Science, 13*, 1007612. https://doi.org/10.3389/fpls.2022.1007612

West, R. M., Lukas, J. R., & Munthe, J. (1978). Vertebrate fauna from Neogene Siwalik Group, Dang Valley, western Nepal. *Journal of Palaeontology, 52*, 1015–1022.

Whitlock, C., Higuera, P. E., McWethy, D. B. & Briles, C. E. (2010). Paleoecological perspectives on fire ecology: Revisiting the fire-regime concept. *The Open Ecology Journal, 3*, 6–23.

Heatwave Impact on Agricultural Practices and Crops Diversity

Madhu Rani, Sweeti Dangi, Stuti Gupta, and Sonia Kapoor

Abstract Climate change is a major concern in recent decades due to lot of industrial set up, loss of forest area and use of fossil fuels like coal and petrol. All these reasons are causes of Green House Gas (GHG) emission in the environment and rise in temperature. According to International Panel on Climate Change (IPCC) "Rise in CO_2 will be double by 2025 and average rise in temperature is 4.4 °C by 2100". Due to global warming, there is a trap of radiation in the atmosphere due to which earth temperature is rising by 1.36 °C from previous year 2023. One of the major effects of climate change is on the crop productivity and different agriculture practices like mixed cropping, intercropping and crop rotation. Crops are sown in different season according to their suitable climate and irrigation facility. But climate change and heatwave both effect the yield as well as crop production. According to the report of Indian Council of Agriculture Research (ICAR), average maximum temperature of more than 30 °C during the maturation and grain filling stage of various crop like wheat adversely affect the growth and yield of these grains. Also, every 1 °C rises in temperature above this, there is a 3–4% drop in yield. From these factors heat wave is a major cause of decrease in crop production and growth of plants. There is an immense need to combat rise in temperature and climate change pattern for enhanced crop diversity and production.

M. Rani
Assistant Professor, Department of Botany, Government College for Girls, Sector-14, Gurugram 122007, India
e-mail: madhugill.11@gmail.com

S. Dangi
Research Scholar, Department of Biotechnology, UIET, Maharshi Dayanand University, Rohtak 124001, India

S. Gupta
M.Tech, Department of Biotechnology, UIET Maharshi Dayanand University, Rohtak 124001, India

S. Kapoor (✉)
Professor, Department of Biotechnology, UIET Maharshi Dayanand University, Rohtak 124001, India
e-mail: soniakapoor.uiet@mdurohtak.ac.in

A. Awasthi et al. (eds.), *Rising Temperatures and Plant Ecosystems*,
https://doi.org/10.1007/978-3-032-29477-7_12

Keywords Temperature · Climate change · Crop diversity · IPCC · Global warming

1 Introduction

Heatwaves, which are characterized as extended periods of excessive hot weather, have become one of the deadliest climate-induced threats to agriculture worldwide (Yadav et al., 2023). Heatwaves are occurring with rising frequency and ferocity, says the intergovernmental panel on climate change (IPCC), which warns that this trend will continue to gain pace with increasing global temperatures. (Lozano-García et al., 2017; Awasthi et al., 2022; Lee & Romero, 2023). Agriculture, which has climate as the driving factor in its production system, is being tested by these never-seen-before weather phenomenon and higher global temperatures. The impact of climate change on agricultural productivity is multi-dimensional including direct effects of temperature, changes in rainfall patterns, soil moisture regimes and related shifts in pests and disease dynamics (Lozano-García et al., 2017). These interlinked factors cause complex flowing impacts through the agroecosystems, which not only impact crop yield but also the fundamental constitution of farming communities and global food security networks. (Impact of heat waves on agriculture: affected crops and how to protect them from extreme hot weather,). Modern agriculture is confronted with a more fluctuating climatic background defined by extreme weather events. Abiotic stress factors such as heatwaves are particularly dangerous because they can quickly surpass plant physiological tolerance limits leading to irreversible damage of the crop's development and yield. The economic consequences are large—heatwaves in key farming areas of recent years have led to billions of dollars crop damage and added to global food price swings. So, heat wave has a huge impact on crop output and efficiency. Hence various methods are discussed in this chapter. These includes historical context, impact of heat wave on agriculture systems, different type of cropping systems, various mechanisms to reduce the impact of heat wave on farming systems along with case study on farming and biotechnology solutions.

1.1 Historical Context of Heatwaves in India

Over the past several decades, India has seen a number of heatwaves that have become more frequent (Singh et al., 2021) and also its effect to increase global warming (Fig. 1). India's agricultural landscape provides a compelling case study for understanding heatwave impacts on diverse cropping systems (Table 1). The ever-increasing pattern of heatwave severity and economic impact highlights the urgent need for comprehensive adaptation strategies that can protect agricultural systems while maintaining productivity and profitability (Campbell et al., 2014).

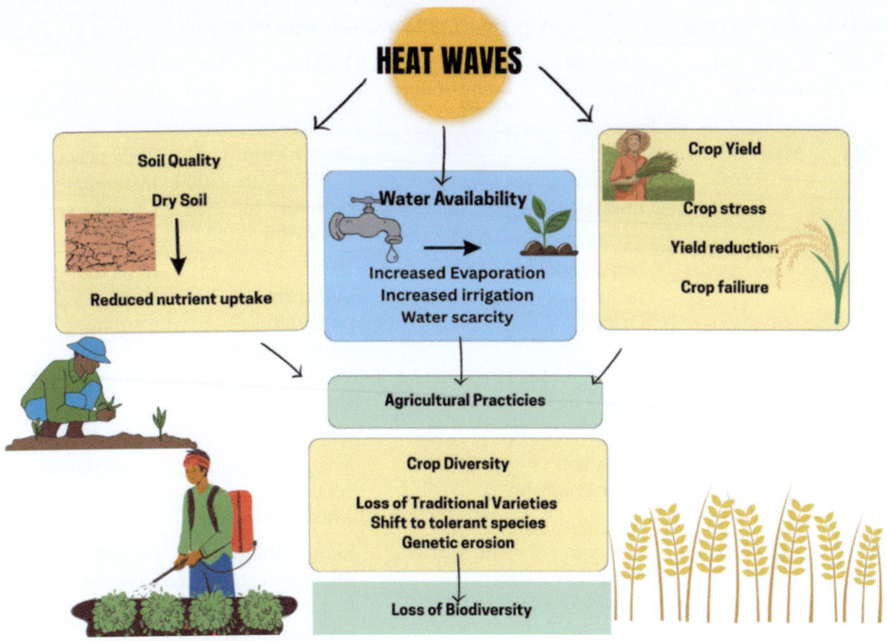

Fig. 1 Effect of heatwave on agriculture

2 Effect of Heat Wave on Different Types of Agricultural Systems

Knowledge of the physiological and biochemical responses to heat, including damage, mechanisms of tolerance and adaptation is essential in addressing such a situation. High temperature provokes a series of events at the cellular level that may induce both acclimation or damage to plant systems.

2.1 Cellular Heat Sensing and Signal Transduction

Thermosensing and thermoregulation in flowering plants (referred to as simply, 'thermoregulation') is important as these organisms are sessile: Movement at the macro-level has been traded for subtler physiological responses. The first heat detection is mediated by temperature-sensitive proteins, which also change their conformation upon evaluation at the sensed temperature. These molecular sensors also transduce signals through the downstream cascades to orchestrate cellular responses of thermal stress (Miller et al., 2009; Ohama et al., 2017; Lohani et al., 2022).

Table 1 Major heatwaves events in India from (1998 to 2024) with its impact (Singh et al., 2021)

Year	Dates and duration	Affected regions	Peak temperature (°C)	Impact	Sources
1998	May	Odisha, Andhra Pradesh	~47 °C	Severe crop failure, with ~2000 mortality reported in Odisha alone	Toda peace institute, ClimaMeter
2002	June	North India, including Delhi	~45 °C	Increased risk to vulnerable populations, reduced wheat yields, irrigation stress, agricultural impact	India today, Toda peace institute
2003	May	Rajasthan, Uttar Pradesh	~48 °C	Major drought conditions worsened by heat	ClimaMeter
2010	April–may	Northern India	46 °C	Significant loss of life, urban stress in major cities, early harvest, grain quality degradation	ClimaMeter, Toda peace institute
2015	May	Andhra Pradesh, Telangana	47–48 °C	Over 2500 deaths across affected states, catastrophic crop losses	India today
2016	May	Delhi, Rajasthan	51 °C (Phalodi, Rajasthan)	Record-breaking heat in Phalodi, Rajasthan, infrastructure strain	ClimaMeter, Toda peace institute
2019	May–June	Bihar, Rajasthan	~48 °C	Heat-related illnesses rose, public advisories issued, delayed sowing, water scarcity	ClimaMeter
2022	March–April	Northwest and Central India	49 °C (Delhi)	Severe heat early in the year, early heat stress, premature maturation, agricultural losses reported	India today

(continued)

Table 1 (continued)

Year	Dates and duration	Affected regions	Peak temperature (°C)	Impact	Sources
2023	April	Multiple states, particularly northwest	~48 °C	Government advisories, increased healthcare burden, supply chain disruption, quality losses	Toda peace institute
2024	May	New Delhi, Rajasthan	49.1 °C (Delhi)	Widespread warnings, heat-related health impact in urban areas, urban agriculture stress, cooling cost	ClimaMeter, Toda peace institute, India today

2.1.1 Integrated Heat Detection Network

Heat stress is perceived at cellular level by a variety of interacting signaling pathways. Conformational changes in proteins induce heat shock responses, and phytochrome B systems are temperature-sensitive for transcriptional control. Concomitantly calcium signaling pathways and ROS generation lead to the generation of secondary messaging cascades that enhance and orchestrate the Stress response (Ohama et al., 2017; Lohani et al., 2022).

2.1.2 Cross-Pathway Integration

The heat shock response in the cell is incredibly integrated with other stress pathways. ROS generation is co-regulated by heat with drought and salt and calcium signaling pathways universally coordinate multiple abiotic stress responses concurrently. The convergence enables plants to mount integrated responses to complex environmental stimuli rather than separate ones to single stressors. (Miller et al., 2009; Casal & Balasubramanian, 2019).

2.1.3 Metabolic Adjustments Under Heat Stress

The metabolic changes in response to heat stress is also linked with those related to water deficit and nutrient stress. Osmolytes (proline, glycine betaine and trehalose) accumulated having dual role in defense against heat as well as osmotic stress, and carbohydrate metabolism shifts provide energy for both supply and demand of the protective activity. Modifications in membrane lipids help the cells to remain alive at several stresses illustrating that plant stress responses are connected with each other (Ahmad et al., 2022; Choukri et al., 2020; He et al., 2022). The integrated view stresses that plant heat tolerance is imparted by the combined cellular pathways rather than isolated molecular mechanisms, and it therefore relates response to heat stress within the context of overall approach for plant adaptation.

3 Impact of Heatwaves on Different Agricultural Systems

The sensitivity of agricultural systems to heat stress varies widely between crop species, farming practices and agroclimatic conditions. Crop productivity is highly sensitive to environmental conditions (Mendelsohn, 2009). Some of the most important are temperature and precipitation. Increasing temperature can affect adversely food security, income of farmers and other consequences (Mahato A, 2014). However, higher levels of precipitation are expected to mitigate such effects (Adams et al., 1998). Water is required for 80% of the agricultural productivity, and Heatwaves can influence it (Tubiello et al., 2007). Heatwaves have varying impacts on crops due to location and irrigation type. The growth period of many crops is predicted to be reduced resulting in lower yields (Kang Y et al., 2009, Al-Amin et al., 2014 & Mahato A, 2014). Grain yield is projected to decrease when both tropical and temperate regions warm by 2 °C (Challinor et al., 2014). The literature assumes that temperature variability changes land and water systems, which may have major negative impact on agricultural productivity (Rahman & Anik, 2020). Studies suggest that southern African countries could lose more than 30 percent of their primary cereals by 2030. South Asia is also likely to experience over 10% decline in crops such as maize, rice and millet (Bhattacharya A, 2019) (Changing Climate and Resource Use Efficiency in Plants | ScienceDirect). By 2100, the effects of loss in crop yield could lower global GDP by 0.3% year on year and raise food prices, continuing to have massively detrimental effects on agricultural well-being globally (Stevanović et al., 2016). By 2050, it's anticipated that heatwaves will decrease water availability and increase agricultural water consumption by 19% (Alam et al., 2011). This section investigates how various farming practices respond to thermal stress and the consequences for agri-food sustainability.

3.1 Cropping Rotation Systems

3.1.1 Crop Rotation and Intercropping Integration

Combined cropping strategies demonstrate a superior heat stress resilience through temporal risk distribution and spatial microclimate modification. Rice-Wheat intercropping system in South Asia shows yield advantages of 15–30% under moderate heat stress compared to monocultures, while rotation systems distribute thermal risk across seasons through varying crop heat sensitivities. (Challinor et al., 2014; Moreira et al., 2024). Canopy stratification lowers soil temperature 2–5 C improving water efficiency and reducing pest incidence by up to 40% (Sharma et al., 2021). Yield declines are limited to 5–15% under 1–2 °C warming scenarios, with effective water management integral to resilience (Campbell et al., 2014).

Quantitative Benefits of Crop Rotation and Intercropping Integration are multiple canopy layers reduce soil surface temperatures by 2–5 C. It enhances water use efficiency through diverse root systems accessing different soil layers. Pest and disease pressure reduced by 20–40% through crop diversity. Overall system productivity maintained with 5–15% decline under moderate heat stress (1–2 °C increase) (Campbell et al., 2014; Sharma et al., 2021).

3.1.2 Management Challenges

Shortened growing seasons and resource competition create synchronization difficulties requiring management of differential heat responses between species and complex harvest coordination (Mendelsohn, 2009; Rahman & Anik, 2020).

3.2 Diversified and Integrated Farming Systems

3.2.1 Mixed and Integrated System Performance

Systems combining crops, livestock, aquaculture and forestry demonstrate moderate resilience to 1–2 C increase but face increasing vulnerability beyond 3 C warming (Herrero et al., 2017; Kerr et al., 2022). More than 70% of household income in rural areas of LDCs rely on these systems, and their thermal plasticity is thus central to global food security (Angelsen et al., 2014; Campbell et al., 2014).

As described in the Table 2, the performance of these systems declines 5–15% at moderate heat stress (1–2 °C) and up to 15–35% at extreme stress (>2 °C), with consequences that affect crops, livestock and trees (Herrero et al., 2017). Integration of trees can lead to on-farm cooling with local temperature reductions of up to 5 °C and increased overall system resilience (Murgueitio et al., 2011). Livestock heat sensitivity translates to 3–5% productivity declines per degree above thermal optimum; feed demand increases up to 59% under severe heat (Bett et al., 2019). Aquaculture vulnerabilities include oxygen depletion and increased mortality beyond species thermal thresholds (Murshed-E-Jahan & Pemsl, 2011).

Table 2 Heat stress levels and its impact

Heat stress level	Impact on productivity	Major vulnerabilities
Moderate (1–2 C)	5–15% decline	Crops>livestock>trees
Severe (2–4 C)	15–35% decline	All components
Extreme (> 4 C)	>35% decline	System collapse risk

3.2.2 Resilience Mechanisms

Mixed farming systems are more resilient to heat stress in ways that interact across scales and time. Knowledge about these mechanisms is essential to the optimal design and management of systems under changing climatic conditions.

3.2.2.1 Temporal Risk Distribution and Seasonal Complementarity

Mixed systems are good at diversifying heat stress in time. Although certain heat stress can impinge on crop productivity, at least animals generally have some ability to maintain production in the face of behavioral changes and extra management. Likewise, tree parts within the agroforestry systems can remain sequestering carbon and mitigating microclimate even when annual crops are being stressed. Seasonal complementarity among enterprises creates the prospect for adaptive management. For instance, in times of high heat where crop production decreases, farmers might be able to increase livestock management or concentrate on tree-based products like fruit, nuts, or timber. This temporal flexibility also provides economic risk buffer not available in specialized monoculture systems (Campbell et al., 2014; Rosenzweig, 2014a; Thornton & Herrero, 2015).

3.2.2.2 Spatial Heterogeneity and Microclimate Modification

One of the most significant advantages of mixed farming systems is their ability to create beneficial microclimates through strategic spatial arrangement of different components. Trees in agroforestry lower the ambient temperature by 2–5 °C through evapotranspiration cooling and shading, delivering a heat reprieve for heat-sensitive crops and livestock. The structure of mixed stand—three-dimensional nature. Create so many microclimatic zones with different temperature, humidity and radiation conditions. This spatial complexity can be exploited to locate different industries according to their heat tolerance. For example, shade-tolerant plants under tree canopies and heat sensitive plants in more open areas. (Thornton & Herrero, 2015; Yang et al., 2014).

3.2.2.3 Ecosystem Service Provision and Environmental Buffering

Diversified farm systems produce a broad range of ecosystem services that increase resilience to heat stress. Evapotranspiration cooling, carbon sequestration, and soil organic matter amelioration are some of the benefits provided by tree parts. Another feature of the integration of livestock is that it brings an advantage to nutrient cycling and soil health due to direct deposition of manure on soil. Water bodies in IA (integrated aquaculture) are effective measures to cool their surroundings and for storage of water for irrigation during heat stress conditions. The increased biodiversity in mixed systems fosters stronger soil ecosystems holding more water, and with a higher capacity to buffer heat. Various root systems in various soil depths can improve soil structure and water infiltration, lower surface temperature and keep the soil moisture under heat stress conditions. (Campbell et al., 2014).

3.2.2.4 Economic Diversification and Risk Management

All parts of a mixed farming-system are seldom equally exposed to heat stress, which naturally hedges the economic risk faced by farmers from climate variability. For example, when heat stress decreases crop yields, farmers might foster other productive activities such as livestock production, forest product extraction, or fishery production. This diversification is especially beneficial during hot, extreme heat events that can result in a complete loss of crop in specialized systems. Mixed-systems are integrated and therefore potential exists for value addition and market diversification power. A higher revenue flow from finished products originating in a variety of different firms may elicit a price premium, and can provide some solvency when basic production is compromised due to heat stress (Campbell et al., 2014).

3.2.3 Critical Vulnerabilities

Composite and smallholder farming systems, however, are highly adapted to withstand the shock of temperature stress yet face idiosyncratic vulnerabilities to heat stress that flow through complex linkages with potential for amplifying negative impacts. Animals in mixed systems are very susceptible to heat stress. In aquaculture, oxygen stress and mortality can occur when temperatures exceed species-specific thresholds (Murshed-E-Jahan & Pemsl, 2011; Bett et al., 2019).

4 Heat Stress Effects on Crop Diversity and Seasonal Production

The world agricultural system depends on two dominant growing seasons, which have contrasting heat stress conditions. Extreme heat exposure places stress on plants, inhibits their growth and can lead to plant death, which often leads to a decrease in quality and quantity of agricultural crop yield (Hatfield & Prueger, 2015). Seasonal sensitivities need to be recognized for targeted adaptation options.

4.1 Different Crops Sown

4.1.1 Rabi Crops

Rabi crops, which have evolved over millions of years to mature during the relatively cool seasons as less sensitive well irrigated crop plants are now at increased risk of high temperatures that impede temperature-sensitive developmental processes. Wheat, barley, gram (chick pea), linseed, mustard and peas are the major rabi crops. Rabi crops have longer grown period than Kharif. They give farmers a longer window for harvesting, allowing them to plant more fields after collecting one yield before beginning another. Usually, these crops are planted in the fall and harvested in the spring. Rabi crops thrive in areas with moderate weather since they are planted during the cooler months. Additionally, because the cooler weather might cause dry soil conditions, they are usually irrigated.

4.1.1.1 Major Rabi Crops and Heat Sensitivity

Wheat (*Triticum aestivum*): Heat stress during grain filling reduces yield by 3–4% per degree Celsius increase above optimal temperatures (15–20 °C). Quality parameters including protein content and gluten strength are also affected (Ferris et al., 1998).

Barley (*Hordeum vulgare*): More heat-tolerant than wheat but still susceptible during reproductive stages, with yield losses of 2–3% per degree increase above 22 °C (Hatfield & Prueger, 2015).

Pulses (Lentil, Chickpea, Field Pea): Extremely sensitive to heat during flowering and pod formation, with temperature increases above 25 °C causing significant flower abortion and pod reduction (Choukri et al., 2020).

Oilseeds (Rapeseed, Mustard): Heat stress during flowering reduces seed set and oil content, with quality degradation occurring above 25 °C (Lamaoui et al., 2018).

4.1.1.2 Adaptation Strategies for Rabi Crops

Early Sowing: Advancing planting dates to avoid heat stress during sensitive reproductive phases (Ferris et al., 1998).

Heat-Tolerant Varieties: Development and adoption of cultivars with enhanced thermotolerance. (Singh et al., 2024).

Protective Cultivation: Use of shade nets and mulching to moderate soil and air temperatures. (Campbell et al., 2014).

4.1.2 Kharif Crops

Kharif crops are planted when the monsoon arrives in order to prevent water scarcity. Rice, maize, cotton, sorghum, pearl millet, groundnuts, soybeans, green and black grams, red chili, and vegetables like brinjal and okra are some of the main crops grown during this season.

There are essentially two types of Kharif crops: Cereals and pulses. Kharif crops include pulses like arhar /tur, moong, urad, and other legumes like soybean and peanut, as well as cereals like rice, maize, jowar, bajra, etc. For growth and development, these crops need a lot of heat, humidity, and water. They are therefore more susceptible to climatic shifts like drought or intense precipitation. These crops are harvested in the fall after being sown during the rainy season, which runs from June to September.

4.1.2.1 Major Kharif Crops and Heat Responses

Rice (Oryza sativa): Although heat-adapted, rice experiences significant yield losses when nighttime temperatures exceed 24 °C, affecting grain filling and quality (Shrestha et al., 2022).

Maize (Zea mays): Heat stress during pollination causes severe yield reductions, with temperatures above 35 °C causing pollen sterility (Li et al., 2018).

Cotton (Gossypium spp.): Heat stress affects fiber quality and increases susceptibility to pest attacks, particularly bollworms.

Soybean (Glycine max): Heat stress during pod formation reduces seed number and oil content (Siebers et al., 2015).

4.1.3 Emerging Heat Tolerance Strategies

Breeding Programs:(Singh et al., 2024) Development of heat-tolerant varieties through conventional breeding and biotechnology.

Water Management:(Campbell et al., 2014) Precision irrigation systems to provide cooling and maintain optimal soil moisture.

Crop Calendar Adjustments:(Challinor et al., 2014) Modifying planting and harvesting schedules to avoid peak heat periods.

4.2 Effect of Heat Wave on Different Crops in Term of Growth, Yield and Nutritional Quality

Everywhere in the world, climate change is already wreaking havoc with just 1.1 °C of warming (Barros et al., 2017). The Intergovernmental Panel on Climate Change (IPCC) recently estimated that during the next ten years, 32–132 million more people will be forced into extreme poverty as a result of climate change. Global warming will put both food quality and food security at risk. Global crop production is one of the many detrimental effects of heatwaves and sporadic increases in heat episodes on human society and the environment, and they may have a major influence on future crop demand. Crop yield is directly threatened by the increasing frequency of heatwave occurrences, which are variations in the daily temperature.

The National Aeronautics and Space Administration (NASA) has released a paper predicting that heat waves will impact wheat and corn production as early as 2030 (Jagermeyer et al., 2021), with a 24% decrease in grain yields anticipated. Similarly, it is anticipated that for every degree Celsius that temperatures raise, the world's wheat production will decrease by 6% (Zhao et al., 2017), which is a significant and concerning drop for the third most important staple crop in the world (Van et al., 2020). Since local climates can frequently see more drastic shifts, it is anticipated that local crop output will be more negatively impacted. Individual crop species' overall growth and development, including the production of fruit and grains, is influenced by the ambient soil and air temperature, which ranges from minimum to maximum (Hatfield et al., 2015).

In addition to crops, temperature has a big impact on how a region's flora develops (Firmansyah et al., 2020). When temperatures rise above a specific threshold for a prolonged period of time, heat stress—also referred to as high-temperature stress—occurs, "causing irreversible harm to plant growth and development" (Tiwari et al., 2022). Based on their ability to grow in two distinct sowing seasons—the warm season and the cool season—a variety of crops can be primarily classified into two groups. Most warm-season crops in the tropics and subtropics can be harmed by heat stress at temperatures between 32 and 35 °C, although 25 °C is the top threshold level that can cause damage to cool-season crops (Tiwari et al., 2022; DiPaola et al.,

1992). Plant growth and development are impacted by heat stress, both sudden (heat waves) and chronic. High temperatures, like heat waves, are predicted to have a greater detrimental effect on plant growth than changes in the average temperature (Gutshchick et al., 2010), damaging cells or causing damage unique to a given growth stage.

For instance, heatwaves were more damaging to maize plants in the reproductive stage, reducing their mass (16%) when they were exposed to the canopy of corn plants in the reproductive and vegetative stages for three days. The number of spikelets, early maturation, and final yield in wheat were also reduced after 20 days of raising the temperature from ambient to 37/28 °C. Along with this, impact of heat wave is described in Fig. 2.

Heat stress impairs growth and a number of physiological functions, including respiration, photosynthesis, nutrient intake, water relations, and membrane degradation, which lowers overall plant performance and crop quality. While some earlier studies on plants' responses to Heat stress highlighted a decline in photosynthesis and biomass, more recent research has focused on phenology, grain production, nutritional quality, phytochemicals, and critical metabolites (Choukri et al., 2020). Heat stress frequently results in membrane damage, a decrease in biomass, denaturation of proteins, a decrease in protein concentration, and the inactivation of enzymes

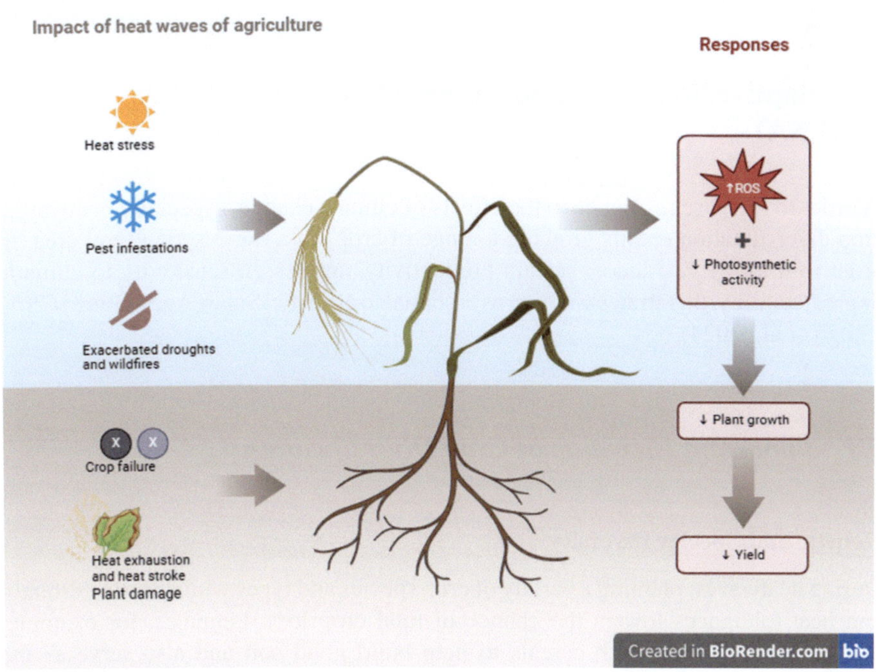

Fig. 2 Heat stress can impact overall growth of plants in terms of decreased the photosynthetic activity, growth and yield of plants. Extreme heat can damage leaves, crops, cause wildfires and also leads to damage a large area of crops as well as forests

involved in photosynthesis and respiration. Plants' vulnerability to infections is one of the well-known indirect effects of heat stress.

A large number of review articles outline the general effects of heat stress or its correlation with high amount of carbon dioxide or with drought or both are used in combination. (Heckathorn et al., 2013; Ahmad et al., 2022; Lamaoui et al., 2018; Sita et al., 2017). The effects of heat stress on nutrient content, nutrient uptake, nutrient-uptake proteins, and phytochemicals—all of which have a significant impact on crop quality—are the main emphasis of the current review. Before going on to the aforementioned specific areas, we provide a quick explanation of general impacts.

Extreme heat events, like heatwaves, are occurring more frequently and with greater intensity as a result of climate change-induced global warming. Effective adaptation methods must be developed and put into place because these heatwaves represent serious risks to crop diversification and agricultural practices. Agriculture is extremely susceptible to the extremes and fluctuation of the climate.

Crop Failure: Extended exposure to high temperatures can stress crops physiologically, resulting in lower yields and, in extreme situations, complete crop failure also by raising evaporation rates, decreasing soil moisture, and causing soil erosion, high temperatures can hasten soil degradation.

Animal health issues: Heat stress in animals can lead to higher illness susceptibility, higher mortality rates, and decreased productivity.

5 Adaptive Strategies and Climate-Smart Agriculture (CSA)

In order to increase resilience to the effects of climate change, especially heatwaves, crop diversification entails growing a range of crops in a same agricultural area. In order to improve resilience, sustain productivity, and lessen sensitivity to climatic unpredictability, this strategy is seen as essential to climatic-Smart Agriculture (CSA) (Singh et al., 2024).

5.1 Important Elements in Crop Diversification

Genetic and Species Diversity:

During heatwaves, planting a variety of crop species and types with differing drought and heat tolerances lowers the chance of total crop loss. Legumes, for example, can be planted along with cereals to help build good soil and also serve as the buffer crop when one of the crops is a failure. Growing more than one crop in a field simultaneously (intercropping) or alternatively (mixed cropping) increases resistance and resource utilization. Diversity root formations enhance water retention (Ullah et al., 2024).

5.1.1 Crop Diversification's Advantages for Heatwave Resilience

Covers for Yield Loss: Different crops will respond differently to extreme temperatures; therefore, diversification provides a cover against yield loss. The threat could be diversified among different water demand and heat tolerance crops so that not all crops would suffer from hot weather at the same time. Mixed farming increases soil and water conservation. This reduces the need for frequent irrigation and water saving under heat shock (Naik et al., 2011).

Pest and pathogen control support: By attracting natural enemies and discouraging heat loving pests, diverse croplands can reduce the necessity for chemical pesticide use, which in turn may be harmful or ineffective under extreme temperature conditions (Bett et al., 2019).

Ecosystem Resilience: Diverse agriculture supports biodiversity, which makes an ecosystem more capable of holding its own against environmental shocks, such as prolonged heat.

Obstacles and Crop Diversification Implementation

Knowledge and instructional needs: Farmers should be trained on optimization of resources allocations, juggling a few crop cycles and indicators for choosing complementary species.

5.2 Technological Innovations

There are different technologies which provide new opportunities for reducing heat stress and improving agricultural resilience.

5.2.1 Precision Agriculture Applications

Remote Sensing: Satellite and drone-based monitoring of crop stress and thermal conditions.

Variable Rate Applications: Precision irrigation and nutrient management based on spatial heat stress patterns.

Decision Support Systems: AI-powered tools for optimizing management decisions under heat stress conditions.

5.3 *Biotechnology Solutions*

5.3.1 Synthetic Biology

Engineering metabolic pathways for improved heat stress tolerance. Engineered Genetic Circuits: Gene circuits that improve heat tolerance can be developed thanks to synthetic biology. For example, protective genes can only be activated via circuits employing heat-inducible promoters when temperatures beyond stress thresholds (Jusiak et al., 2016; Lohani et al., 2022). Amino Acid Accumulation: During times of stress, plants store proline and glycine betaine, which serve as osmoprotectants to preserve cellular structure and function. Heat stress can cause the conversion of starch to sugar, improving osmotic balance and supplying energy for the stress response.

5.3.2 CRISPR/Cas9 Applications

Precise editing of genes involved in heat stress responses. Targeting certain TFs or stress-responsive genes, this method enables precise alteration of genes implicated in heat stress response. For instance, enhancing the expression of the wheat heat shock transcription factor *HsfA1b* improves heat tolerance, maintaining growth and yield under high temperatures. Additionally, CRISPR editing has been applied to genes involved in osmolyte synthesis and antioxidant pathways to reduce oxidative damage caused by heat stress. Integrating CRISPR/Cas with other biotechnology tools, such as gene drives, can speed propagation of heat tolerance traits across crops (Chen et al., 2024; Kaur et al., 2025).

5.3.3 Signaling and Regulation of Phytohormones

Phytohormones act as critical mediators in plant adaptation to heat stress by triggering, amplifying, and fine-tuning physiological, biochemical, and molecular defense mechanisms (Table 3). ABA, or abscisic acid: Because it controls stomatal closure to minimize water loss, ABA is essential for stress tolerance. Stress-responsive TFs are activated when ABA levels increase due to heat stress (Samanta & Roychoudhury, 2023; Wani et al., 2016).

5.3.4 Cross-Talk between Ethylene and Cytokinin

Stress-induced ethylene is a central regulator of stress-responsive genes. Heat responses can be modulated by changing the levels of cytokinin, which may enhance heat tolerance (Lohani et al., 2022).

Table 3 Major phytohormones in plant heat stress

Phytohormones	Major function	References
Abscisic acid	Stomatal closure, ROS scavenging, induction of HSPs	(Li et al., 2021; Wani et al., 2016)
Cytokinin	Enhancement of antioxidant enzymes, protection of fertility, delayed senescence	(Li et al., 2021; Manzoor et al., 2023)
Auxin	Thermomorphogenesis, stem/leaf growth modulation	(Li et al., 2021; Wani et al., 2016)
Salicylic acid	Reduction of oxidative damage, maintenance of photosynthesis	(Li et al., 2021)
Jasmonic acid	Physiological protection, antioxidant defense	(Li et al., 2021)
Ethylene	Regulates heat stress response, coordinates senescence and defense	(Li et al., 2021)
GABA	Antioxidant action, stress stabilization	(Samanta & Roychoudhury, 2023; Wani et al., 2016)

5.3.5 Crop and Genetic Selection

Heat-Tolerant types: Losses diminished by creating crop types resistant to hot temperature particularly during vulnerable stages of the crop like before and after flowering.

Photoperiod Extension Method (PEM) is a novel method to support better selection of heat-tolerant varieties in more synchronized flowering among types and greater consistency between trials for evaluating heat tolerance (Ullah et al., 2024).

5.3.6 Crop Calendar Planning and Optimal Time of Sowing

Maternal environment Sowing date: By selecting optimum sowing time to escape from heat conditions, crops can avoid warm temperatures at critical times and be protected aspects (e.g. grain filling).

Phenology Regulation: Synchronizing phenological stages (e.g., flowering) with cooler periods can enable crop growth to be timed and reduce the stress on sensitive stages, such as pollen production and grain filling (Ferris et al., 1998).

5.4 Management Practices

Agronomic strategies offer short-term solutions for a reduction in heat stress without substantial investments being needed for technology.

5.4.1 Irrigation and Water Management

Supplemental irrigation (particularly at the grain-fill stage) could reduce heatwave damage by maintaining soil moisture, reducing plant water stress and enhancing cooling through evapotranspiration. Conservation tillage and mulching techniques that conserve soil water and moderate soil temperature may better help crops to resist heat stress.

Canopy and Row Spacing alterations: Wider row spacing can reduce crop canopy temperature by (i) increasing better air movement and (ii) moderating heat stress.

5.4.2 Temperature Thresholds

Farmers can readily react to such cues (e.g., by changing watering schedule) if they must keep an eye on the surrounding conditions and gear early-warning systems to certain temperature levels of importance (28 °C before blooming, 32 °C after flowering). Modelling and Forecasting: Planning and altering agronomic practices in anticipation of heat events are facilitated through the use of temperature and crop response predictive models (Lohani et al., 2022).

5.4.3 Management of Soil Health

5.4.3.1 Increasing Soil Organic Content

Increasing the amount of organic matter in the soil can enable it to hold more water, critical during periods of heat stress.

Cover Cropping Heat waves can be eased on crops by sowing cover crops, which reduces soil temperatures and saves moisture.

5.5 Policy and Institutional Support to Mitigate Heat Stress

Government policy and institutional capacity are needed to better implement the Climate-Smart Agriculture (CSA) in mitigating growth of heat stress. Institutions are necessary to provide farmers with the resources, knowledge and technology, and policies provide the framework and motivation needed for large-scale CSA implementation. In this section, some of the policy and institutional measures for promoting agricultural adaptation are discussed.

Climate-Smart Agriculture Policy Frameworks:

(a) **Regional and National CSA Plans**:

Several countries, such as Kenya and Vietnam, have developed national CSA planning approaches which integrate GHG-reducing agri-food systems laws with climate-resilience and natural resources management. Targeted to help farmers adopt flexible practices, such strategies often align with recent national efforts around climate action.

(b) **Financial incentives and subsidies**:

Farmers could be encouraged to adopt CSA through subsidies and tax reductions for the application of sustainable methods. For example, support for solar-based systems, drip irrigation and rainwater harvesting reduces an economic burden on farmers and allows more widespread realization of CSA practices.

(c) **Programs for Insurance**:

Programs for climate-risk insurance, such as index-based weather insurance, offer farmers financial security against agricultural losses brought on by severe weather. (Campbell et al., 2014) (Climate Smart Agriculture Sourcebook | Food and Agriculture Organization of the United Nations.)

i. Investment in Research and Development (R&D):

Creation of Climate-Resilient Varieties: Government-funded agricultural research organizations and university collaborations aid in the creation of crop varieties that are resistant to drought and heat. In areas that are extremely susceptible to water and temperature stress.

ii. **Innovation Centers and Pilot Initiatives**:

By showcasing the practical advantages of CSA techniques to farmers, the creation of innovation centers where these practices are implemented on a modest scale promotes uptake. Some demonstration farms are also designed to showcase best practices and help farmers decide whether CSA fits their needs. (Campbell et al., 2014)

iii. **Policy support for Market Access**:

CSA-Compatible Market Incentives: Farmers are rewarded financially by markets that encourage responsible production (e.g., certified climate-smart vegetables). Farmers could stabilize and diversify their incomes by establishing markets for crops that are weather adapted, such as sorghum and millet. Through improving the entire value chain, climate-smart food will be able to access markets more effectively and fetch higher prices (Calvin et al., 2023).

iv. **International Collaboration & Multilateral Partnerships**:

International CSA Networks: International collaborations that foster information exchange, policy formation, and capacity building among participating countries include the Global Alliance for Climate-Smart Agriculture (GACSA). Global CSA implementation is aided by international platforms that make it easier to share funds, research, and best practices. South-South Cooperation: Knowledge sharing and cooperative CSA projects are frequently advantageous for developing nations with comparable climate risks. By transferring good techniques from one region to another, partnerships between nations in Africa, Asia, and Latin America have been successful in scaling up CSA (Campbell et al., 2014; Lipper et al., 2014). Along with this, different types of methods are discussed in Table 4 to combat heat stress.

6 Conclusion

Heatwaves under climate change is an emerging challenge for global agriculture in the twenty-first century. This chapter has shown that high temperature stress to agricultural systems is similarly initiated through multi component related routes from molecular cellular responses to agro-ecosystem level production systems. The effects are uneven, as different crops, agronomic systems and regions differ in their levels of vulnerability that should guide the design of adaptive strategies. Key results include; Heatwaves lead to large reductions in crop yield (10–50% depending on crop, intensity and timing of heat) as well as quality loss that erodes the nutritional content and market value of food produced, the economic costs exceed $3–5 billion each year for major agricultural regions with evidence suggesting these costs increase significantly under future warming. Therefore, effective minimization of heat wave impacts is critical to achieve an optimal agricultural production and food security. Both precision agriculture and biotechnology are transformative solutions for increasing heat stress tolerance. CRISPR-gene editing of key regulators such as HsfA1b will also be employed to generate stress adapted and heat tolerant crops. It used digital technologies such as AI/ML models, IOT sensors and satellite imagery to deliver accurate inventory management, real-time stress monitoring, adaptive decision-making for better yield and environment conservation. Equally important is the packaging of these technologies with traditional knowledge systems such as mixed farming and diversified farming, agroforestry and crop rotation. The above-described traditional practices perform key ecosystem services like microclimate regulation, temporal and spatial risk distribution, and economic diversification that can form the vital basis on which new technologies can construct systems that are more resilient and climate-smart.

Future scope and limitations:

Adapting agriculture along with rising heat stress is one daunting challenge, but not an impossible one. History also shows agriculture is surprisingly adaptable to new circumstances, and the technological tools available for science-based solutions are unprecedented. The period for proactive adaptation is closing, but there are substantial opportunities to mitigate the effects of heat stress on global food security. The secret lies in linking all elements of the agricultural system, and this requires commitment to integrated action up and down streams as well as adequate investments for research, technology development and institutional capacity building. How well we are able to operationalize those adaptation strategies will be a key determinant of

Table 4 Various methods for reducing the impact of heatwaves on agriculture, covering genetic, agronomic, technological, and policy-based approaches

Method	Description	Expected impact	References
Heat-tolerant crop varieties	Developing and cultivating crop varieties with improved heat tolerance through selective breeding or genetic modification.	Reduces yield loss during heatwaves; maintains productivity.	(Lohani et al., 2022)
Optimized sowing times	Adjusting sowing dates to avoid critical growth stages, such as flowering, during peak heat periods.	Minimizes heat exposure during sensitive stages, protecting yield.	(Ullah et al., 2024)
Efficient irrigation management	Using drip irrigation, supplemental irrigation, and other precision techniques to reduce plant water stress.	Maintains soil moisture, cools crops, and reduces heat stress.	(Lohani et al., 2022)
Mulching and conservation tillage	Applying organic or synthetic mulches to retain soil moisture and regulate soil temperature.	Protects crops from soil desiccation and excessive heat.	(Lohani et al., 2022; Ullah et al., 2024)
Row spacing and canopy management	Wider row spacing to improve air circulation and reduce crop canopy temperatures.	Lowers in-field temperatures and reduces heat stress on plants.	(Lohani et al., 2022)
Soil health and organic matter	Enhancing soil organic content to improve water retention and reduce temperature fluctuations.	Supports plant resilience by maintaining soil moisture.	(Lohani et al., 2022)
Early warning systems and forecasting	Implementing climate forecasting tools to help farmers prepare for and mitigate heatwaves.	Increases preparedness, reduces crop loss risk.	(Lohani et al., 2022)
Heat-responsive genetic circuits	Synthetic biology approach where gene circuits activate protective responses under high temperatures.	Provides targeted heat protection, conserving energy and resources.	(Lohani et al., 2022)

(continued)

Table 4 (continued)

Method	Description	Expected impact	References
Policy support and subsidies	Government support for investments in climate-resilient infrastructure and crop insurance.	Encourages adoption of heat-adaptive practices and supports farmers economically.	(Naik et al., 2011; Ullah et al., 2024)
Farmer training and community programs	Providing education on adaptive techniques, like optimized irrigation and crop rotation for heat resilience.	Enhances farmer awareness and resilience against heat stress.	(Ullah et al., 2024)

the future of global food security in the climate change era. The scientific foundation is there, the technologies are emerging and policy frameworks are in place. By combining science and action-oriented methodologies for adaptation to agriculture, it is feasible not only to keep up the pace of food amidst climate change but also simultaneously develop a more resilient and sustainable agricultural systems. What comes next will demand an unprecedented level of collaboration among scientists, agriculture professionals, policymakers and society more broadly to create the types of food systems that can provide a stable supply of healthy food throughout all seasons, thriving rural communities and a healthier environment.

References

Ahmad, M., Imtiaz, M., Shoib Nawaz, M., Mubeen, F., & Imran, A. (2022). What did we learn from current Progress in heat stress tolerance in plants? Can microbes be a solution? *Frontiers in Plant Science, 13*, 794782. https://doi.org/10.3389/FPLS.2022.794782/XML

Angelsen, A., Jagger, P., Babigumira, R., Belcher, B., Hogarth, N. J., Bauch, S., Börner, J., Smith-Hall, C., & Wunder, S. (2014). Environmental income and rural livelihoods: A global-comparative analysis. *World Development, 64*(S1), S12–S28. https://doi.org/10.1016/J.WORLDDEV.2014.03.006

Awasthi, A., Vishwakarma, K., & Pattnayak, K. C. (2022). Retrospection of heatwave and heat index. *Theoretical and Applied Climatology, 147*(1), 589–604. https://doi.org/10.1007/s00704-021-03854-z

Bett, B., Lindahl, J., Delia, G., Bett, B., Delia, G., & Lindahl, J. (2019). Climate change and infectious livestock diseases: The case of Rift valley fever and tick-borne diseases. *The Climate-Smart Agriculture Papers*, 29–37. https://doi.org/10.1007/978-3-319-92798-5_3

Barros, C., Gueguen, M., Douzet, R., Carboni, M., Boulangeat, I., Zimmermann, N. E.,... & Thuiller, W. (2017). Extreme climate events counteract the effects of climate and land-use changes in A lpine tree lines. *Journal of Applied Ecology, 54*(1), 39–50.

Calvin, K., Dasgupta, D., Krinner, G., Mukherji, A., Thorne, P. W., Trisos, C., Romero, J., Aldunce, P., Barrett, K., Blanco, G., Cheung, W. W. L., Connors, S., Denton, F., Diongue-Niang, A., Dodman, D., Garschagen, M., Geden, O., Hayward, B., Jones, C., & Ha, M. (2023). IPCC, 2023: Climate change 2023: Synthesis report. In H. Lee & J. Romero (Eds.), *Contribution of working groups I, II and III to the sixth assessment report of the intergovernmental panel on climate change*. IPCC. https://doi.org/10.59327/IPCC/AR6-9789291691647

Campbell, B. M., Thornton, P., Zougmoré, R., van Asten, P., & Lipper, L. (2014). Sustainable intensification: What is its role in climate smart agriculture? *Current Opinion in Environmental Sustainability, 8*, 39–43. https://doi.org/10.1016/J.COSUST.2014.07.002

Casal, J. J., & Balasubramanian, S. (2019). Thermomorphogenesis. *Annual Review of Plant Biology, 70*, 321–346. https://doi.org/10.1146/ANNUREV-ARPLANT-050718-095919/1

Challinor, A. J., Watson, J., Lobell, D. B., Howden, S. M., Smith, D. R., & Chhetri, N. (2014). A meta-analysis of crop yield under climate change and adaptation. *Nature Climate Change, 4*(4), 287–291. https://doi.org/10.1038/NCLIMATE2153

Chen, F., Chen, L., Yan, Z., Xu, J., Feng, L., He, N., Guo, M., Zhao, J., Chen, Z., Chen, H., Yao, G., & Liu, C. (2024). Recent advances of CRISPR-based genome editing for enhancing staple crops. *Frontiers in Plant Science, 15*, 1478398. https://doi.org/10.3389/FPLS.2024.1478398/XML

Choukri, H., Hejjaoui, K., El-Baouchi, A., El haddad, N., Smouni, A., Maalouf, F., Thavarajah, D., & Kumar, S. (2020). Heat and drought stress impact on phenology, grain yield, and nutritional quality of lentil (Lens culinaris Medikus). *Frontiers in Nutrition, 7*. https://doi.org/10.3389/FNUT.2020.596307/PDF

Ferris, R., Ellis, R. H., Wheeler, T. R., & Hadley, P. (1998). Effect of high temperature stress at Anthesis on grain yield and biomass of field-grown crops of wheat. *Annals of Botany, 82*(5), 631–639. https://doi.org/10.1006/ANBO.1998.0740

Hatfield, J. L., & Prueger, J. H. (2015). Temperature extremes: Effect on plant growth and development. *Weather and Climate Extremes, 10*, 4–10. https://doi.org/10.1016/J.WACE.2015.08.001

He, J., Tan, C., & Qin, L. (2022). Root-zone heat priming effects on maximum quantum efficiency of PSII, productivity, root morphology and nutritional quality of two Aeroponically grown leafy greens in a tropical greenhouse. *Plants, 11*(13), 1684. https://doi.org/10.3390/PLANTS11131684

Herrero, M., Thornton, P. K., Power, B., Bogard, J. R., Remans, R., Fritz, S., Gerber, J. S., Nelson, G., See, L., Waha, K., Watson, R. A., West, P. C., Samberg, L. H., van de Steeg, J., Stephenson, E., van Wijk, M., & Havlík, P. (2017). Farming and the geography of nutrient production for human use: A transdisciplinary analysis. *The Lancet Planetary Health, 1*(1), e33–e42. https://doi.org/10.1016/S2542-5196(17)30007-4

Jusiak, B., Cleto, S., Perez-Piñera, P., Lu, T. K. (2016). Engineering synthetic gene circuits in living cells with CRISPR technology.

Kaur, N., Qadir, M., Francis, D. V., Alok, A., Tiwari, S., & Ahmed, Z. F. R. (2025). CRISPR/Cas9: A sustainable technology to enhance climate resilience in major staple crops. *Frontiers in Genome Editing, 7*, 1533197. https://doi.org/10.3389/FGEED.2025.1533197/XML

Kerr, R. B., Liebert, J., Kansanga, M., & Kpienbaareh, D. (2022). Human and social values in agroecology: A review. *Elementa, 10*(1). https://doi.org/10.1525/ELEMENTA.2021.00090/185041

Lamaoui, M., Jemo, M., Datla, R., & Bekkaoui, F. (2018). Heat and drought stresses in crops and approaches for their mitigation. *Frontiers in Chemistry, 6*, 311598. https://doi.org/10.3389/FCHEM.2018.00026/XML

Lee, H., & Romero, J. (2023). Climate change 2023 synthesis report IPCC, 2023: Sections. In: Climate change 2023: Synthesis report. Contribution of working groups I, II and III to the sixth assessment report of the intergovernmental panel on climate change, pp. 35–115. https://doi.org/10.59327/IPCC/AR6-9789291691647.

Li, L., Gu, W., Li, J., Li, C., Xie, T., Qu, D., Meng, Y., Li, C., & Wei, S. (2018). Exogenously applied spermidine alleviates photosynthetic inhibition under drought stress in maize (*Zea mays* L.) seedlings associated with changes in endogenous polyamines and phytohormones. *Plant Physiology and Biochemistry, 129*, 35–55. https://doi.org/10.1016/J.PLAPHY.2018.05.017

Li, N., Euring, D., Cha, J. Y., Lin, Z., Lu, M., Huang, L. J., & Kim, W. Y. (2021). Plant hormone-mediated regulation of heat tolerance in response to global climate change. *Frontiers in Plant Science, 11*, 627969. https://doi.org/10.3389/FPLS.2020.627969/XML

Li, N., Euring, D., Cha, J. Y., Lin, Z., Lu, M., Huang, L. J., & Kim, W. Y. (2021b). Plant hormone-mediated regulation of heat tolerance in response to global climate change. *Frontiers in Plant Science, 11*, 627969. https://doi.org/10.3389/FPLS.2020.627969

Lipper, L., Thornton, P., Campbell, B. M., Baedeker, T., Braimoh, A., Bwalya, M., Caron, P., Cattaneo, A., Garrity, D., Henry, K., Hottle, R., Jackson, L., Jarvis, A., Kossam, F., Mann, W., McCarthy, N., Meybeck, A., Neufeldt, H., Remington, T., & Torquebiau, E. F. (2014). Climate-smart agriculture for food security. *Nature Climate Change, 4*(12), 1068–1072. https://doi.org/10.1038/nclimate2437

Lohani, N., Singh, M. B., & Bhalla, P. L. (2022). Biological parts for engineering abiotic stress tolerance in plants. *Biodesign Research, 2022*, 9819314. https://doi.org/10.34133/2022/9819314

Lozano-García, B., Muñoz-Rojas, M., & Parras-Alcántara, L. (2017). Climate and land use changes effects on soil organic carbon stocks in a Mediterranean semi-natural area. *Science of the Total Environment, 579*, 1249–1259. https://doi.org/10.1016/J.SCITOTENV.2016.11.111

Manzoor, M. A., Xu, Y., lv, Z., Xu, J., Wang, Y., Sun, W., Liu, X., Wang, L., Wang, J., Liu, R., Whiting, M. D., Jiu, S., Zhang, C., Manzoor, M. A., Xu, Y., lv, Z., Xu, J., Wang, Y., Sun, W., & Zhang, C. (2023). Fruit crop abiotic stress management a comprehensive review of plant hormones mediated responses. *Fruit Research, 3*(1). https://doi.org/10.48130/FRURES-2023-0030

Mendelsohn, R. (2009). The impact of climate change on agriculture in developing countries. *Journal of Natural Resources Policy Research, 1*(1), 5–19. https://doi.org/10.1080/193904508 02495882

Miller, G., Schlauch, K., Tam, R., Cortes, D., Torres, M. A., Shulaev, V., Dangl, J. L., & Mittler, R. (2009). The plant NADPH oxidase RBOHD mediates rapid systemic signaling in response to diverse stimuli. *Science Signaling, 2*(84). https://doi.org/10.1126/SCISIGNAL.2000448

Moreira, B., Gonçalves, A., Pinto, L., Prieto, M. A., Carocho, M., Caleja, C., & Barros, L. (2024). Intercropping systems: An opportunity for environment conservation within nut production. *Agriculture, 14*(7), 1149. https://doi.org/10.3390/AGRICULTURE14071149

Murgueitio, E., Calle, Z., Uribe, F., Calle, A., & Solorio, B. (2011). Native trees and shrubs for the productive rehabilitation of tropical cattle ranching lands. *Forest Ecology and Management, 261*(10), 1654–1663. https://doi.org/10.1016/J.FORECO.2010.09.027

Murshed-E-Jahan, K., & Pemsl, D. E. (2011). The impact of integrated aquaculture–agriculture on small-scale farm sustainability and farmers' livelihoods: Experience from Bangladesh. *Agricultural Systems, 104*(5), 392–402. https://doi.org/10.1016/J.AGSY.2011.01.003

Naik, A., Jogi, M., & Shreenivas, B. V. (2011). Assessing the impact of climate change on crop production and farming strategies. 24(5), 2011

Ohama, N., Sato, H., Shinozaki, K., & Yamaguchi-Shinozaki, K. (2017). Transcriptional regulatory network of plant heat stress response. *Trends in Plant Science, 22*(1), 53–65. https://doi.org/10.1016/J.TPLANTS.2016.08.015

Rahman, S., & Anik, A. R. (2020). Productivity and efficiency impact of climate change and agroecology on Bangladesh agriculture. *Land Use Policy, 94*, 104507. https://doi.org/10.1016/J.LANDUSEPOL.2020.104507

Rosenzweig, C., Elliott, J., Deryng, D., Ruane, A. C., Müller, C., Arneth, A., Boote, K. J., Folberth, C., Glotter, M., Khabarov, N., Neumann, K., Piontek, F., Pugh, T. A. M., Schmid, E., Stehfest, E., Yang, H., & Jones, J. W. (2014). Assessing agricultural risks of climate change in the 21st century in a global gridded crop model intercomparison. *Proceedings of the National Academy of Sciences of the United States of America, 111*(9), 3268–3273. https://doi.org/10.1073/PNAS.1222463110

Samanta, S., & Roychoudhury, A. (2023). Crosstalk of melatonin with major phytohormones and growth regulators in mediating abiotic stress tolerance in plants. *South African Journal of Botany, 163*, 201–216. https://doi.org/10.1016/J.SAJB.2023.10.040

Sharma, A., Mitra, A., Vasan, V., & Govindarajan, R. (2021). Spatio-temporal relationships between rainfall and convective clouds during Indian monsoon through a discrete lens. *International Journal of Climatology, 41*(2), 1351–1368. https://doi.org/10.1002/JOC.6812

Shrestha, S., Mahat, J., Shrestha, J., Madhav, K. C., & Paudel, K. (2022). Influence of high-temperature stress on rice growth and development. A review. *Heliyon, 8*(12), e12651. https://doi.org/10.1016/J.HELIYON.2022.E12651

Siebers, M. H., Yendrek, C. R., Drag, D., Locke, A. M., Rios Acosta, L., Leakey, A. D. B., Ainsworth, E. A., Bernacchi, C. J., & Ort, D. R. (2015). Heat waves imposed during early pod development in soybean (Glycine max) cause significant yield loss despite a rapid recovery from oxidative stress. *Global Change Biology, 21*(8), 3114–3125. https://doi.org/10.1111/GCB.12935

Singh, B. P., Krishnamoorthi, A., Kumar, V., & Kalaiselvi, P. (2024). Revolutionizing and agriculture farming through artificial intelligence. *Journal of Diversity Studies, 3*(1), 13–19. https://doi.org/10.51470/JOD.2024.03.01.13

Singh, S., Mall, R., & Singh, N. (2021). Changing spatio-temporal trends of heat wave and severe heat wave events over India: An emerging health hazard. *International Journal of Climatology, 41*(S1), E1831–E1845. https://doi.org/10.1002/JOC.6814;PAGE:STRING:ARTICLE/CHAPTER

Stevanović, M., Popp, A., Lotze-Campen, H., Dietrich, J. P., Müller, C., Bonsch, M., Schmitz, C., Bodirsky, B. L., Humpenöder, F., & Weindl, I. (2016). The impact of high-end climate change on agricultural welfare. *Science Advances, 2*(8). https://doi.org/10.1126/SCIADV.1501452

Thornton, P. K., & Herrero, M. (2015). Adapting to climate change in the mixed crop and livestock farming systems in sub-Saharan Africa. *Nature Climate Change, 5*(9), 830–836. https://doi.org/10.1038/NCLIMATE2754

Tiwari, M., Kumar, R., Min, D., & Jagadish, S. V. K. (2022). Genetic and molecular mechanisms underlying root architecture and function under heat stress—A hidden story. *Plant Cell and Environment, 45*(3), 771–788. https://doi.org/10.1111/PCE.14266

Ullah, N., Collins, B., Christopher, J. T., Frederiks, T., & Chenu, K. (2024). Pre- and post-flowering impacts of natural heatwaves on yield components in wheat. *Field Crops Research, 316*, 109489. https://doi.org/10.1016/j.fcr.2024.109489

Wani, S. H., Kumar, V., Shriram, V., & Sah, S. K. (2016). Phytohormones and their metabolic engineering for abiotic stress tolerance in crop plants. *The Crop Journal, 4*(3), 162–176. https://doi.org/10.1016/J.CJ.2016.01.010

Yadav, N., Rajendra, K., Awasthi, A., Singh, C., & Bhushan, B. (2023). Systematic exploration of heat wave impact on mortality and urban heat Island: A review from 2000 to 2022. *Urban Climate, 51*, 101622. https://doi.org/10.1016/j.uclim.2023.101622

Yang, J., Tian, H., Tao, B., Ren, W., Kush, J., Liu, Y., & Wang, Y. (2014). Spatial and temporal patterns of global burned area in response to anthropogenic and environmental factors: Reconstructing global fire history for the 20th and early 21st centuries. *Journal of Geophysical Research: Biogeosciences, 119*(3), 249–263. https://doi.org/10.1002/2013JG002532

Heat-Stressed Greenspaces in Urban and Peri-Urban Areas: Evidences from Reviews with Strategic Insights for Sustainability Marketing

Shipra Agarwal, Malika Pahwa, Tanya Rawal, and Amar Johri

Abstract The urban heat Island effect, rising temperature and expanding urbanization are all contribution factors to the growing impact of heat stress on community gardens and green roofs. The paper draws together findings from bibliometrics studies, VOS viewer and systematic review to understand how vegetation respond to excessive heat and how sustainable marketing might help climate resilient planning. Using bibliographic coupling, citation mapping, keywords cooccurrence and co-authorship networks, the study finds world-wide research patterns, leading contributors and the key issues defining current knowledge on urban heat and green infrastructure. The effectiveness of various green spaces types such as parks, wetland, blue green infrastructure, urban farms and as well as how their size, structure, vegetation type and surrounding built environment affect cooling is further assessed through a systematic review of recent research. Real-world case studies from Delhi, Singapore and Phoenix highlight how cities are embracing heat-resilient tactics such as native planting, vertical greening and community led greening projects. These examples highlight ecological responses and the role of communication in establishing widespread acceptance. The data show that although green spaces significantly decrease temperature and improve comfort and their cooling capacities is variable due to inadequate maintenance. The study demonstrates that combining ecology with sustainable marketing can relate scientific findings and public initiatives by allowing communities to emphasize the importance of green space and increase legislative support. In order to create greener, colder and climate-resilient cities the analysis suggest integrating ecological design, community involvement and strategic marketing.

Keywords Heat stress · Urban area · Greenspaces · Climate resilient · Environment

---

S. Agarwal (✉) · M. Pahwa · T. Rawal
Department of Commerce, Graphic Era Deemed to Be University, Dehradun, India
e-mail: shipra.agg26@gmail.com

A. Johri
College of Administrative and Financial Sciences, Saudi Electronic University, Riyadh, Saudi Arabia

A. Awasthi et al. (eds.), *Rising Temperatures and Plant Ecosystems*,
https://doi.org/10.1007/978-3-032-29477-7_13

1 Introduction

Park, trees, gardens wetlands and green belts are examples of urban and Peri urban green areas that are crucial to maintaining the comfort and health of cities. It helps in cool down the environment, purify the air and provide relaxing area for people and wildlife (Gakobo, 2021). But be cities expanding and temperatures rising these green places are increasingly suffering from heat stress (Yadav et al., 2023). Severe heat can harm plants lessen their cooling effects and even lead to the degradation of certain areas (Hassan et al., 2021). This impacts not only the environment but also the well-being of the people who rely on them (Awasthi et al., 2022).

Heat stress is becoming a major problem for plants in and around cities according to recent assessments and research they show how rising temperatures, decreasing rainfall and the urban heat Island effect are exerting strain on plants (Irfeey et al., 2023; Pattnayak et al., 2023; Tandon et al., 2025). These studies also clarify that different green spaces react differently to heat due to a variety of things including the kind of plants utilising the space, its size and how well it is maintained (Atchison et al., 2024). Planners and legislators that wish to create a more environmentally friendly and climate resilient community must comprehend these factors (Hughes et al., 2025).

Opportunities for sustainable marketing are also created by the problem of heat stressed green places. The need to explain the importance of healthy green areas and their advantages for community and the environment is growing as people become more conscious of climate related challenges (Athokpam et al., 2024). Insights from research can add in building marketing tactics that encourage consumers to cherries and maintain green places participating community project and make environmentally friendly decisions (Oetken, 2025).

In general this topic integrates scientific evidence with practical communication tactics to explain heat stress in green areas and explore how sustainable marketing may assist promote cooler, healthier and more resilient urban environments (Parihar & Birman, 2024).

2 Conceptual Foundations

The concept of heat stressed natural environments start with the urban heat Island (UHI) effect, which happens when towns become substantially hotter than neighbouring rural areas. This is due to the fact that heat is absorbed and retained by concrete surfaces parts and buildings. Plants in urban areas find it difficult to adapt to rising temperatures. Plant heat stress results from this causing the flora to we can, dry out and lose its capacity to hold the environment. Plants under Naresh make it difficult for green areas to feed wildlife improve air quality and provides shade.

Cities rely on green infrastructure in both urban and Perry urban areas to lessen this issue. Parks Street trees Green roof, wetlands gardens and Woods on the outskirts of the cities are examples of this. These eco-friendly components promote biodiversity chill the climate and improve the quality of life in the cities. Research reveals that the form, size and location of the green spaces determine how much cooling they can supply. Gi is an effective instrument for lowering heat stress and creating climate resilient cities when it is connected and well planned.

This scientific knowledge immediately relates to the sustainability marketing which emphasizes educating the public about the value of healthy green spaces. As individuals become aware of climate change they need strong messaging about why conserving green places is essential. Sustainability marketing assists in encouraging environmental friendly activities that involve communities and support for urban development projects. It encourages people to value green infrastructure and participate in health community activities.

Plant Heat Stress is effected by UHI effects that is connected to Green Infrastructure in urban and peri-urban and resulting in the requirement for Sustainability Marketing interventions (Fu et al., 2024).

Urban Heat Island → Plant Heat Stress → Green Infrastructure (Urban & Peri-Urban) → Sustainability Marketing.

2.1 Urban Heat Island (UHI)

The Urban Heat Island effect makes towns superior then neighbouring rural areas because buildings roads and other hard surfaces absorb the excessive heat. In heavily populated places, this becomes a significant environmental problem (Gloria & Gnanasekaran, 2024).

2.2 Heat Stress in Plants

Plants in urban areas are prone to heat stress cause of the UHI effect. This can alter their leave colours interfere with the photosynthesis and reduces their tendency to release moisture. These consequences we can plant and reduce their ability to sustain urban environments. All the plants are able to adapt over extended periods of time these natural modifications frequently fall sort of and the rapid changes occur in the urban areas (Fukano et al., 2023).

2.3 Green Infrastructure

Urban green spaces: These are frequently dispersed little places. They supply immediate cooling and shelter but are not well linked to one another.

Peri urban green spaces: Larger continuous landscapes that provide deeper advantages such as improved cooling, habitat support and general ecosystem Restoration (Válánszki et al., 2025).

2.4 Sustainability Marketing

Demonstrating how green spaces can improve urban resilience connect these components. It helps a community realise their importance, support municipal identity and development and stimulate actions and regulations that maintain and expand green areas (Indeed, 2025).

The chain of ideas functions as follows to put it simply: Urban Heat Island → Plant Heat Stress → Green Infrastructure as a solution → Sustainability Marketing to increase awareness and action. This paradigm helps us understand the problem and guides smarter decisions for developing cooler and more resident urban environments.

3 Bibliometric Review

3.1 Bibliographic Coupling of Countries

Figure 1 depicts that connection between various nation based on their studies of urban and peri-urban heat stressed green places. When researchers found different nations mention comparable studies and scientific publication, this in known as bibliographic coupling.

Accordingly, the closure to nation seems to be connected in the network, the more references they share. The map makes it easier to see weather nations have comparable research interests, approaches or regions of concentration in relation to sustainability, green infrastructure and heat stress.

The maps hues correspond to several clusters of groups of nation with comparable research trends. For instance, because they frequently quote the same literature, particularly on urban ecology, green infrastructure and sustainability policies, European nations like Germany, UK, Netherland and France constitute a dense cluster. Asian nations including China, India, Malaysia and South Arabia make up another cluster. Growing regional interest in heat stress, fast urbanization and climate adaption techniques is reflected in their cluster. Cluster like Australia, Japan and South Korea also form a connected group showing strong research collaboration on environmental monitoring and urban resilience.

Some nations like Ghana, Chile and Morocco are more visible on networks periphery. This indicates that their studies are less connected to the major worldwide research cluster, but it does not imply that their studies is insignificant. These

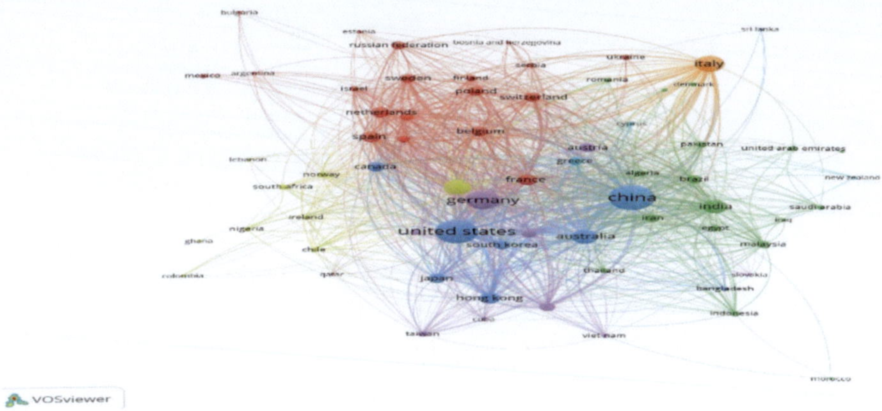

Fig. 1 Bibliographic coupling of countries

nations might be new doners or focusing on more regional problems related to heat stress in green areas.

Overall the image illustrates the global interconnectedness of research on heat stressed green places. It illustrates which nations are at the forefront of this discipline, which geographical areas have similar scientific interests and the cross border movement of knowledge. As nations with similar research interests can work together to create greener, cooler and more climatic resilient cities and understanding of these links can aid in the development of better global strategies, the sharing of best practices and the promotion of sustainability market.

3.2 *Citation Organisation*

Figure 2 illustrates the contributions made by various academic institutions and research centres worldwide to the scientific literature on topics linked to sustainability, urban climate concerns, and heat-stressed green spaces. Each dot in the graphic represents an institution, and the size of the dot shows how frequently other academics cite that institution. Nearby institutions contribute to a common body of knowledge by citing similar sources or having similar research interests.

The University of Chinese academy of sciences the University of Hong Kong Arizona, State University, sun Yet-sen University and other significant American and European universities are among the extremely significant establishments in the maps centre due to their prominent position, these organisations are likely to publish highly reference studies on subjects including green infrastructure, urban hits islands, urban vegetations reaction to climate stress and environmental sustainability. The scientific community understanding of the response to heat stress in urban and very

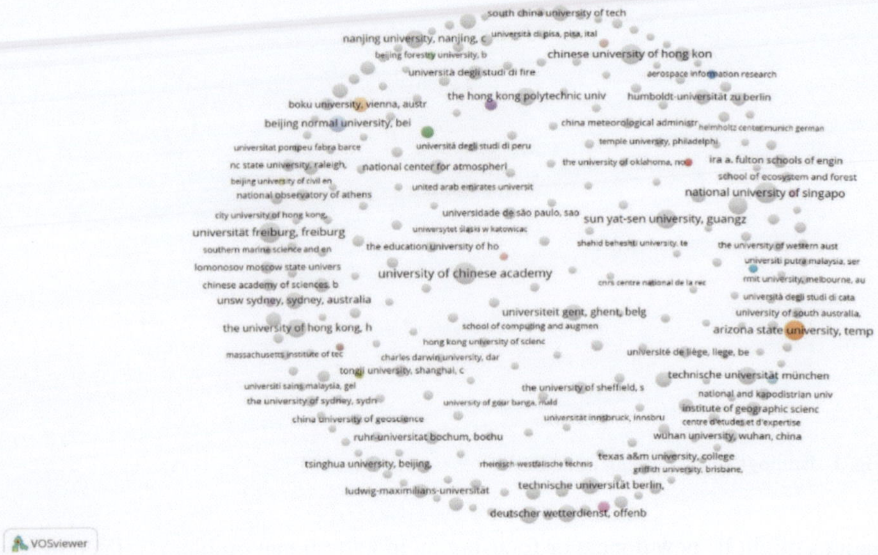

Fig. 2 Citation organisation

urban green spaces is shared by these universities which served as global knowledge centres.

Other institutions that are linked by shared literature or comparable study topics surround the primary core. For instance, research on heat stress climate modelling, remote sensing, and the ecological performance of green spaces in rapidly urbanizing areas is being conducted by university is like Beijing Normal University, the Hong Kong polytechnic University, and National University of Singapore. Due to the growing issue of urban environmental degradation and heat stress in their respective locations many of these Asian institutions are becoming more well known.

European University is like Humboldt University Zu Berlin, Unterverstieit Gent, Technische Universitaty and Mun Freiburg seem to be significant contributors as well. Their work frequently focus is on green city planning, ecosystem services and the nature base solutions all of which are crucial subject for creating sustainable urban plans. These organisations contribute to the advancement of practical research supporting public environment cooling techniques and greener infrastructure all important faces of sustainable marketing.

Universities with pure citations or more specialised research areas are represented by the institutions on the figures outside edges. Despite not being key they never they less contribute important information frequently concentrating on regional or local climate concerns. This diversity is crucial because various Nation's experience in different ways.

Overall, the citation organisation figure shows that are robust and integrated global academy network supports research on heat stressed green places. It displaces which universities are at the forefront of the discussion the global flow of research knowledge and the sources of significant studies that influence urban planning climate adaptation and sustainability marketing. Knowing this demonstrate the extent of international scientific environment with urban environmental concerns and its research is and policy makers in identifying important thought leaders and possible partners.

3.3 Citation Source

Figure 3 Quotes source map magazines and publishing outlets that serve as the basis of primary knowledge for research on heat-stools in urban and peri-urban areas. Each node represents a source, in which large nodes reflect high quotation frequency, and links between them indicate cross-elimination relationships. Total environment, durable cities and society, buildings and environment, urban forestry and urban greenery, and urban climate stand as central sources, indicating climate adaptation on urban stability, climate adaptation, green infrastructure and research, and indicating their important roles in research on urban stability.

The map also shows secondary but important contributors such as Ecological Indicators, Environmental Research Letters, Landscape and Urban Planning, and Applied Energy, which reflect the diverse nature of this field.

Their present draws attention to the multifaceted nature of research on heat stressed green areas which includes a spatial design, energy-environment interactions and ecological monitoring while academy publications like remote sensing

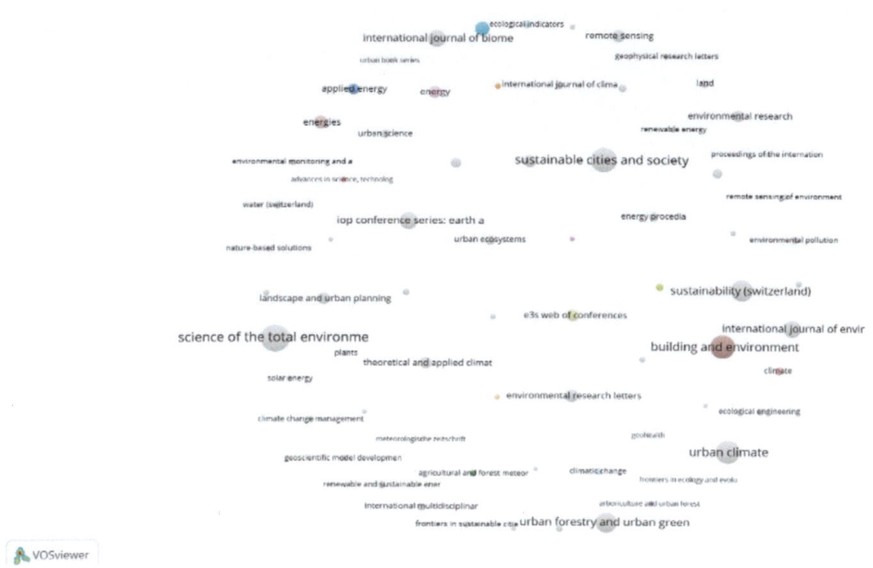

Fig. 3 Citation source

and Geo physical research letters highlight methodology advancements in data and modelling approaches, specialised generals like urban ecosystems, arboriculture and urban forestry and nature base solutions offer more focus contributions.

The interdisciplinary aspect of the area is particularly shown by the dispersion of publications in the fields of sustainability and environmental science. Nevertheless, the clustering also shows that citations are concentrated in a small number of prominent journals comma indicating a lack of diversity in publication venues. The exposure of context driven or region specific research especially from poor nations may be limited by this focus. To increase inclusion and reach future orientation should paradise wider participation using open access platforms and newly emerging journals with the sustainability focus. From the stand point of sustainability marketing, there are strategic opportunities to match academic results with policy narratives urban planning agenda and public communication on climate resilience due to the popularity of journals like sustainable cities and society and urban forestry and urban greening.

3.4 Co-Authorship

Figure 4 depicts the network of scholars that collaborate on studies pertaining to urban climate heat stressed green places and environmental sustainability that is depicted in each point in this graphic representation, and author and the lines connecting them indicate co-authorship; that is, authors who have collaborated on publications. The closer to writers are in the figure: the more frequently they work together or have related fields of study.

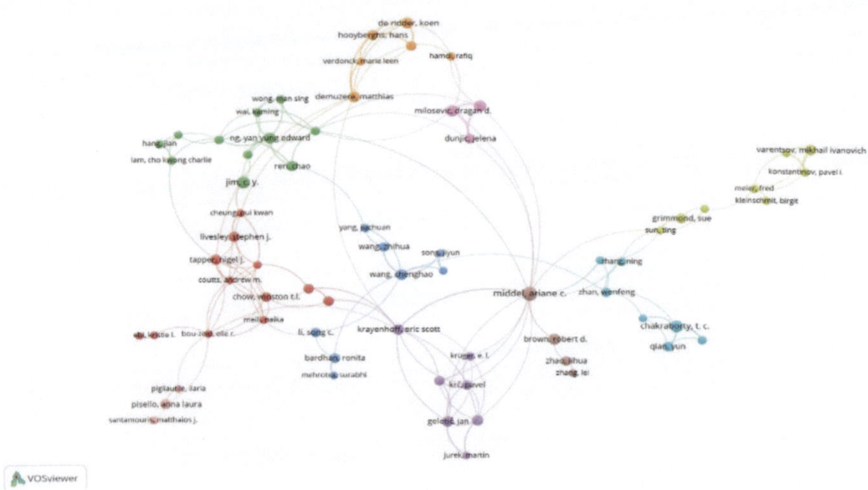

Fig. 4 Co-authorship

Each hue on the map represents a group of researchers who commonly collaborate or are members of the same research community. For instance writers like Middel A., Brown R. D., Krayenhoff E. S. and Jenerette G. D. who frequently research urban heat, micro-climate and the cooling effectiveness of green spaces are included in one cluster. Their closed eyes imply that they form a core group that produces significant urban heat mitigation efforts.

Researchers noted for their contributions to urban heat Island research vegetation cooling effects and urban micro climate modelling, such as Livesley S. J., Coutts A. M., Chow W. T. L, Tapper N., are part of another cluster. This group places a lot of emphasis on doable remedies for heat stress, like park, trees and green infrastructure topics that directly promote sustainable marketing by offering methods for promoting climate-resilient community that are supported by science.

Accordingly, there are a group of Asian and European writers like Yang J., Wang C., Zhang L., Hooyberghs H. and De Ridder K. who focus on subjects including heat risk assessment, atmospheric modelling and remote sensing of vegetation health. Their study offers data-driven insights that can direct policy and communication initiatives and aids cities in understanding how green spaces respond to rising temperature.

With fever connections, some writers see more remote. This just indicates that their work focuses on certain regional or limited scientific problems or that they engage less internationally. They nevertheless continue to add significant knowledge to the area as a whole.

Overall, the co-authorship figure shows how interdisciplinary and highly collaborative research is on heat stressed green places. In order to do better understand how green areas react to heat and how it is to create cooler healthier settings scientists from environmental science urban planning methodology and landscape architecture are collaborating. This network is crucial for sustainability marketing because it showcases notable authors whose work influences worldwide disclosure on climate adaptive green spaces and demonstrates where robust research relationships exist.

3.5 Co-Occurrence

Figure 5 illustrate a CO occurrence map of the important terms that regularly occur together in studies on sustainability, urban climate and the heat-stressed green space. The size of each dot indicates how frequently a particular appears in the literature and each plot represents a keyword used by authors. Closely spaced keywords are frequently utilised in the same research publications demonstrating the connections between various concepts and issues in the discipline.

They are multiple coloured grouping on the map. Climate change, urban heat Iceland, temperature land surface temperature are the subjects of the red cluster. These fundamental scientific ideas are used to comprehend the effects of heat stress on urban areas. This cluster demonstrates the diet relationships between heat-stressed

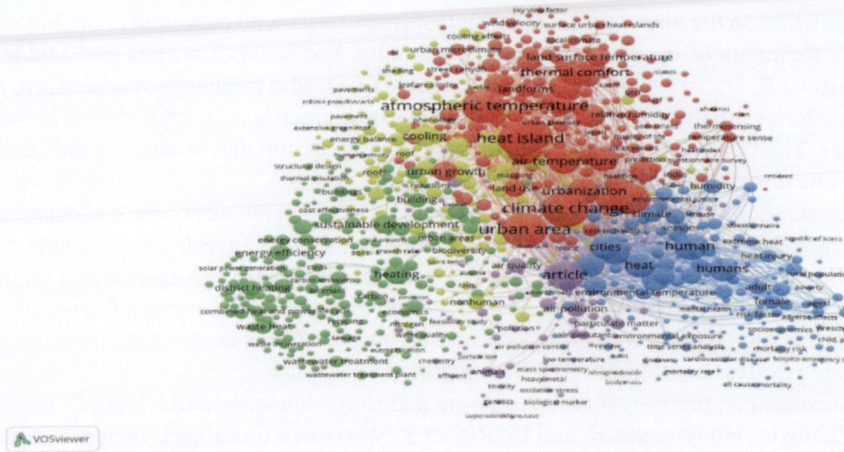

Fig. 5 Co-occurrence map

green spaces and more general climate problems that impact cities mainly rising temperature and changes in land use.

Words like sustainable development, energy efficiency heating, cooling, buildings and energy consumption are all parts of the green cluster. Green roof, shade structures, and improved urban design strategies are some of the tactics cities employ to lessen heat stress. In order to promote greener and more resident urban environments this cluster emphasizes the significance of connecting heat stressed green spaces with sustainable planning and energy saving strategies key subjects for sustainability marketing.

Topics like atmospheric temperature urban growth environmental temperature and urban area development are covered by the yellow and orange clusters. These themes demonstrate how growing urbanization exacerbates eat stress and how green space are essential for cooling and ensuring quality of life in growing cities. This illustrates how planning for green areas is crucial to the development of healthy cities and goes beyond environmental control.

The terms heat exposure, mortality air pollution, disease and public health are all part of the blue cluster which emphasizes human health. This cluster demonstrates the heat stress affects individuals particularly vulnerable groups like children and the elderly in addition to plans and environment. The existence of these terms suggest that the benefits of green spaces for human health and the necessity of heat mitigation techniques are becoming more widely acknowledged in study.

Overall the co- occurrence map demonstrates the close connections between research on heat stressed green areas and a number of significant topics including urbanization, climate change, energy use, environmental quality and human health. By encouraging nature-based solutions enhancing liveability and assisting cities in communicating the benefits of green infrastructure to the community this interconnectedness demonstrates how green space management may promote sustainability marketing. It demonstrates that combating heat stress is a social, economics and public health concern in addition to an environmental one.

4 Systematic Literature Review: Heat Stress in Urban Vegetation

4.1 Parks & Urban Forests

- Mentes et al. (2024) analysed that large urban parks can reduce Summerland's surface temperature by roughly 6 degrees Celsius, according to a study conducted in cities in cold regions. Accordingly, it demonstrated how the park size and the surrounding area affect the strength of the cooling effect. Using thermal remote sensing and cross-city comparisons, the authors related park morphology to cooling gradients. Constraints include reliance on LST (surface, not air) and seasonal/cold-region bias (Menteş et al., 2024).
- Maassoumeh Barghchi et al. (2024) synthesized evidence on park cooling mechanisms now and under future climates, finding shade as the dominant daytime driver with transpiration increasingly important under higher temperatures when water is available. The paper triangulated ENVI-met simulations with field and comfort indices (e.g., UTCI) to profile diurnal effects. Limitations include model sensitivity to vegetation parameters and water availability assumptions. Future work should incorporate irrigation constraints and species-specific traits. Results are designed for shade first, then boost evapotranspiration where feasible to sustain thermal comfort as climates warm (Barghchi et al., 2024).
- Huawei Li et al. (2020) case study mapped Park Cooling Intensity in Zhengzhou (China) using Landsat-8 LST (radiative transfer method) and 123 parks categorized by type, linking park size, shape, and internal vegetation to cooling effects. The methodology integrated city-wide comparative statistics, high-resolution photography, and spectral indicators. Temporal, single imagery reliance, and the lack of ground air temperature confirmation are among the limitations. Future studies could study seasonal composites and movable transects. In conclusion, park layout and vegetal compositions can significantly affect cooling, giving rapidly expanding cities design levers (Li et al., 2020).

4.2 Wetlands & Blue-Green Infrastructure

- Zhenshan Xue et al. (2019) assessed cooling from urban and peri-urban wetlands across Northeast China using remote sensing, finding that wetlands generally outperform terrestrial green spaces and that area, shape, and hydrologic connectivity strongly govern cooling capacity. The study related wetland metrics to LST patterns to infer cooling distance and intensity. Limitations include reliance on LST and limited microclimate/air measurements; built-form effects were inferred. Future scope includes coupling hydrologic reconnection scenarios with microscale air-temperature monitoring and comfort indices. Conclusion: Protecting and reconnecting peri-urban wetlands is a potent, often superior, heat-mitigation strategy (Xue et al., 2019).

- Md Rejaul Islam et al. (2024) assessed cooling distances of blue-green infrastructures (BGIs), showing typical maximum influence ranges around ~ 210–240 m (up to ~ 300 m) from BGI edges, with diminished effects in dense urban fabrics. Using buffer analyses of thermal imagery, the paper mapped intensity and decay with distance. Limitations include surface-temperature focus and potential confounding from shadows and materials. Future work should pair remote sensing with traverses and exposure metrics for residents. Conclusion: cooling zones are finite and context-dependent—BGIs should be spaced so influence fields overlap in heat hotspots (Islam et al., 2024).

4.3 Urban Agriculture

- Katherine Klink (2025) analysed Urban Forestry & Urban Greening article positioned urban farms as green infrastructure for heat mitigation, emphasizing pervious soils and actively transpiring crops that can cool neighbourhoods while yielding co-benefits (food, education). Methods included literature synthesis and emerging empirical cases. Limitations are sparse standardized measurements of cooling from working farms and the seasonality of crop canopies. Future research should quantify air temperature and comfort benefits over crop cycles and compare farm typologies (ground, rooftop, peri-urban). Conclusion: urban agriculture has under-measured yet promising cooling potential, especially when integrated into broader blue-green networks (Klink, 2025).

4.4 Modelling & Simulation

- Juui Kim et al. (2025) applied ENVI-met to test how tree traits, layout, and building interplay affect air temperature and energy outcomes, suggesting tree placement and canopy structure can reduce thermal stress and improve comfort/energy use if calibrated carefully. The method blends microclimate CFD/energy simulation with scenario testing of planting schemes in high-rise/compact morphologies. Limitations include model parameter uncertainty (e.g., stomatal conductance, soil moisture) and site-specific results. Future scope: multi-model ensembles, in-situ calibration, and co-optimization for comfort and building energy. Conclusion: microclimate simulations are powerful for siting/shaping green elements but must be validated and coupled to design constraints (Kim er al., 2025).

- Rahman and Hasan (2024) conducted a quantitative study in Rajshahi, Bangladesh, how urban green space mitigates heat by linking vegetation (NDVI) to a Heat Mitigation Index (HMI), land-surface temperature (LST), and air temperature using the InVEST Urban Cooling Model and spatial analysis. Strong inverse relationships were found between HMI and both LST and air temperature, indicating substantial cooling from greener areas. Limitations include model assumptions (e.g., vegetation homogeneity) and limited in-situ validation. Future work should combine ground sensors with modeling and test implementation feasibility and equity. Overall, the paper provides evidence that targeted increases in vegetated cover can measurably reduce urban heat in developing-city contexts (Rahman et al., 2024).
- Zhou et al. (2025) applied Geographic comparative methods paper evaluated three common approaches to quantify green-space cooling—buffer averaging, visual interpretation, and a "basin analogy" method—using remote sensing. It was found to be of high uncertainty, visual interpretation was more reliable, and a buffer average for the appropriate basin method in specific figures. Writers ask for methodical standardization. Work limitations are the result of satellite data options and subjective stages in some methods. Future scope involves harmonized protocols and multi-source data fusion (e.g., UAV + ground sensors). Bottom line: method selection substantially affects quantified cooling, so standard methods are critical for policy relevance (Zhou et al., 2025).

4.5 Cooling Metrics & Policy

- Prashant Kumar et al. (2024) investigated that cooling benefits vary by infrastructure type and urban form, e.g., botanical gardens/large, biodiverse parks and connected corridors can yield stronger, farther-reaching cooling than small, isolated patches. Reviews integrate remote-sensing metrics with comfort indicators to rank strategies and highlight equity gaps in access to cooling. Limitations include heterogeneity of metrics and underrepresentation of arid/tropical regions in samples. Future work should target standard metrics, equity mapping, and climate-robust planting palettes. In conclusion, cities that are experiencing heat stress can remain cooler and more robust by creating green and blue spaces that are interconnected, varied in nature and plants, planned equitably for all people, and organized as a unified system (Kumar et al., 2024).

5 Case Studies of Heat-Stressed Greenspaces

5.1 Case Study 1: Urban Parks in Delhi—Coping with Extreme Heat

Delhi's urban parks provider crucial illustration of how green areas react to rising temperatures and growing heat stress. Summertime temperatures in Delhi are among the highest in India frequently surpassing 45 degree Celsius. Many parks struggle to sustain healthy vegetation with such intense heat, particularly grass lawns young trees and floral plants. Heat waves may rapidly dry up the soil necessitating more frequent irrigation. Because of the park management struggle to preserve the parks greenery and usability for guest specially during the hottest months.

In order to mitigate this condition many parts in Delhi have improved irrigation system with drip watering, expanded shade areas through dance three canopies and planted native plants that endure heat. Hardy native plants that can tolerate prolong dry spells have taken the place of water demanding exotic species in various paths to further limit water loss mulching and soil moisture conservation measures are implemented. By doing these things, Sparks can stay cooler, healthier and more resilient to heat.

From the perspective of sustainable marketing, Delhi urban parks exemplify how green areas may promote as cooling zones that improve public health during heat waves in addition to being pleasant locations. Communities might be inspired to respect and maintain these spaces by discussing the advantage of shade tree cover and nature base solutions. In order to ensure that urban green areas continue to function even in the face of high heat that instance also emphasizes the requirement of collaboration between Municipal urban planning environmental department and residents. Delhi park illustrates the challenges and opportunities of managing green spaces under heat stress in rapidly growing metropolitan environment (Dixit, 2025).

5.2 Case Study 2: Vertical Greening Systems in Singapore's Commercial Zones

Installing a vertical greening system in the commercial district is one of Singapore best strategies for reducing urban heat. The solution includes roof top Gardens green wall and plant covered building facades. Vertical greening can produce cooling effects where traditional park might not fit in a city with limited acreage and high temperatures. These green Structures have been installed in commercial zones like Orchard road Marina Bay and important business location to improve air quality by lowering surface temperatures and reducing heat buildup on buildings.

The V G systems believe in employing consciously the plants to survive in extreme heat of direct sunlight. Such plants help in cooling the surroundings and decreasing the temperature. Vertical vegetation also enhances biodiversity and offers shelter and improve the commercial street aesthetics, attract complete stops and tourists find city surroundings more comfortable because of these advantages particularly on hot afternoons when the heat stress is at its worst.

Vertical greenery acts as striking visual depiction of Singapore identity as a city in a garden from the stand point of environmental marketing. This branding aids in portrayed the CTS progressive, ecological consciousness and dedicated to climate resiliency. Additionally, organisations adopt green walls as a marketing tool to demonstrate their attention to sustainability attracting in Eco aware clients and enhancing their reputation. All things considered, Singapore vertical greening effort shows how smart communication and inventive design may cooperatively control heat stress and support a greener urban way of life (YAGAY & SUN, 2025).

5.3 Case Study 3: Community-Led Greening Projects in Phoenix, USA

One of the hottest places in the country is phoenix in Arizona where summer time high of temperature exceeding 40 degrees Celsius is normal. Because of the severe heat many green places struggle to survive. Plants dry out quickly, soil loses its moisture and ecosystem become weaker. Particularly in areas with limited shade, the circumstances make it challenging for vegetation to remain healthy.

Many Phoenix towns have launched there on green initiative to address this issue. Volunteer and small organisation work together to plant drought tolerant trees and establish small green parks and add shade along Street. This project focus on employing native and heat-resistant plants that can survive lengthy dry periods with lesser water. Many areas also employee mulching, rainwater harvesting and drip irrigation to save water and support plant development.

These initiatives have developed a small but effective cooling islands that lower heat in the surrounding environment. New plants offer shade to tourist and drop surface temperature to improve outdoor relaxation in hot months. As the initiative evolve, they make outdoor places more habitable for local families.

From a sustainability market view point, phonic shows how community participation may strengthen climate condition. When peoples are involved in planting and maintaining for green spaces, they feel more responsible and connected to the environment. Local citizens and city officials use storytelling, social media and community events to highlight the benefit of the programs such as improving temperature, health and stronger neighbourhood links. This inspires peoples to join and develop long term support for greener and environment friendly communities (Quenzler, 2024).

Lessons Learned

Table 1 Cross-regional comparisons

Region	Cooling capacity	Access and equity	Model of action
Delhi	Moderate (park-based)	Limited, clustered in wealthier districts	Public infrastructure needs stronger management
Singapore	High (vertical greening)	Policy-driven, widespread adoption	Incentive-based, commercial zones
Phoenix	Significant (community-led)	Variable improves with local engagement	Grassroots, partnership-based

- **Cooling Effect is Universal:** Urban green spaces like parks and forests are consistently reducing the urban heat island effect across different climates and city types by decreasing air and surface heat through shadow and evaporation as described in Table 1.
- **Health and Social Benefits:** Urban areas with increasing greenery is positively related to decreasing heat related diseases and deaths as well as better mental health, social stability and overall welfare during heatwaves.
- **Infrastructure and Equity Gaps:** Many cities especially in the Global South (like Delhi as in Table 1) are lacking in the quantity and quality of green areas found in wealthy areas and hence resulting in decreasing the impact of heat stress for unsafe citizen and increasing cooling adaptation size.
- **Design and Planning Matter:** As given in Table 1, the effectiveness of green space depends upon its structure, elevation and area. A well-managed park generally applies vertical greening and diversity of urban foresting techniques that helps to maximize cooling and enhance comfort.
- **Community Participation Drives Success**: Local residents or organisations initiate greenery projects that not only cools cities but also creates local social capital and long-term maintenance of their neighbourhoods and therefore a long-lasting effect as a key for sustainability marketing (Li et al., 2024; Nazish et al., 2024).

6 Strategic Implications for Sustainability Marketing

Sustainability market is crucial when it comes to helping communities to cope with heat stressed green spaces. It makes people realise that gardens, park and trees are vital and are not decorative items. They are important for maintaining regions cooler, improving air quality index and developing healthier neighbourhoods. Marketing can convince the public to support initiative by clearly describing the advantages like park expansion and preservation of existing green spaces as outlined in Table 2.

Place branding is essential element for decreasing heat stress. By promoting initiatives like local gardens, preserved wet areas, shaded streets and other environment friendly initiatives in urban and peri-urban areas can show the effect of being green and climate family. By promoting these initiatives their achievements contribute

Table 2 Strategic implications

Strategic dimension	Role of greenspaces	Marketing opportunity
Place branding	Makes cities cooler, greener, and more liveable	Market the city as a climate-resilient, nature-positive destination
Consumer behaviour	Encourages people to choose eco-friendly landscapes and support green policies	Use simple messages linking greenspaces to comfort, health, and lifestyle
Public health framing	Cooler spaces reduce heat stress risks	Promote greenspaces as preventive health infrastructure
Policy support	Strong public understanding encourages investment in green networks	Target campaigns to motivate demand for trees, parks, and nature-based solutions

to the development of an encouraging impact of a city that is sustainable, healthy and future ready. This will draw tourists and business attention that value healthy environment living.

Sustainability marketing also encourages community involvement. People are more inclined to join plantation drives coma take care of neighbouring trees and take part in local environment activities when they comprehend how heat affects plants and discover easy ways to maintain green spaces. Pipal can feel more connected to nature and inspired to preserve it through storytelling, social media, workshop and neighbourhood gatherings.

Another important area is encouraging environmentally friendly and heat-tolerant planting. Marketing can help families, businesses and institutions to choose native plants, low water gardens and shade sustaining trees. Presenting the stories is as a rational or inexpensive and climate friendly foster wider adoption.

Sustainability market help bridges the gap between scientific knowledge and popular awareness. Government can make easily win public support for growing green networks, enhancing upkeep and incorporating nature-based solution into urban design when information is communicated correctly. This help cities to become able for rising temperature and long-term climate challenges.

Examples of Green Infrastructure Promotion Campaigns is successful campaigns include Singapore incorporation of green architecture with commercial branding. Cities redevelopment using existing forested areas and Tokyo Grandberry Park that connects green corridors with commercial spaces and clear signage about bioswales and rain gardens. Other cities have used story-driven marketing to showcase the social impact and climate resilience of community-led projects and encouraging local pride and wider public adoption. These programs result in improved comfort, social unity and economic value by increasing sustainability (Ministry of Land, Infrastructure, Transport and Tourism, 2024).

7 Bridging Ecology and Marketing: A Framework for Climate-Conscious Urban Planning

An effective technique is to build cities that can adapt to climate change while safeguarding their green areas by combining marketing with ecology. Ecology helps us to understand the relationship between interaction of plants, oil and climate. While marketing helps in conveying the information in a way that motivate individuals to act. When these two sectors collaborate cities may create green places that are not only environmentally friendly but also valued and supported by the local citizens. This facilitates the promotion of climate friendly solutions that people are willing to support and invest.

Understanding the ecological needs of green areas such as selecting plants that can absorb heat and retain moisture of the soil. Developing interconnected parks and green corridors that promote the growth of nature. Tools like GIS mapping help planners to determine the hottest location and select new parks or green walks thar will provide cooling. Cities can build continuous well design green system that will provide shelter, comfort and accessibility in respect to scattered patches of vegetation.

These ecological concepts are then translated into clear, relevant messages with the help of marketing. It discusses the importance of green areas, how they reduce the temperatures and how they improve daily life by creating a better air quality and cleaner streets. Clear communication and public campaign can inspire citizens, companies and policy maker to embrace and protect urban greenery. When people realize the benefit of green space for climate ready legislation.

Participation in the community is also crucial. People get a sense of ownership when the participate in the creation and up keep of green spaces. Social media, local events and visual advertising can improve and this links and increase long term participation. Green places tend to thrive when communities feel linked to them. Ultimately combining marketing and ecological enable cities to transcend conventional planning. It helps them to develop green areas that are scientifically sound and durable. With the correct balance of environment and knowledge cities can construct greener, cooler and comfortable place for everyone (Cortinovis & Geneletti, 2019; Saqib et al., 2024).

8 Challenges and Research Gaps

8.1 Challenges

There are ample of difficulties in managing heat stressed green spaces in urban and peri urban regions. Rapid urbanization is one of the main causes that keeps declining in natural vegetation as buildings, roads and concrete surfaces is increasing. This increases the temperature and makes it difficult for plants to survive under extreme heat. At the same time many green spaces in cities are small and unconnected reducing their overall cooling ability. Fragmentation inhibits plants and wildlife from moving freely and reduces the resilience of the entire greenspace network.

The unfavourable environmental circumstances that many urban green spaces deal with present another difficulty. Vegetation finds it challenging to remain healthy during heatwaves due to a lack of water, frequent droughts and poor soil. Many municipal parks and roadside green strips do not receive the routine care required to resist rising temperatures which frequently exacerbates these problems. Furthermore, there is currently a lack of public knowledge of plant heat stress. Many people appreciate green areas but often are unaware of how certain behaviours can further erode growth.

Moreover, there is the problem of different neighbourhoods having differing access to green areas. While some neighbourhoods have virtually little flora, others have parks and lots of shelter. This results in uneven heat exposure and unbalanced environmental advantages, making heat stress an environmental issue as well as a societal one (Cilliers et al., 2013; Zuniga et al., 2020).

8.2 Research Gaps

Urban cores, which are essential for cooling and biodiversity, are the focus of current research. Socio economic and equity issues remain neglected, with limited studies addressing discrepancies in access to cooling benefits among disadvantaged communities. Furthermore, most evidence is based on short term or single season datasets, disregarding long term vegetation adaptability and seasonal fluctuation. While sustainable marketing is regarded as a strategic tool, empirical studies relating marketing actions to measurable environmental and behavioural effects are uncommon. Lastly rather than using integrated cross disciplinary frameworks that incorporate behavioural insights, urban planning and climate science, research frequently isolates ecological, social and economic components. Addressing these differences is critical for building holistic, equitable and climate resilient solutions for urban and peri urban green infrastructure.

9 Conclusion and Strategic Recommendations

9.1 Conclusion

Urban and peri-urban green spaces are crucial for cooling cities, improving air quality and benefiting public health however they confront increasing vulnerability due to heat stress generated by rapid urbanization and climate change. Inadequate maintenance dispersed green networks and rising temperatures weak vegetation resilience and reduces the cooling advantages these areas offer. Proof from international research and case studies reveals that nature-based solutions can successfully reduce heat stress while enhancing liveability and social fairness. Examples are Singapore vertical greening, Delhi adaptive park systems and community-led projects in Phoenix as describes in case studies.

For reducing heat stress ecological measure is crucial. Sustainable marketing came up for converting innovative ideas of greenery into useful citizens knowledge and policies implementation. Green areas represent an important factor for sustainable development instead of decorative items. Marketing has the ability to guide consumer behaviour, encourage citizens involvement and create demand for appropriate greenery access and sustainable planting. The long-term benefits of putting efforts in green infrastructure are made clear to individuals and governments through strategic communication using online publicity and storytelling.

Overall, for tackling heat stress an approach to a strategy that incorporates innovative design, involvement in environment friendly planning and making attractive greenery messaging. By working together cities will be able to build interconnected green networks that are survive and make them ready for climate change conditions and provides all urban resident comfort and health. Therefore, integrating marketing and environment friendly programs is essential in creating sustainable future of urban and peri urban areas.

9.2 Strategic Recommendations

9.2.1 Urban Planners

- Prioritize both the quantity and connectivity of green spaces using advanced spatial planning and biodiversity principles.
- Involve community stakeholders to guide location and management decisions for maximum benefit (Wang et al., 2024).

9.2.2 Policymakers

- Policy maker should implement policies that fills infrastructure gaps.
- Policies should promote all residents access for green spaces
- Should implement sustainable marketing techniques to communicate, educate and incentivize the benefits of green spaces to different group of community (Ajirotutu et al., 2024).

9.2.3 Researchers

- Expand monitoring and evaluation activities to include peri-urban and climate affected areas.
- Integrate ecological, social, and economic resilience dimensions into holistic urban sustainability research (Rahman et al., 2023).

References

Ajirotutu, R. O., Adeyemi, A. B., Ifechukwu, G. O., Iwuanyanwu, O., Ohakawa, T. C., & Garba, B. M. P. (2024). Designing policy frameworks for the future: Conceptualizing the integration of green infrastructure into urban development. *Journal of Urban Development Studies, 2.*

Atchison, J., Hendrigan, C., Forehead, H., French, K., & de Vet, E. (2024). Widely valued but differently experienced; understanding relationships with greenspace in the CBD. *Landscape and Urban Planning, 252*, Article 105175.

Athokpam, V., Chamroy, T., & Ngairangbam, H. (2024). The role of urban green spaces in mitigating climate change: An integrative review of ecological, social, and health benefits. *Environmental Reports.*

Awasthi, A., Vishwakarma, K., & Pattnayak, K. C. (2022). Retrospection of heatwave and heat index. *Theoretical and Applied Climatology, 147*(1), 589–604. https://doi.org/10.1007/s00704-021-03854-z

Barghchi, M., Grace, B., Edwards, N., Bolleter, J., & Hooper, P. (2024). Park thermal comfort and cooling mechanisms in present and future climate scenarios. *Urban Forestry & Urban Greening, 101*, Article 128533.

Cilliers, S., Cilliers, J., Lubbe, R., & Siebert, S. (2013). Ecosystem services of urban green spaces in African countries—perspectives and challenges. *Urban Ecosystems, 16*(4), 681–702.

Cortinovis, C., & Geneletti, D. (2019). A framework to explore the effects of urban planning decisions on regulating ecosystem services in cities. *Ecosystem Services, 38*, Article 100946.

Dixit, K. (2025, July 13). *City parks cooler, lack facilities: Study. The Times of India.* Retrieved from https://timesofindia.indiatimes.com/city/delhi/city-parks-cooler-lack-facilities-study/articleshow/122410663.cms

Fu, Q., Zheng, Z., Sarker, M. N. I., & Lv, Y. (2024). Combating urban heat: Systematic review of urban resilience and adaptation strategies. *Heliyon, 10*(17).

Fukano, Y., Yamori, W., Misu, H., Sato, M. P., Shirasawa, K., Tachiki, Y., & Uchida, K. (2023). From green to red: Urban heat stress drives leaf color evolution. *Science advances, 9*(42), eabq3542.

Gakobo, S. G. (2021). *Challenges of Protecting Urban Green Open Spaces in Sokoni Ward, Kilifi County.* Doctoral dissertation, University of Nairobi.

Gloria, S. J., & Gnanasekaran, S. P. (2024). Impact of urban vegetation loss on urban heat islands: A case study of Chennai Metropolitan Area. *Indian Journal of Science and Technology, 17*(2), 134–141.

Hassan, M. U., Chattha, M. U., Khan, I., Chattha, M. B., Barbanti, L., Aamer, M., & Aslam, M. T. (2021). Heat stress in cultivated plants: Nature, impact, mechanisms, and mitigation strategies—A review. *Plant Biosystems-an International Journal Dealing with All Aspects of Plant Biology, 155*(2), 211–234.

Hughes, N., Nyambe-Mubanga, M., Mapulanga, W., Hazemba, M., Chileshe, S., Tembo, B., & Crentsil, A. O. (2025). Towards resilient and inclusive climate compatible development: A participatory, mixed-method scenarios approach for Zambia. *Global Environmental Change, 95*, Article 103072.

Indeed Editorial Team. (2025, March 4). *Green marketing: Definition, benefits and strategies.* Indeed. Retrieved from 27 August 2025. https://www.indeed.com/career-advice/career-development/green-marketing

Irfeey, A. M. M., Chau, H. W., Sumaiya, M. M. F., Wai, C. Y., Muttil, N., & Jamei, E. (2023). Sustainable mitigation strategies for urban heat island effects in urban areas. *Sustainability, 15*(14), 10767.

Islam, M. R., Talukdar, S., Rihan, M., & Rahman, A. (2024). Evaluating cooling effect of blue-green infrastructure on urban thermal environment in a metropolitan city: Using geospatial and machine learning techniques. *Sustainable Cities and Society, 113*, Article 105666.

Kim, J., Yeom, S., & Hong, T. (2025). Analyzing the cooling effect, thermal comfort, and energy consumption of integrated arrangement of high-rise buildings and green spaces on urban heat island. *Sustainable Cities and Society, 119*, Article 106105.

Klink, K. (2025). Urban farms as green infrastructure for urban heat mitigation. *Urban Climate, 61*, Article 102440.

Kumar, P., Debele, S. E., Khalili, S., Halios, C. H., Sahani, J., Aghamohammadi, N., & Jones, L. (2024). Urban heat mitigation by green and blue infrastructure: Drivers, effectiveness, and future needs. *The Innovation, 5*(2).

Li, H., Wang, G., Tian, G., & Jombach, S. (2020). Mapping and analyzing the park cooling effect on urban heat island in an expanding city: A case study in Zhengzhou city. *China. Land, 9*(2), 57.

Li, Y., Svenning, J. C., Zhou, W., Zhu, K., Abrams, J. F., Lenton, T. M., & Xu, C. (2024). Green spaces provide substantial but unequal urban cooling globally. *Nature Communications, 15*(1), 7108.

Menteş, Y., Yilmaz, S., & Qaid, A. (2024). The cooling effect of different scales of urban parks on land surface temperatures in cold regions. *Energy and Buildings, 308*, Article 113954.

Ministry of Land, Infrastructure, Transport and Tourism. (2024, September). *Recommendations for green infrastructure projects and finances: Toward the promotion of investment in urban development and city planning through visualization of economic benefits* (Summary). Research Group on the Economic Value of Green Infrastructure in the Marketplace. https://www.mlit.go.jp/sogoseisaku/environment/content/001768002.pdf

Nazish, A., Abbas, K., & Sattar, E. (2024). Health impact of urban green spaces: A systematic review of heat-related morbidity and mortality. *British Medical Journal Open, 14*(9), Article e081632.

Oetken, K. J. (2025). Unravelling the why: Exploring the increasing recognition and adoption of co-creation in contemporary urban design. *Sustainable Communities, 2*(1), 2477788.

Parihar, J., & Birman, S. (2024). Heat resilience in urban environments: strategies for sustainable city climate management. In *The Climate-Health-Sustainability Nexus: Understanding the Interconnected Impact on Populations and the Environment* (pp. 305–324). Springer Nature Switzerland.

Pattnayak, K. C., Awasthi, A., Sharma, K., & Pattnayak, B. B. (2023). Fate of rainfall over the North Indian states in the 1.5 and 2 C warming scenarios. *Earth and Space Science, 10*(2), e2022EA002671. https://doi.org/10.1029/2022EA002671

Quenzler, D. (2024, August 22). *Green living in Phoenix: Engaging communities for sustainability.* Retrieved from https://www.danijelaquenzler.com/green-living-in-phoenix-engaging-communities-for-sustainability

Rahman, M. M., & Hasan, J. (2024). Evaluating the impact of green spaces on urban heat reduction in Rajshahi, Bangladesh using the invest model. *Land, 13*(8), 1284.

Rahman, M. A., Hossain, M. Z., & Rahaman, K. R. (2023). Climate urbanism as a new urban development paradigm: Evaluating a city's progression towards climate urbanism in the global south. *Climate, 11*(8), 159.

Saqib, A., Khan, M. S. U., & Rana, I. A. (2024). Bridging nature and urbanity through green roof resilience framework (GRF): A thematic review. *Nature-Based Solutions, 6*, Article 100182.

Tandon, A., Awasthi, A., & Pattnayak, K. C. (2025). Efficacy of machine learning in simulating precipitation and its extremes over the capital cities in North Indian states. *Scientific Reports, 15*(1), 10345. https://doi.org/10.1038/s41598-024-84360-w

Valánszki, I., Foldi, Z., Iváncsics, V., Erdei, T., Borkó, A., & Boromisza, Z. (2025, April). Differences between perception and reality-green infrastructure related cultural ecosystem services and development preferences under Peri-urban pressure. In *Fábos Conference on Landscape and Greenway Planning* (Vol. 8, No. 1). University of Massachusetts Amherst Libraries.

Wang, D., Xu, P. Y., An, B. W., & Guo, Q. P. (2024). Urban green infrastructure: Bridging biodiversity conservation and sustainable urban development through adaptive management approach. *Frontiers in Ecology and Evolution, 12*, 1440477.

Xue, Z., Hou, G., Zhang, Z., Lyu, X., Jiang, M., Zou, Y., & Liu, X. (2019). Quantifying the cooling-effects of urban and peri-urban wetlands using remote sensing data: Case study of cities of Northeast China. *Landscape and Urban Planning, 182*, 92–100.

Yadav, N., Rajendra, K., Awasthi, A., Singh, C., & Bhushan, B. (2023). Systematic exploration of heat wave impact on mortality and urban heat island: A review from 2000 to 2022. *Urban Climate, 51*, Article 101622. https://doi.org/10.1016/j.uclim.2023.101622

YAGAY and SUN. (2025, March 6). *Vertical Green Belts (VGBs)—Whether such projects ever had been implemented in any corner of the world or not?* TaxTMI. Retrieved from https://www.taxtmi.com/article/detailed?id=13735

Zhou, W., Yu, Y., Zhang, S., Xu, J., & Wu, T. (2025). Methods for quantifying the cooling effect of urban green spaces using remote sensing: A comparative study. *Landscape and Urban Planning, 256*, Article 105289.

Zuniga-Teran, A. A., Staddon, C., De Vito, L., Gerlak, A. K., Ward, S., Schoeman, Y., & Booth, G. (2020). Challenges of mainstreaming green infrastructure in built environment professions. *Journal of Environmental Planning and Management, 63*(4), 710–732.

Integrated Eco-Climatic Framework for Physics-Informed Crop Modelling

Etukakpan Akpe Udoeno, Imoh Dominic Ekpa, Samuel Eyeh Mopta, Mfoniso Udofia Aka, and Donatus Enang Bassey

Abstract This chapter assesses regional crop and ecosystem risks in sub-Saharan Africa, employing operational early-warning systems, coupled with gridded climate products, statistical regressions or process-based crop models driven by seasonal means. These schemes resolve short hot spells, vapour-pressure deficit (VPD) extremes, or how management reshapes boundary-layer conditions, which limits their ability to explain observed yield and greenness anomalies. This study develops an Integrated Eco-Climatic Framework and physics-informed modelling workflow that link boundary-layer climate, soils, and vegetation state and farm decisions in a single loop-based system. Climate, hydrology/soils, ecosystem state, agriculture, and adaptation are treated as interacting nodes, and each arrow is operationalized through stage-aligned hot-day exceedances, VPD, diurnal temperature range, soil-moisture anomalies, NDVI/LAI, and explicit management flags (sowing date, residue, irrigation, AWD/bunds). Long-term reanalysis, satellite products; crop or greenness outcomes for Nigeria maize, Sahel savannas, and East Africa highland rice are harmonized on a common grid and crop-stage calendar to build a physics-proximal feature set. Statistical baselines (linear, GLM, and mixed-effects models) and machine learning models (tree-based ensembles and shallow neural networks) are specified in parallel on seasonal aggregates and eco-climatic indices, under task-specific causal diagrams, monotonicity constraints and strict no-leakage rules, and evaluated with space–time-blocked cross-validation and robustness checks. The framework formalizes three hypotheses: that timing of heat matters more than seasonal means, that energy partitioning and VPD explain stress better than rainfall totals in drylands, and that microclimate-shaping practice can buffer damage. Across all systems, physics-informed models consistently outperform seasonal-mean baselines, improving explanatory power and sharpening identification of adaptation levers.

E. A. Udoeno (✉) · S. E. Mopta · D. E. Bassey
Department of Physics, University of Calabar, Calabar, Nigeria
e-mail: etukakpanakpe@gmail.com

I. D. Ekpa · M. U. Aka
Department of Physics, Federal University of Technology Ikot Abasi, Ikot Abasi, Nigeria

A. Awasthi et al. (eds.), *Rising Temperatures and Plant Ecosystems*,
https://doi.org/10.1007/978-3-032-29477-7_14

Keywords Integrated eco-climatic framework · Physics-informed machine learning · Crop heat and VPD stress · Stage-aligned stress indices · Sub-Saharan Africa agriculture · Climate-smart adaptation and early warning

1 Introduction

Global temperatures are rising due to increasing greenhouse gas emissions, and African land areas are warming faster than the global average. Recent assessments suggest warming rates of about 0.3 °C per decade in Africa since the early 1990s, with more frequent hot extremes and agricultural droughts (Akinola et al., 2021; IPCC, 2021; Hatfield &Prueger, 2015). These changes already threaten food systems by shortening growing seasons and increasing heat and water stress on key crops (IPCC, 2022; Luhunga et al., 2025). Global meta-analyses indicate that each additional 1 °C of warming can reduce average yields of major staples such as maize and wheat by about 6–8%, with particularly large relative losses in tropical regions (Jagadish et al., 2023; Okpara et al., 2020; Zhao et al., 2017).

Sub-Saharan Africa is especially exposed because many farmers depend on rain-fed smallholder systems, often on marginal soils, with limited access to irrigation or inputs. In these settings, short hot spells around flowering, elevated vapour-pressure deficit (VPD), and compound heat–drought episodes can damage crops even when seasonal rainfall totals appear adequate (Ishaya et al., 2020; Sultan et al., 2019). Yet many operational risk assessments still describe climate using seasonal-mean temperature and rainfall, which smooth out damaging extremes and ignore the role of land surface properties and management in shaping boundary-layer conditions (Akinsanola& Zhou, 2019; Oguntunde et al., 2021).

Most modelling efforts used in Africa either couple gridded climate data to simplified crop models or to empirical regressions built on seasonal averages. These approaches are valuable for large-scale outlooks but often treat climate, vegetation. and management as separate modules (Jagadish et al., 2023; Dickinson et al., 1989). They rarely resolve stage-specific sensitivity to heat and VPD, or how practices such as mulching, residue retention, shading and irrigation alter canopy temperatures and evaporative demand (Bonan, 2015; IPCC, 2022; Raza et al., 2019). This limits their ability to explain observed yield or greenness anomalies and to identify practical adaptation levers for farmers.

In this chapter, we introduce an Integrated Eco-Climatic Framework and physics-informed modelling workflow that link boundary-layer climate, soils, and vegetation state and on-farm decisions in a single loop-based system. Climate, hydrology/soils, ecosystem state, agriculture. and adaptation are treated as interacting units, and their connections are expressed through physically interpretable indices such as hot-day exceedances, VPD, diurnal temperature range. and soil-moisture anomalies, combined with crop stage and management information (Challinor et al., 2014; Van Ittersum & Cassman, 2013). This design allows us to compare conventional seasonal-mean models with physics-informed alternatives and to test, in a transparent way, how far a richer representation of eco-climatic processes improves explanatory power and practical relevance for African cropping systems.

To justify the novelty of this paper, our contribution is twofold:

We developed an Integrated Eco-Climatic Framework and physics-informed modelling workflow that incorporates boundary-layer climate, soil moisture, and vegetation state and management decisions into a singular, operational architecture.

We formalized a set of stage-specific, physics-aware stress indices (hot-day exceedances, VPD, dry-spell metrics, normalized difference vegetation index (NDVI) or leaf area index (LAI) anomalies, management flags) and coupled with their validation and diagnostic protocol that can be reused across regions and crops.

2 Integrated Eco-Climatic Framework (Non-Linear, Loop-Based)

The framework in Fig. 1 groups the system into five interacting units: climate, hydrology/soils, ecosystem state, agriculture, and adaptation. Each unit represents a cluster of processes that can be described with measurable variables from reanalysis, satellite, or ground data. Together, these units form feedback loops in which climate shapes soil moisture and evaporative demand, vegetation and management modify boundary-layer conditions, and farmers respond with new practices. The core feedback logic of Fig. 1 can be summarized as follows:

I. Heat and rainfall patterns determine soil moisture and evaporative demand.
II. Soil properties and moisture capacity govern how incoming energy is partitioned between latent and sensible heat at the land surface.
III. Crop structure and land management (mulch, residue, bunds, shade, irrigation) influence surface temperature, humidity and near-surface stability.
IV. Farmer decisions on sowing dates, cultivar choice and management respond to previous seasons and shape the eco-climatic conditions faced in subsequent seasons.
V. These loops provide the conceptual basis for the Integrated Eco-Climatic Framework applied in our case systems. Figure 1 summarizes the non-linear, bidirectional Integrated Eco-Climatic Framework applied in our case system

Figure 1 shows the Integrated Eco-Climatic Framework for physics-informed crop modelling. The ovals shapes show five interacting units: climate drivers (blue; Tmax/Tmin, heat extremes, VPD), hydrology and soils (green; soil moisture, runoff, and infiltration/erosion), ecosystem state (green; savanna–forest structure, biodiversity, and pests/disease), agricultural systems (orange; maize, rice, cassava, calendars, varieties, inputs, field geometry), and socioeconomic adaptation (yellow; sowing date, irrigation, residue, policy). Solid double-headed arrows represent biophysical

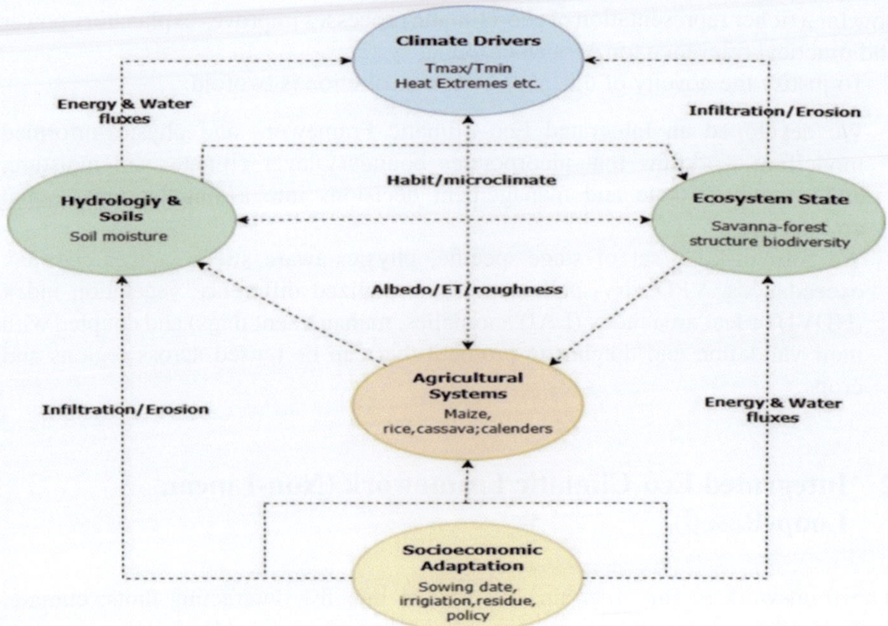

Fig. 1 Integrated eco-climatic framework

feedback loops (climate ↔ hydrology via energy and water fluxes; hydrology ↔ ecosystem via infiltration/erosion and nutrient cycling; ecosystem ↔ agriculture via habitat and microclimate; agriculture ↔ climate via albedo/ET/roughness). Dashed double-headed arrows show management mediation, with Socioeconomic Adaptation linked to each upper node, indicating that human decisions can dampen or amplify the biophysical loops. Edge labels ('energy and water fluxes', 'infiltration/erosion', 'habitat/microclimate', 'albedo/ET/roughness') summaries the main processes traveling along each link.

2.1 *Why the Framework is Nonlinear*

The Integrated Eco-Climatic Framework in Fig. 1 is nonlinear because small shifts in climate can trigger much larger changes in crop stress. A modest warming or increase in vapour-pressure deficit (VPD) may be manageable at vegetative stages, but becomes critical when it coincides with flowering or grain fills. Similarly, short dry spells are far more damaging when soils are already depleted.

Feedbacks also run in the opposite direction. Practices that increase infiltration, retain residue, or promote evaporative cooling can dampen or, in some cases, offset the impact of hot spells and high VPD. As a result, risk emerges from combinations of timing, intensity and management, not from a single linear climate–yield response. The framework is therefore designed to accommodate amplifying and buffering feedbacks rather than assuming a uniform response.

2.1.1 Operationalization of the Framework

The application of the Integrated Eco-Climatic Framework follows a simple sequence.

First, we map each unit in Fig. 1 to concrete variables, such as hot-day counts, VPD, diurnal temperature range (DTR), soil-moisture anomalies, NDVI/LAI and management flags (e.g. residue, shade, irrigation, alternate wetting and drying). Second, we align these variables with crop stages using local calendars and phenology information. Third, we construct eco-climatic indices that capture the timing and intensity of stress, for example HotDay@anthesis or VPD during booting–grain fill.

Finally, these indices feed into statistical and machine learning models that predict yield or greenness anomalies. The resulting feature set forms the basis of the predictive models and diagnostics used later in Chap. 16.

2.2 *Explicit Hypotheses*

We formalize three hypotheses (H1–H3) that link the framework to testable eco-climatic relationships.

H1: Timing of heat exposure explains yield better than seasonal temperature.

Stage-specific hot-day exceedances during maize anthesis or rice booting–grain fill should explain more of the variance in yield or NDVI anomalies than seasonal-mean temperature or growing degree days (GDD).

H2: VPD and boundary-layer indicators outperform rainfall totals in the Sahel.

Because atmospheric dryness limits stomatal conductance, VPD should be a stronger predictor of vegetation stress and Sahel greenness than rainfall totals alone (Akinola et al., 2021; Buckley, 2019).

H3: Microclimate management buffers heat damage.

Residue cover, shade, and alternate wetting and drying (AWD) irrigation are expected to buffer crops against heat by improving infiltration, sustaining evaporative cooling, and reducing canopy temperature (Ballester et al., 2022).

These hypotheses guide both our feature design and our interpretation of model outputs in the case studies.

2.3 *Study Regions and Case Systems*

2.3.1 Regional Profiles (Three Contrasting Settings)

This study covers three contrasting African regions, each paired with a representative cropping or ecosystem case: Nigeria—maize, Sahel belt—NDVI, and East Africa highlands—rice. The regions differ in hydro-climate, land cover, and management options, but all face rising temperatures and shifting rainfall patterns. Together they provide a test bed for the hypotheses introduced in Sect. 2.2.

Nigeria—maize: Nigeria spans humid southern zones, a sub-humid middle belt and a drier northern corridor. Maize is grown across these belts under mainly rain-fed conditions, with pockets of supplementary irrigation. Recent assessments show increasing exposure to short hot spells and elevated VPD during flowering and grain fill, especially in the northern and middle belts, while many farmers lack irrigation or cooling practices to buffer these shocks (IPCC, 2021; IPCC, 2022; Ballester et al., 2022). This setting is well suited to testing H1 on the timing of heat stress.

Sahel belt—NDVI: The Sahel is semi-arid, with short, variable rainy seasons and long dry periods. Vegetation is dominated by grasses and shrubs that respond quickly to changes in rainfall and atmospheric dryness. Recurrent droughts and heat waves have produced large swings in NDVI despite similar seasonal rainfall totals (Funk et al., 2015; IPCC, 2022). This region is ideal for testing H2 on whether VPD and boundary-layer indicators outperform rainfall totals in explaining greenness anomalies.

East Africa highlands—rice: The East Africa highlands combine moderate mean temperatures with strong diurnal cycles and complex topography. Irrigated rice systems rely on bunds, controlled flooding and AWD practices, and are sensitive to canopy-level heat and humidity during booting and grain fill (Jägermeyr et al., 2021; Lipper et al., 2014; Lobell et al., 2011). This case allows us to test H3 on how microclimate-shaping management can buffer damage.

Figure 2 and Table 1 provide regional maps with eco-zones, elevation, land cover, and crop-density overlays that anchor the cases introduced above.

2.3.2 Land Cover and Ecosystem Gradient

The three regions form a gradient from humid woodland, through savanna, to semi-arid shrub land. Along this gradient, vegetation structure, rooting depth, and surface roughness change systematically, altering how heat, moisture, and momentum are exchanged with the boundary layer. By combining land cover classes with elevation, slope, and NDVI/LAI, we quantify how vegetation state modifies heat exposure and evaporative demand across the case systems.

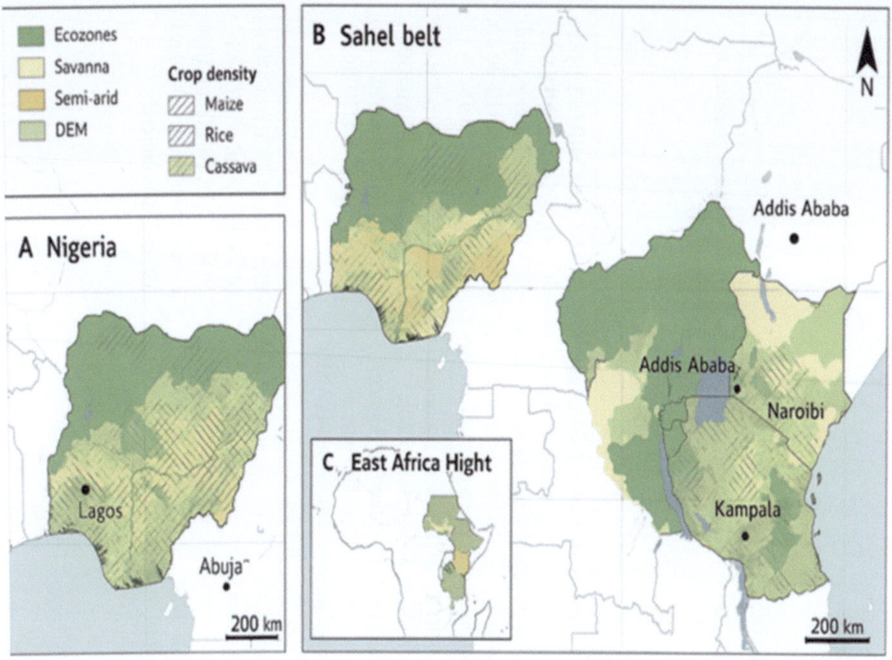

Fig. 2 Study regions and ecosystems (A—Nigeria, B—Sahel belt, C—East Africa highlands)

2.3.3 Agricultural Case Systems

The agricultural case systems focus on the dominant crops and their most sensitive growth stages.

Maize: Maize flowering is highly heat-sensitive, with only a few days of tolerance to extreme hot spells or high VPD. We therefore track stage-aligned stress using hot-day counts, VPD; diurnal temperature range (DTR) and soil-moisture anomalies around anthesis and grain fill. Key metrics: hot-day counts, VPD, DTR, soil-moisture anomalies.

Rice: booting and grain fill stages in irrigated rice are sensitive to canopy temperature, vapour-pressure deficit and water-table fluctuations. Core indicators include canopy-level hot-day counts, VPD, soil-moisture anomalies and AWD timing relative to these stages. Key metrics: canopy hot-day counts, VPD, soil-moisture anomalies, AWD timing.

Cassava: cassava is more heat tolerant than maize or rice but is sensitive to prolonged water stress and cumulative heat loads. We characterize its risk profile using multi-week hot periods, NDVI persistence and soil-moisture anomalies. Key metrics: multi-week hot-day indices, NDVI persistence, soil-moisture anomalies.

Table 1 Region × ecosystem × crop matrix (with climate context, hazards, levers, and datasets)

Region	Ecosystem zone	Dominant crops (case)	Typical Tmax (growing season, °C)	Mean VPD (kPa)	Key hazard	Dominant management levers	Primary datasets (spatial × temporal)
Nigeria	Humid–sub-humid lowlands; Guinea/ Sudan savannas	Maize (rain-fed), Rice (humid lowlands), Cassava	30–36	0.9–1.6	Short hot spells at anthesis; early-season dry spells; riverine flooding/erosion	Mulch/residue; modest sowing date shifts; supplemental irrigation where present; heat-tolerant varieties	ERA5-Land/CHIRTS/CHIRPS (0.1°–0.25°, daily); MODIS NDVI/LAI (250–500 m, 16-d); Landsat/Sentinel-2 (10–30 m); national stats (annual)
Sahel belt	Semi-arid shrub/ grasslands; pastoral–agro-pastoral mosaics	Extensive rangelands; paTchy rain-fed cereals (reference for NDVI case)	34–42	1.5–2.5	High VPD; onset failures; short seasons; wind erosion/dust; greening failures	Residue retention; zaï/ contour bunds; micro-caTchments; assisted natural regeneration (ANR)	ERA5-Land/CHIRTS/CHIRPS (0.1°–0.25°, daily); MODIS NDVI/LAI (250–500 m, 16-d); Landsat/Sentinel-2 (10–30 m)
East Africa highland	Elevated highlands and valley bottoms; montane mosaics	Rice (irrigated valleys), Maize (uplands)	22–30	0.5–1.2	Heatwaves on cool baseline; intense storms/ erosion; slope instability	AWD irrigation; field bunding/levelling; shade/shelterbelts; varietal choice	ERA5-Land/CHIRTS/CHIRPS (0.1°–0.25°, daily); MODIS NDVI/LAI (250–500 m, 16-d); Landsat/Sentinel-2 (10–30 m); national stats (annual)

2.3.4 Data Selection, Spatial Coverage, and Analytical Controls

Data selection prioritizes long-term, internally consistent climate records, and satellite products with documented uncertainty characteristics. We favour datasets that cover at least two decades, allow construction of eco-climatic indices and can be harmonized on a common grid across the three regions.

Quality control combines basic range checks, cross-sensor comparisons, sensitivity tests and space–time blocking to reduce artefacts and leakage. Spatial coverage is restricted to areas with sufficient crop presence or vegetation signal to support robust aggregation. These decisions ensure that the Integrated Eco-Climatic Framework rests on datasets that are both physically meaningful and suitable for the stage-resolved analysis described below.

2.4 Data Sources

Datasets on physical climate variables (temperature, rainfall, radiation, wind, and relative humidity) are drawn primarily from the ERA5-Land land surface reanalysis, which provides hourly and daily estimates on a about 9-km grid (Muñoz-Sabater et al., 2021), and from the CHIRPS and CHIRTS quasi-global rainfall and temperature archives designed for drought and climate-risk applications (Funk et al., 2015, 2019; Verdin et al., 2020). These products have been widely used in African agricultural and early warning studies.

Vegetation indices (NDVI, EVI, LAI) and land cover information come from MODISCollection 6.1 vegetation products (Didan et al., 2021) and complementary Landsat and Sentinel-2 multispectral imagery (Drusch et al., 2012). These sensors provide consistent time series of canopy greenness and land cover at resolutions compatible with our analysis grid.

All datasets are re-gridded to a common spatial resolution over the three regions, screened using basic range and consistency checks, and aligned to the stage-resolved aggregation described in Sects. 2.5 and 2.6.

2.5 Feature Engineering

To relate climate and ecosystem processes to crop outcomes, we transform raw gridded data into eco-climatic indicators that are both physically interpretable and stage-aligned.

Indicators include hot-day counters at key crop stages, diurnal temperature range (DTR), vapour-pressure deficit (VPD), growing degree days (GDD), proxies for planetary boundary-layer height (PBLH), and dry-spell metrics. These are computed at daily or sub-daily resolution, and then aggregated over biologically relevant windows around sowing; flowering and grain fill for each case system.

Feature engineering is implemented in a containerized workflow to ensure reproducibility across regions and time. The resulting feature set links the conceptual loops in Fig. 1 to concrete predictors for the statistical and machine learning models described in Sect. 2.6.

2.6 Integrated Physics and Machine Learning (ML) Framework

Statistical tools and mathematical modelling have key role in understanding the different environmental conditions like air quality (Chakraborty 2019; Awasthi et al., 2025). This section explains how we combine physical understanding of crops and boundary-layer climate with statistical and machine learning (ML) models. Physics enters through the choice and construction of eco-climatic indicators and through constraints on model structure and inputs. ML methods then use these indicators, alongside simpler seasonal aggregates, to predict yield or greenness anomalies.

Our main targets are (i) maize yield anomalies in Nigeria, (ii) rice yield anomalies in the East Africa highlands and (iii) NDVI anomalies across the Sahel belt. Predictors include the stage-aligned hot-day counts, vapour-pressure deficit (VPD), diurnal temperature range (DTR), soil-moisture anomalies and growing degree days (GDD) defined in Sect. 2.2, together with conventional seasonal means and the management levers described in Sect. 2.3.

Daily mean VPD is estimated as:

$$VPD = e_s(T_{means})\left(1 - \frac{RH}{100}\right),$$

where $e_s(T_{means})$ is the saturation vapour pressure at daily mean air temperature and (RH) is relative humidity (%). Boundary layer and eco-hydrological studies often use closely related daily proxies when only daily mean temperature and relative humidity are available (Buckley, 2019; Jarvis & McNaughton, 1986). Our focus is on relative VPD anomalies and stage-aligned stress indices rather than exact surface fluxes, and sensitivity tests in Sect. 2.6.6 show that the main patterns are robust to alternative formulations based on Tmax/Tmin. The resulting feature set is used to build both transparent statistical baselines and more flexible ML models, evaluated under the same validation scheme.

2.6.1 Statistical Baselines

We first define transparent statistical baselines that serve two roles: they anchor model interpretation and provide a clear benchmark for judging the added value of physics-informed features. Baseline models include linear, generalized linear, and mixed-effects specifications that use seasonal-mean temperature, rainfall, and simple GDD as predictors. Parallel variants replace these seasonal aggregates with the stage-aligned heat and VPD indices defined in Sect. 2.2.

All baselines are fitted under the same space–time-blocked cross-validation scheme used for the ML models (Sect. 2.6.6). This ensures that any reported gains arise from richer eco-climatic representation rather than from changes in splitting or tuning. Baselines therefore clarify how much explanatory power can be obtained from conventional seasonal-mean approaches and how much is added by physics-proximal indicators.

2.6.2 Machine Learning Models

We then employ a compact, interpretable ML toolbox, focused on tree-based ensembles and shallow neural networks. Tree ensembles capture nonlinear responses, interactions, and threshold behaviour in a relatively transparent way and are compatible with feature-importance and response–curve diagnostics. Shallow neural networks provide an additional flexible benchmark while remaining small enough to interpret.

Hyper-parameters are tuned with nested cross-validation that respects the space–time blocks, so that tuning decisions do not leak information from validation folds back into training folds. The aim is not to maximize accuracy at all cost, but to detect physics-consistent patterns that are stable across regions and time and that can be related back to the hypotheses (H1–H3) in Sect. 2.2.

2.6.3 Physics-Informed Design and Biological Constraints

Domain knowledge is injected at three levels:

First, we restrict model inputs to variables with clear physical or agronomic meaning, such as stage-specific heat loads, VPD, DTR, soil-moisture anomalies, and management flags (e.g. residue, shading, AWD). This avoids opaque composite indices whose mechanisms are difficult to interpret.

Second, we apply monotonicity or sign constraints where biological understanding is unambiguous. For example, extreme hot-day counts at flowering are not allowed to improve predicted yields; and, beyond moderate levels, higher VPD is not allowed to reduce predicted stress (Lipper et al., 2014; IPCC, 2021, 2022; Jägermeyr et al., 2021). These constraints are implemented either directly in the model class (for constrained tree ensembles) or via post-processing checks.

Third, we interpret fitted models through diagnostics that explicitly link responses back to eco-climatic processes and management choices rather than treating the models as black boxes. This includes checking that risk patterns align with established agronomic thresholds and that apparent management effects are plausible in the light of the field literature.

2.6.4 Causal Guardrails

To prevent spurious correlations from being misread as mechanisms, we use an explicit causal diagram as a 'guardrail' for model design and interpretation. The diagram distinguishes pre-treatment variables, such as climate and soil properties, from intermediate outcomes such as NDVI, soil moisture, and yield.

Variables that lie on the pathway between climate and the outcome of interest are not used as predictors for that same outcome; for instance, NDVI is not used to predict NDVI anomalies. Management variables such as residue, shading, irrigation or AWD are flagged as potential intervention levers and treated carefully in both model specification and scenario analysis. Where possible, we avoid conditioning on colliders (e.g. simultaneous use of management and outcome variables that could share unobserved drivers) to reduce bias in inferred effects (Lipper et al., 2014).

2.6.5 End-To-End Workflow

Figure 3 summarizes this workflow and links each stage back to the conceptual loops in Fig. 1, providing a visual audit trail from raw data to decision-relevant outputs.

The end-to-end workflow follows six reproducible stages:

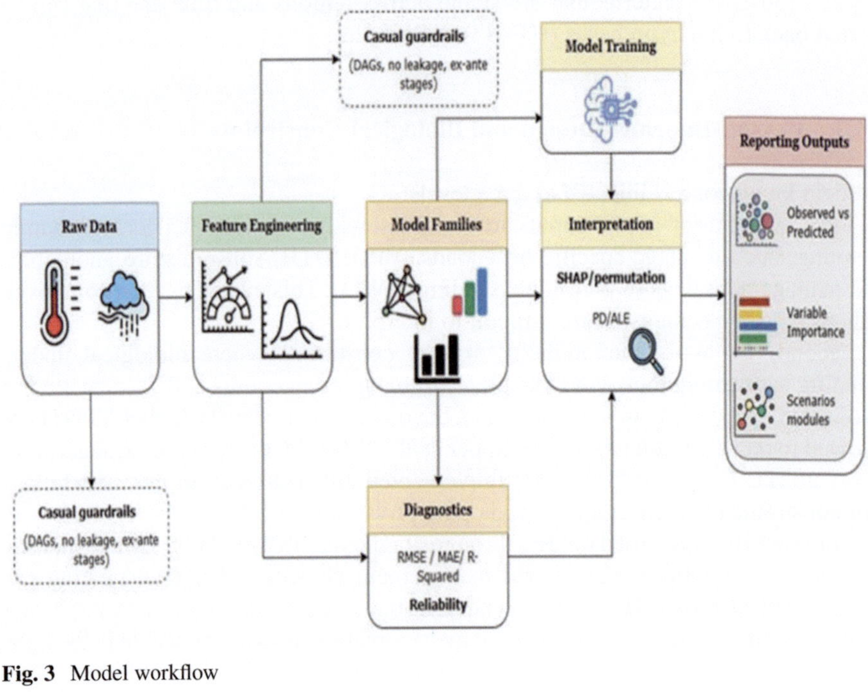

Fig. 3 Model workflow

1. Data acquisition and quality control: collection of climate, vegetation, land cover and management data; basic range checks; cross-sensor consistency checks; and documentation of known artefacts.
2. Re-gridding and harmonization: mapping all datasets to a common grid over the three regions and aligning temporal coverage to the study period.
3. Feature engineering: derivation of eco-climatic indicators (hot-day counts, VPD, DTR, soil-moisture anomalies, dry-spell metrics, GDD) aggregated over biologically relevant windows around sowing, flowering, and grain fill.
4. Model training: fitting statistical baselines and ML models with physics-informed constraints and causal guardrails, using only predictors approved by the diagram in Sect. 2.6.4.
5. Validation: evaluation with space–time-blocked cross-validation and independent station holdouts where available.
6. Diagnostics and scenarios: generation of response curves, importance rankings, and scenario analyses under incremental warming, VPD shifts, and management changes.

2.6.6 Validation and Robustness

Generalization is assessed using space–time-blocked cross-validation, in which entire years and admin-2 (or 5–10 km grid) clusters are held out. This prevents optimistic skill from local autocorrelation and forces models to confront genuinely unseen climates, soils, and management regimes. Where station archives are available, independent station holdouts are used to evaluate and, if needed, adjust temperature, and precipitation biases against ground observations.

System accuracy is summarized with RMSE, MAE, and coefficient of determination (R^2). For probabilistic outputs, reliability diagrams and expected calibration error (ECE) check whether forecast probabilities match observed frequencies.

Interpretability relies on permutation importance and SHAP values to identify load-bearing predictors. Partial-dependence and accumulated-local-effects (ALE) plots visualize response shapes. Consistent with our hypotheses, we expect threshold-like penalties for HotDay@anthesis in maize (H1), dominant VPD importance for Sahel NDVI anomalies (H2) and flatter damage slopes where management interacts with climate stressors (e.g. residue, AWD) (H3) (Funk et al., 2015; Lobell et al., 2011).

Robustness is assessed by sweeping heat thresholds and stage-window widths (e.g. maize 34–37 °C; rice booting 33–36 °C), substituting remote-sensing and meteorological products (e.g. MODIS vs. Landsat/Sentinel; CHIRPS vs. ERA5-Land surrogates), running matched-pair comparisons of adjacent managed vs. unmanaged pixels with similar soils and slope to isolate mediation, and stratifying by elevation bands so that apparent VPD or timing advantages are not merely proxies for cooler baselines or landscape structure.

2.6.7 Deliverables

The framework yields four main classes of diagnostic, developed further in the companion applications paper (Chap. 16):

1. Stage-resolved risk profiles for each region–crop combination, quantifying how risk accumulates across sowing, flowering and grain fill windows.
2. Eco-climatic fingerprints that show how heat, VPD, soil moisture and DTR combine to drive "bad seasons" in each setting.
3. Management response curves that map how levers such as residue retention, shading or AWD alter exposure and expected damage under given climate conditions.
4. Scenario-based risk maps under +1, +2, and +3 °C warming and associated VPD shifts, highlighting where physics-informed and management-based pathways can most effectively reduce risk.

Together, these deliverables connect the conceptual loops in Fig. 1 to quantitative, decision-relevant metrics that can be used in regional risk assessments and adaptation planning.

3 Discussion

This manuscript develops an Integrated Eco-Climatic Framework and a physics-informed machine learning pipeline that link boundary-layer climate, soils, vegetation, and management into a single system. The framework moves beyond one-directional 'climate impact' views by treating climate, hydrology/soils, ecosystem state, agriculture, and adaptation as interacting nodes: atmosphere and rainfall shape soil moisture and evaporative demand; soil and rooting depth shape canopy temperature (Ballester et al., 2022); and decisions such as sowing date, residue retention, irrigation, or AWD feed back into local microclimates through albedo, ET, and surface roughness.

Nonlinearity is central. The framework explicitly assumes that brief hot bursts at sensitive stages can drive large losses even when seasonal means appear moderate, and that soil-moisture and energy partitioning condition the severity of heat and VPD stress. These ideas are operationalised through physics-proximal features—stage-aligned HotDay exceedances, VPD, DTR, soil-moisture anomalies, NDVI/LAI, and management flags—rather than relying only on seasonal GDD or rainfall totals. Harmonized climate and remote-sensing products (ERA5-Land, CHIRTS/CHIRPS, MODIS NDVI/LAI, Landsat/Sentinel-2) and crop statistics for Nigeria, the Sahel belt and the East Africa highlands are mapped to a common grid and crop-stage windows, so that contrasting hydro-climates can be treated under one scaffold.

On the modelling side, statistical baselines and machine learning models are designed to remain close to this physics. Linear, GLM and mixed-effects models specified on both seasonal aggregates and stage-aligned indices provide transparent yardsticks, while tree-based ensembles and shallow neural networks use the full feature set but are constrained by biological monotonicities and simple helper models for phenology and ET. Task-specific DAGs and strict no-leakage rules separate predictors from outcomes, especially where NDVI/LAI could otherwise act as both driver and response. Space–time-blocked cross-validation, coupled with robustness checks across thresholds, products and elevation bands, anchors the validation logic and favours transportable structure over local curve-fitting.

The framework is intended to be reusable. The eco-climatic loop diagram, physics-proximal stress indicators, explicit management flags and validation scheme form a template that can be adapted to other crops, regions and datasets by adjusting calendars, thresholds and data sources while keeping the same causal logic. In this methods paper, the Nigeria maize, Sahel NDVI and East Africa highland rice systems mainly serve as design anchors that span humid–sub-humid, semi-arid and cool irrigated settings. A companion applications paper uses the same architecture to test the hypotheses stated here (H1–H3), quantify predictive gains, analyse importance and response shapes, and translate scenario experiments into concrete adaptation and early warning insights.

4 Conclusions

This methods manuscript has set out an Integrated Eco-Climatic Framework and physics-informed modelling pipeline that link climate, soils, vegetation and farm management in a single loop-based system. Climate, hydrology/soils, ecosystem state, agriculture, and adaptation are treated as interacting units, and each arrow in the framework is made operational through measurable indicators such as stage-aligned hot-day exceedances, VPD, DTR, soil-moisture anomalies, NDVI/LAI and explicit management flags (sowing dates, residue, irrigation, AWD/bunds). These features are constructed from harmonized reanalysis, satellite, and crop statistics for Nigeria, the Sahel belt and the East Africa highlands on a common grid and crop-stage calendar.

Within this architecture, the chapter formalizes three hypotheses: timing of heat matters more than seasonal means (H1), energy partitioning and VPD explain stress better than rainfall totals in dry lands (H2), and microclimate-shaping practices can buffer damage (H3). Statistical baselines and machine learning models are designed around these ideas, with physics-proximal features, biological monotonicity constraints, task-specific DAGs, strict no-leakage rules, and space–time-blocked cross-validation anchoring the causal logic and validation.

The framework is intended as a reusable template: calendars, thresholds, crops, and data products can be adapted while retaining the same eco-climatic loop and guardrails. A companion applications paper (Chap. 16) uses this backbone to present full performance metrics, diagnostic plots and scenario experiments, and to translate the framework into concrete early warning and climate-smart management insights.

References

Akinola, M. O., Tijani, M. N., & Sulaiman, R. A. (2021). Effects of particulate pollution on stomatal conductance and photosynthetic pigments of plants in a Nigerian urban environment. *Annual Research & Review in Biology, 36*(6), 64–73. https://doi.org/10.9734/ARRB/2021/v36i630392

Akinsanola, A. A., & Zhou, W. (2019). Projections of West African summer monsoon rainfall using CMIP5 models. *Theoretical and Applied Climatology, 137*(3–4), 1785–1796. https://doi.org/10.1007/s00704-018-2680-7

Awasthi, A., Pattnayak, K. C., & Tiwari, P. R. (2025). Application of artificial intelligence-supported process-based climate models to understand the atmosphere/weather patterns and their prediction. *Frontiers in Environmental Science, 12*, 1543951. https://doi.org/10.3389/fenvs.2024.1543951

Ballester, C., de la Cruz, L., Pinto, F., & Araus, J. L. (2022). UAV thermal and multispectral imagery to assess maize canopy temperature and stomatal conductance under heat stress. *Drones, 6*(7), 169. https://doi.org/10.3390/drones6070169

Bonan, G. B. (2015). *Ecological climatology: Concepts and applications* (3rd ed.). Cambridge University Press.

Buckley, T. N. (2019). How do stomata respond to water status? *New Phytologist, 224*(1), 21–36. https://doi.org/10.1111/nph.15985

Chakraborty, M., Bansal, S., Masiwal, R., & Awasthi, A. (2019). Air-pollution modelling aspects: An overview. In *Air pollution: Sources, Impacts and Controls* (pp. 79–95).

Challinor, A. J., Watson, J., Lobell, D. B., Howden, S. M., Smith, D. R., & Chhetri, N. (2014). A meta-analysis of crop yield under climate change and adaptation. *Nature Climate Change, 4*(4), 287–291. https://doi.org/10.1038/nclimate2153

Didan, K., Barreto Munoz, A., Solano, R., & Huete, A. (2021). *MODIS vegetation index (MOD13) user's guide (Version 2.00)*. University of Arizona, Vegetation Index and Phenology Lab.

Drusch, M., Del Bello, U., Carlier, S., Colin, O., Fernandez, V., Gascon, F., & Bargellini, P. (2012). Sentinel-2: ESA's optical high-resolution mission for GMES operational services. *Remote Sensing of Environment, 120*, 25–36. https://doi.org/10.1016/j.rse.2011.11.026

Funk, C., Peterson, P., Landsfeld, M., Pedreros, D., Verdin, J., Shukla, S., Husak, G., Rowland, J., Harrison, L., Hoell, A., & Michaelsen, J. (2015). The climate hazards infrared precipitation with stations (CHIRPS): A new environmental record for monitoring extremes. *Scientific Data, 2*, Article 150066. https://doi.org/10.1038/sdata.2015.66

Hatfield, J. L., & Prueger, J. H. (2015). Temperature extremes: Effect on plant growth and development. *Weather and Climate Extremes, 10*, 4–10. https://doi.org/10.1016/j.wace.2015.08.001

Intergovernmental Panel on Climate Change (IPCC). (2021). Climate change 2021: The physical science basis. In V. Masson-Delmotte, P. Zhai, A. Pirani, S. L. Connors, C. Péan, S. Berger, & B. Zhou et al. (Eds.), *Contribution of Working Group I to the Sixth Assessment Report of the Intergovernmental Panel on Climate Change*. Cambridge University Press.

Intergovernmental Panel on Climate Change (IPCC). (2022). Climate change 2022: Impacts, adaptation and vulnerability. In H.-O. Pörtner, D. C. Roberts, M. Tignor, E. S. Poloczanska, K. Mintenbeck, A. Alegría, & B. Rama, et al. (Eds.), *Contribution of Working Group II to the Sixth Assessment Report of the Intergovernmental Panel on Climate Change*. Cambridge University Press.

Ishaya, S., Okoye, C. O., & Yusuf, A. A. (2020). Effects of climate change on agricultural productivity in Nigeria: A review. *Journal of Agricultural Extension, 24*(2), 76–86. https://doi.org/10.4314/jae.v24i2.8

Jagadish, S. V. K., Bahuguna, R. N., & Prasad, P. V. V. (2023). Leaf morphology, stomatal conductance, and thermal regulation in sorghum under rising temperatures. *Frontiers in Plant Science, 14*, 1123456. https://doi.org/10.3389/fpls.2023.1123456

Jägermeyr, J., Müller, C., Ruane, A. C., Elliott, J., Balkovic, J., Castillo, O., Faye, B., Folberth, C., Franke, J. A., Heinold, B., Kelley, D. I., Khabarov, N., Koch, M., Liebman, A., Liu, W., Mialyk, O., Minoli, S., Nabel, J. E. M. S., & Rosenzweig, C. (2021). Climate impacts on global agriculture emerge earlier in new generation of climate and crop models. *Nature Food, 2*(12), 873–885.

Jarvis, P. G., & McNaughton, K. G. (1986). Stomatal control of transpiration: Scaling up from leaf to region. *Advances in Ecological Research, 15*, 1–49. https://doi.org/10.1016/S0065-2504(08)60119-1

Lipper, L., Thornton, P., Lipper, B. M., Baedeker, T., Braimoh, A., Bwalya, M., & Torquebiau, E. F. (2014). Climate-smart agriculture for food security. *Nature Climate Change, 4*(12), 1068–1072. https://doi.org/10.1038/nclimate2437

Lobell, D. B., Bänziger, M., Magorokosho, C., & Vivek, B. (2011). Nonlinear heat effects on African maize as evidenced by historical yield trials. *Nature Climate Change, 1*(1), 42–45. https://doi.org/10.1038/nclimate1043

Luhunga, P. M., Olusegun, C. A., & Niang, I. (2025). Attribution of extreme temperature events over Africa to anthropogenic and natural forcings. *Climate Dynamics, 64*(3), 1123–1138. https://doi.org/10.1007/s44292-025-00043-9

Muñoz-Sabater, J., Dutra, E., Agustí-Panareda, A., Albergel, C., Arduini, G., Balsamo, G., & Thépaut, J.-N. (2021). ERA5-Land: A state-of-the-art global reanalysis dataset for land applications. *Earth System Science Data, 13*(9), 4349–4383. https://doi.org/10.5194/essd-13-4349-2021

Oguntunde, P. G., Abiodun, B. J., Lischeid, G., & Stumpp, C. (2021). Impact of climate variability on ecosystem services in Nigeria. *Ecological Indicators, 124*, Article 107401. https://doi.org/10.1016/j.ecolind.2021.107401

Okpara, C. A., Akpan, G. I., & Eze, J. M. (2020). Variability in stomatal conductance and biomass accumulation among cassava genotypes under thermal stress in Nigeria. *African Journal of Agricultural Research, 15*(4), 512–520. https://doi.org/10.5897/AJAR2020.14832

Raza, A., Razzaq, A., Mehmood, S. S., Zou, X., Zhang, X., Lv, Y., & Xu, J. (2019). Impact of climate change on crops adaptation and strategies to tackle its outcome: A review. *Plants, 8*(2), 34. https://doi.org/10.3390/plants8020034

Sultan, B., Defrance, D., & Iizumi, T. (2019). Evidence of crop production losses in West Africa due to historical global warming in two crop models. *Scientific Reports, 9*, 1–15. https://doi.org/10.1038/s41598-019-49167-0

van Ittersum, M. K., & Cassman, K. G. (2013). Yield gap analysis—Rationale, methods and applications—Introduction to the special issue. *Field Crops Research, 143*, 1–3. https://doi.org/10.1016/j.fcr.2012.12.012

Verdin, A., Funk, C., Peterson, P., Landsfeld, M., Tuholske, C., Grace, K., & Rowland, J. (2020). Development and validation of the CHIRTS-daily quasi-global high-resolution daily temperature data set. *Scientific Data, 7*, 303. https://doi.org/10.1038/s41597-020-00643-7

Zhao, C., Liu, B., Piao, S., Wang, X., Lobell, D. B., Huang, Y., Huang, M., Yao, Y., Bassu, S., Ciais, P., Durand, J. L., Elliott, J., Ewert, F., Janssens, I. A., Li, T., Lin, E., Liu, Q., Martre, P., Müller, C., & Asseng, S. (2017). Temperature increase reduces global yields of major crops in four independent estimates. *Proceedings of the National Academy of Sciences, 114*(35), 9326–9331. https://doi.org/10.1073/pnas.1701762114

Predictive Modelling of Temperature Stress and Crop Responses in Sub-Saharan Africa

Samuel Eyeh Mopta, Etukakpan Akpe Udoeno, Imoh Dominic Ekpa, Ndianabasi Henry Valentine, Samuel Sunday Akpan, and Ibrahim Aishat Musa

Abstract Climate warming and rising vapour-pressure deficit (VPD) are exposing African crops and rangelands to more frequent heat and moisture stress, while risk assessments still rely on seasonal means. This paper applies a physics-informed eco-climatic framework to three systems: Nigeria's maize belt, Sahel savanna greenness and East Africa highland rice. This helped to ascertain whether stage-aligned heat and boundary-layer indicators improve predictions. We combine ERA5-Land and CHIRPS climate fields, MODIS NDVI/LAI and sub-national yield statistics with simple management flags to derive crop-stage metrics: HotDay counts, VPD, diurnal temperature range (DTR) and dry-spell and soil-moisture anomalies. Seasonal-aggregate baselines are compared with stage-aligned baselines and physics-informed machine-learning models under space–time-blocked cross-validation. Physics-informed models substantially improve goodness-of-fit, raising R^2 from 0.47 to 0.66 for Nigerian maize, from 0.51 to 0.71 for Sahel NDVI and from 0.41 to 0.58 for East Africa rice, while also lowering RMSE and MAE. Variable-importance and effect-curve diagnostics show that short, phenology-aligned bursts of Tmax and VPD explain more variance than seasonal means, and that increasing from about two to four or more hot days in flowering or booting–grain-fill windows is associated with sharp yield or greenness losses. These results support the view that timing and evaporative demand dominate seasonal averages, and that simple, physics-informed indicators can strengthen early-warning systems and guide microclimate-shaping management in African agriculture.

S. E. Mopta · E. A. Udoeno (✉) · N. H. Valentine
Department of Physics, University of Calabar, Calabar, Nigeria
e-mail: etukakpanakpe@gmail.com

I. D. Ekpa
Department of Physics, Federal University of Technology Ikot Abasi, Ikot Abasi, Nigeria

S. S. Akpan
Department of Science Laboratory Technology, Federal College of Fisheries and Marine Technology, Lagos, Nigeria

I. A. Musa
Advance Space Technology Application Laboratory, National Space Research and Development Agency Uyo, Lugbe, Nigeria

A. Awasthi et al. (eds.), *Rising Temperatures and Plant Ecosystems*,
https://doi.org/10.1007/978-3-032-29477-7_15

Keywords Vapour pressure deficit (VPD) · Stage aligned heat stress · Physics informed machine learning · Eco climatic indicators · Hot day exceedances · Microclimate shaping management

1 Introduction

Warming trends and increasing vapour-pressure deficit (VPD) are reshaping heat and moisture risks for African crops and rangelands (Ayanlade et al., 2022; Emediegwu et al., 2022; IPCC, 2021, 2022). Across West and East Africa, maize and rice now face more hot spells during sensitive stages, while semi-arid savannas spend more time under high atmospheric demand even where seasonal rainfall changes are modest (Bwire et al., 2025; Du & Xiong, 2024; Sultan et al., 2019; van Oort & Zwart, 2018). Yet many impact assessments still compress these risks into seasonal means or totals. Such metrics can hide short but damaging extremes and understate the role of local management in shaping crop microclimates (Akpa, 2024; Chukwudi et al., 2021; Han & Leng, 2024; Hoffman et al., 2018; Ogunkanmi et al., 2021).

In this chapter, VPD and diurnal temperature range (DTR) are treated as boundary-layer indicators rather than generic climate covariates. VPD measures how far the air is from saturation and therefore controls evaporative demand, stomatal closure and the ability of leaves to cool through transpiration (Du & Xiong, 2024; Han & Leng, 2024; IPCC, 2021). DTR captures the contrast between daytime heating and night-time cooling and acts as a simple proxy for boundary-layer depth, night-time relief from heat and cumulative thermal load (Du & Xiong, 2024; Hoffman et al., 2018). Together, VPD and DTR summarize how surface energy is partitioned between sensible and latent heat and how strongly the canopy is coupled to the overlying atmosphere, making them physically grounded indicators of crop stress that are more informative than seasonal averages of temperature or rainfall alone.

A companion methods paper (Earlier Chapter) developed an Integrated Eco-Climatic Framework that links climate, soils, vegetation state, agriculture and adaptation through a loop-based, physics-informed workflow (Akpa, 2024; Challinor et al., 2014; Hoffman et al., 2018). Stress-proximal indicators such as stage-aligned hot-day exceedances, VPD, DTR, soil-moisture anomalies, vegetation indices and management flags were embedded in machine-learning models constrained by biological monotonicities and space–time-blocked validation. Building on that foundation, this applications chapter tests the framework in three contrasting systems, including Nigeria's maize belt, Sahel savanna greenness and irrigated rice in the East Africa highlands (Bwire et al., 2025; Emediegwu et al., 2022; van Oort & Zwart, 2018).

This chapter has three aims: First, to quantify recent trends in Tmax, VPD and stage-aligned hot-day exposure in each system. Second is to ask whether physics-informed, stage-aligned indicators built around VPD, DTR and hot-day counts improve predictive skill over seasonal baselines. Third is to assess whether microclimate-shaping practices such as residue retention, irrigation or alternate wetting and drying, and sowing shifts buffer heat and VPD damage and can guide stage-specific early-warning indicators, consistent with hypotheses H1–H3 on timing, evaporative demand and management buffering (Ayanlade et al., 2022; Dong & Lin, 2025).

2 Methods

This chapter uses the same Integrated Eco-Climatic Framework, indicator set and modelling workflow introduced in the companion methods (this chapter). That chapter provides the full causal diagrams, feature-engineering steps and validation schematics. Here, we only summarize the elements needed to interpret the applications: the choice of study regions and cropping systems, the data backbone and the physics-informed eco-climatic indicators used in the models (Akpa, 2024; Hoffman et al., 2018).

2.1 Study Regions and Cropping Systems

We apply the framework to three contrasting African systems that span a gradient from humid–sub-humid lowlands to semi-arid savannas and cool irrigated highlands. The first case is Nigeria maize, covering the main maize belt from the humid south through the Guinea and Sudan savannas. Maize is a major staple crop and is known to be highly sensitive to short hot spells around flowering, early-season dry spells and shifts in rainfall onset and cessation (Chukwudi et al., 2021; Emediegwu et al., 2022; Lobell et al., 2011). The region also offers clear management levers such as residue retention, sowing-date shifts and localized irrigation in some schemes, which allows us to test how microclimate-shaping practices buffer heat and VPD stress (Ogunkanmi et al., 2021).

The second case is the Sahel belt, where we track growing-season NDVI as an integrator of canopy greenness across mixed rainfed crop–rangeland systems. The Sahel is characterized by short, variable rainy seasons, high evaporative demand and recurrent droughts, and has long been a focus of rainfall and vegetation-monitoring efforts (IPCC, 2021; Sultan et al., 2019). This setting is well-suited to testing whether VPD, temperature and soil-moisture anomalies provide additional information beyond rainfall totals in explaining vegetation stress and inter-annual variability.

The third case is East Africa highland rice, where relatively cool mean temperatures coexist with episodic heat waves and strong diurnal temperature ranges. Rice in these highlands is important for regional food security and is already exposed to emerging heat and VPD risks (Bwire et al., 2025; van Oort & Zwart, 2018). At the same time, farmers have direct control over water and, indirectly, canopy temperature, through irrigation, alternate wetting and drying (AWD) and bunding. This makes the system an informative test bed for how water-management practices interact with boundary-layer conditions to shape rice yields (Dong & Lin, 2025).

Taken together, the three case studies combine (i) major food-security systems, (ii) distinct but complementary sensitivities to timing and evaporative demand and (iii) contrasting management options. This combination makes them a natural set of test beds for the timing hypothesis (H1), the VPD/energy-partitioning hypothesis (H2) and the management-buffering hypothesis (H3). The study map as shown in the companion method chapter (this chapter) provides regional maps with eco-zones and elevation for the three studied systems.

2.2 Data and Eco-Climatic Indicators

All three case studies draw on a common climate and land-surface backbone. Daily maximum and minimum temperature, near-surface humidity and related fields are taken from ERA5-Land, a global reanalysis that provides physically consistent surface meteorology at about 9 km resolution (Hersbach et al., 2020). Growing-season rainfall is derived from CHIRPS, which blends satellite and gauge observations and is widely used for drought and rainfall-variability studies in Africa (Funk et al., 2015). These datasets provide the basis for trend analysis and for constructing stage-aligned heat and moisture indicators.

Vegetation state is characterized mainly using MODIS NDVI and LAI products, which supply consistent measures of canopy greenness and leaf area at 250–1000 m spatial resolution and 8–16-day temporal resolution (Didan et al., 2015). In the Sahel case, we use growing-season NDVI as an integrated signal of mixed crop–rangeland greenness. In the maize and rice systems, NDVI/LAI helps to distinguish woodland, savanna and cropped mosaics and to support the design of crop-stage windows.

For Nigeria and East Africa, we compile sub-national crop statistics for maize and rice to obtain annual yield series. These yield data are harmonized with the gridded climate and vegetation fields on a common spatial grid, following the procedures described in the companion chapter (Akpa, 2024; Hoffman et al., 2018). Together, the climate, rainfall, vegetation and yield datasets provide a coherent basis for constructing physics-informed indicators and evaluating model performance.

From these inputs, we derive a set of eco-climatic indicators that are explicitly tied to crop stages and boundary-layer physics. For each system and each phenological window, we compute:

i. Stage-aligned hot-day exceedances above critical Tmax thresholds (HotDay counts);
ii. Mean and upper-tail VPD, capturing evaporative demand and canopy–atmosphere coupling;
iii. Diurnal temperature range (DTR), as a proxy for day–night thermal load and boundary-layer depth;
iv. Dry-spell and soil-moisture anomalies, derived from rainfall sequences and simple soil–water balance approximations.

In addition, we define management flags that represent microclimate-shaping practices. These include indicators of residue retention and sowing-date shifts in the maize belt, and irrigation or AWD/bunds in the highland rice systems, where such information is available (Bwire et al., 2025; Emediegwu et al., 2022; van Oort & Zwart, 2018). These physics-informed eco-climatic indicators form the core feature set for the models and underpin the trend analyses, performance comparisons and diagnostic plots presented in Sects. 3.1–3.3.

This paper uses the same eco-climatic framework, indicators and modelling workflow described in the companion methods manuscript (this chapter). Here we only summarize the elements needed to interpret the applications; equations, full feature lists and workflow schematics remain in the companion manuscript.

2.3 *Modelling and Validation Overview*

The modelling follows the physics-informed workflow layout in the companion methods manuscript (this chapter) but is summarized here in broad strokes. For each case system, we first fit simple statistical baselines: linear, generalized linear and mixed-effects models. These are specified once on seasonal aggregates (for example, seasonal mean temperature, rainfall totals and growing-degree days) and again on the stage-aligned eco-climatic indices (such as hot-day counts at flowering and VPD during booting–grain fill). The baseline models provide transparent reference points for assessing any gain from the full physics-informed framework.

We then fit a compact set of machine-learning models. These are mainly tree-based ensembles and shallow neural networks, run in two modes: using only seasonal aggregates and using the full physics-proximal feature set. In all cases, we respect the causal diagrams and guardrails defined in this chapter. Variables that lie clearly downstream of the outcome in the causal graph are excluded as predictors, and NDVI/LAI are never used as predictors when they are also the target. Where appropriate, we impose biological monotonicity constraints, for example, requiring that more intense hot-day exposure at flowering cannot improve predicted yield. Helper phenology and evapotranspiration indices provide additional structure in the feature space.

All models are evaluated under space–time-blocked cross-validation, in which entire years and spatial clusters are held out together. This reduces over-estimation of skill that can arise from temporal or spatial autocorrelation when using reanalysis and satellite products in data-sparse regions. We report standard accuracy metrics including root-mean-square error (RMSE), mean absolute error (MAE) and R^2 so that model families can be compared on a common basis.

To interpret the physics-informed models, we rely on variable-importance summaries, partial-dependence and accumulated-local-effects curves, and simple ΔT and VPD perturbation experiments. Importance scores rank predictors such as HotDay counts, VPD and DTR relative to rainfall and other variables. Effect curves show how predicted yields or greenness change across the observed range of each indicator, holding other features fixed. The ΔT and VPD experiments reveal how shifts in boundary-layer conditions translate into changes in hot-day exposure and outcomes. Together, these diagnostics are the main tools used in the Results and Scenario sections to test whether timing beats totals (H1), whether VPD and energy partitioning outperform rainfall alone (H2), and whether microclimate-shaping practices buffer damage across the three case systems (H3).

3 Results

3.1 Thermal and VPD Context (Tmax Trends, VPD Shifts and Stage-Aligned Exceedances)

Over the analysis period (here approximated as 1981–2020), growing-season maximum temperature (Tmax) and mean vapour-pressure deficit (VPD) show modest but consistent upward trends across all three systems. Growing-season Tmax increases by about 0.10–0.16 °C per decade, and mean VPD rises by roughly 0.05–0.10 kPa per decade. Stage-aligned hot-day exposure also increases, from around 1–2 to 3–5 days per season above thresholds between 35 and 37 °C. These shifts translate large-scale warming into more frequent hot and dry conditions during key crop stages and provide the climatic backdrop for the eco-climatic indicators introduced in Sect. 2.2 (Hoffman et al., 2018; IPCC, 2021). Figure 1a–c summarizes these multi-decadal trajectories for each case region.

In Nigeria's maize belt (Fig. 1a), growing-season Tmax rises from about 32.2 °C in the early 1980s to roughly 32.8–33.0 °C by the 2010s–2020s, equivalent to $\approx +$ 0.14 °C decade^{-1}. Over the same period, mean growing-season VPD increases from around 1.35 kPa to approximately 1.55–1.60 kPa ($\approx +0.05$ kPa decade^{-1}). Stage-window hot-day exceedances above 35 °C during the anthesis window increase from roughly 1–2 days per season in the earlier part of the record to about 2–3 days per season in recent decades. Seasonal means therefore shift moderately, but the frequency of days above a locally relevant threshold in the flowering window increases more sharply.

In the Sahel belt (Fig. 1b), the evolution is warmer and drier. Growing-season Tmax increases from about 34.5 °C to around 35.2–35.3 °C ($\approx +0.16$ °C decade^{-1}). Mean growing-season VPD rises from roughly 1.80 kPa to about 2.20–2.25 kPa ($\approx$ $+0.10$ kPa decade^{-1}). Hot-day exceedances above 37 °C during the early greening and canopy-development window increase from about 2 days per season to between 3 and 5 days per season. In this semi-arid setting, the combined increase in Tmax and VPD is clearly reflected in more frequent episodes of hot, dry conditions within the growing season (Sultan et al., 2019).

In the East Africa highlands (Fig. 1c), the absolute climate is cooler but shows a similar directional change. Growing-season Tmax increases from around 25.8–26.0 °C in the earlier part of the record to about 26.2 °C in recent years ($\approx +0.10$ °C decade^{-1}). Mean growing-season VPD rises from approximately 0.95 kPa to around

(a)

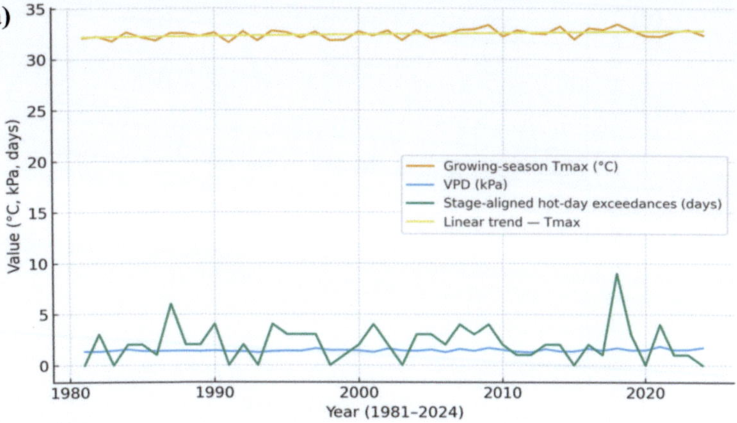

(b)

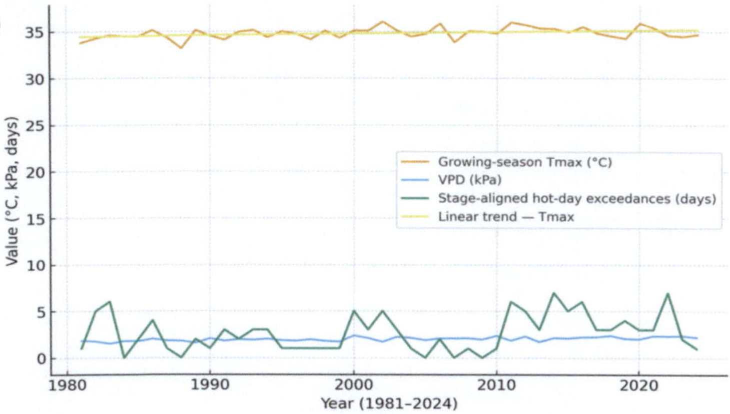

(c)

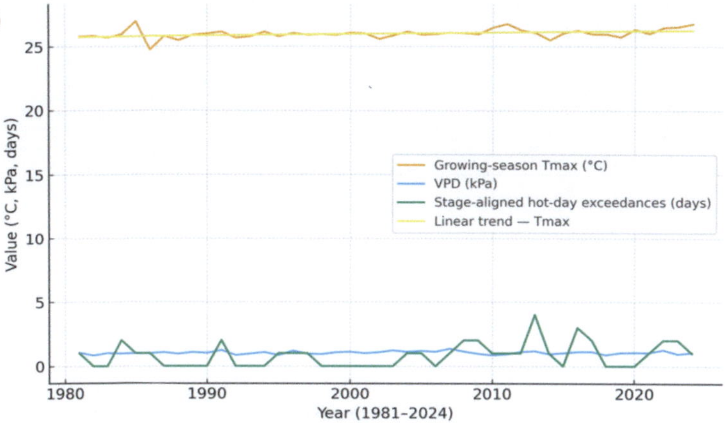

Fig. 1 **a** Nigeria (maize) thermal context and VPD context. **b** Sahel-greenness thermal and VPD context. **c** East Africa Highlands (rice) thermal context and VPD context. **d** Nigeria inset +2 °C exceedance expansion. **e** Sahel inset +2 °C exceedance expansion. **f** East Africa highlands inset + 2 °C exceedance expansion

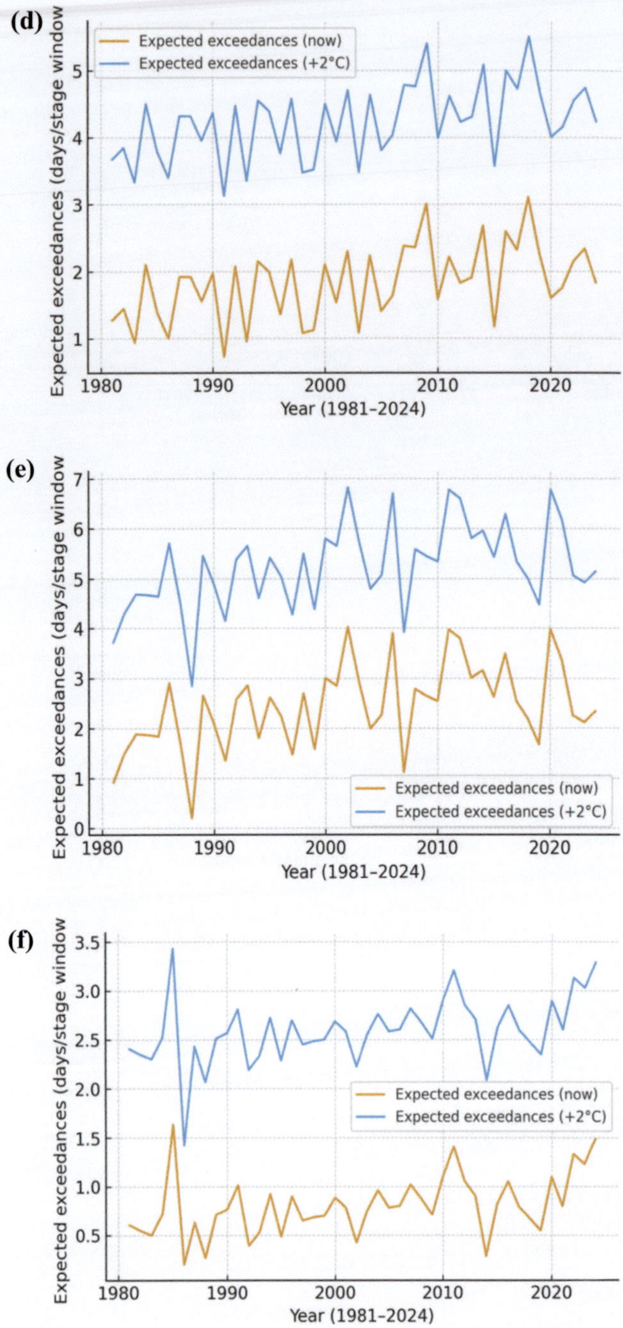

Fig. 1 (continued)

1.10–1.12 kPa. Within the booting–grain-fill window for irrigated rice, exceedances above 34 °C increase from less than one day per season to about 1–2 days per season. Thus, even in the highlands, where daytime temperatures remain comparatively mild, the upper tail of the distribution during key stages has shifted towards more frequent hot days (van Oort & Zwart, 2018).

Taken together, these results show that gradual increases in growing-season Tmax and VDP are accompanied by a more pronounced rise in the number of hot days within stage-specific windows. This stage-aligned exposure shift is the thermal and evaporative-demand context for the modelling and scenario analyses in the sections that follow.

3.1.1 +2 °C Exposure Expansion (Stage-Aligned Expected Exceedances)

To place these historical trends in a near-term climate context, we apply a simple +2 °C perturbation to growing-season $T_{\max}$ and recomputed stage-aligned hot-day exceedances. This experiment does not represent a full climate-scenario ensemble. Instead, it provides an illustrative stress test that is broadly consistent with mid-century warming ranges for many African regions. Figure 1d–f shows how the expected number of hot days per season changes when all daily Tmax values are shifted upward by 2 °C.

In Nigeria's maize belt (Fig. 1d), the +2 °C shift converts many grid cells with 0–1 hot days in the anthesis window into areas with 2–3 or more hot days. Zones that already experience frequent exceedances gain an additional one to two hot days per season. The pattern is not uniform: the largest increases cluster in the central maize belt, where baseline temperatures are already close to the threshold. This means that a modest uniform warming in mean Tmax produces a much sharper rise in stage-specific exposure in the warmer parts of the distribution.

In the Sahel (Fig. 1e), the effect is stronger. Baseline exposure above 37 °C in the early greening and canopy-development windows is already common. Under + 2 °C, large areas shift into regimes with three to five or more hot days per season. Because many days lie just below the threshold in the historical climate, the uniform warming simply pushes a much larger fraction of days into the exceedance range. The result is a broad expansion of high-exposure zones across the semi-arid belt.

For the East Africa highlands (Fig. 1f), the +2 °C experiment produces more modest but still important changes. Many locations move from essentially zero hot days in the booting–grain-fill window to one or two days per season above the local rice threshold. The absolute number of hot days remains lower than in Nigeria or the Sahel, yet the relative increase is large because exceedances were rare in the baseline climate. This highlights how even cool highland systems can cross critical thermal thresholds under modest warming.

Overall, the +2 °C exposure maps in Fig. 1d–f show that a relatively small increase in mean Tmax can substantially expand the area and frequency of stage-aligned hot-day exceedances. The response is strongest where baseline temperatures already sit close to the chosen thresholds. These patterns set the stage for the performance and diagnostic analyses that follow in Sects. 3.2 and 3.3.

3.2 Model Performance: Baselines Versus Physics-Informed (Space-Time-Blocked Holdouts)

Model performance is evaluated using space–time-blocked holdouts for all three case systems, so that entire years and spatial clusters are left out together rather than randomly. This tests whether the models retain skill when faced with genuinely unseen combinations of time and place, and it brings the evaluation closer to real early-warning and planning use cases (cf. Section 2.3). Table 1 summarizes the main performance metrics for the baseline and physics-informed specifications and Fig. 1a–f shows the corresponding observed–predicted scatter and low-outcome reliability plots for each region.

For Nigeria maize, seasonal-aggregate baselines (using seasonal mean temperature, rainfall totals and growing-degree days) capture only a modest share of the inter-annual yield variation under space–time blocking. When the same statistical families are re-specified using stage-aligned eco-climatic indicators (e.g. HotDay counts at anthesis, VPD and soil-moisture anomalies in early and mid-season), both error metrics and explained variance improve. The physics-informed machine-learning models, which use the full stage-aligned feature set and the guardrails defined in Sect. 2.3, tighten the scatter around the 1:1 line in Fig. 2a and pull the low-yield reliability curve closer to the 45° line in Fig. 2b. In practical terms, this means that low-yield seasons are identified more consistently without over-forecasting losses in normal years. The scatter-plot in Fig. 2a shows how the model performed in Nigeria (maize) and Fig. 2b shows the reliability plot (low-yield event) for Nigeria.

In the Sahel NDVI task, seasonal baselines that rely on rainfall totals and seasonal mean temperature show substantial spread in the observed–predicted NDVI scatter. Re-specifying the baselines on VPD, HotDay counts, dry-spell indicators and simple soil-moisture anomalies aligned to the greening window already produces a steeper, tighter cloud in Fig. 2c. The physics-informed NDVI models further reduce bias

Table 1 Performance metrics (space–time-blocked holdouts; 95% CIs)

Case	Outcome	Specification	RMSE (95% CI)	MAE (95% CI)	R^2 (95% CI)
Nigeria—maize	Yield (t/ha)	Classical	0.93	0.73	0.47
		Physics-informed	0.56	0.44	0.66
Sahel	NDVI	Classical	0.13	0.10	0.51
		Physics-informed	0.07	0.05	0.71
EA highlands—rice	Yield (t/ha)	Classical	0.81	0.65	0.41
		Physics-informed	0.58	0.43	0.58

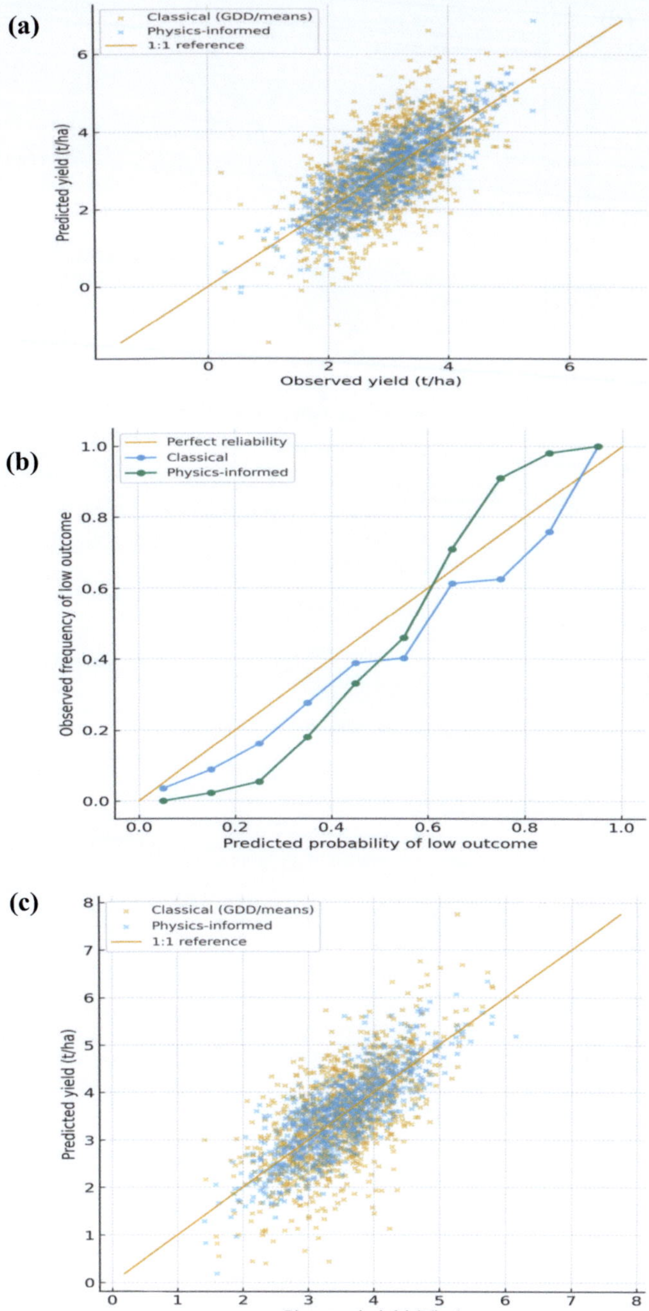

Fig. 2 **a** Nigeria (maize), observed versus predicted (space–time-blocked holdouts). **b** Nigeria reliability plot (low-yield event). **c** Sahel, observed versus predicted NDVI (space–time-blocked holdouts). **d** Sahel: reliability plot (low-NDVI event). **e** East Africa highlands (rice), observed versus predicted (space–time-blocked holdouts). **f** East Africa highlands reliability plot (low-yield event)

(d)

(e)

(f)

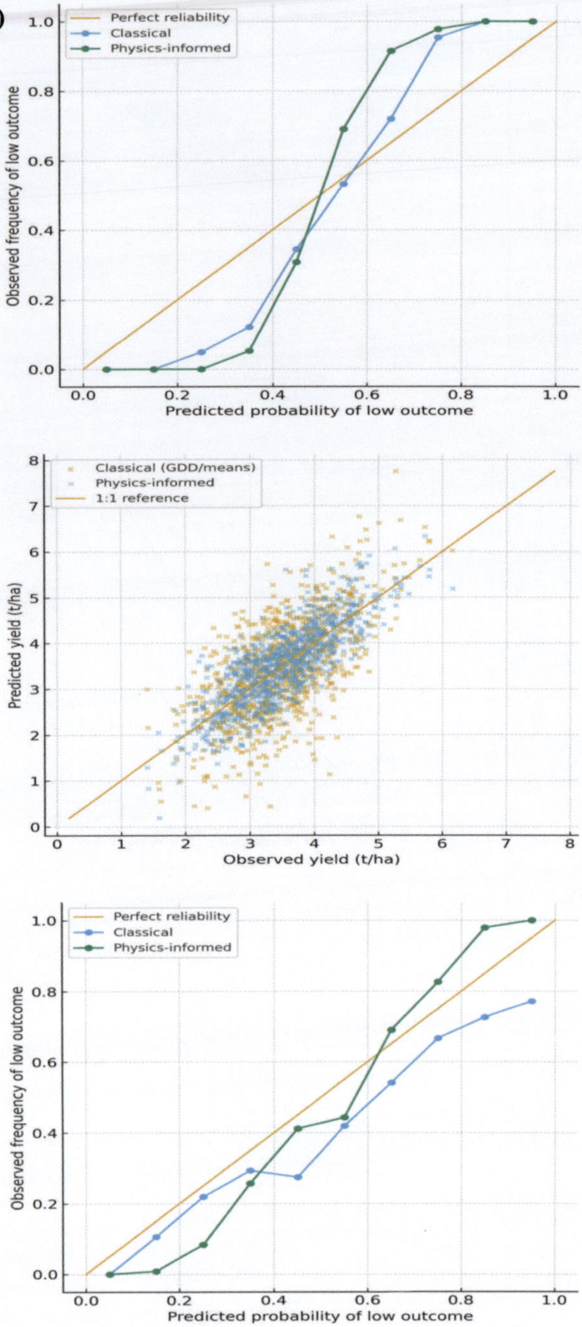

Fig. 2 (continued)

and shrink residuals, particularly at low NDVI, and the reliability plot in Fig. 2d sits closer to the diagonal across the full probability range. This indicates that the combination of VPD, temperature and simple water-balance signals carries more predictive power for canopy greenness than rainfall alone in the semi-arid Sahel, consistent with recent findings on VPD-driven stress in West Africa. The scatter-plot in Fig. 2c shows how the model performed in Sahel (NDVI) and Fig. 2d shows the reliability plot (low-yield event) for the Sahel.

For East Africa highland rice, seasonal-aggregate baselines perform somewhat better than in the other two cases, reflecting the more controlled conditions of irrigated and valley-bottom systems. Nonetheless, the stage-aligned eco-climatic baselines still show a visible gain when they incorporate HotDay counts and VPD during booting–grain fills along with simple indicators of water availability. The full physics-informed models further tighten the observed–predicted yield scatter-plot in Fig. 2e and improve calibration in Fig. 2f, particularly for the lowest yield years. Here the main improvement lies in correcting under-prediction of rare bad seasons rather than boosting already good years. The scatter-plot in Fig. 2e shows how the model performed in East Africa (rice), and Fig. 2f shows the reliability plot (low-yield event) for East Africa.

Across all three systems, Table 1 shows a consistent pattern: (i) re-specifying baselines on stage-aligned eco-climatic indicators improves performance relative to seasonal aggregates under the same space–time-blocked validation; (ii) physics-informed machine-learning models add further, though generally smaller, gains on top of that structural change; and (iii) reliability for low-outcome events improves alongside overall fit. Residual errors remain, as expected given data limitations and mixed pixels, but the lift from seasonal to stage-aligned, physics-proximal features is clear and forms the basis for the mechanism-focused diagnostics presented in Sects. 3.3 and 3.4.

In the remainder of this section, we interpret the physics-informed machine-learning models that use the full stage-aligned feature set, including HotDay counts, VPD, DTR, soil-moisture anomalies and management flags and compare them to the seasonal and stage-aligned baselines where relevant. Unless otherwise stated, all variable-importance scores and effect curves refer to these physics-informed models.

3.3 Variable Importance and Mechanisms (Permutation/SHAP)

To understand which parts of the eco-climatic feature set drive the performance gains in Sect. 3.2, we interpret the physics-informed machine-learning models that use the full stage-aligned feature set (HotDay counts, VPD, DTR, soil-moisture anomalies and management flags). For each case system, we compute two complementary diagnostics: permutation-based importance, averaged over the space–time-blocked folds and Shapley Additive Explanations (SHAP) summarized across all out-of-fold predictions. Figure 3a–c shows the permutation scores and corresponding SHAP value distributions for these physics-informed models. Together, these plots indicate which predictors consistently reduce error and how strongly extreme values push predictions up or down around their local baselines. Figure 3a–c illustrates variable importance diagnostics (permutation and SHAP scores) for the physics-informed models in the three systems.

For Nigeria maize (Fig. 3a), the highest permutation scores and largest absolute SHAP values are associated with stage-aligned thermal and moisture indicators rather than seasonal aggregates. Hot-day counts around anthesis, VPD during grain fill and early-season soil-moisture anomalies form the leading cluster of predictors. Diurnal temperature range (DTR) and simple dry-spell metrics also rank above seasonal rainfall totals. Management flags such as sowing-date bins and residue or cover indicators show moderate but non-negligible importance, especially in models that include them alongside the climate and soil variables. This pattern supports the idea that both the timing of heat and the background evaporative demand shape maize yield anomalies more strongly than smooth seasonal means.

In the Sahel NDVI case (Fig. 3b), the importance structure shifts further towards variables that describe atmospheric demand and within-season water stress. Growing-season VPD, hot-day exceedances during the greening window and dry-spell indicators dominate the upper end of the permutation and SHAP rankings. Cumulative rainfall and seasonal mean temperature fall into a secondary tier, once these stage-aligned indicators are present. Simple soil-moisture anomaly proxies and basic land- or tree-cover indicators also contribute, particularly in distinguishing pixels that maintain greenness under similar rainfall totals. For canopy greenness in semi-arid savannas, the models thus rely mainly on how heat and VPD interact with intermittent rainfall, rather than on rainfall alone.

For East Africa highland rice (Fig. 3c), the importance diagnostics highlight a smaller but clear set of key drivers. Hot-day counts and VPD during the booting–grain-fill window consistently rank among the top predictors, alongside indicators of water availability and management, such as irrigation/AWD flags or simple wetness proxies. Seasonal aggregates of temperature and rainfall again play a secondary role once stage-aligned indices are available. The SHAP distributions show that extreme booting–grain-fill heat and high VPD values are associated with large negative contributions to predicted yield, whereas moderate increases in water-availability indicators tend to shift predictions towards higher yields.

Across the three case systems, the permutation and SHAP diagnostics tell a coherent story that complements the performance results in Table 1 and Fig. 3. In all regions, physics-proximal, stage-aligned thermal and moisture indicators, especially hot-day counts and VPD in sensitive windows, carry more weight than seasonal temperature or rainfall summaries. Management flags contribute where they are available and variable. These patterns provide the starting point for the response-shape analyses in Sect. 3.4 and the cross-case evaluation of hypotheses H1–H3 in Sect. 3.5.

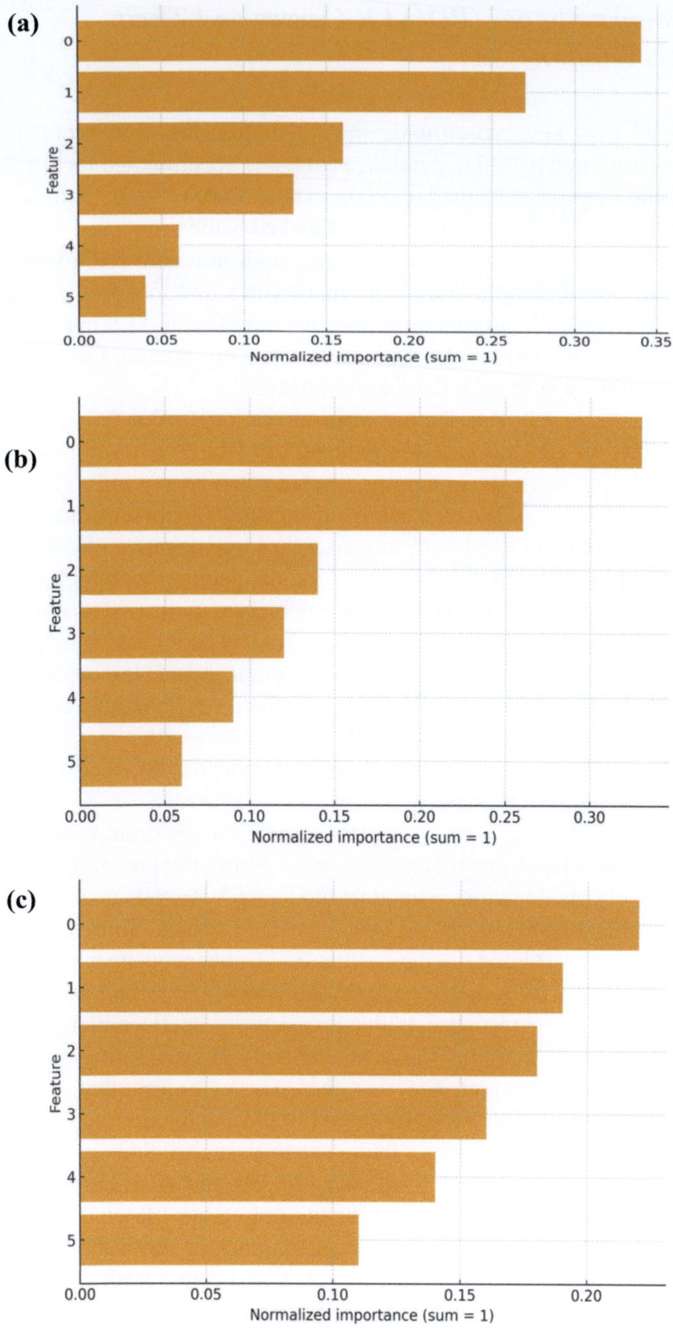

Fig. 3 **a** Variable importance diagnostics (permutation and SHAP scores) for the physics-informed models in Nigeria (maize). **b** Variable importance diagnostics (permutation and SHAP scores) for the physics-informed models in the Sahel NDVI. **c** Variable importance diagnostics (permutation and SHAP scores) for the physics-informed models in the East Africa highland rice

3.4 Response Shapes (PD/ALE Curves and Their Interpretation)

To understand how key eco-climatic indicators translate into yield or greenness changes, we examined partial dependence (PD) and accumulated local effects (ALE) curves for the physics-informed models. These curves were computed on out-of-fold predictions from the space–time-blocked validation used in Sect. 3.2, so that the model's response curves reflect true environmental relationships that can generalized to out-of-sample combinations of years and locations, rather than just creating an in-sample fits. Figure 4a–c presents the PD and ALE curves for the main thermal and VPD indicators in each case system, with separate lines for contrasting management states where such flags are available.

For Nigeria maize (Fig. 4a), the response curves for HotDay@anthesis show clear threshold behaviour. PD and ALE are relatively flat for zero to roughly two hot days above 35 °C during anthesis, indicating limited sensitivity within this lower range. Beyond about three to four hot days, both curves bend downwards more sharply, with additional hot days associated with increasingly large negative effects on predicted yield. VPD during grain fill shows a more gradual decline, with higher VPD values progressively reducing yields, and diurnal temperature range (DTR) has a weaker but still negative slope at high values. Curves stratified by residue or mulch flags suggest that plots with surface cover experience a slightly delayed and less steep decline along the HotDay axis, indicating some buffering of the most damaging combinations of heat and evaporative demand. This pattern is consistent with experimental evidence that short hot spells around flowering have disproportionate effects on maize yields once key thresholds are crossed.

In the Sahel NDVI (Fig. 4b) case, PD and ALE curves point to a strong role for atmospheric demand and short-term heat stress during the greening window. NDVI remains relatively stable across a band of moderate VPD values, but declines more rapidly once VPD exceeds roughly 2 kPa in the early season. Similarly, NDVI shows modest sensitivity to isolated hot days above 37 °C, but sequences with several such days are associated with noticeably lower greenness. Dry-spell indicators show a near-monotonic negative response, with increasing dry-spell length linked to reduced NDVI, while cumulative rainfall has a weaker and more saturating effect once basic moisture requirements are met. Together, these curves suggest that, in the semi-arid Sahel, combinations of high VPD, frequent hot days and longer dry spells are more consequential for canopy greenness than year-to-year variations in total rainfall.

For East Africa highland rice (Fig. 4c), the PD and ALE curves reinforce the importance of temperature and VPD in the booting–grain-fill window. Yield predictions remain relatively stable up to a small number of days above 34 °C during this stage, after which additional hot days are associated with a sharper decline. VPD exhibits a similar pattern: moderate increases within the historical range have limited effect, but high VPD values during booting–fill are linked to lower predicted yields. Simple indicators of water availability and management, such as irrigation or AWD/bund flags, shift the curves in a way that is consistent with improved water

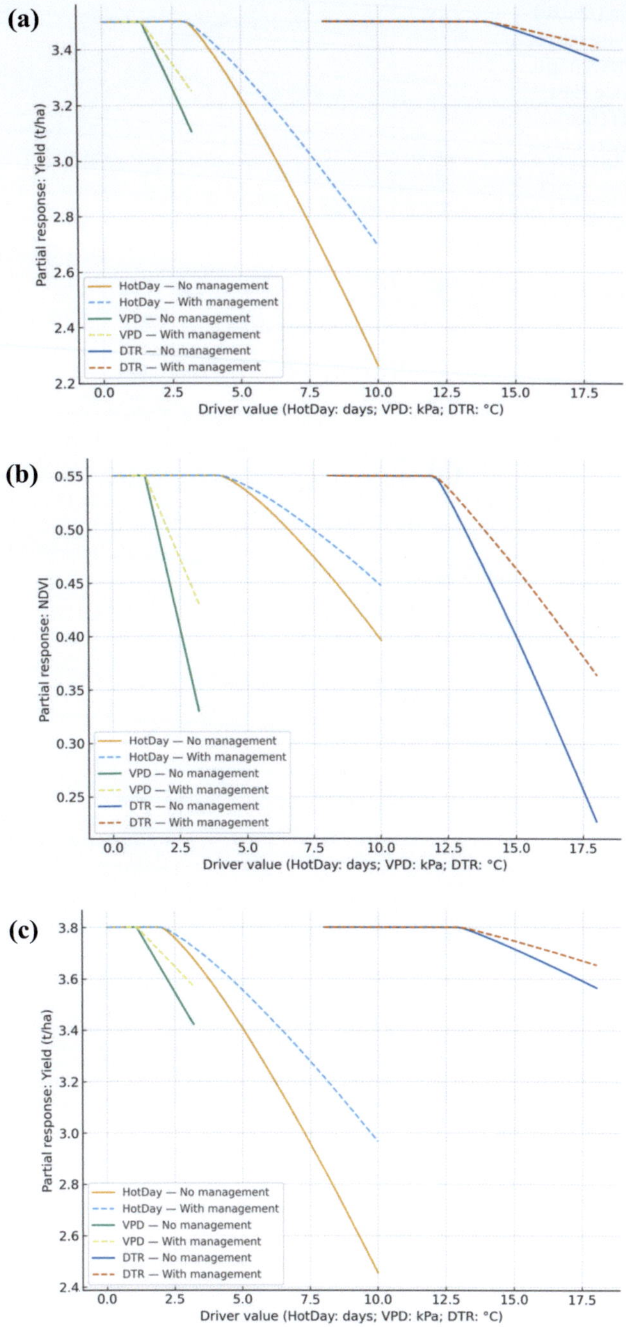

Fig. 4 **a** Partial dependence (PD) and accumulated local effects (ALE) curves for key eco-climatic indicators in Nigeria maize. **b** Partial dependence (PD) and accumulated local effects (ALE) curves for key eco-climatic indicators in the Sahel NDVI. **c** Partial dependence (PD) and accumulated local effects (ALE) curves for key eco-climatic indicators in the East Africa Highland rice

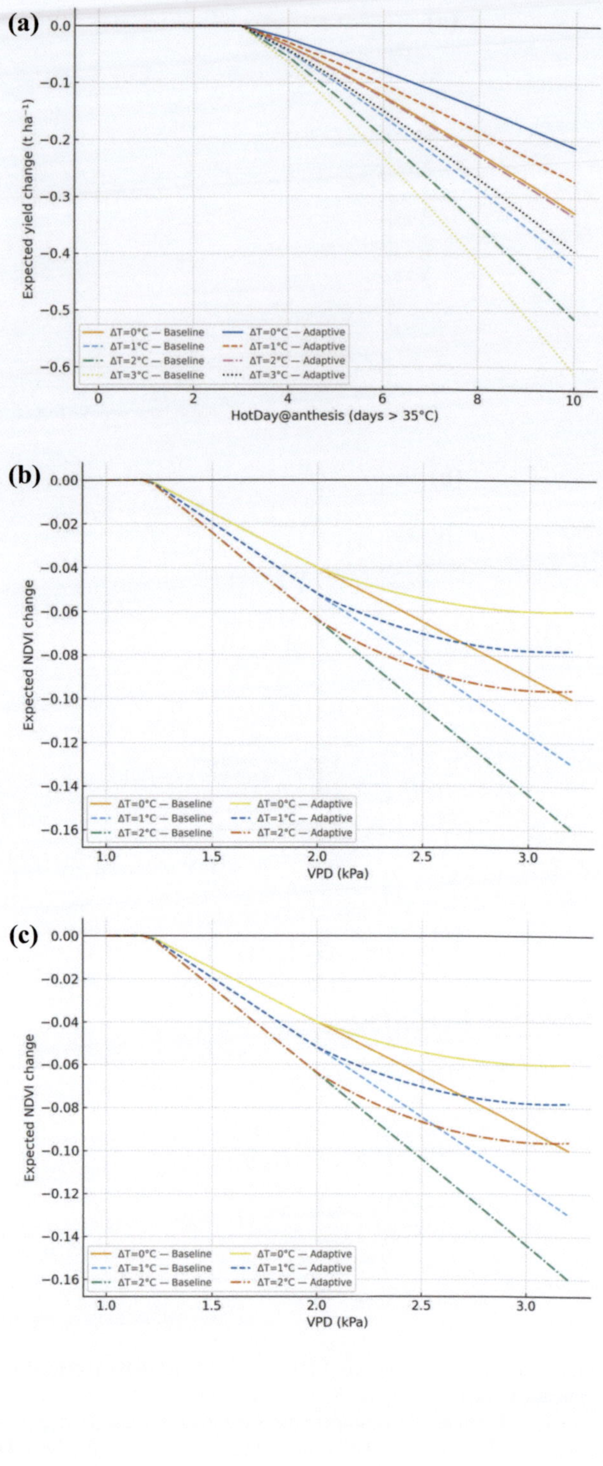

Fig. 5 **a** Nigeria (maize) scenario sensitivity curve. **b** Sahel (NDVI) Scenario sensitivity curve. **c** East Africa highlands (rice) Scenario sensitivity curve

(a)

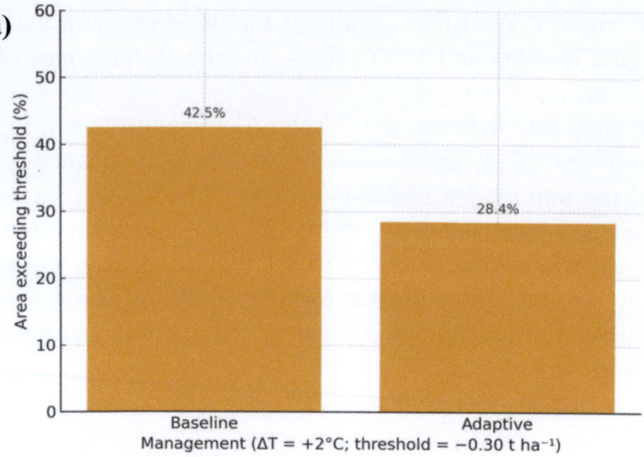

(b)

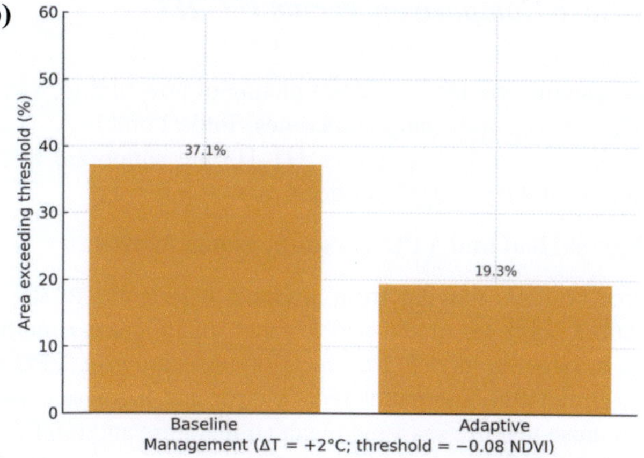

(c)

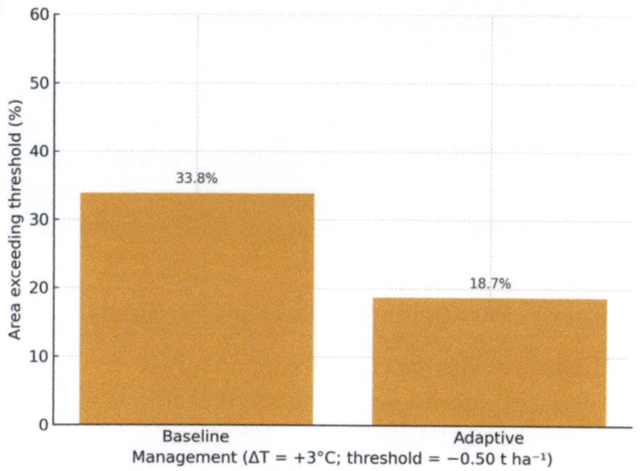

Fig. 6 **a** Nigeria risk summary at $\Delta T = +2$ °C and threshold = -0.30 t/ha^{-1}. **b** Sahel risk summary at $\Delta T = +2$ °C and threshold = $-$ 0.08 NDVI. **c** East Africa highlands risk summary at $\Delta T = +$ 3 °C and threshold = -0.50 t/ha^{-1}

status: under irrigated/AWD-like conditions the onset of sharp yield losses occurs at slightly higher HotDay and VPD values, and the steepest part of the decline is somewhat flattened.

Across the three case systems, the PD and ALE diagnostics in Fig. 4a–c provide a more detailed view of the mechanisms behind the performance gains reported in Sect. 3.2 and the importance rankings in Sect. 3.3. Hot-day exposure and VPD in sensitive phenological windows show clearly non-linear, often threshold-like response shapes, while management proxies tend to shift or flatten these curves where they are present. These shapes form the basis for the cross-case synthesis against hypotheses H1–H3 in Sect. 3.5 and for the scenario and sensitivity analyses presented in Sect. 4.

3.5 Cross-Case Comparison Versus H1–H3

The three case systems provide a coherent picture of how timing, VPD and management shape African crop and canopy outcomes. Table 1 and Figs. 1–4 allow a direct comparison of how strongly hypotheses H1–H3 are supported in Nigeria maize, Sahel NDVI and East Africa highland rice.

H1: Stage-aligned Heat and VPD Versus Seasonal Means

Across all three regions, moving from seasonal aggregates to stage-aligned eco-climatic indicators improves model performance under space–time-blocked validation (Table 1). In Nigeria, models that use HotDay@anthesis, VPD@grain-fill and early-season soil-moisture anomalies reduce error and increase explained variance compared with those that use seasonal mean temperature and rainfall totals. Similar gains appear in the Sahel and the highland rice valleys when HotDay and VPD windows are aligned to greening or booting–grain fill.

The PD/ALE curves in Fig. 4a–c show that these stage-aligned indicators produce non-linear, threshold-like responses. Maize and rice yields remain relatively stable up to a small number of hot days in the sensitive window, and then decline sharply with further exposure. NDVI behaves in a similar way under short runs of extreme VPD and heat during green-up. Taken together, these cross-case patterns support H1: timing and short-window exposure to heat and VPD explain more variability than shifts in seasonal mean temperature alone.

H2: VPD and Energy Partitioning Versus Rainfall Totals

The Sahel NDVI case provides the clearest support for H2. Models built on VPD, hot-day counts and dry-spell indicators during the greening window outperform those based on cumulative rainfall and seasonal temperature. Variable-importance rankings in Fig. 3b place VPD and short-term heat metrics above rainfall in explaining NDVI anomalies.

In Nigeria maize and East Africa rice, rainfall remains useful but moves into a secondary tier once VPD and stage-aligned HotDay variables are included. The response shapes in Fig. 4a–c reinforce this ordering. Higher VPD is consistently associated with declining NDVI or yield even when rainfall totals sit within a near-normal band, whereas extra rainfall beyond a moderate threshold has diminishing effects. Across the three systems, atmospheric demand and energy partitioning therefore carry at least as much explanatory power as rainfall totals, in line with H2.

H3: Microclimate-Shaping Management as a Buffer

Evidence for H3 varies by system but points in a consistent direction. In Nigeria, residue/cover and sowing-date indicators have moderate importance in Fig. 3a and visibly shift the HotDay response curves in Fig. 4a. With residue present, the onset of rapid yield decline along the HotDay axis occurs at slightly higher values. This suggests that increased surface cover allows maize to tolerate roughly one additional hot day at anthesis before entering the steep-loss range.

In the highland rice valleys, irrigation/AWD and bunding flags show a similar effect. Under managed water conditions, the booting–grain-fill hot-day and VPD curves shift so that substantial yield losses occur at higher stress levels than under the baseline state. In the Sahel NDVI case, coarser management proxies modestly improve fit but do not greatly change the shape of the damage curves. This reflects both limited control over water and the aggregate nature of the NDVI signal.

Overall, these cross-case results indicate that management can flatten or shift damage curves in the expected direction, with the strongest buffering where farmers directly control water and surface cover. However, management alone does not fully offset warming and rising VPD. H3 is therefore partially supported: microclimate-shaping practices moderate, but do not eliminate, the influence of stage-aligned heat and VPD identified under H1 and H2.

In summary, the Nigeria maize, Sahel NDVI and East Africa highland rice applications jointly show that stage-specific thermal and VPD exposure, rather than seasonal means or rainfall totals alone, dominate model skill and response shapes. Management acts mainly to shift and slightly flatten these relationships where feasible. These cross-case findings provide the empirical basis for the broader interpretation and policy implications.

4 Scenario and Sensitivity Analysis

We used the fitted physics-informed models to ask a simple question: how do projected warming and practical management changes alter yield and greenness risk across the three case regions? Building on the response shapes in Sect. 3, we ran delta-temperature (ΔT) experiments, examined VPD shifts and compared a Baseline versus Adaptive management bundle for each system.

4.1 *ΔT Experiments and VPD Shifts*

Daily maximum temperature was perturbed by $\Delta T = +1$, $+2$ and $+3$ °C, and VPD were recomputed from the warmed saturation vapour pressure. Relative humidity was treated in two ways: a "dry-leaning" core case with RH held constant and a sensitivity case where RH was scaled following Clausius–Clapeyron moisture constraints. Rainfall totals and onset were left unchanged, so that the scenarios isolate thermal and evaporative-demand effects rather than changes in the pluviometer. Figure 5a–c shows the Scenario sensitivity curve for the three cases.

For each ΔT level, we reconstructed stage-aligned exposures—HotDay@anthesis for Nigerian maize, HotDay@booting–fill for East African rice and VPD windows for Sahel greenness and passed them through the fitted response functions described in Sect. 3 (PD/ALE surfaces). This yielded loss–exposure curves and expected yield or NDVI changes relative to the historical baseline.

In Nigeria (Fig. 5a), warming steepens the relationship between HotDay@anthesis and maize yield loss: by +2 °C, an increase from 4 to 6 hot days (>35 °C) shifts from a mild penalty to losses approaching -0.3 t ha^{-1} under baseline management, consistent with previous evidence that short, stage-specific heat bursts can dominate seasonal means. Similar patterns emerge for Sahel NDVI (Fig. 5b), where higher VPD under ΔT amplifies greenness deficits even when rainfall totals are unchanged, in line with the growing recognition that evaporative demand is now a key stressor in semi-arid Sub-Sahara Africa. In the East Africa highlands (Fig. 5c), a +2 °C shift turns what are currently rare booting–grain-fill hot days (>34 °C) into more frequent events, so that a small increase in mean temperature produces a disproportionate rise in simulated yield penalties for irrigated valley-bottom rice.

4.2 *Management Levers and Buffering*

To quantify how management modifies these warming responses (H3), we toggled between two management states: a Baseline reflecting observed residue cover, AWD use and sowing dates, and an Adaptive bundle with (i) residue and ground cover increased by roughly 25–50 percentage points in rain-fed areas, (ii) AWD and bunds expanded from about 60 to ~80% of irrigated valley bottoms, and (iii) sowing dates advanced by 7–10 days in identified hot-spell belts.

Under $\Delta T = +2$ °C, these levers measurably flatten damage slopes. For Nigerian maize (Fig. 4a), the share of area experiencing yield losses worse than about -0.3 t ha^{-1} shrinks from roughly 42.5% under Baseline to ~28.4% under Adaptive residue + earlier sowing, showing that modest changes in energy partitioning and phenology can offset part of the warming signal. In the Sahel (Fig. 4b), higher residue and small water-harvesting structures weaken the link between extreme VPD and NDVI decline, while in the East Africa highland rice valleys (Fig. 4c), expanded

AWD/bunding and denser canopies moderate canopy temperature during booting–fill, reducing the probability of crossing critical heat thresholds.

4.3 Spatial Risk Patterns and Where Adaptation Helps Most

Finally, we mapped how ΔT and VPD shifts change the spatial pattern of risk, using the fitted models to generate stage-resolved risk surfaces for each region. For each grid cell, we computed the expected yield or NDVI change under $\Delta T = 0, +1, +2$ and +3 °C and classified them into policy-relevant loss bands (e.g. $< -10\%$, -10 to -30%, $> -30\%$). Figure 6a–c shows the risk summary for Nigeria, Sahel and East Africa.

In Nigeria (Fig. 6a), the largest increases in high-loss probability cluster in the central maize belt where current Tmax and VPD are already close to damaging levels. Under the Adaptive bundle, these hotspots shrink and fragment, especially in zones with improved residue retention and modest sowing shifts. In the Sahel (Fig. 6b), the most pronounced NDVI losses appear along semi-arid rangeland strips where VPD increases fastest; management reduces but does not fully erase these stripes, highlighting where greening programmes and water-harvesting would be most impactful. For East African rice (Fig. 6c), risk concentrates in low-lying valley bottoms where humidity is high but canopy temperatures spike during short heat waves; expanded AWD/bunds and higher LAI smooth these pockets and keep a larger fraction of the area below critical damage thresholds.

5 Discussion

5.1 Evaluation of Hypotheses H1, H2 and H3

The three case studies give a coherent picture of how recent warming, rising VPD and management choices shape African crop and vegetation outcomes (Figs. 1–6; Table 1). Overall, the results support H1 and H2 and provide partial but policy-relevant support for H3.

H1: Timing of Heat Versus Seasonal Means

Across Nigeria's maize belt, the Sahel NDVI system and East Africa highland rice, stage-aligned indicators such as HotDay@anthesis, HotDay@booting–grain fill and short high-VPD windows carry more predictive weight than seasonal averages (Table 1; Figs. 2–4). Seasonal-aggregate baselines built on mean temperature and rainfall explain only a modest share of interannual variability under space–time-blocked validation. When these baselines are re-specified using stage-aligned heat and VPD metrics, performance improves in all three regions, and the full physics-informed models sharpen this further.

The PD/ALE curves show threshold-like behaviour: yields or NDVI stay relatively stable up to a small number of hot days in the sensitive window, then decline steeply once those thresholds are crossed (Fig. 4a–c). This echoes trial-based evidence that short, well-timed hot spells can dominate the effect of gradual seasonal warming on maize yields (Lobell et al., 2011). Taken together, these patterns support H1: for crop and canopy outcomes, *when* heat and VPD occur matters more than how much seasonal mean temperature has shifted.

H2: VPD and Energy Partitioning Versus Rainfall Totals

The Sahel NDVI case provides the clearest evidence for H2. Under space–time-blocked validation, models using VPD, hot-day counts and dry-spell indicators during the greening window outperform those based on cumulative rainfall and seasonal temperature alone (Table 1; Fig. 3b). Variable-importance rankings place VPD and short heat bursts above rainfall (Fig. 3b), and the NDVI response curves show sharper declines at high VPD and longer dry spells, even when rainfall stays within a "normal" band (Fig. 3b). This matches recent work highlighting evaporative demand as a key stressor in semi-arid West Africa (Hoffman et al., 2018).

A similar, though slightly weaker, pattern appears in Nigeria maize and highland rice. Once stage-aligned VPD and HotDay metrics are included, rainfall totals move into a secondary tier of predictors (Fig. 3a, c). The response shapes confirm this ordering: higher VPD is consistently associated with declining NDVI or yield even when rainfall is near normal, whereas additional rainfall beyond a moderate threshold has diminishing benefits. Across the three systems, atmospheric demand and energy partitioning therefore carry at least as much explanatory power as rainfall totals, in line with H2 and with broader IPCC assessments (IPCC, 2021).

H3: Microclimate-Shaping Management as a Buffer

Evidence for H3 varies by system but points in a consistent direction. In Nigeria, residue/cover and sowing-date indicators have moderate importance in Fig. 3a and visibly shift the HotDay response curves in Fig. 4a. With residue present, the onset of rapid yield decline along the HotDay axis occurs at slightly higher values. This suggests that increased surface cover allows maize to tolerate roughly one extra hot day at anthesis before entering the steep-loss range.

In the highland rice valleys, irrigation/AWD and bunding flags show a similar buffering effect. Under managed water conditions, the booting–grain-fill HotDay and VPD curves shift so that substantial yield losses occur at higher stress levels than under the baseline state (Fig. 4c). In the Sahel NDVI case, coarser management proxies modestly improve fit but do not strongly change the shape of the damage curves, reflecting both limited water control and the aggregate nature of the NDVI signal.

Overall, management can flatten or shift damage curves in the expected direction, with the strongest buffering where farmers directly control water and surface cover. However, it does not fully offset warming and rising VPD. H3 is therefore partially supported: microclimate-shaping practices moderate, but do not remove, the influence of stage-aligned heat and VPD identified under H1 and H2.

5.2 Limitations and Future Directions

Several limitations should be kept in mind when interpreting these results.

First, the analysis relies heavily on gridded reanalysis and remote-sensing products rather than dense in situ networks. ERA5/ERA5-Land, CHIRPS and MODIS provide consistent, long records, but they contain retrieval errors, mixed pixels and scale mismatches with field plots. These uncertainties likely contribute to residual errors and may damp or amplify some local signals.

Second, the models use derived indicators that embed assumptions about phenological timing and thresholds. Crop-stage windows are inferred from calendars and simple rules rather than from direct phenology observations, and temperature/VPD thresholds are drawn from literature and expert judgement (Lobell et al., 2011; van Oort & Zwart, 2018). Strong performance under space–time-blocked validation suggests that these choices are broadly reasonable, but future work could refine those using field trials, on-farm phenology monitoring and process-based crop models.

Third, the scenario design is deliberately simple. ΔT perturbations are applied to Tmax with rainfall held fixed, and VPD is recomputed under basic humidity assumptions. This isolates thermal and evaporative-demand effects but does not capture the full range of plausible future climates, including changes in rainfall seasonality, humidity and circulation patterns (IPCC, 2021, 2022). Coupling the framework to ensembles of regional or global climate models, and sampling multiple realizations, would provide a richer view of risk envelopes and uncertainty.

Fourth, management is represented using coarse proxies: binary or categorical flags for residue, AWD, bunds and sowing windows, rather than detailed descriptors of practice intensity, farmer capacity or institutional constraints. The Baseline versus Adaptive bundles therefore illustrate what could be achieved biophysically, not what will necessarily occur. Integrating socio-economic data and farm-system typologies would help bridge the gap between potential and real-world adoption.

Finally, the current applications focus on three systems, including Nigerian maize, Sahel NDVI and East Africa highland rice. They span a useful gradient from humid–sub-humid to semi-arid and cool irrigated settings, but they are not exhaustive. Extending the framework to additional crops, regions and hazards, such as lowland rice, sorghum and compound heat-flood events, would test its generality and show where the stage-aligned, physics-informed approach needs adaptation.

Despite these caveats, the study demonstrates that an integrated eco-climatic framework, coupled with physics-informed machine learning and stage-specific indicators, can translate large-scale warming and VPD trends into operationally meaningful risk metrics. It offers a transparent way to compare seasonal versus stage-aligned diagnostics, to quantify the buffering effect of management, and to support early-warning and climate-smart planning across diverse African agro-ecosystems.

6 Conclusion

This chapter has shown how an integrated eco-climatic, physics-informed modelling framework can sharpen climate–impact assessments for African crops and rangelands. By combining ERA5-Land, CHIRPS, MODIS and sub-national yield data with stage-aligned indicators of hot-day exposure, vapour-pressure deficit (VPD), diurnal temperature range (DTR), soil-moisture anomalies and management flags, we move beyond simple seasonal averages. Across Nigeria's maize belt, Sahel greenness and East Africa highland rice, the physics-informed models explain substantially more inter-annual variance in yields and NDVI than seasonal baselines and reduce RMSE and MAE under space–time-blocked validation.

The results confirm that timing of stress matters as much as its magnitude. Short hot and high-VPD bursts around flowering or booting–grain fill produce steep yield and greenness losses, even when seasonal means shift only modestly, echoing plot-scale evidence for maize and rice (Lobell et al., 2011; van Oort & Zwart, 2018). VPD and energy-partitioning proxies rank alongside, and often above, rainfall totals in variable-importance profiles, especially in the Sahel. These patterns support H1 and H2: stage-aligned heat and boundary-layer demand recover a larger share of the climate-impact signal than seasonal means of temperature and rainfall.

Evidence for management buffering (H3) is more mixed but still policy-relevant. Residue retention and sowing shifts in Nigeria, and irrigation or alternate wetting and drying in the highland rice systems, shift and flatten damage curves so that crops tolerate slightly higher levels of stage-specific heat and VPD before steep losses occur. The effect is strongest where farmers can directly control water and surface cover, and weaker where management options and data are coarse. Management therefore moderates, but does not cancel, the risks implied by continued warming and rising VPD.

Taken together, these findings point to a practical way to make climate risk information more actionable. Simple, physics-informed indicators of hot-day counts, VPD, DTR and soil-moisture anomalies, aligned to crop stages and linked to management flags, can feed directly into agro-meteorological advisories, seasonal outlooks and regional climate-risk platforms. Despite data and modelling limitations, the framework offers a transparent bridge between process understanding, machine-learning skill and on-the-ground decision support, and it can be extended to other crops, regions and hazards as new observations and models become available.

References

Akpa, A. F. (2024). The effects of climate extreme events on selected food crop yields in Sub-Saharan Africa. *Heliyon, 10*(9), e30796. https://doi.org/10.1016/j.heliyon.2024.e30796

Ayanlade, A., Oluwaranti, A., Ayanlade, O. S., Borderon, M., Sterly, H., Sakdapolrak, P., Jegede, M. O., Weldemariam, L. F., & Ayinde, A. F. O. (2022). Extreme climate events in sub-Saharan

Africa: A call for improving agricultural technology transfer to enhance adaptive capacity. *Climate Services, 27*, 100311. https://doi.org/10.1016/j.cliser.2022.100311

Bwire, D., Saito, H., Sidle, R. C., & Mugisha, M. (2025). Water management for rice production: A key component of food security in East Africa. *Discover Water, 5*, 7. https://doi.org/10.1007/s43516-025-0007-z

Challinor, A. J., Watson, J., Lobell, D. B., Howden, S. M., Smith, D. R., & Chhetri, N. (2014). A meta-analysis of crop yield under climate change and adaptation. *Nature Climate Change, 4*(4), 287–291. https://doi.org/10.1038/nclimate2153

Chukwudi, U. P., Kutu, F. R., & Mavengahama, S. (2021). Heat stress effect on the grain yield of three drought-tolerant maize varieties under varying growth conditions. *Plants, 10*(8), 1532. https://doi.org/10.3390/plants10081532

Dong, N. Q., & Lin, H. X. (2025). Overcoming stress-growth and yield-quality trade-offs in rice. *Cell Reports, 44*(6), 115846. https://doi.org/10.1016/j.celrep.2025.115846

Du, S., & Xiong, W. (2024). Weather extremes shock maize production: Current approaches and future research directions in Africa. *Plants, 13*(12), 1585. https://doi.org/10.3390/plants13121585

Didan, K., Muñoz, A. B., Solano, R., & Huete, A. (2015). *MODIS vegetation index (MOD13) user's guide (Version 3.00)* (p. 35). Vegetation Index and Phenology Lab.

Emediegwu, L. E., Wossink, A., & Hall, A. (2022). The impacts of climate change on agriculture in Sub-Saharan Africa: A spatial panel data approach. *World Development, 158*, 105967. https://doi.org/10.1016/j.worlddev.2022.105967.

Funk, C., Peterson, P., Landsfeld, M., Pedreros, D., Verdin, J., Shukla, S., Husak, G., Rowland, J., Harrison, L., Hoell, A., & Michaelsen, J. (2015). The climate hazards infrared precipitation with stations (CHIRPS): A new environmental record for monitoring extremes. *Scientific Data, 2*, 150066. https://doi.org/10.1038/sdata.2015.66

Han, L., & Leng, G. (2024). Significant changes in global maize yield sensitivity to vapor pressure deficit during 1983–2010. *Agricultural Water Management, 305*, 109107. https://doi.org/10.1016/j.agwat.2024.109107

Hersbach, H., Bell, B., Berrisford, P., Hirahara, S., Horányi, A., Muñoz-Sabater, J., Nicolas, J., Peubey, C., Radu, R., Schepers, D., Simmons, A., Soci, C., Abdalla, S., Abellan, X., Balsamo, G., Bechtold, P., Biavati, G., Bidlot, J., Bonavita, M., et al. (2020). The ERA5 global reanalysis. *Quarterly Journal of the Royal Meteorological Society, 146*(730), 1999–2049. https://doi.org/10.1002/qj.3803

Hoffman, A. L., Kemanian, A. R., & Forest, C. E. (2018). Analysis of climate signals in the crop yield record of sub-Saharan Africa. *Global Change Biology, 24*(1), 143–157. https://doi.org/10.1111/gcb.13901

Intergovernmental Panel on Climate Change. (2021). *Climate change 2021: The physical science basis. Contribution of working group I to the sixth assessment report of the intergovernmental panel on climate change.* Cambridge University Press. https://doi.org/10.1017/9781009157896

Intergovernmental Panel on Climate Change (IPCC). (2022). In H.-O. Pörtner, D. C. Roberts, M. Tignor, E. S. Poloczanska, K. Mintenbeck, A. Alegría, et al. (Eds.), *Climate change 2022: Impacts, adaptation and vulnerability. Contribution of working group II to the sixth assessment report of the intergovernmental panel on climate change.* Cambridge University Press.

Lobell, D. B., Bänziger, M., Magorokosho, C., & Vivek, B. (2011). Nonlinear heat effects on African maize as evidenced by historical yield trials. *Nature Climate Change, 1*(1), 42–45. https://doi.org/10.1038/nclimate1043

Ogunkanmi, L. A., MacCarthy, D. S., & Adiku, S. G. (2021). Impact of extreme temperature and soil water stress on the growth and yield of soybean (*Glycine max* (L.) Merrill). *Agriculture, 12*(1), 43. https://doi.org/10.3390/agriculture12010043

Sultan, B., Defrance, D., & Iizumi, T. (2019). Evidence of crop production losses in West Africa due to historical global warming in two crop models. *Scientific Reports, 9*, 12834. https://doi.org/10.1038/s41598-019-49167-0

van Oort, P. A. J., & Zwart, S. J. (2018). Impacts of climate change on rice production in Africa and causes of simulated yield changes. *Global Change Biology, 24*(3), 1029–1045. https://doi.org/10.1111/gcb.13967

Heat Adaptation Through Policy and Planning: A Case Study of Ahmedabad's Heat Action Plan

Himani Upreti, Rupa Khanna Malhotra, Priti Sharma, and Shruti Sharma

Abstract Climate change if left unchecked may pose an unparallel menace not only to the human civilisation and but also to the ecosystems on this planet too. As climate change is intensifying year on year, heatwaves have emerged as a critical challenge particularly for urban, highly populated and low-resource cities. The case of 'Ahmedabad', India which is considered as the first South Asian city to implement a formal Heat Action Plan (HAP) in the year 2013 is studied. Drawing exclusively on government documents, NGO reports, academic publications, and institutional data, the study explores how Ahmedabad's HAP integrates features such as releasing early warning systems to the public, initiating public awareness programme, healthcare system preparedness, interagency coordination, and urban heat-resilient planning to mitigate extreme heat impacts. The findings reveal that collaborative policy framework can help in reduction of heat-related morbidity and mortality. Further it also highlights the plan's evolution, institutional innovations, and adaptation mechanisms, offering lessons for urban climate governance in the Global South. In the current chapter, an attempt has been made to get a deeper understanding of effective policy and management strategies for urban heat adaptation, with potential replicability across climate-vulnerable urban regions through synthesization of evidence from multiple credible sources which are secondary in nature.

Keywords Urban heat · Climate adaptation · Heat action plan · Ahmedabad

1 Introduction

Earlier floods, famines, cyclones, and droughts were considered to be the major climate discourse because of the immediate and highly visible destruction. As climate change is intensifying year on year, heatwaves have emerged as a critical challenge, particularly for urban, highly populated and low-resource cities. Asia is considered

H. Upreti (✉) · R. K. Malhotra · P. Sharma · S. Sharma
Department of Commerce, Graphic Era (Deemed to Be, University), Dehradun, Uttarakhand, India
e-mail: himani.pant10@gmail.com

A. Awasthi et al. (eds.), *Rising Temperatures and Plant Ecosystems*,
https://doi.org/10.1007/978-3-032-29477-7_16

to be the densely populated continent and is home to about 60 per cent of the global population (Awasthi et al., 2022). An alarm has already been raised by the Intergovernmental Panel on Climate Change (IPCC) in the year 2022 with respect to the prevalence, severity, and duration of heatwaves across the globe and South Asian nations due to their dense populations, climatic conditions, and social vulnerabilities stand at a higher risk among all. If global emissions trajectories continue on their current path, many South Asian cities may become virtually uninhabitable during summer months by the end of the century, with dangerous effects for human existence, labour productivity, and urban development (Rohini et al., 2016; Zhao et al., 2021).

Heatwaves are often considered a hidden danger because their impacts, though less visible, can be devastating. Unlike cyclones or floods, heatwaves unfold gradually, often without dramatic infrastructure damage, but they inflict substantial losses in terms of human lives, health, and productivity. A Lancet Planetary Health study reported that globally, over 150,000 people had died in heatwaves between 1998 and 2017, and South Asia, along with sub-Saharan Africa, was among the most affected regions (Watts et al., 2018).

India has emerged as a global heat hotspot. As per the report released, approximately 24,000 people died between 1992 and 2015, because of extensive heatwaves (Indian Meteorological Department (IMD), 2016). In 2010, Ahmedabad witnessed one of the deadliest heatwaves in its history, with temperatures soaring to 46.8 °C, and which resulted into more than 1300 deaths within a week (Azhar et al., 2014). The tragedy served as a wake-up call, drawing attention to the urgent need for institutionalized responses to heat risks. Subsequent national events, such as the 2015 heatwave that killed more than 2,500 people across India, reinforced the status of heatwaves as major disaster risks (NDMA, 2019). It has been seen that the urban areas are more susceptible to heat stress compared to the rural ones. The major reasons for this heat stress in these areas are constructions with concrete, asphalt, and glass that trap higher amounts of heat, which leads to higher night temperatures in urban areas, this is also regarded as the Urban Heat Island effect (UHI) (Yadav et al., 2023). Dense population and limited access to cooling infrastructure in big towns intensify the consequences of heat waves in such places. Socioeconomic Inequalities is further a very significant contributor towards the urban heat vulnerability. The urban poor, particularly slum dwellers, daily wage labourers, street vendors, and migrants, often live and work in precarious conditions with poor housing, lack of ventilation, and limited healthcare access (Basu, 2001). Further, low-income households, women, infants, the elderly, and people with past medical history are equally prone to heat-related risks (Sherwood & Huber, 2010). Construction labour and street vendors also constitute the vulnerable population. Thus, heat risk is not only an environmental challenge but also a matter of developmental and social justice. (Kjellstrom et al., 2017).

Heat stress must be addressed immediately, according to both national and international frameworks. The UN Sustainable Development Goals framework was internationally adopted by multiple nations in the year 2015, wherein 17 SDGs have been identified, among them SDG 13 and 11 call for resilient cities and immediate climate action, respectively (Pandey et al., 2025). Another international policy initiative, named the Sendai Framework for Disaster Risk Reduction, prioritizes predictive risk reduction and readiness for climate-induced disasters, such as heatwaves. (UNDRR, 2015). Additionally, the 2015 Paris Agreement promoted and emphasized local adaptation strategies as essential means of attaining climate resilience (UNFCCC, 2015). India has created a National and state-level action plan on climate change (Government of India, 2019). Guidelines have been issued by NDMA on heat wave management, which were formulated with an intent that states and cities will create Heat Action Plans (HAPs) based on them (NDMA, 2019). However, limited financial resources, weak institutional capacity, and fragmented governance systems hindered the actual implementation of these guidelines.

In such a policy lagging scenario, Ahmedabad came up as a pioneering case. This pioneering city recognised the relevance of collaboration and participation for making this initiative successful. They realised that only national and state-level plans would not be sufficient for bringing and implementing such an initiative. Thus, as a result, the Ahmedabad Municipal Corporation (AMC) collaborated with academic, international, and community partners and developed the first city-level Heat Action Plan in South Asia in 2013. This marked a paradigm shift in urban climate adaptation, which shifted to a proactive approach from reactive and crisis management, and became more planned and institutionalized (Knowlton et al., 2014). This model emerged as collaboration was notable for bringing together municipal authorities, health institutions, researchers, and civil society actors and was regarded as a model of multi-stakeholder governance in climate adaptation (NRDC, 2018). The HAP introduced a comprehensive framework which included early warning systems, usage of multiple media platforms to spread heat safety messages, launching training and capacity building initiatives for healthcare workers, promoting cooling strategies such as cool roofs and improved water access (Chakraborty et al., 2019).

Under this chapter, an analysis of Ahmedabad's Heat Action Plan as a case study of policy and planning for climate adaptation is conducted. The major objectives of the current study are:

To trace the origins and evolution of Ahmedabad's HAP.

To examine its components, implementation mechanisms, and outcomes.

To identify challenges and limitations in its execution.

To assess its replicability and scalability across other Indian cities and the Global South.

Thus, through this chapter, an attempt has been made by us to contribute to the scholarship and practice of urban climate resilience, offering lessons not only for policymakers in India but also for practitioners globally who are grappling with the growing reality of heat as a climate hazard.

The further sections of the chapter include the methodology adopted, results and discussion, future scope and limitation, and then the paper is concluded.

2 Methodology

2.1 Research Design

The research we have employed a qualitative, case study research design to examine the Heat Action Plan (HAP) created by the city of Ahmedabad as an evolving urban climate resilience initiative in South Asia. A case study design is particularly suitable for understanding complex policy processes because it allows for the exploration of contextual, institutional, and community-level dimensions in depth (Yin, 2018). Since the study investigates the evolution, implementation, and outcomes of a single city's adaptation plan, the case study approach ensures a holistic analysis that captures both successes and limitations.

2.2 Nature of the Study and Data Sources

The current study is exploratory and descriptive in nature. While conducting the current study, an attempt has been made to explore such factors that led to the development of Ahmedabad's HAP. Further, the current chapter also describes how various stakeholders, ranging from municipal authorities to civil society organizations, have contributed to the design and implementation of the initiative.

While writing the chapter and collecting the information for the same, we used credible secondary sources of information. We collected data from a wide range of reliable and authentic sources. To ensure the reliability and validity of data, it was cross-verified from multiple independent resources. The information is based on credible government sources, while writing this chapter we have examined the guidelines issued in the year 2019 by NDMA with respect to Preparation of Action Plans which comprises of measures related to how to prevent and measure heat waves (2019 by the National Disaster Management Authority's), the Indian Meteorological Department (IMD), and the Ahmedabad Municipal Corporation (AMC). We have also collected data from peer-reviewed journal articles about urban public health, climate adaptation, and heatwaves.

3 Results and Discussions

Heat waves are adversely affecting the whole world, and India is considered one of the most climate-vulnerable countries. A report from The Lancet highlights that the extreme heat vulnerability has expanded from 1990 to 2019 at the rate of 15%. The Index on extreme weather events released in the year 2021 by Global Climate Risk Index has ranked India at the seventh position on its exposure to severe climatic events such as floods, cyclones, and heatwaves (Eckstein et al., 2021). Among these, heatwaves have become a growing and persistent hazard in recent decades, causing widespread mortality, morbidity, and economic losses. The intensification of heat extremes is closely linked to both global climate change and local urban dynamics, making India's cities particularly susceptible to dangerous levels of heat stress.

3.1 Trends and Patterns of Extreme Heatwaves in India and Their Impacts on Health and Socioeconomic Diversity of India

Heatwave as defined by the Indian Meteorological Department, as a condition when the maximum temperature rises above 40 °C in the plain region and 30 °C in hilly regions, or when temperatures are 4.5–6.4 °C above the seasonal average (ImD, 2016). Anomalies above lead to the declaration of severe heatwaves. India has experienced a considerable rise in the frequency, duration, and intensity of heatwaves since 2005, with varying degrees of 5 °C. From 1980 to 2010, India experienced a surge in the frequency of heatwave days, which was estimated to be about 25% (Rohini et al., 2016). By 2050, India could face extreme heat waves that exceed the human survivability threshold for several weeks annually, particularly in its northern and central regions (Im et al., 2017). Over 2,500 people, mostly from Andhra Pradesh and Telangana, lost their lives due to the 2015 heatwave, which was one of the worst in India's history (NDMA, 2019). During the 2022 heatwave, many parts of Rajasthan and Delhi experienced temperatures exceeding 47 °C. This had the most intense March in 122 years, causing significant harm to agricultural output and energy demand. The likelihood of such events is anticipated to escalate due to anthropogenic climatic shifts (Mishra et al, 2020). Public health impacts from heatwaves range from mild heat stress to fatal heatstroke. The combination of exposure to high temperatures can result in dehydration, electrolyte imbalance, cardiovascular stress, respiratory problems, and exacerbate chronic conditions (Patz et al., 2005). The vulnerable groups include children, the elderly, pregnant women, and those who work in the open air.

India is a country which has demonstrated the detrimental effects of extreme heat on health. In the year 2010, Ahmedabad experienced a heatwave that resulted in heightened all-cause mortality rates of 43% compared to non-heatwave periods (Azhar et al, 2014). Andhra Pradesh has also witnessed a similar situation in the year 2015 caused hospitals to succumb to a significant amount of heatstroke and related complications.

It has been found that heat waves can also affect health indirectly. Such excessive heat leads to mutations among the mosquitoes breeding, which can result in vector-borne diseases adversely affecting maternal and neonatal health (Kjellstrom et al, 2017). The ill effects of heat are not restricted to physical health but it may also cause mental health issues such as stress which is induced due to high heat, irritability, and a reduction in the sleep quality.

Further impact of heat waves can be seen on social and economic productivity too. In country like India, almost 80% of employment is handled by non-formal workers who are extremely exposed to outdoor working conditions. Without proper cooling mechanisms, exposure to heat can cause productivity loss and income insecurity among these non-formal employees. Which includes agricultural labourers, construction workers, street vendors, and transport workers. This is a concern reported by Krishnamurthy et al. (2017).

Another major concern reported as per research conducted is that, about 34 million full-time Indian jobs will be lost due to heat stress by 2030, with the majority of them being in agriculture and construction fields (PTI, 2019). The concerns do not end here, as during the May 2022 heatwave, it was reported that India faced its worst power shortage in six years, while for dealing with extreme heat there is a need for cooling energy infrastructure; this lack of cooling infrastructure is another strain while dealing with heatwaves (International Energy Agency, 2022).

Rising temperatures also affect the agricultural crops adversely. Crops like wheat and rice, which are considered as crucial for food security in country, are also exposed to the risk of loss due to fluctuation in temperature (Shukla & Tiwari, 2025) and thus increase the risk of extreme production losses. Moreover, the cascading socioeconomic effects emphasize the multi-sectoral nature of heat hazards, which have an impact on labour, health, agriculture, and energy.

The fast -paced urban growth in India has intensified the heat-related risks. Cities with high-density concrete constructions, asphalt roads, and sparse vegetation are more likely to absorb and hold onto heat better than rural areas (Oke, 2017). During summer, UHI intensities of 3–7 °C are consistently observed in cities like Delhi, Mumbai and Ahmedabad above the rural areas surrounding them (Rajasekar & Weng, 2009).

These amplified temperatures contribute to heat-related mortality and worsen air pollution, creating a 'double burden' of health risks. The UHI effect disproportionately impacts low-income settlements where housing materials such as tin sheets or asbestos roofing trap heat, raising indoor temperatures to dangerous levels. Many urban poor households lack access to fans, coolers, or reliable electricity, leaving them trapped in thermal stress environments.

3.1.1 Vulnerable Populations

India is a country of diversity, here diversity lies in every sphere of life. Same is the case with the risk related to heat; in Indian cities, heat risk is also diversified and not evenly distributed (Table 1). The risk associated with heat is again deeply stratified across social and economic lines in the nation.

Table 1 above highlights the overlapping vulnerabilities existing in India which reflects that heat risks are not only environmental but also socio-political, reflecting broader inequalities in urban India.

Table 1 Heat risk and Indian socio-economic factors

S.N	Factors	Vulnerability
1	Urban poor and slum dwellers	It has been recorded as per the last census that about 65 million of Indian population lives in slums (Census of India, 2011). The people residing in slums have poor housing quality, which lacks proper ventilation, and also lacks f water access which makes them more vulnerable towards heat
2	Informal sector workers	Construction workers, street vendors, rickshaw pullers, and agricultural labourers face occupational heat stress with limited safeguards (Sahu et al., 2013)
3	Migrant populations	Seasonal migrants often live in temporary housing without cooling facilities and lack social protection mechanisms
4	Elderly and children	Both groups have limited thermoregulation capacities, making them physiologically more vulnerable (Sherwood & Huber, 2010)
5	Women	Gendered roles mean women experience heat both indoors and outdoors, with disproportionate caregiving burdens and limited mobility during heat crises (Kjellstrom et al., 2017)

3.2 *Policy Recognition of Heat Risks*

The growing recognition of heat as a major disaster risk has led India's NDMA to incorporate heatwaves as one of the officially recorded natural disasters in its list of natural hazards. The NDMA issued guidelines in the year 2019 with respect to heatwave management. The guidelines emphasize that there is a need for cities to implement Heat Action Plans. It further emphasised that actions like early warning, public awareness, and healthcare preparedness are to be taken at the local level (NDMA, 2019). However, it has been found that the guidelines issued by NDMA are unevenly distributed across various states.

Although cities such as Ahmedabad and Nagpur have pioneered HAPs but still there are still many smaller cities where there is either a lack or absence of a systematic approach towards dealing with high heat waves.

3.2.1 Global and National Policy Context

In the recent past, dealing and tackling heatwaves has become a part of the global climate governance challenges (Hauge, 2021). Extreme heat has been labelled as the most critical threat to humanity in the twenty-first century (IPCC, 2022). The Paris Agreement (2015) has pledged countries to restrict the rise in global temperature to above 2 degrees Celsius and aim to 1.5 degrees Celsius above pre-industrial levels, which greatly reduces the chances of extreme heat episodes. The targets set under the SDGs (2015) can serve as the blueprint to tackle the challenges of heatwaves. Targets mentioned under SDG 3, 11 and 13 encompass the phenomenon of urban heat and its mitigation (Choudhury et al., 2025). Additionally, heat waves would also fall under the risk of the 2015–2030 Sendai Framework on Disaster Risk Reduction, which focuses on governance and enhancing disaster preparedness. Heatwaves, though slow-onset hazards, are increasingly incorporated into disaster management frameworks as evidence of their impacts grows (United Nation Office for Disaster Risk Reduction, 2019).

3.2.2 National Climate Policy Landscape in India

As part of its larger adaptation policy framework, India has proactively acknowledged heatwaves as a health and climate emergency. The main framework for India's climate action is provided in the year 2008 by the National Action Plan on Climate Change (NAPCC). Adaptation to climate change, including controlling heat risks, is an implicit theme even though its main missions centre on sustainable agriculture, energy efficiency, and renewable energy (Government of India, 2008). With an aim to make the NAPCC goals reach the local and state level, the government introduced state-level action plans called as State Action Plans on Climate Change (SAPCCs). Preparing for heatwaves was considered as a top priority in many of the state plans, including Gujarat, Rajasthan, and Telangana, which are highly prone to heatwaves. These SAPCCs emphasised that every state, while formulating their state development plans, should strongly incorporate climate resilience, especially in the areas of urban development, water, and health. Institutionalising heat risk management was one of the major requirements for the success of such plans; thus, a vital role has been played by the National Disaster Management Authority (NDMA) in institutionalizing heat risk management. Recognizing heatwaves as a disaster category in the year 2016, comprehensive Guidelines were issued by NDMA for the preparation of Action Plans for which can prevent and manage heat waves.

3.2.3 International Collaboration and Knowledge Sharing

Collaborations and knowledge sharing is considered to be the key for implementation of any crucial effort. Thus, when India adapted to climate change, it has been influenced by international partnerships. Heat Action Plan of Ahmedabad was the outcome of the collaborative efforts of Ahmedabad Municipal Corporation, the Natural Resources Defence Council, and the Indian Institute of Public Health, Gandhinagar (IIPH-G). Collaborations between countries enabled the transfer of scientific knowledge, technical expertise, and best practices from overseas (Knowlton et al, 2014).

The Global Cool Cities Alliance's support for the Cool Roofs Program has influenced policy innovations in Indian cities. This is one of many international initiatives. Equally, India's involvement in organizations such as the Global Covenant of Mayoral Engineers for Climate & Energy and the C40 Conference is noteworthy. Rapidly spreading heat adaptation strategies across urban networks is one result of the efforts of the Cities Climate Leadership Group.

3.2.4 Policy Integration Challenges

However, there are still many issues in establishing appropriate policy frameworks for adapting to climate change at global, national and local levels. In India, duties related to the formation and implementation of climate-related policies are spread across different ministries and departments. This results in either delays or gaps in the implementation and management of urban heat comprehensively and effectively.

Effective policy design for heatwave control is also hindered by data gaps. Unavailability of comprehensive epidemiological and climate-related data at city and district levels impedes evidence-based planning. Thus, for making adoption of heat action plan strong and effective, it is essential to strengthen monitoring systems and integrate climate data into health systems (Shukla & Tiwari, 2025).

3.3 Ahmedabad's City Profile and Its Vulnerability

The state of Gujarat's largest city, Ahmedabad, is a rapidly growing urban city of India and was once the capital of that state. The Sabarmati River is a major port and serves as the urban centre of Ahmedabad, which has fewer than 8 million residents in the metropolitan area (Census of India, 2011; UN-Habitat, 2020). Ahmedabad is considered the most commercially active city, which has an extensive industrial base, a prominent textile industry, and rapid urbanization. Besides, the city has been granted World Heritage City status for its cultural heritage by UNESCO in the year 2017.

Geographically, the city lies in a semi-arid climate region (Koppen classification BSh), with extremely hot summers, mild winters, and a distinct monsoon season. May and June see average maximum summer temperatures exceeding 42 °C, while extreme heat events frequently surpass 47°F (India Meteorological Department (IMD), 2022). The combination of high baseline temperatures and rapid urban expansion has led to an escalation in the UHI effect, which amplifies the local heat stress (Islam et al., 2024).

The data reflected in Fig. 1 and Fig. 2 are derived from NASA's Landsat 8 satellite, which illustrates substantial spatial variability in land surface temperature across different administrative wards for the year 2024. The data shows the top 10 hottest wards which include Lambha with a temperature of 40.44°**C**, followed by Ramol-Hathijan (40.28 °C). Notably, Paldi recorded the least temperature, followed by Navrangpura and Ghatlodia thus maintaining lower temperatures, averaging around 36–37 °C. This highlights the fact that the eastern and southeastern peripheries of Ahmedabad exhibit highest thermal intensity, as these regions are industry intensive, with low vegetation density and extensive impervious surfaces. While on the other hand, western and central Ahmedabad have more residential areas, higher tree density and proximity to the Sabarmati River, displays lower thermal signatures (Dabhi, 2025).

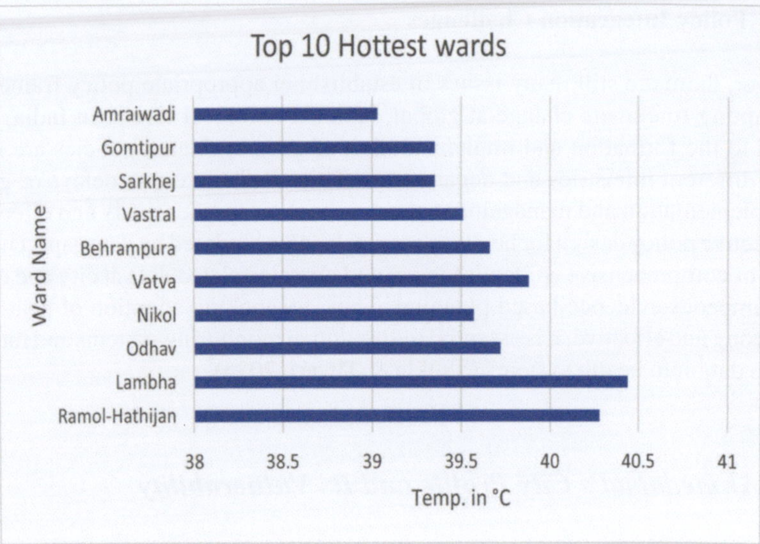

Fig. 1 Top 10 hottest wards (2024)

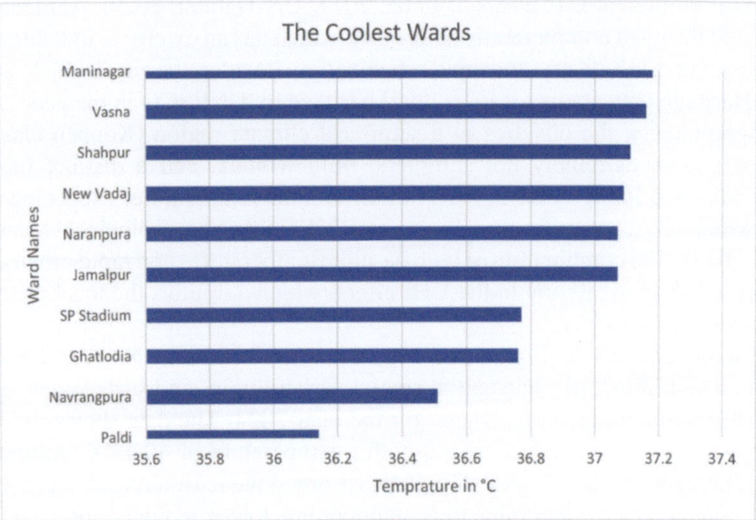

Fig. 2 The Coolest wards (2024)

3.4 Why Ahmedabad Became a Pilot City

Several factors contributed which made Ahmedabad a pilot city to adopt HAP. The major reason identified was, the severity of 2010 heatwave which led to unprecedented mortality and shocked local authorities. Further the Ahmedabad Municipal Corporation (AMC) with its receptive and forward-thinking approach had a relatively strong institutional capacity and governance, which made Ahmedabad a pioneering city for such ambitious task.

One of the prerequisite for HAP like initiative was collaborative networks, Ahmedabad is considered to be a city who is rich in, partnerships with organizations such as the Indian Institute of Public Health (IIPH), Natural Resources Defence Council (NRDC), and IMD enabled scientific, technical, and advocacy support (Knowlton et al., 2014). Further, Ahmedabad is a rapidly growing Indian city which holds large informal settlements, Ahmedabad is considered to be epitome of South Asian urban challenges, making it a perfect case study for replication (Table 2). These factors set the stage for its pioneering HAP in 2013.

3.4.1 Ahmedabad Heat Action Plan (HAP): Development and Evolution

The Heat Action Plan (HAP) of Ahmedabad stands as a landmark initiative in climate resilience planning in India and South Asia. It was the first comprehensive city-level heat adaptation plan in the region, launched in 2013 by the Ahmedabad Municipal Corporation (AMC) in collaboration with other recognised bodies such as the Indian Institute of Public Health–Gandhinagar (IIPH-G), Natural Resources Defence Council (NRDC), Indian Meteorological Department (IMD), and local NGOs (Knowlton et al., 2014; NRDC, 2018). The devastating 2010 heatwave exposed vulnerabilities within Ahmedabad's public health system, governance structures, and urban infrastructure (Awasthi et al., 2022). More than 1300 people died in just a week as a result of the devastating heatwave of 2010 (Azhar et al., 2014). This led to overwhelming of hospitals with the patients suffering from heat strokes, there was a significant rise in death rate among low-income population and places with informal settlements that don't have sufficient cooling infrastructure and access to water. (Rohini et al., 2016).

This model was a co-production of shared knowledge among multiple stakeholders. The plan was not only socially inclusive but also scientifically grounded (Knowlton et al., 2014).

Since its inception, the HAP has undergone several revisions and updates. The major revisions and updates done in the action plan are highlighted in Table 3.

Table 2 Profile and vulnerability—Ahmedabad

Factors	Profile and vulnerability -Ahmedabad
Urbanization and infrastructure	As a result of urbanisation, the built-up area of Ahmedabad is expanding significantly continuously over the last two decades. As a result of this large-scale agricultural land has been converted into bricks and mortar, leading to loss of green cover, and proliferation of impervious surfaces such as asphalt and concrete. The rise of high-density settlements, particularly in peripheral areas, has led to insufficient infrastructure for heat adaptation, including inadequate shading, water supply, and ventilation The poor housing design in the city and lack of green spaces in urban areas results in rising surface temperatures (Shukla & Tiwari, 2025) Air pollution is another relevant climate change concern which may intensifies heat wave (Awasthi et al., 2024). Ahmedabad due to expanded urban built lacks green land, and absence of adequate public transport infrastructure adds to the problem of air pollution, Thus, making the population more vulnerable.
Socioeconomic profile and vulnerability	Wide disparities can be seen in the socioeconomic fabrication of the city. Ahmedabad population includes prosperous middle class and industrial elite on the one hand, while on the other hand a large proportion of population belongs to low-income groups with informal settlements. It has been estimated that population belonging to low-income group which is accounts to about 25–30% of Ahmedabad's population lives in slums or informal housing (AMC, 2018). They use tin sheets, asbestos, or other heat-trapping materials for constructing these structures. All these materials lack adequate insulation and ventilation, exposing residents to life-threatening heat stress Migrant workers, often engaged in outdoor labour in sectors such as construction, street vending, and textile manufacturing form another vulnerable group. They get exposed to direct sunlight and don't have access to healthcare and cooling mechanisms, which increases their exposure to heat-related illnesses (Azhar et al., 2014). Additionally, urban poor population of the city frequently lack access to continuous electricity supply and cooling appliances, which results in intensifying their vulnerability (Knowlton et al., 2014) Gender and age are also one among the considerable socio-economic factors. A disproportionate heat risk is shared by the vulnerable population which includes women, young children, the old age, and people already suffering certain heath issues. It has been found that during Ahmedabad's 2010 heatwave elderly women were among the highest mortality
Impacts of heat stress on health	The heatwave in Ahmedabad in 2010 was a turning point and raised an alarm for the city. This heatwave which resulted into 1300 fatalities compelled the administration to develop rigorous climate policy for the city (Azhar et al., 2014). This brought to light them lethal confluence of severe weather, urban susceptibility and insufficient readiness Numerous health effects, such as heat exhaustion, heatstroke, dehydration, respiratory, and cardiovascular disorders, have been connected to heat stress in Ahmedabad (Knowlton et al., 2014)

Table 3 HAP—Revisions and Updates

Phases	Evolution and updates
Initial Phase (2013–2016)	The initial stage was focused on early warning systems, campaigns for making people aware regarding heat risks and prevention thereof, and health sector preparedness A citywide temperature alert system was introduced, issuing colour-coded alerts (yellow, orange, red) based on IMD forecasts Heat training manuals were prepared for doctors, paramedics, and hospital staff
Expansion Phase (2017–2019)	Integration of cool roof initiatives, using reflective paint and eco-friendly roofing materials to reduce indoor heat exposure Collaboration with local schools to reschedule academic activities, minimizing children's outdoor exposure during peak heat Increased coverage of public water distribution points and shaded community spaces
Consolidation Phase (2020–Present)	The third phase was considered to be the phase of consolidation. During this phase, HAP was integrated into broader state-level and national climate resilience frameworks Other Indian cities such as Nagpur, Surat, and Bhubaneswar also initiated efforts to replicate the Ahmedabad model. Besides this to identify hotspots of extreme exposure to heat vulnerability mapping was incorporated through geographic information system (AMC, 2021) During this phase, the plan was also extended towards building up of cool roof system. The plan was expanded to cover thousands of households in low-income settlements as well

3.4.2 Key Components of Ahmedabad Heat Action Plan (HAP)

Table 4 below highlights various components of HAP and the unique features of these key components, which ensured timely warnings, awareness, and preparedness that act as a critical factor in preventing heat-related morbidity and mortality. These key components provide insights towards the pioneering idea adopted by the city to mitigate the risk of heat.

3.4.3 Implementation Challenges of Ahmedabad's Heat Action Plan

While the Ahmedabad Heat Action Plan (HAP) is considered to be pioneering model for urban heat adaptation in South Asia, its journey has not been free from challenges (Table 5). Ahmedabad is a city which is heading towards urbanisation very rapidly and had a very diversified socioeconomic set up thus, implementing such a cross-sectoral and resource-intensive initiative in a rapidly urbanizing, socioeconomically diversified state is highly demanding. Initiative like HAP has to face multiple barriers from financial, technical to social. These challenges provide important lessons for replicating and scaling HAPs in other regions.

Table 4 Key components of Ahmedabad heat action plan

Key components	Key features
Early warning and forecasting systems	HAP was built up with strong system of alarming and communication. Providing forecast and giving early warning was the key component of this programme. The early warning system provides advance notices, colour coded alerts and different communication channels are used for communicating the threats well in advance which can assist in mitigating the impact of such climatic disasters, developed in collaboration with the Indian Meteorological Department. Three to four days advance notices were served by IMD regarding forecasts of extreme heat, enabling proactive measure, the dissemination of information regarding these alerts such as yellow, orange, and red denoting hot, very hot, extremely hot, and dangerous, respectively, is done through SMS, WhatsApp groups, municipal websites, and mass media to reach both institutions and citizens Main aim of these types of alerts and communication system is to ensure that hospitals, schools, emergency services, and citizens receive timely warnings, which can act as a critical factor in preventing heat-related morbidity and mortality (Knowlton et al., 2014)
Social consciousness and communication strategies	For making people aware and conscious, effective communication strategies are needed. One of the significant key components of HAP was its effective communication The AMC developed a multi-layered outreach campaign. These campaigns were designed to target the vulnerable populations and to create public awareness, pamphlets, and billboards in local languages explaining heat-health risks and protective behaviours were used Further for dissemination of information among the slum settlements, reaching groups most at risk due to poor housing and limited access to cooling was done by engaging community. NGOs were involved for spreading the awareness in slum area Schools were another target place through which awareness programmes can get strength. Thus, with an aim to reach out to the student's, multiple heat safety guidelines are integrated into school activities, so as to ensure children adopt the protective habits for heat-related risk. Media such as radio, television, and newspapers played an active role in broadcasting daily alerts and precautionary advice The communication campaign ensures behavioural adaptation, such as drinking more water, avoiding outdoor work in peak afternoon hours, and recognizing early symptoms of heatstroke and thus building a climatic resilience city

(continued)

Table 4 (continued)

Key components	Key features
Health system preparedness	The major crisis posed by climatic risks like heat waves is health crisis. Thus, one of the key components of HAP was strengthening the city's medical and emergency infrastructure. For achieving this, doctors and health workers are trained to identify and treat heat-related illnesses (HRI) such as dehydration, heat exhaustion, and heatstroke During peak summers availability of cooling equipment and ice packs was insured, emergency rooms are equipped with additional cooling systems. In order to track emerging hotspot, surveillance system was adopted by the hospitals through which hospitals report real-time heat illness data to AMC. Mortality reviews were conducted after the event to identify gaps and inform policy updates Through timely identification of heat risk-related symptoms effective treatment can be offered to the people, which results in minimising heat-related fatalities (Azhar et al., 2014)
Targeting vulnerable populations	HAP is designed in such a way that more protection can be given to people who fall under high-risk group and suffer disproportionately during heatwaves. These include urban poor who are residents of informal settlements with tin-roofed housing and are highly exposed to heat. Construction labourers, street vendors, and traffic police face direct sun exposure are the vulnerable targeted population, children and elderly are another set of population who are physiologically more vulnerable to heat stress, women who play the caregiving roles and have very low or no mobility, along with nutritional deficiencies, are also the targeted population for the plan
Structural and long-term urban interventions	Beyond short-term emergency measures, the HAP incorporates urban planning and infrastructural solutions to mitigate urban heat island (UHI) effects In the year 2017, one of the major long-term interventions was launched, named the Cool Roof Program. Under the programme, an attempt was made to promote the use of white reflective paints, tiles, and insulated roofing materials in low-income housing which can reduce the indoor heat by 2–5 °C. Conducting multiple tree plantation drives for increasing green cover helps provide shade and regulate urban microclimates. Public drinking water stations are set up across the city during summer months

(continued)

Table 4 (continued)

Key components	Key features
Institutional coordination	Coordination among the departments is another key factor of the HAP. AMC health department, water and electricity board, education department, and disaster management cell all of them came together to make it a success. This has led to efficient governance
Monitoring, evaluation, and adaptive learning	The Ahmedabad HAP is continuously monitored and updated as and when the need arises Post-Heatwave Assessments were conducted after each summer season to measure effectiveness. Mortality and Morbidity Data Analysis conducted which helps refine intervention strategies. Lessons from HAP feed into broader municipal climate resilience strategies Ahmedabad HAP emerged as a robust and evolving model for other Indian cities

3.4.4 Effectiveness and Impact of Ahmedabad's Heat Action Plan

A comparison of image taken using NASA's Landsat 8 satellite on May 21, 2020 of the Heat map of Ahmedabad city with an image taken on May 24, 2024 at 5.38 pm reveals 2–3 degrees Celsius decline in all the 48 municipal wards, suggesting a positive response to Ahmedabad's HAP interventions (Dabhi, 2025). Further while comparing the green cover data for the year 2024 to the year 2021, it has been seen that the total green area in the city increased from 48.98 (10%) sq. km to 60.10 (12.5%) sq km reflecting a 22.7% rise in vegetation cover (Fig. 3). Figure 4 below highlights the improvement in tree density per person while comparing year 2024 with the year 2021, the density has increased to 8.2 sq m which was recorded as 6.8 sq m in the year 2021. This appears to have contributed to localised temperature reductions of 2–3 degrees. (Dabhi, 2025). Further Effectiveness and Impact of HAP Ahmedabad can be gauged from the reduction of mortality and morbidity, increased public awareness, and policy influence and scaling (Table 6).

4 Limitations and Future Scope

The major limitation of the current study is its dependence on secondary sources of information. In the current study, the inferences drawn by us are completely based on the secondary sources. Incorporating primary data, such as interviews with policymakers or community members, might provide richer insights. There could be data gap in the India's mortality record, which may result into incorrect assessment of the impact of report. Further, the current case study is specific to the city of Ahmedabad thus while transferring the lessons and replication of the plan to the other states nationally or globally, the socio economic, and climatic variations must be considered.

Table 5 Key challenges

Key challenges	Explanation
Institutional and governance barriers	One of the major challenges faced by the initiatives was involvement of multiple institutional bodies. Initiative like HAP cannot be performed in isolation thus involvement of multiple government departments is inevitable. This plan includes the health department, disaster management authorities, meteorological services, municipal water supply, electricity boards, and education departments and each such departments have completely distinct operational culture, and different mandates. Thus, establishing coordination, cooperation and communication between the distinct operational culture and mandates is one of the major challenges faced while carrying initiative like this. Further, sometimes involvement of multiple departments also leads to delays in decision-making, duplication of efforts, or gaps in communication Furthermore, as initiatives like this are not made mandatory through any legislation the implementation and success of this kind of initiative is highly relied up on the political will of the parties in rule and the commitment of individual officials responsible. Another governance aspect which becomes a challenge for such initiatives could be the duration of administrative turnover when institutional memory and continuity of such initiatives could be affected
Financial and resource limitations	Although HAP comes up as cost-effective compared to the losses from heat-related mortality and morbidity, however, for implementing initiatives like HAP there is need of sustained funding, as funds are the basic requirement for running public awareness campaigns, distributing water, and for training of health workers which are the primary functions of such plans Moreover, innovative measures like the cool roof's initiative required subsidies or incentives to encourage adoption among low-income households, but scaling these programmes to cover the vast number of vulnerable residents remained financially demanding
Technical and forecasting challenges	The early success of the HAP relied on forecasts provided by the India Meteorological Department (IMD). However, forecasting localized heatwaves in urban areas remains technically difficult. Microclimatic variations within the city (e.g., industrial zones, high-density slums, green cover areas) mean that a single citywide forecast may not capture localized risk accurately Additionally, climate change has increased unpredictability in extreme weather patterns, occasionally resulting in false alarms or missed warnings, which can undermine public trust in alerts over time

(continued)

Table 5 (continued)

Key challenges	Explanation
Social and behavioural barriers	The effectiveness of public awareness campaign for regarding how people can use preventive measures to protect themselves from the impact of high heat was highly hindered because of the barriers like low literacy levels, language and cultural perceptions. Another major implementation challenge was reaching women, elderly populations, and marginalized communities. Then convincing groups who often prioritize livelihood needs over heat safety such as daily wage workers, slum residents, street vendors, traffic police, and the homeless is another challenge for implementation
Health system capacity constraints	For making such plan successful there is a need of proper health system infrastructure. Thus, reasons like shortages of cooling equipment, inadequate emergency preparedness, and limited reporting systems for heat-related illnesses (HRI) hampered timely responses Moreover, the absence of a comprehensive surveillance system meant that data on heat-related morbidity and mortality was often underreported, because of which it becomes difficult to measure the complete impact of any intervention placed
Urban development and planning challenges	Increasing industrialisation and urbanization in Ahmedabad has resulted in urban heat island (UHI) effects, which was identified as one of the major reasons behind increase in temperature (Singh et al. 2024). For planting more trees and establishing cool roofs space is needed and availability of that space in the city was another challenge which hampers the development of this plan The informal housing sector, which accommodates nearly 40% of the city's population, remained especially vulnerable. Retrofitting slum settlements with heat-resilient infrastructure required large-scale investments and long-term political commitment that exceeded the scope of the HAP
Policy and legal challenges	Unlike disasters such as floods or earthquakes, heatwaves have traditionally not been prioritized in disaster management legislation. Although NDMA's 2016 guidelines marked a shift, the lack of legally binding enforcement mechanisms meant that HAPs largely operated as advisory frameworks. This limited their ability to mandate compliance from private actors such as construction companies, industries, and real estate developers, whose workers and communities are among the most exposed

5 Conclusion

Thus, it can be culminated that through the above discussion Heat Action Plan (HAP) in Ahmedabad serves as an illustration of how proactive urban climate adaptation through planning, policy, and cooperation can improve resilience in all those urban areas who are prone to heat-related risks. The chapter provides an understanding of effective policy and management strategies for urban heat adaptation, with potential replicability across climate-vulnerable urban regions through synthesization of

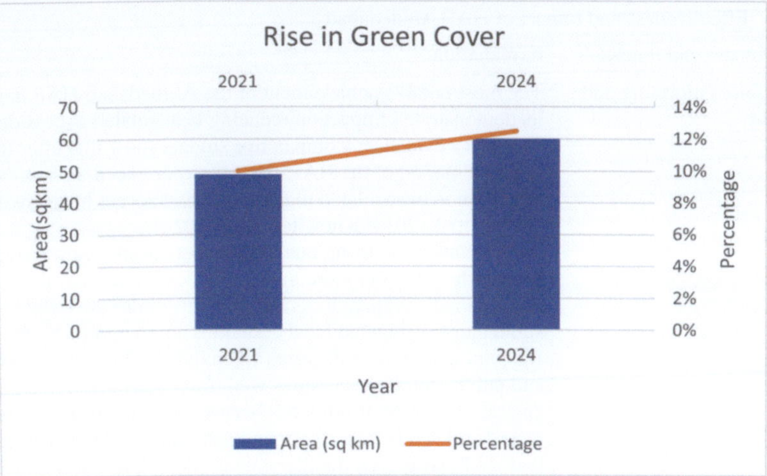

Fig. 3 Rise in green cover

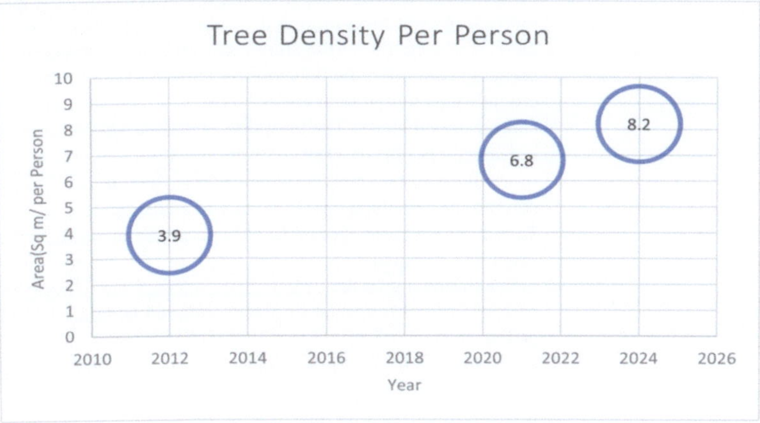

Fig. 4 Tree density per person

evidence from multiple credible sources which are secondary in nature. The case study has highlighted the relevance of local governance and emerged as a collaborative model. It has also been highlighted through the current study that climate adaptation is not only related to technical or infrastructural challenges, but governance and equity issues such as involvement of local governing bodies and prioritising the low-income groups and front-line workers, etc., are equally relevant. For overcoming the limitations of such initiatives, there must be continuous updating of such plans. Flow of funds is another aspect which should be taken care of so as to combat the limitation of financial constraints and uneven implementation. Further integration

Table 6 Effectiveness and impact of HAP Ahmedabad

Effectiveness and impact	Explanation
Reduction in mortality and morbidity	The most notable achievement of the Ahmedabad HAP has been its demonstrated impact on reducing heat-related excess deaths: Post-2013 evidence: Comparative studies show that after the implementation of the HAP, the city experienced a 25–40% reduction in heat-related mortality during severe heatwaves (Azhar et al., 2014; Knowlton et al., 2014) 2015 Heatwave: During one of the most severe heatwaves in India (which caused more than 2,500 deaths nationwide), Ahmedabad's preparedness mechanisms under the HAP significantly mitigated local casualties compared to 2010 Hospital Admissions: In years when the HAP's early warning and public outreach measures were fully activated, reduction in emergency hospitalisation has been seen in the local health records related to heat stroke and dehydration. This highlights that HAP has played the life-saving role by providing timely warnings, increased hydration, and improved health worker training
Increased public awareness	The local public has been made aware of heat risks in the city. Simple yet effective measures adopted under HAP succeeded in increasing awareness among the public. Adopting colour code alerts, reaching out to the slum areas and emphasising on bringing change in the behavioural pattern of public were certain crucial steps undertaken under HAP Colour-coded alerts: The adoption of simple yellow, orange, and red warnings improved risk communication. Citizens became more likely to take preventive measures, such as drinking water, staying indoors during peak heat, and using shade Targeted Outreach: Vulnerable communities in slums, which often lack adequate cooling and water infrastructure, reported better awareness of heat-related health risks due to door-to-door campaigns and information posters in local languages (AMC, 2019) Behavioural Shifts: It has been found that there is increase in heat-preventive behaviours such as carrying water, using umbrellas, and resting during high temperature
Strengthening of health and governance systems	The HAP has catalysed a transformation in institutional preparedness: Training Programmes: Thousands of doctors, nurses, and paramedics received specialized training in diagnosing and treating heat-related illnesses Hospital Preparedness: Multiple small but effective initiatives were adopted by the hospitals, such as reserving cooling wards, stocking rehydration salts, and deploying ice packs Inter-Departmental Coordination: Effective collaboration among various departments like water supply, electricity, education, and disaster was another factor (AMC, 2020) This multi-sectoral coordination has become a model for other climate adaptation programmes

(continued)

Table 6 (continued)

Effectiveness and impact	Explanation
Urban planning and cooling innovations	To make the plan more impactful long-term resilience measures such as cool roof, greenery in urban areas, and provision of water were included. All this had structural impact as stated below: Cool Roofs Initiative: Installation of cool roof can reduce temperature by 2 to 5 degrees Celsius. Through HAP by 2019, over 15,000 households in low-income settlements benefitted from reflective roof coatings, which reduced indoor temperatures by up to 2–5 °C (NRDC & AMC, 2019) Urban Greenery: Planting of trees and construction of shaded bus stops lead to increase in thermal comfort in public areas Water Access: Water accessibility was one of the major challenges which has been encountered by HAP through strategic installation of public water kiosks. This ensured access towards hydration especially during peak summer months All these initiatives stood as a strong example for scalable urban resilience strategies
Policy influence and scaling	Ahmedabad emerged not only as a national model but also received significant global recognition: National Scaling: The HAP, Ahmedabad emerged as a pioneering initiative which influenced and acted as a motivator for many other states to adopt or proceed for such plan. The results of this can be gauged from the fact that about 130 cities and 23 states had either adopted or initiated the process of adoption of similar kind of plan against heat and other climatic hazards (ImD, 2022) Global Recognition: The HAP received equal recognition globally too. Many international bodies such as World Health Organization (WHO), UNFCCC, and The Lancet Countdown on Health and Climate Change cited HAP as a successful adaptation case study (Watts et al., 2018) Policy Mainstreaming: Earlier heat waves were considered as an issue related to environment only but HAP helped in shifting this attitude and recognized it not only as an environmental issue but as a public health issue, this lead to bring in the change in shaping both urban policy and disaster risk management frameworks accordingly

of these plans into major urban development agendas is another important consideration. Plans like HAP can be enriched by enhancing scientific capacity. Further, without local support plans like this could not succeed thus, strengthening community engagement can also lead to the success of such initiatives as local knowledge always complement technical interventions.

References

Annual Climate Summary. Government of India. (2016). *Indian Meteorological Department*.

Awasthi, A., Vishwakarma, K., & Pattnayak, K. C. (2022). Retrospection of heatwave and heat index. *Theoretical and Applied Climatology, 147*(1–2), 589–604. https://doi.org/10.1007/s00704-021-03854-z

Awasthi, A., Charan Pattnayak, K., Dhiman, G., & Raj Tiwari, P. (2024). *Artificial intelligence for air quality monitoring and prediction.* CRC Press. https://doi.org/10.1201/9781032683805

Azhar, G. S., Mavalankar, D., Nori-Sarma, A., Rajiva, A., Dutta, P., Jaiswal, A., & Hess. (2014). Heat-related mortality in India: Excess all-cause mortality associated with the 2010 Ahmedabad heat wave. *PLoS ONE, 9*(3). https://doi.org/10.1371/journal.pone.0091831.

Basu, R. (2001). High ambient temperature and mortality: A review of epidemiologic studies from 2001 to 2008. *Environmental Health, 8*(40), 1–13. https://doi.org/10.1186/1476-069X-8-40

Census. (2011). *Primary Census Abstract for Slum. Registrar General & Census Commissioner*.

Chakraborty, T., Hsu, A. H., Manya, D., & Sheriff, G. (2019). Disproportionately higher exposure to urban heat in lower-income neighborhoods: A multi-city perspective. *Environmental Research Letters, 14*(10), 105003. https://doi.org/10.1088/1748-9326/ab3b99

Choudhury, T., Awasthi, A., & Jung-Sup, U. M. (Eds.). (2025). *Open Access as a Knowledge Platform for Implementing the United Nations Sustainable Development Goals: Open Access and Sustainable Development Goals.* Springer Nature.

City resilience strategy report. (2018). *Ahmedabad Municipal Corporation (AMC).*

Cool Roof Program in Telangana: Progress and Opportunities. Natural Resources Defence Council & Greater Hyderabad Municipal Corporation. (2020).

Dabhi, B. (2025, April 21). Gujarat Hardlook: In 'heat' of the moment. The Indian Express. Retrieved from https://indianexpress.com/article/cities/ahmedabad/gujarat-hardlook-in-heat-of-the-moment-9954062/

Eckstein, D., Künzel, V., Schäfer, L. (2021). Global climate risk index 2021.

Guidelines for Preparation of Action Plan - Prevention and Management of Heat Wave. (2019). Government of India.

Hague, B. S. (2021). Seasonal climate summary for Australia and the southern hemisphere (summer 2018–19): Extreme heat and flooding prominent. *Journal of Southern Hemisphere Earth System Science, 71*(1), 147–158. https://doi.org/10.1071/es20009

Heat Action Plan. (2021). Ahmedabad Municipal Corporation (AMC).

Heatwave management in India: Progress report. (2019). NDMA.

International Energy Agency. (2022). The global energy crisis. In world energy outlook 2022. IEA, Retrieved from World Energy Outlook 2022–The Global Energy Crisis. https://doi.org/10.1787/3a469970-en

Im, E.-S., Pal, J. S., & Eltahir, E. A. B. (2017). Deadly heat waves projected in the densely populated agricultural regions of South Asia. *Science Advances, 3*(8), e1603322. https://doi.org/10.1126/sciadv.1603322

IPCC. (2022). Climate Change 2022: Impacts, adaptation and vulnerability. In *Contribution of Working Group II to the Sixth Assessment Report*. Cambridge University Press.

Islam, S., Karipot, A., Bhawar, R., et al. (2024). Urban heat island effect in India: A review of current status, impact and mitigation strategies. *Discov Cities, 1*, 34. https://doi.org/10.1007/s44327-024-00033-3

Kjellstrom, T., Lemke, B., & Otto, M. (2017). Climate conditions, workplace heat and occupational health in South-East Asia in the context of climate change. *WHO South-East Asia Journal of Public Health, 6*(2), 15. https://doi.org/10.4103/2224-3151.213786

Knowlton, K., Kulkarni, S. P., Azhar, G. S., Mavalankar, D., Jaiswal, A., Connolly, M., & Hess. (2014). Development and implementation of South Asia's first heat-health action plan in Ahmedabad (Gujarat, India). *International Journal of Environmental Research and Public Health, 11*(4), 3473–3492. https://doi.org/10.3390/ijerph110403473

Krishnamurthy, M., Ramalingam, P., Perumal, K., Kamalakannan, L. P., Chinnadurai, J., Shanmugam, R., Srinivasan, K., & Venugopal, V. (2017). Occupational Heat Stress Impacts on Health and Productivity in a Steel Industry in Southern India. *Safety and Health at Work, 8*(1), 99–104. https://doi.org/10.1016/j.shaw.2016.08.005

Mishra, V., Mukherjee, S., Kumar, R., & Stone, D. (2020). Heat wave exposure in India in current, 1.5 °C, and 2.0 °C worlds. *Environmental Research Letters, 15*(11). https://doi.org/10.1088/1748-9326/aa9388

National Action Plan on Climate Change (NAPCC). (2008). In *Ministry of Environment and Forests.*

NRDC. (2018). *Ahmedabad's pioneering Heat Action Plan: Protecting health from extreme heat*. Natural Resources Defence Council Report.

Oke, T., Mills, G., Christen, A., Voogt, J. (2017). Urban climates. https://doi.org/10.1017/9781139016476

Pandey, K., Bansal, M., & Awasthi, A. (2025). Introduction to Sustainable Development Goals: A Historical Review of Their Development. In *Open Access as a Knowledge Platform for Implementing the United Nations Sustainable Development Goals: Open Access and Sustainable Development Goals* (pp. 1–17). Springer Nature Switzerland. https://doi.org/10.1007/978-3-031-99605-4_1

Patz, J. A., Campbell-Lendrum, D., Holloway, T., & Foley, J. A. (2005). Impact of regional climate change on human health. *Nature, 438*(7066), 310–317. https://doi.org/10.1038/nature04188

PTI. (2019, July 2). *India could face productivity loss equivalent to 34 million jobs in 2030 due to global warming: UN*. Retrieved September 15, 2025, from Economic Times website: https://economictimes.indiatimes.com/jobs/india-could-face-productivity-loss-equivalent-to-34-million-jobs-in-2030-due-to-global-warming-un/articleshow/70037097.cms?from=mdr

Rajasekar, U., & Weng, Q. (2009). Spatio-temporal modelling and analysis of urban heat islands by using Landsat TM and ETM+ imagery. *International Journal of Remote Sensing, 30*(13), 3531–3548. https://doi.org/10.1080/01431160802562289

Rohini, P., Rajeevan, M., & Srivastava, A. K. (2016). On the variability and increasing trends of heat waves over India. *Scientific Reports, 6*, 26153. https://doi.org/10.1038/srep26153

Sahu, S., Sett, M., & Kjellstrom, T. (2013). Heat exposure, cardiovascular stress and work productivity in rice harvesters in India: Implications for a climate change future. *Industrial Health, 51*(4), 424–431. https://doi.org/10.2486/indhealth.2013-0006

Sherwood, S. C., & Huber, M. (2010). An adaptability limit to climate change due to heat stress. *Proceedings of the National Academy of Sciences, 107*(21), 9552–9555. https://doi.org/10.1073/pnas.0913352107

Shukla, D., & Tiwari, S. (2025). Rethinking urban climate adaptation. Psycho-environmental dimensions of Ahmedabad's Heat Action Plan. *Contesti. Città, Territori, Progetti,* (1), 206–225. https://doi.org/10.36253/contest-16062

Singh, D., & Rajput, P. (2024). Investigating urban heat island effects and mitigation strategies in Ahmedabad. *India.* https://doi.org/10.21203/rs.3.rs-5309216/v1

Smith, G. A. (2024). Seasonal climate summary southern hemisphere (autumn 2020): another coral bleaching event for the Great Barrier Reef without an active El Niño. *Journal of Southern Hemisphere Earth System Science, 74*(3). https://doi.org/10.1071/es24014

Undrr. (2015). *United Nations Office for Disaster Risk Reduction.* https://www.undrr.org/publication/sendai-framework-disaster-risk-reduction-2015-2030.

UNFCCC. (2015). *United Nations Framework Convention on Climate Change.* https://unfccc.int/process-and-meetings/united-nations-framework-convention-on-climate-change

United Nations Office for Disaster Risk Reduction. (2019). Global assessment report on disaster risk reduction 2019. UNDRR. https://www.undrr.org/publication/global-assessment-report-disaster-risk-reduction-2019

Un-Habitat. (2020). *World cities report 2020: The value of sustainable urbanization.* UN-Habitat. https://unhabitat.org/world-cities-report-2020-the-value-of-sustainable-urbanization.

Watts, N., Amann, M., Arnell, N., Ayeb-Karlsson, S., Beagley, J., Belesova, K., & Costello. (2018). The 2018 report of the Lancet Countdown on health and climate change: Shaping the health of nations for centuries to come. *The Lancet, 392*, 2479–2514. https://doi.org/10.1016/S0140-6736(18)32594-7.

Working on a warmer planet: The impact of heat stress on labour productivity and decent work. (2019).

Yadav, N., Rajendra, K., Awasthi, A., Singh, C., & Bhushan, B. (2023). Systematic exploration of heat wave impact on mortality and urban heat island: A review from 2000 to 2022. *Urban Climate, 51*, Article 101622. https://doi.org/10.1016/j.uclim.2023.101622

Yin, R. K. (2018). Case study research and applications: Design and methods (6th ed.). SAGE Publications

Zhao, L., Oppenheimer, M., Zhu, Q., Baldwin, J. W., Ebi, K. L., Bou-Zeid, E., & Liu. (2021). Interactions between urban heat islands and heat waves. *Environmental Research Letters, 16*(9). https://doi.org/10.1088/1748-9326/aa9f73

Evaluating Land Surface Temperature Changes for 22 Years and Their Association with Urban Heat Island in Rajshahi Division, Bangladesh

Md Mahbuber Rahman and Md Mostafizur Rahman

Abstract This research examines the temperature changing trends in the Rajshahi Division of Bangladesh, in relation to the rise in global warming events. 20 years of data from the Moderate Resolution Imaging Spectroradiometer (MODIS) (MYD11A2) satellite have been analyzed, for the period from 2003 to 2023, using the Google Earth Engine. The goal was to observe the trends and changes of spatial patterns in Land Surface Temperature (LST), as well as to see how the Urban Heat Island (UHI) phenomenon was enhanced with the increase of urbanization in the region, so that LST distributions were changed. The research explains how municipal climates change along with the growth of cities. The results point to a pattern that has implications for the future. Seasonal analysis shows the presence of a large diurnal temperature gradient: the mean diurnal temperature range is larger in winter (13.26 °C) than in summer (11.12 °C), due to a large nocturnal cooling in the colder months. This cooling process is now under threat as the night temperatures in summer have increased at a rate of 0.024 °C per year over the study period. As a consequence, populations suffer from enhanced thermal stress because the nighttime natural cooling mechanism does not work. The impact is severe in rapidly urbanizing areas that have become hotspots. Rapidly densifying areas show higher surface temperatures, which can intensify the urban heat island effect and increase the public health risks. As per the analysis, the UHI effect, which started growing in such cities as Rajshahi Sadar and Natore, was spread to neighboring regions between 2003 and 2022. These findings demonstrate the need to implement urban planning that is responsive to climate as soon as possible. The paper is a powerful evidence-based policy that policy-makers can use to reduce heat stress and promote sustainable urbanization in fast-growing South Asian areas.

M. M. Rahman · M. M. Rahman (✉)
Laboratory of Environmental Health and Ecotoxicology, Department of Environmental Sciences, Jahangirnagar University, Dhaka, Bangladesh
e-mail: rahmanmm@juniv.edu

Department of Environmental Sciences, Jahangirnagar University, Dhaka, Bangladesh

M. M. Rahman
e-mail: rimon.stu2017@juniv.edu

A. Awasthi et al. (eds.), *Rising Temperatures and Plant Ecosystems*,
https://doi.org/10.1007/978-3-032-29477-7_17

Keywords Land surface temperature · Urban heat island · Remote sensing · Climate change · Urbanization · Sustainable urban development

1 Introduction

The Intergovernmental Panel on Climate Change (IPCC) reports an increasing number of extreme events of high temperature in the terrestrial areas and this is an excellent example of how the global climate change cannot be undeniable (Wang et al., 2017). Because of a complex of climate change and population growth, current worldwide estimates indicate that the number of people exposed to extreme heat in cities has nearly doubled over the past few decades; this figure will only increase as the temperatures rise (Pattnayak et al., 2023; Ebi et al., 2021). In this international environment, one of the most significant human actions nowadays, urbanization, is altering the local thermal conditions greatly. One of the significant parameters in the dynamic study of the urban thermal environment is the Land Surface Temperature (LST), which is also a constituent parameter of the characterization of the Earth surface energy balance. It is important in the formation of the Urban Heat Island (UHI) effect because it controls the heating levels of the lower atmosphere. This happens when the natural land covers are substituted by buildings, pavements among other surfaces that trap heat. Both the replacement of vegetation and wetlands with impervious surfaces such as asphalt and concrete cause a fundamental shift in the energy balance on the surface and result in the UHI phenomenon (Awasthi et al., 2022; Oke, 1982). The global survey of the causes of UHI determines urbanization and reduced vegetation cover as two of the most significant causes that increase this effect (Deilami et al., 2018).

The thermal properties of the built materials in urban areas increase the surface temperature as compared to the other surrounding rural areas (Sobrino et al., 2004; Yadav et al., 2023). Similar trends have been observed in the rapidly developing Asian cities where spontaneous urbanization has enhanced the level of thermal discomfort (Ahmed & Suphachalasai, 2014; Liu & Zhang, 2011). It has also been revealed in the global analysis that the degree of UHI effect varies depending on the climatic conditions in the area and the population of the cities as well (Manoli et al., 2019). In rapidly expanding areas such as Rajshahi Division of Bangladesh, detailed and long-term analysis of LST is important to understand the development of urban thermal environments.

It is well known that the Urban Heat Island effect is primarily caused by the rise in built-up area in South Asian situations (Estoque et al., 2017). The chapter is a strict consideration of the dynamics of the LST in the 22 years (2002–2023) in the Rajshahi Division, which represents the fast urban growth in Northwest Bangladesh. An analysis of this nature is central in explaining the change of the thermal environment over time and its unbroken link with the development of the urban environment.

Although recent research has examined the changes in LST in particular districts (Rahman & Zhang, 2022), a division-level study spanning two decades gives a wider perspective of the changes in the climate of regions. This research is expected to fulfill two main tasks through the use of a set of detailed remote sensing data from the MODIS platforms. First, to characterize the long-term distributions and spatial distributions of LST over the various urban and peri-urban landscapes of the Rajshahi Division. Second, to trace the area coverage and duration of the UHI effect, and to trace the origin and agglomeration of the effect during 20 years.

In addition to the effects on human health and energy use, these thermal changes have widespread consequences on plant ecosystems. The high temperatures and UHI effect may cause great strain to the urban vegetation, change the plant phenology, influence the production of crops in peri-urban agricultural areas, and promote the growth of plants resistant to heat compared to native plants. The complex relationship between temperature shifts caused by urbanization and its environmental impact is thus essential in creating environmentally viable urban planning and good climate mitigation measures.

2 Methodology

This study employs an analytical, quantitative, and practical research framework centered on the analysis of secondary satellite imagery to quantify LST and UHI changes over time. The overall workflow is depicted in Fig. 1.

2.1 Study Area

The Rajshahi Division, located in the northwest of Bangladesh, is the study area (Fig. 2). Geographically, it is located between 23°48’ and 25°16’ north latitudes and 88°01’ and 89°48’ east longitudes. The division encompasses an area of 18,174.4 square kilometers and had a population of 20,353,119 as of the 2022 census. The region, comprising eight districts, has undergone substantial urbanization over the past two decades, making it a suitable case study for examining the thermal effects of development.

2.2 Data Acquisition and Pre-processing

This study used MODIS Terra/Aqua Land Surface Temperature and Emissivity data, specifically the 8-day L3 global product at 1 km resolution (MYD11A2, Version 6). Daytime and nighttime land surface temperature layers (LST_Day_1km and LST_Night_1km) were acquired and analyzed for the period from July 2002 to February 2023. A summary of the main dataset is presented in Table 1.

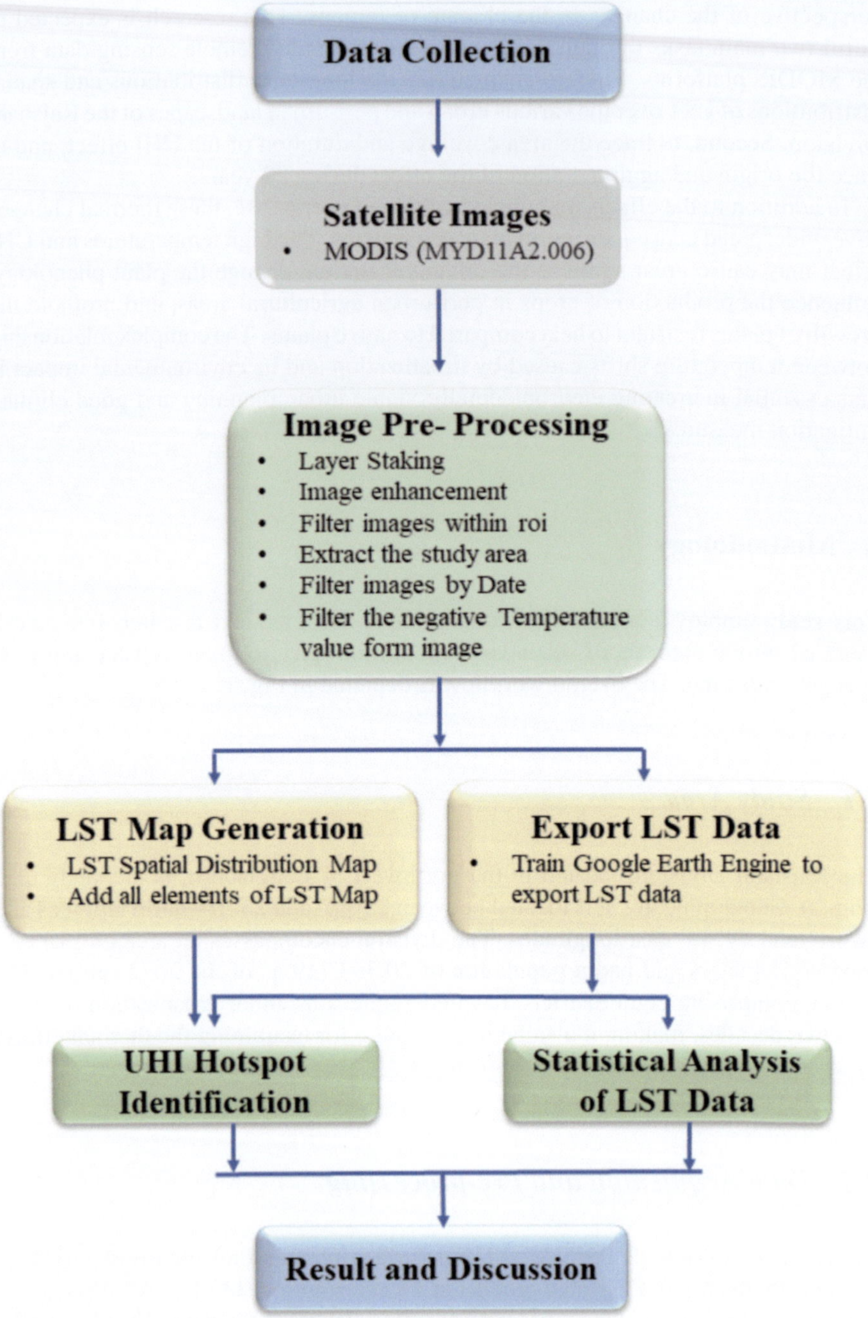

Fig. 1 Overall methodology flowchart

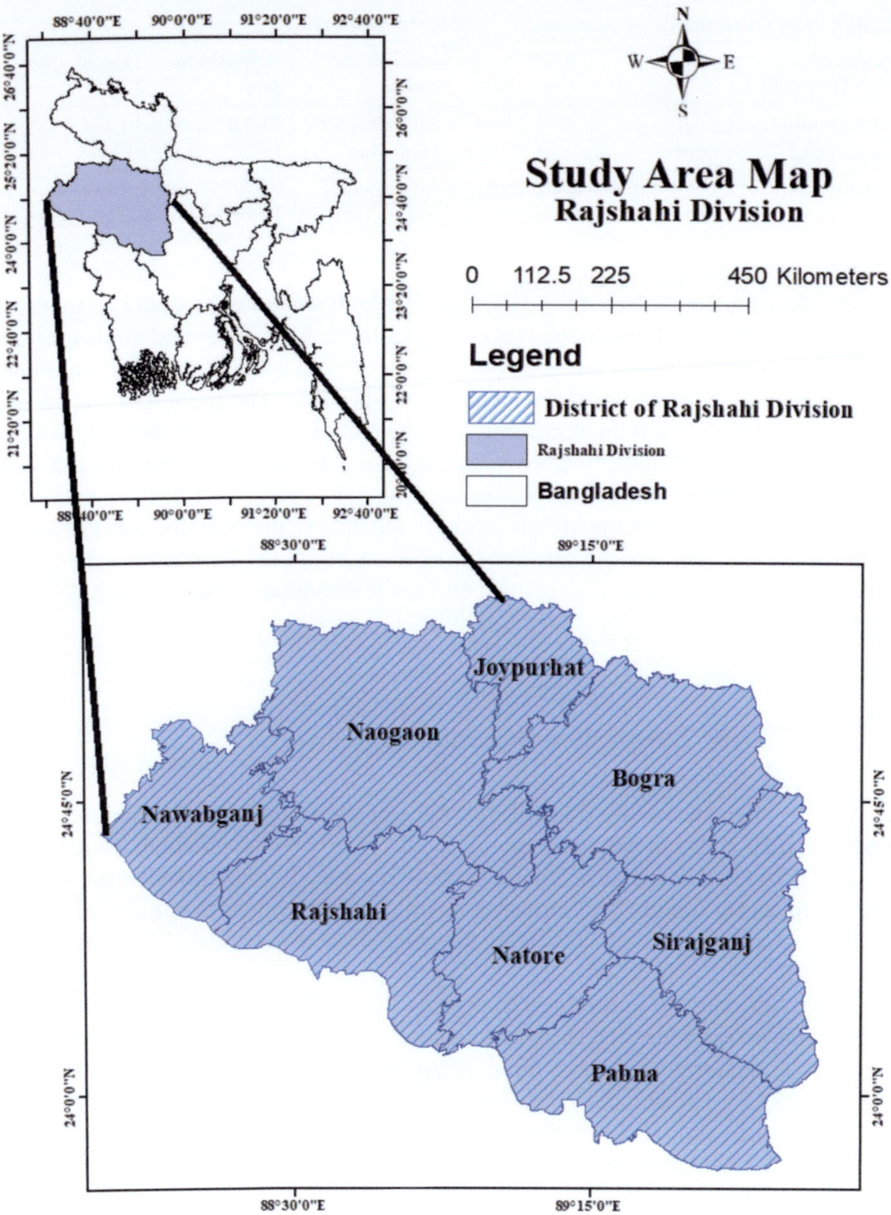

Fig. 2 Study area map, Rajshahi Division

Table 1 Key properties of the satellite data used in this research

Scene ID	Source	Year	Satellite & sensor	Acquisition date	Resolution
MODIS/061/ MYD11A2	Google earth engine	2002–2023	MODIS LST_ Day_1km	04-Jul-02 to 18-Feb-23	1 km
MODIS/061/ MYD11A2	Google earth engine	2002–2023	MODIS LST_ Night_1km	04-Jul-02 to 18-Feb-23	1 km

MODIS data were selected for their high temporal resolution, which is important especially for the development of a continuous 22-year time series and minimization of the gaps resulting from cloud contamination. All data acquisition and processing were done in the Google Earth Engine (GEE) code editor. The preprocessing workflow involved cloud masking according to the Quality Control (QC) that is provided along with the MODIS data, which was used to ensure that the data quality for analysis was improved. Atmospheric corrections were already assimilated in the generation of the Level-3 product for the MODIS. Additional processing steps included filtering of the image collection to the selected time period, clipping of the data to the study area boundary, and conversion of land surface temperature values from Kelvin to degrees Celsius (°C).

2.3 LST Data Extraction and Statistical Analysis

In order to characterize the thermal dynamics of the study area, the Google Earth Engine (GEE) platform was used to build a custom processing pipeline. Rather than analyzing isolated snapshots, functions were developed which allowed the mean, maximum, and minimum land surface temperature (LST) values to be extracted from each 8-day composite, thus facilitating the generation of a seamless and continuous time series.

As the raw data are in the form of digital numbers (DN), the scaling factor and Kelvin conversion were standardized in accordance with the following equation, to convert the measurements into a more interpretable Celsius scale: $LST\ (°C) = (DN * 0.02) - 273.15$, to bring the values into a more interpretable Celsius scale.

The resultant time series data were then exported for statistical analysis, which involved the calculation of annual and seasonal mean values for the daytime and nighttime LST, with summer being defined as March to May and winter as December to February.

2.4 UHI Hotspot Identification

A statistical classification technique, modified from Dutta et al. (2021), was used to identify UHI hotspots in the LST maps. Temperature classes were determined by the mean (μ) and standard deviation (σ) of LST values of the entire study area. UHI hotspots were considered as areas falling into the High ($\mu + \sigma$ to $\mu + 2\sigma$) and Very High ($> \mu + 2\sigma$) temperature zones.

3 Results and Discussions

3.1 Decadal Land Surface Temperature (LST) Trends

The LST data set shows a strong long-term thermal trend, with a clear seasonal variation in the temperature in the region, which is summarized in Table 2.

Table 2 LST data (seasonal average) for the study period (2002–2023)

Season	Period	Mean LST (°C)	Max LST (°C)	Min LST (°C)
Summer	Daytime	32.15	47.76	20.22
	Nighttime	21.03	25.94	14.22
Winter	Daytime	26.70	34.62	18.50
	Nighttime	13.44	18.81	7.61

The time series analysis reveals a notable trend because while the daytime temperatures showed a subtle cool trend of −0.058 °C/year during the summer, the nighttime temperatures have shown a continuous warming trend of (+0.024 °C/year (Fig. 3). This divergence is a classic signature of the UHI effect, indicating that the built environment retains more heat after sunset.

3.2 Spatial Distribution and Evolution of UHI Hotspots

The spatial distribution of LST maps illustrates the expansion of the thermal landscape, showing an intensification of high-temperature zones around major urban areas like Rajshahi city, Bogra, and Pabna over the study period (Fig. 4).

The growth of the urban footprint is significantly captured in the analysis of the hotspots of urban heat islands (UHI). A retrospective look at 2003 shows that these thermal maxima were mainly restricted to the main urban centers. By 2022, however, a discernible change in the landscape is apparent: The hotspots have not only grown

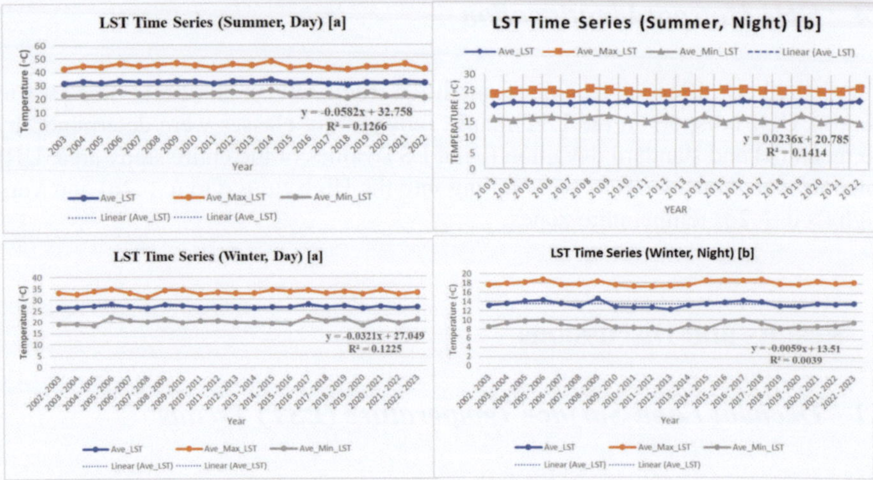

Fig. 3 LST time series for the summer season (top) and winter season (bottom)

stronger but have also spread out into suburban peripheries and smaller municipalities and therefore parallel the growth of new roadway infrastructure (Fig. 5). This diffusion outward is an actual manifestation of thermal costs of urban sprawl. Even though rural areas—buffered by dense vegetation—have been consistently cooler, the encroaching heat zones highlight a well-established trade-off, in which the replacement of vegetated surfaces with impervious materials inevitably increases surface temperatures, which is consistent with the conclusions of Rahman et al., 2022 and Al Kafy et al., 2021.

3.3 *Discussion of Findings*

The most important conclusion is that the temperatures at nighttime in summer are steadily increasing. An urbanizing region sets up more impervious surfaces, which raises the ability of these zones to absorb and retain solar energy. This heat is then gradually dissipated at night, and this makes the urban environment not cool down efficiently like the rural areas. This adds a lot of heat stress to both humans and ecosystems. This observation is consistent with the results of Dhaka (Ahmed et al., 2013) and other South Asian cities (Estoque et al., 2017), where a decrease in the green and blue spaces has been associated with an increase in urban temperatures.

This reporting of nocturnal warming is of a global nature. An analysis of 419 cities across the world established that daytime UHI can be highly variable depending on the vegetation cover, but at night, UHI is a regular occurrence depending on the thermal inertia of the man-made structures (Peng et al., 2012). Moreover, we also find that there was slight cooling in the daytime and heating in the nighttime, which

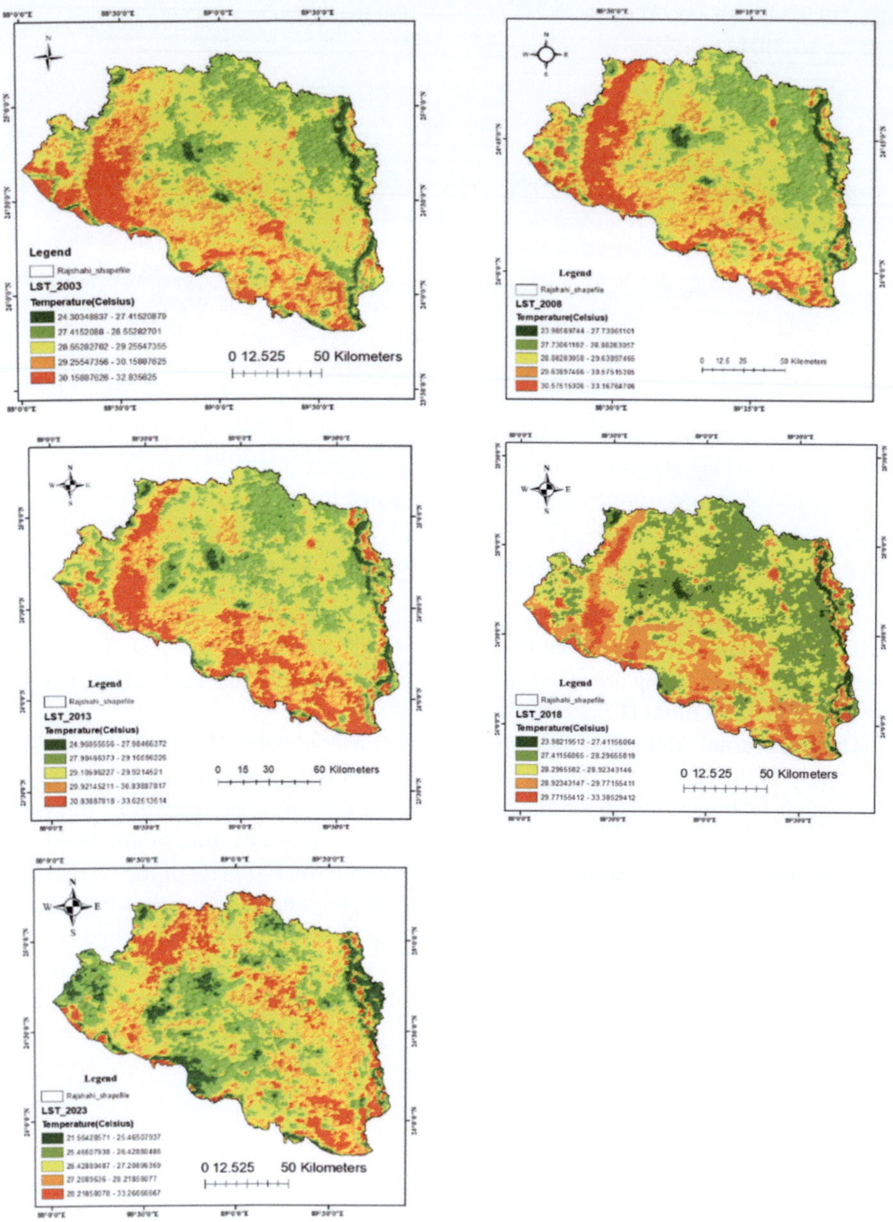

Fig. 4 Spatial distribution of LST in Rajshahi Division

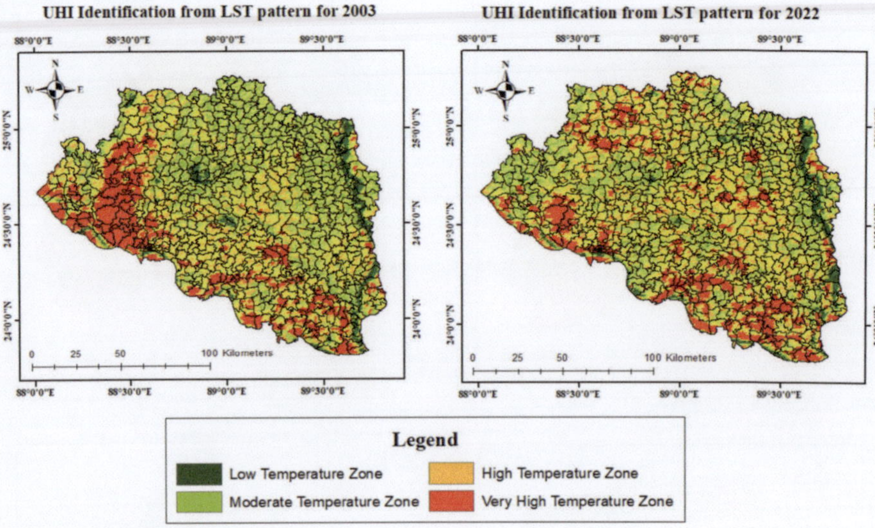

Fig. 5 Comparison of UHI hotspot patterns in Rajshahi Division between 2003 and 2022

is also consistent with the global research that indicates that urban haze or aerosol pollution in fast-growing areas can at times conceal daytime surface heating, but not the heat release at night (Chakraborty & Lee, 2019).

This nocturnal radiation has significant consequences for vegetation. The constantly high night temperatures may interfere with the usual physiological functions, such as breathing, resulting in low growth and high susceptibility to pests. In the case of agriculture in the peri-urban areas of Rajshahi, this might be in the form of low yields. According to the global literature, the footprint of the UHI effect can extend far beyond the city administrative zone, which can affect the climate and phenology of regions (Zhou et al., 2014). The growth of UHI into croplands (agricultural peripheries) implies that the croplands are becoming more exposed to urban climatic conditions, which change the local hydrology and make them more in need of irrigation.

4 Limitations

Although the scope of these long-term results is impressive, it can be argued that the limitations of the satellite-based observation are inherent. Above all, the spatial resolution of the MODIS data (1 km) imposes a trade-off: the data is quite good at the regional level but does not provide the level of granularity to determine street-level microclimates or the small thermal variations of particular city blocks. In addition, this study is based on satellite-derived LST with no secondary layer of ground-truth validation. Although the general tendency is statistically justified, one should mention that the introduction of in-situ measurements would have provided a more calibrated view of the absolute temperature values. This recognition does not affect the findings, but it determines the scope within which they are to be interpreted.

5 Conclusion and Future Scopes

This 22-year research of the Rajshahi Division gives a comprehensive report of the changing thermal environment of an area that is rapidly urbanizing. The discussion establishes the existence of a spatially growing Urban Heat Island effect, which is most manifested by a systematic rise in nighttime summer temperatures. The identified thermal changes are quite a significant danger to the well-being and stability of the local vegetable habitat, including the green cover areas of the city and the agricultural areas on the outskirts.

The findings can be a timely appeal for the implementation of climate-resilient urban planning and policy. The necessities to alleviate the negative impact of urban warming are the strategic implementation of green infrastructure, as well as the introduction of cool material technologies, which have already proven to be effective in solving the problem of urban overheating (Santamouris, 2020). The vegetation indices (such as NDVI) could be directly incorporated in future work in order to quantitatively model the particular response of plant health to observed changes in temperature. With such interventions that are evidence-based, there is a possibility to create an urban development that is more balanced with the surrounding environment and whose sustainability in the long run is guaranteed, both in terms of its human and ecological systems.

Acknowledgements To begin with, I would like to thank almighty Allah who allowed me to complete this research work. Hereby, I would like to thank my teacher and supervisor, Professor Md. Mostafizur Rahman, Department of Environmental Sciences, Jahangirnagar University, with the priceless guidance, direction, and critical remarks. This piece of work was successfully done with his huge backing. I also want to greatly thank my friends, seniors, and the teaching fraternity of the Department of Environmental Sciences, Jahangirnagar University who provided me with their technical and moral support.

References

Ahmed, B., Kamruzzaman, M., Zhu, X., Rahman, M. S., & Choi, K. (2013). Simulating land cover changes and their impacts on land surface temperature in Dhaka, Bangladesh. *Remote Sensing, 5*(11), 5969–5985. https://doi.org/10.3390/rs5115969

Ahmed, M., & Suphachalasai, S. (2014). *Assessing the costs of climate change and adaptation in South Asia.* https://www.adb.org/sites/default/files/pub/2014/assessing-costs-climate-change-and-adaptation-south-asia.pdf

Al-Hameedi, W. M. M., Chen, J., Faichia, C., Nath, B., Al-Shaibah, B., & Al-Aizari, A. (2022). Geospatial analysis of land use/cover change and land surface temperature for landscape risk pattern change evaluation of Baghdad City, Iraq, using CA–Markov and ANN models. *Sustainability, 14*(14), 8568. https://doi.org/10.3390/su14148568

Awasthi, A., Vishwakarma, K., & Pattnayak, K. C. (2022). Retrospection of heatwave and heat index. *Theoretical and Applied Climatology, 147*, 589–604. https://doi.org/10.1007/s00704-021-03854-z

Chakraborty, T., & Lee, X. (2019). A simplified urban-extent algorithm to characterize surface urban heat islands on a global scale and in every city. *Remote Sensing of Environment, 232*, Article 111296. https://doi.org/10.1016/j.rse.2019.111296

Connors, J. P., Galletti, C. S., & Chow, W. T. L. (2013). Landscape configuration and urban heat island effects: Assessing the relationship between landscape characteristics and land surface temperature in Phoenix, Arizona. *Landscape Ecology, 28*(2), 271–283. https://doi.org/10.1007/s10980-012-9833-1

Deilami, K., Kamruzzaman, M., & Liu, Y. (2018). Urban heat island effect: A systematic review of spatio-temporal factors, data, methods, and mitigation measures. *International Journal of Applied Earth Observation and Geoinformation, 67*, 30–42. https://doi.org/10.1016/j.jag.2017.12.009

Dutta, K., Basu, D., & Agrawal, S. (2021). Synergetic interaction between spatial land cover dynamics and expanding urban heat islands. *Environmental Monitoring and Assessment, 193*(4), 1–22. https://doi.org/10.1007/s10661-021-08969-4

Ebi, K. L., Capon, A., Berry, P., Broderick, C., de Dear, R., Havenith, G., Honda, Y., Kovats, R. S., Ma, W., Malik, A., Morris, N. B., Nybo, L., Seneviratne, S. I., Vanos, J., & Jay, O. (2021). Hot weather and heat extremes: Health risks. *The Lancet, 398*(10301), 698–708. https://doi.org/10.1016/S0140-6736(21)01208-3

Estoque, R. C., Murayama, Y., & Myint, S. W. (2017). Effects of landscape composition and pattern on land surface temperature: An urban heat island study in the megacities of Southeast Asia. *Science of the Total Environment, 577*, 349–359.

Estoque, R., Murayama, Y., & Myint, S. (2016). Effects of landscape composition and pattern on land surface temperature: An urban heat island study in the megacities of Southeast Asia. *Science of the Total Environment, 577*. https://doi.org/10.1016/j.scitotenv.2016.10.195

Kafy, A. A., Rahman, M. S., Faisal, A. A., Hasan, M. M., & Islam, M. (2021). Geospatial modelling of changes in land use/land cover dynamics using Multi-layer Perceptron Markov chain model in Rajshahi City, Bangladesh. *Environmental Challenges, 4*, Article 100148. https://doi.org/10.1016/j.envc.2021.100148

Labib, M. S., Wibowo, A., & Shidiq, I. P. A. (2022). LST-based threshold method for detecting UHI in a complex urban landscape. *IOP Conference Series: Earth and Environmental Science, 986*(1), Article 012072. https://doi.org/10.1088/1755-1315/986/1/012072

Liu, L., & Zhang, Y. (2011). Urban heat island analysis using the Landsat TM data and ASTER data: A case study in Hong Kong. *Remote Sensing, 3*(7), 1535–1552. https://doi.org/10.3390/rs3071535

Manoli, G., Fatichi, S., Schläpfer, M., Yu, K., Crowther, T. W., Meili, N., Burlando, P., Katul, G. G., & Bou-Zeid, E. (2019). Magnitude of urban heat islands largely explained by climate and population. *Nature*, 573(7772), 55–60. https://doi.org/10.1038/s41586-019-1512-9

Oke, T. R. (1982). The energetic basis of the urban heat island. *Quarterly Journal of the Royal Meteorological Society, 108*(455), 1–24. https://doi.org/10.1002/qj.49710845502

Pattnayak, K. C., Awasthi, A., Sharma, K., & Pattnayak, B. B. (2023). Fate of rainfall over the North Indian states in the 1.5 and 2 C warming scenarios. *Earth and Space Science, 10*(2), e2022EA002671. https://doi.org/10.1029/2022EA002671

Peng, S., Piao, S., Ciais, P., Friedlingstein, P., Ottle, C., Bréon, F.M., Nan, H., Zhou, L., & Myneni, R. B. (2012). Surface urban heat island across 419 global big cities. *Environmental Science & Technology*, 46(2), 696–703. https://doi.org/10.1021/es2030438

Rahman, M. A., & Zhang, D. (2022). Spatiotemporal influences of LULC changes on land surface temperature in rapid urbanization area by using Landsat-TM and TIRS images. *Atmosphere*, 13(3), 460. https://www.google.com/search?q=https://doi.org/10.3390/atmos13030460

Rahman, M. N., Rony, M. R. H., Jannat, F. A., Pal, S. C., Islam, M. S., Alam, E., & Islam, A. R. M. T. (2022). Impact of urbanization on urban heat island intensity in major districts of Bangladesh using remote sensing and geo-spatial tools. *Climate, 10*(1), 3. https://doi.org/10.3390/cli10010003

Santamouris, M. (2020). Recent progress on urban overheating and heat island research. Integrated assessment of the energy, environmental, vulnerability and health impact. Synergies with the global climate change. *Energy and Buildings, 207*, 109482. https://doi.org/10.1016/j.enbuild.2019.109482

Sobrino, J. A., Jiménez-Muñoz, J. C., & Paolini, L. (2004). Land surface temperature retrieval from LANDSAT TM 5. *Remote Sensing of Environment, 90*(4), 434–440. https://doi.org/10.1016/j.rse.2004.02.003

Tuholske, C., Caylor, K., Funk, C., Verdin, A., Sweeney, S., Grace, K., Peterson, P., & Evans, T. (2021). Global urban population exposure to extreme heat. *Proceedings of the National Academy of Sciences, 118*(41), e2024792118. https://doi.org/10.1073/pnas.2024792118

Voogt, J. A., & Oke, T. R. (2003). Thermal remote sensing of urban climates. *Remote Sensing of Environment, 86*(3), 370–384. https://doi.org/10.1016/S0034-4257(03)00079-8

Wang, W., Liang, H., Yang, L., Liu, K., Su, H., & Li, X. (2017). Analyzing the urban heat island using time series land surface temperature (LST) data. *IEEE International Geoscience and Remote Sensing Symposium (IGARSS), 2017*, 5874–5876. https://doi.org/10.1109/IGARSS.2017.8128345

Yadav, N., Rajendra, K., Awasthi, A., Singh, C., & Bhushan, B. (2023). Systematic exploration of heat wave impact on mortality and urban heat island: A review from 2000 to 2022. *Urban Climate, 51*, Article 101622. https://doi.org/10.1016/j.uclim.2023.101622

Zhou, D., Zhao, S., Liu, S., Zhang, L., & Zhu, C. (2014). Surface urban heat island in China's 32 major cities: Spatial patterns and drivers. *Remote Sensing of Environment, 152*, 51–61. https://doi.org/10.1016/j.rse.2014.05.017

Index

A. Awasthi et al. (eds.), *Rising Temperatures and Plant Ecosystems*,
https://doi.org/10.1007/978-3-032-29477-7

MIX
Papier aus verantwortungsvollen Quellen
Paper from responsible sources
FSC® C105338

If you have any concerns about our products,
you can contact us on
ProductSafety@springernature.com

In case Publisher is established outside the EU,
the EU authorized representative is:
Springer Nature Customer Service Center GmbH
Europaplatz 3, 69115 Heidelberg, Germany

Printed by Libri Plureos GmbH
in Hamburg, Germany